Mechanisms of Primary Energy Transduction in Biology

Chemical Biology

Editor-in-Chief:
Tom Brown, *University of Oxford, UK*

Series Editors:
Kira J. Weissman, *Lorraine University, France*
Sabine Flitsch, *University of Manchester, UK*
Nick J. Westwood, *University of St Andrews, UK*

Titles in the Series:
1: High Throughput Screening Methods: Evolution and Refinement
2: Chemical Biology of Glycoproteins
3: Computational Tools for Chemical Biology
4: Mass Spectrometry in Chemical Biology: Evolving Applications
5: Mechanisms of Primary Energy Transduction in Biology

How to obtain future titles on publication:
A standing order plan is available for this series. A standing order will bring delivery of each new volume immediately on publication.

For further information please contact:
Book Sales Department, Royal Society of Chemistry, Thomas Graham House, Science Park, Milton Road, Cambridge, CB4 0WF, UK
Telephone: +44 (0)1223 420066, Fax: +44 (0)1223 420247,
Email: booksales@rsc.org
Visit our website at www.rsc.org/books

Mechanisms of Primary Energy Transduction in Biology

Edited by

Mårten Wikström
University of Helsinki, Finland
Email: Marten.Wikstrom@Helsinki.Fi

ROYAL SOCIETY OF CHEMISTRY

THE QUEEN'S AWARDS FOR ENTERPRISE: INTERNATIONAL TRADE 2013

Chemical Biology No. 5

Print ISBN: 978-1-78262-865-1
PDF ISBN: 978-1-78801-040-5
EPUB ISBN: 978-1-78801-363-5
ISSN: 2055-1975

A catalogue record for this book is available from the British Library

The Royal Society of Chemistry is a charity, registered in England and Wales, Number 207890, and a company incorporated in England by Royal Charter (Registered No. RC000524), registered office: Burlington House, Piccadilly, London W1J 0BA, UK, Telephone: +44 (0) 207 4378 6556.

For further information see our web site at www.rsc.org

Printed in the United Kingdom by CPI Group (UK) Ltd, Croydon, CR0 4YY, UK

Preface

Bioenergetics is at the heart of biochemistry, biophysics and molecular biology, and encompasses the key reactions by which the processes of life harness the energy of sunlight and foodstuffs and transduce it into adenine triphosphate (ATP), the currency of energy that is found throughout the three kingdoms of life. As they have a common evolutionary origin, the majority of these basic processes are virtually the same, or at least very similar, in the mitochondria of eukaryotes as they are in prokaryotes, although the mitochondrial versions often exhibit a higher degree of sophistication and regulation. In all cases, the catalysts involved in these reactions are complex assemblies of membrane proteins that carry out primary energy transduction by mechanisms based on Peter Mitchell's universal electrochemical proton gradient. For a long time, research in this area was hampered by the difficulty of crystallising and solving the structures of membrane proteins, but, as described in this volume, today that problem has been largely overcome, so that the three-dimensional structures of all the major energy transducing protein complexes are known to various degrees of resolution. Bioenergetics is also a unique area of biological research in the sense that the atomic level reaction mechanisms of these complex enzymes have been studied in real time, either in molecular ensembles, or by single molecule techniques. The development of computational chemistry has led to especially valuable comparisons between such real-time experiments and advanced atomic simulations using quantum-chemical and molecular dynamics technologies, as well as combinations of the two.

The volume starts out with a chapter by Moser *et al.* (Chapter 1) that describes their outstanding work to build and engineer bioenergetically relevant structures in the laboratory. This is followed by three chapters by Sazanov (Chapter 2), Zickermann (Chapter 3), and Kaila (Chapter 4) on the

Chemical Biology No. 5
Mechanisms of Primary Energy Transduction in Biology
Edited by Mårten Wikström

Published by the Royal Society of Chemistry, www.rsc.org

respiratory Complex I, which is the part of the respiratory chain that contains the largest remaining mechanistic enigmas at this time despite the crystal structures. Gennis *et al.* (Chapter 5) then review the knowledge on transhydrogenase, an honorary member of the respiratory chain, and Barquera *et al.* (Chapter 6) describe the interesting Na^+-translocating NADH-quinone reductase (NQR) found in several pathogenic bacteria. This is followed by two insightful contributions on the bc_1 complex of the respiratory chain by Crofts *et al.* (Chapter 7) and Osyczka *et al.* (Chapter 8). Here we need to recall that analogues of the bc_1 complex are also part of the reactions of Photosystem II of photosynthetic bacteria and green plants. Then follows a contribution by Ferguson-Miller and Hosler on cytochrome *c* oxidase (Chapter 9), the complex of the respiratory chain that enables it to use O_2 as the terminal oxidant. Their emphasis is on the enigmatic subunit III of this enzyme complex, which is lacking in many bacterial variants of the superfamily of heme-copper oxidases. Swanson complements the description of cytochrome *c* oxidase (Chapter 10) with an incisive account of the computational work that has been performed to better understand the proton-pumping function of this enzyme. Siegbahn (Chapter 11) presents the only account here of the structure and function of the photosynthetic water-splitting complex of cyanobacteria and green plants. The remarkable earlier structural achievements by X-ray crystallography are here complemented by advanced quantum-chemical calculations, which give unique atomic insight into the mechanism of dioxygen formation in photosynthesis. In Chapter 12, Lenaz *et al.* present a critical assessment of the currently very popular concept of supercomplexes of the respiratory chain. Finally, in the concluding Chapter 13, Walker gives a survey of the recent status of research on the universally important ATP synthase, also called F_1F_o-ATPase. Apart from interesting details relevant for the proton-driven rotor of this remarkable molecular machine, which is central for all of bioenergetics, there are discussions of ATP synthase as a possible drug target, and its possible role in providing the mitochondrial permeability transition pore that has historically gained great popularity in cell biology.

Mårten Wikström

Contents

Chemical Biology No. 5
Mechanisms of Primary Energy Transduction in Biology
Edited by Mårten Wikström

Published by the Royal Society of Chemistry, www.rsc.org

CHAPTER 1

Making Maquette Models of Bioenergetic Structures

CHRISTOPHER C. MOSER,* NATHAN M. ENNIST, JOSHUA A. MANCINI AND P. L. DUTTON

University of Pennsylvania, Johnson Research Foundation, Department of Biochemistry and Biophysics, Philadelphia, PA, USA
*Email: moserc@mail.med.upenn.edu

1.1 Unravelling Protein Complexity

Extensive work developing X-ray crystal structures of core natural bioenergetic proteins exposed the arrangement of hundreds to thousands of amino acids that surround and bind a range of different cofactors that support bioenergetic light absorption and respiratory and photosynthetic electron and proton transfer.[1–12] Such atomic resolution offers the promise of a detailed understanding of how these naturally evolved systems function to direct electron and proton transfer to support life, or how they may dysfunction in disease.

Indeed, such structures have provided a good appreciation of how electrons tunnel from one redox centre to another across the intervening electrically insulating protein medium to form chains of redox centres that connect catalytic centres of bond breaking and making. A survey of dozens of individual bioenergetic electron-transfer reactions, both productive and unproductive, using both natural and unnatural redox centres, has shown that a relatively simple formula depending on only three parameters captures the quantum mechanics adequately to provide estimates of electron-tunnelling rates within an order of magnitude.[13,14] By constructing

Chemical Biology No. 5
Mechanisms of Primary Energy Transduction in Biology
Edited by Mårten Wikström

Published by the Royal Society of Chemistry, www.rsc.org

a matrix of electron-tunnelling rates between all redox centres in a protein, we can model the progress of electron transfer to reveal the relative weaknesses and strengths in natural electron-transfer chain design.[15–17]

However, factors other than electron-tunnelling distance can dominate performance, especially at the catalytic centres terminating electron-transfer chains.[18] Here, the complexity of natural bioenergetic protein structures obscures which engineering elements are critical for function and which elements may be incidental remnants of an evolutionary legacy. While mutation of individual residues can verify their functional importance, mutation imperfectly clarifies the residue's role. This is because natural selection has an inherent tendency to create systems in which component parts become progressively interdependent[19,20]—the protein-level equivalent to Muller's genetic ratchet.[21,22] Previously independent amino acid residues came to depend upon one another for continued function. Thus mutation of a residue with only indirect support of catalytic function may devastate function and seem to be of central importance. In addition, any one amino acid in natural proteins tends to support many functions, a molecular expression of Darwin's principle of multiple utility;[23] for example, a particular amino acid may support not only chemical activity, but also protein folding, stability and dynamics.[24,25]

The need to unravel this complexity inspired design and construction of simplified maquettes of natural proteins aiming for intentional simplicity, robustness and independent function. *De novo* designed proteins use first principles of protein folding and are untouched by natural selection. These constructs can be exploited to reverse-engineer natural protein function and rigorously test hypotheses. The basic functional principles resolved in this way may then be used to construct novel protein systems that introduce new functions or surpass natural proteins in robustness and performance.

De novo design and construction of cofactor binding protein maquettes exposes the default properties of cofactors in a protein matrix independent of natural selection and gain insight into how redox proteins may have operated early in the evolution of life.[26,27] Such designs help us to isolate and understand the means by which protein can be engineered to manipulate cofactor properties towards desired function.[28] Furthermore, by independently recapitulating critical properties of natural bioenergetic proteins in designed constructs, we gain confidence in protein engineering principles—the maquette approach is the redox-protein counterpart of Feynman's principle "What I cannot create, I do not understand".[29] This chapter surveys the present state of cofactor functionalized bioenergetic maquette design.

1.2 Robust, Modular, Helix-bundle Scaffold for Electron-transfer Architecture

De novo protein design is highly diverse, including architectures quite unlike natural proteins. For example, some early maquette designs sought to guide protein assembly through a template assisted synthetic protein (TASP) strategy whereby separate helical peptides were site specifically attached to a synthetic cyclic peptide template.[30–32] In this chapter, we will focus on designs that can be expressed in a cell; such designs can benefit from integration with natural cellular biochemistry and offer the possibility of conferring new functions to the organism.

De novo protein designs commonly exploit the fundamental principle of binary patterning of amino acid heptad repeats to promote spontaneous helical bundle association.[33,34] Hydrogen bonding between alpha-helical amide backbone oxygens and nitrogens folds two helical turns for every seven amino acids. Selecting a sequence in which residues on one face of the helix are non-polar, while the remaining are polar, drives helical self-association to bury hydrophobic residues in the bundle core. The structurally resolved two-helix coiled-coil protein tropomyosin[35] was used as an early model, in which heptad positions were labelled sequentially from *a* through *g*, with the *a* and *d* positions associated with core hydrophobic residues. By shifting the hydrophobic : hydrophilic balance from two to three non-polar residues per heptad at positions *a*, *d* and *e*, four-helix bundles form spontaneously.

For example, the heptad repeat LQQLLQX where L is Leu, Q is Gln and X is E (Glu) for one sequential pair of helices and K (Lys) for the other pair, forms a fully antiparallel helical bundle arrangement common in natural helical coiled-coil proteins. High-affinity wrapping of the helices in a left handed coiling is driven by the "knob-into-hole" interactions first predicted by Francis Crick, in which amino acid side chains (knobs) pack into spaces between the side chains on adjacent helices (holes).[36] Connecting the four helices into a single chain with Gly rich loops facilitates independent adjustment of the sequences of each helix, assisting in site-specific binding of redox cofactors and permitting surface residue salt-bridges between helices that stabilize the structure. For example, in the sequence just described, negatively charged E residues in helices one and two salt-bridge with positively charged K residues in helices three and four when helices are threaded in a counter-clockwise pattern, as shown in Figure 1.1a. It is not uncommon to generate helical bundle maquettes that resist thermal unfolding even at 100 °C.[25,37,38]

A single chain also facilitates the use of negative design, in which plausible alternative assemblies of helices into a bundle are destabilized. Negative design was effectively demonstrated by the DeGrado group with single-helix peptides of different charge patterns that assembled into heterotetrameric four-helix bundles only when component peptides with stabilizing charge pairing interactions were mixed with the proper stoichiometry; when component peptides were kept separate, the peptides were unstable.[39,40] In the single chain example of Figure 1.1, the alternative helical threading, in which the positions of helices two and four are swapped (Figure 1.1c) buries hydrophobic residues just as well; however this threading will be electrostatically disfavoured as it pairs the negatively

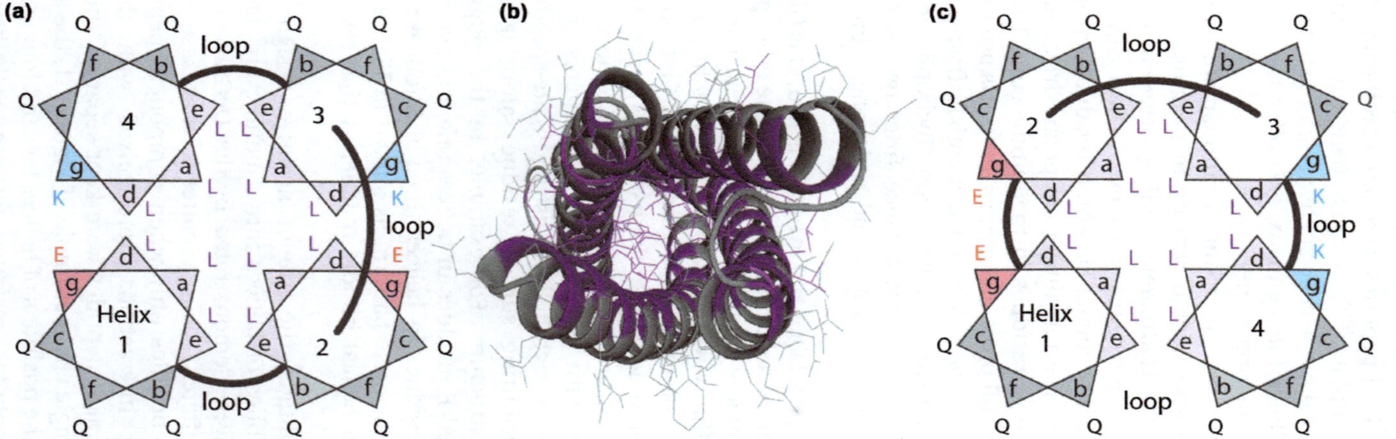

Figure 1.1 Binary patterning of helical heptad repeats drives hydrophobic self-association into a four-helix bundle. (a) seven amino acids completing two turns of an alpha-helix are labelled *a* through *g*. In four-helix bundle maquettes, amino acids at positions *a*, *d* and *e* are non-polar, and the other positions are polar. Helices one through four are connected by flexible loops. (b) Binary patterning leads to spontaneous bundle association forming long non-polar cores, as seen by a maquette crystal structure. Cofactors are anchored at desired positions in these cores. (c) As an example of negative design, this helical threading brings charged groups of similar sign together and is electrostatically disfavoured compared to (a) where nearby charges are of opposite sign.

charged E residues at one helical interface and the positively charged K residues at another interface.[41] Thus surface charge patterning provides both positive and negative design to reliably force a particular helical bundle threading, as verified by X-ray crystallography (Figure 1.1b).

A great advantage of helical coiled-coil designs is adaptable modularity. Helices can be lengthened or shortened by extending or truncating the helical heptad pattern without disturbing the local protein folding. This frees the designer of a coiled-coil maquette to employ electron-tunnelling theory to select the optimal number of cofactors and inter-cofactor distances to support desired electron-transfer function on predetermined timescales. It also promotes compact designs lightening the expression load on the organism. Currently, maquette helices range from five to 13 turns for a length of 26 to 62 Å. The individual helices of a binary patterned four-helix bundle assembly can be extended to include regions of predominantly non-polar residues to facilitate transmembrane insertion into lipid bilayers. In this way a redox cofactor electron-transfer chain can extend from the aqueous phase to cross the low dielectric region of the membrane,[42,43] much like cytochrome bc_1.

Of the 20 natural amino acids commonly coded into proteins, the nitrogen of His and the sulfur of Cys are frequently employed to ligate the metals found in a range of biologically common redox cofactors including hemes, iron–sulfur clusters and light-activated tetrapyrroles. Inserting these residues into the binary patterned sequences anchors cofactors at locations prescribed by electron-transfer theory. For the ligation of metal cations in redox clusters, Cys and His can be used alone or in combination with the carboxylic acids of Asp and Glu.

1.3 Securing Cofactors with Histidines

1.3.1 Hemes

Taking inspiration from heme binding natural four helix bundle proteins cytochrome b_{562} and cyt bc_1, early heme maquette designs bound four hemes per helical bundle through the elementary strategy of placing a pair of His residues at "a" positions separated by four turns on each helix[44] while placing bulkier core Phe residues midway between to separate the helices for heme binding. As the helices self-associate, His residues in the helical core form two clusters at opposite ends of the bundle.[45] Added heme B initially partitions into the hydrophobic bundle core, and then recruits a bis–His ligation of the heme iron.[46] Relatively close packing of hemes in the bundle core leads to conspicuous heme–heme electrostatic interactions—reduction of each heme makes the reduction of the next heme less favourable. This leads to a wide split in heme redox midpoint potentials (E_m values) within the same heme maquette, ranging from −80 to −230 mV. Heme redox potentials are also shifted electrostatically by modulating the charge patterning on the maquette surface.[47]

Heme A and Heme C bind to bis–His sites in maquettes,[48,49] as do many iron-porphyrin variants.[46] For four-helix bundle maquettes with bis–His sites in the core, strong Fe porphyrin binding requires an amphipathic character for the tetrapyrrole peripheral substituents. Polar groups (such as propionates on protoporphyrin IX) should be placed on a single or at least adjacent edges of the macrocycles and non-polar groups on remote edges. This provides a balance between tetrapyrrole water solubility with less aggregation in the aqueous phase, and maquette hydrophobic core solubility for heme partitioning into maquettes before bis–His ligation.[46]

Early heme maquette designs were structurally characterized only in the apo-form by X-ray crystallography.[50] Recently, crystal structures of heme containing maquettes of the coiled-coil design (Figure 1.2) have been achieved to reveal details of the heme structural environment. Despite the intentional absence of sequence similarity between maquettes and natural heme proteins, the maquette heme environment resembles cyt bc_1. This suggests that structure of natural proteins at heme sites may be largely a consequence of the helical bundle fold. One difference between heme in the larger, membrane embedded cyt bc_1 and the smaller and water-soluble coiled-coil maquette is that heme propionates in the maquette are stabilized by access to the aqueous phase towards the end of the bundle. The ~10 nM bis–His heme binding affinity in these coiled-coil designs results in spontaneous incorporation of heme during *in vivo* expression in *E. coli*.

Heme binds to these designs in many different ligation motifs besides bis–His, replicating many of the heme binding geometries seen in natural proteins (Table 1.1).[41] Such ligation alternatives modulate electron transfer properties to support engineering functions beyond electron-transfer chain action. Notably, a myoglobin-like His–Ala version of this maquette binds oxygen to the ferrous state stably for days without auto-oxidation.

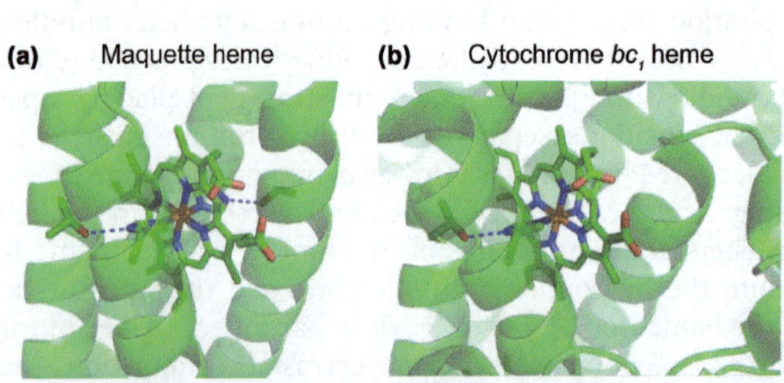

Figure 1.2 Although maquette bundles (a) do not share sequences with natural cytochrome *b* (b), following simple rules such as bis–His ligation stabilized by second shell threonines together with notch glycines to reduce steric bulk near the heme, produces folded structures that are remarkably similar.

Table 1.1 Heme ligation patterns seen in natural proteins can be reproduced in maquettes to confer different redox properties and functions.

Heme Fe ligation	Natural protein	Maquette E_m (mV SHE)	Soret max (nm)
His–Cys	Cystathione β-synthase	−280	424
His–His	Cytochrome *b*	−190	412
Cys–Ala	P450	−190	387
His–Tyr	Bovine liver catalase	−180	407
His–Ala	Myoglobin	−110	416
His–Lys	Cytochrome *f*	−65	414
His–Met	Cytochrome *c*	70	409

1.3.2 Non-iron Tetrapyrroles

As heme binds to bis–His sites stronger than to single His sites, mixing bis–His and mono-His sites within a maquette enables site specific loading with different tetrapyrroles. Light activatable, Zn metallated tetrapyrroles, including porphyrins, chlorins and bacteriochlorins, bind readily to mono-His sites.[51,52] Zn tetrapyrroles are attractive for light-activated electron transfer designs because long-lived triplet excited states form with high yields to facilitate longer-range electron transfer with other redox cofactors. Light-activatable Mg chlorins bind to maquettes as well.[53] Indeed, when expressed in chlorophyll producing algae, natural Mg chlorins can spontaneously insert into maquette His sites. Although not a covalently bound cofactor, carotenoids can also spontaneously insert in the hydrophobic core during maquette expression in algae. The cobalt analogue of heme B (Co protoporphyrin IX) readily binds to maquette potentially providing a means of catalytic reduction of protons to hydrogen.[54]

1.4 Securing Cofactors with Cysteines

1.4.1 Heme C and Bilins

Cys groups inserted near His in a CXXCH motif attack heme B vinyl groups to form thio-ether linked heme C analogous to the natural cytochromes *c*.[49,55,56] Although this reaction takes place *in vitro*, heme ligation is improved through interaction with natural heme lyases during *in vivo* expression. Even though *de novo* maquette sequences are entirely different from the natural substrates of heme lyases, the coupling interaction is efficient.

Similarly, Cys reacts with the vinyls of open chain tetrapyrrole bilins, components of the natural light-harvesting phycobilisome system. As with heme, maquettes interact with natural lyases to enhance binding of a range of bilin types and colours including phycocyanobilin and phycoerythrobilin. Bilin binding efficiency depends upon the positioning of the Cys residue in the maquette sequence. However, when present in sufficient concentrations in the cell, bilins such as biliverdin will also spontaneously associate with maquette hydrophobic cores, even without anchoring Cys residues. Biliverdin ligates to maquettes without the assistance of lyases during expression in mammalian cells, forming an analogue to natural phytochromes.

1.4.2 Iron–Sulfur Clusters

When exposed to iron sulphide under reducing conditions, Cys binds directly to Fe_4S_4 clusters, analogous to the ubiquitous and primeval iron–sulfur redox proteins.[57] A 16 amino acid loop sequence is sufficient to create a ferredoxin analogue.[58] Inserting this sequence as a pair of loops into heme binding maquettes creates a compact six-redox centre ferredoxin-heme maquette.[59] Complementing the Cys loop with three nearby His anchors a Ni(II)–Fe_4S_4 bridged cluster, an analogue to natural carbon monoxide dehydrogenase.[60] Fe_4S_4 clusters also form with Cys residues in the helical core.[61] A maquette beta sheet fold has been used to form a single iron, tetra-thiolate centre as a rubredoxin analogue.[62]

Sulfur chemistry is also readily exploited for site-specific cofactor anchoring in maquettes through means not found in natural proteins. Single Cys residues anchor suitably activated cofactors such as halogen-activated flavin[63,64] or quinone,[65] or maleimide-activated ferrocene[66] or bacteriochlorin.[67]

1.5 Securing Metals with Cys, His or Carboxylates

Maquette His and Cys residues have been used to ligate individual metal ions. The Pecoraro group has engineered three-helix bundle maquettes that tris-thiolate bind Hg(II)[68] or As(III),[69,70] and tris–His bind catalytically active Cu(I/II)[71] or Zn(II).[72] Site specific metal ligation is seen as Hg(II) ions bound preferentially to a tris-thiolate site while Zn(II) ions bound to a tris–histidine site at the other end of the maquette.[72]

Clusters of acidic carboxylate residues Asp and Glu together with His, often bind redox active and inactive metal cations in natural proteins. The four-helix bundle frame of natural bacterio-ferritin inspired the design of a series of metal binding maquettes, some of which are catalytically active.[73] These maquettes bind bivalent metal ions with relative affinities of Mn < Fe < Co < Zn > Cd,[73–75] consistent with the Irving–Williams series.[76] These *de novo* designs have recently been adapted to bind combinations of redox active tetrapyrroles and metal clusters for extended chains of light and electron-transfer active cofactors. X-ray crystal structures closely reproduce the metal ligand arrangement of natural bacterio-ferritin even in the extended designs (Figure 1.6).

1.6 Redox Active Amino Acids

The amino acids Tyr and Trp function as high potential redox centres in natural electron-transfer chains and show similar redox properties in maquette environments.[65,77–82] Maquettes form light-activated flavin-Trp radical pairs that show conspicuous magnetic field sensitivity, reproducing essential elements of cryptochrome.[64] For these non-metallic amino acid redox centres, proton-coupled electron transfer rates may be slower than electron tunnelling and reflect protein dynamics.

Artificial redox cofactor amino acids have been inserted in maquette sequences using a TAG Amber codon and engineered mutant t-RNA synthetase to insert propargyloxyphenylalanine; this group reacts with azido-functionalized redox centres using click chemistry.[66] Artificial redox active amino acids have also been inserted into maquettes through intein semi-synthesis.[83,84]

1.7 Practical Design Rules for Intraprotein Electron Tunnelling

Many natural bioenergetic proteins are composed of catalytic sites, where oxidative and reductive bond making and breaking take place, connected by chains of redox cofactors that relay electrons over distances of four to 14 Å by electron tunnelling through the intervening insulating protein medium.[13] Although the effective height of the electron-tunnelling barrier between co-factors can be altered by making the medium more or less densely packed with single and double bonds, in practice natural selection has not used this means to speed productive electron transfers and slow unproductive, energy wasting electron-transfer reactions. Instead, Nature relies on the exponential dependence of the electron-tunnelling rate with the edge-to-edge distance between cofactors to guide electron transfer by means of proximity: at any step, productive electron transfer is generally shorter than unproductive electron transfer.

The driving force dependence of electron-tunnelling rates is approximately Gaussian.[85] As the driving force ($-\Delta G$) increases from zero, the electron-transfer rate increases. However, it eventually reaches a maximum rate at a driving force defining the reorganization energy (λ). Increasing the driving force still further, into what is called the Marcus inverted region, causes electron-transfer rates to drop. This Marcus rate dependence is captured in eqn (1.1), where the log rate is in units of s^{-1} and ΔG and λ are in units of electron volts (eV).

$$\log_{10} k_{et} \propto -(\Delta G + \lambda)^2/(4\lambda k_B T) \tag{1.1}$$

In this classical Marcus picture, electron-tunnelling rates show conspicuous temperature dependences when driving force does not match reorganization energy. However, working with an extensive collection of quinone electron-transfer reactions in photosynthetic reaction centres over a very wide temperature and free energy range, Gunner and Woodbury demonstrated that quantized nuclear vibrations enforce a more modest temperature dependence.[86,87] The energy of the typical characteristic frequency of vibration coupled to intraprotein electron tunnelling in natural proteins is

$\sim$0.07 eV,[13] larger than the 0.025 eV Boltzmann thermal energy near room temperature.

A survey of many intraprotein electron-tunnelling rates for systems with known distances, driving forces and reorganization energies led to the development of a practical electron-tunnelling rate expression[13] that is used not only to gain insight into natural protein design, but also to design novel electron-transfer proteins.

$$\log_{10}k_{et} = 13 - 0.6(R - 3.6) - 3.1(\Delta G + \lambda)^2/\lambda \quad (1.2)$$

Eqn (1.2) provides an electron-tunnelling rate for an energetically favourable reaction near room temperature in units of s^{-1} when the R is in units of Å, and ΔG and λ are in units of eV. To estimate the rate of an uphill, endergonic reaction, the rate of the reverse exergonic reaction is first calculated using eqn (1.2), and then a rate penalty of one order of magnitude for every 0.06 eV of driving force is imposed. This is consistent with the exergonic–endergonic behaviour described by Woodbury and Gunner.[87]

We combine the ability to estimate electron-tunnelling rates between cofactors with the facility to move maquette cofactors by adjusting the position of cofactor anchoring amino acids in maquette sequences. Designs are vetted by simulating electron-transfer performance, given the constraints of cofactor redox potentials, excited state lifetimes, *etc.*, by solving a system of differential rate equations. Electron-transfer rates are expected to be limited by electron-tunnelling provided there is no diffusive movement or bond making/breaking coupled to the electron transfer, such a proton or hydride transfer.

Figure 1.3 shows an example of an elementary maquette design including a light excitable tetrapyrrole pigment and a tetrapyrrole electron acceptor anchored to histidines in the interior of a four-helix bundle, comparing the differences in optimal design for long-lived charge separation yield originating from the excited singlet or triplet state. The vertical axis adjusts the driving force of electron transfer from the excited pigment state P* to an electron acceptor A. This might be selected by changing the peripheral groups or protein surroundings of the P and A cofactors. The horizontal axis varies the edge-to-edge distance between P and A. This is adjusted by selecting the spacing along a helix of the cofactor anchoring amino acids, typically in increments of one helical turn, or about 5.2 Å. Optimal designs in simple cofactor dyads avoid losses that return the system to the ground state by radiative decay or rapid charge recombination. The figure illustrates that singlet systems should place histidines near three helical turns apart, while triplet systems with longer lifetimes are better if anchoring His are nearly five helical turns apart.

In constructing electron-transfer chains, it is not generally necessary that each successive electron transfer be energetically downhill (exergonic) provided the cofactors are placed close enough to one another that even uphill (endergonic) electron-tunnelling reactions are relatively fast. Indeed, many

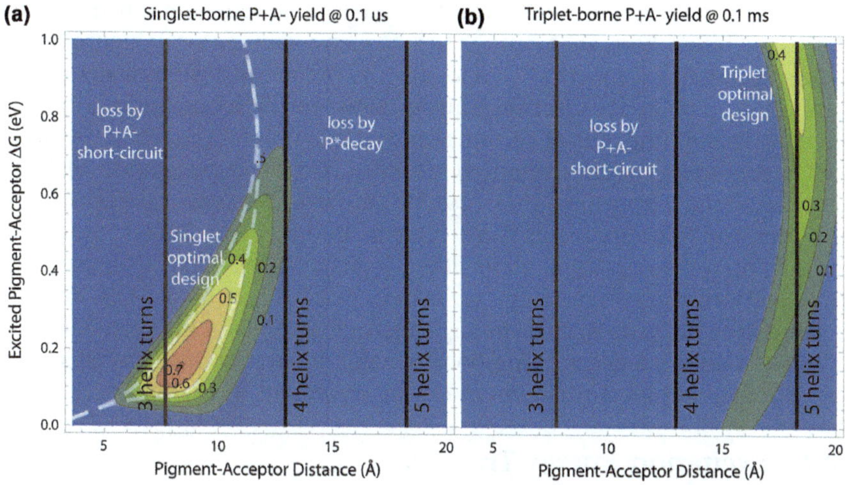

Figure 1.3 Using eqn (1.2), simulated electron-tunnelling kinetics shows that optimal separation of maquette His residues anchoring a light-activated and an acceptor tetrapyrrole, on the one hand, and optimal choice of redox midpoint potentials, on the other hand, is different when charge separation originates from the excited singlet (a) or triplet (b) state. In this simulation, the pigment resembles natural chlorophyll: an ~8 ns lifetime singlet is excited by a red photon at 1.8 eV, while a ~2 ms lifetime triplet phosphorescence energy is 1.3 eV. λ is a moderate 0.7 eV. Optimal designs indicate cofactor ligating His residues should be about three helix turns apart for the singlet and about five turns apart for the triplet.

natural electron-transfer chains show an up and down energy profile.[14] Catalytic bond-making and breaking generally involves pairs of electron transfers. This means that catalytic sites should be designed to accumulate multiple oxidizing or reducing equivalents, for example by placing multiple redox centres in close proximity, so that a relatively slow uphill electron transfer to a catalytic intermediate can be followed by a fast downhill electron transfer to a more stable redox state.

1.8 Bioenergetic Function Examples

1.8.1 Ligand Binding and Transport

In natural proteins, the 6th ligation position of heme iron may bind small ligands, such as molecular oxygen, to either promote electron transfer (such as P450 or cytochrome oxidase) or suppress electron transfer, as in oxygen transport. Natural O_2 transport proteins myoglobin and haemoglobin are ligated by a single His residue with an open 6th coordination site that may be occupied by a water molecule easily displaced by ligands such as O_2, NO or CO. In natural transport proteins such as neuroglobin and cytoglobin, heme is bis-His ligated. One His residue must be displaced to bind O_2, providing a means for these globins to modulate O_2 affinity.[88]

An early single-chain, two heme B bis–His maquette demonstrated the ability to bind both CO and O_2,[89] much like natural neuroglobin. Surprisingly, heme C bis–His[49,56] and His–Cys maquette designs also bind O_2. However, auto-oxidative electron transfer between the heme and O_2 in these initial designs on the tens of seconds timescale is much faster than the hours to days timescale typical in natural globins. The latest generation of heme maquette designs appear to have less mobility and less water access around the heme iron. The net result is oxy-ferrous lifetimes of 1–2 days, much like natural globins. Remarkably, maquette designs generally bind O_2 better than CO, in stark contrast to natural O_2 transport globins for which CO is effective poison. These properties, plus the high thermal stability of maquette designs, make them an attractive starting point for the development of an emergency oxygen transport blood supplement during treatment of trauma.

1.8.2 Excitation Energy Transfer

Maquettes bind a wide variety of light activated Zn metallated tetrapyrroles, including porphyrins, chlorins and bacteriochlorins (Figure 1.4).[51]

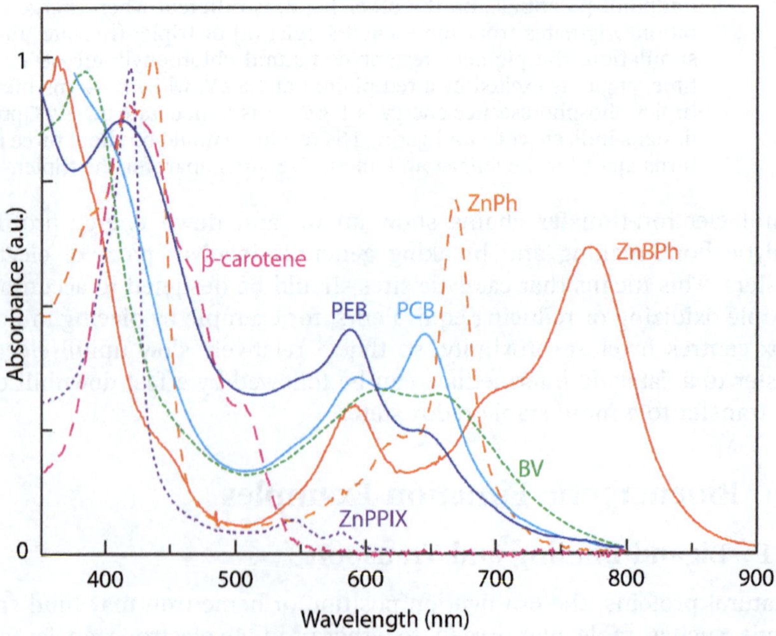

Figure 1.4 Light active cofactors in maquettes have absorbances tuneable across the visible and near IR spectrum. Zn metallated porphyrin (ZnPPIX), chlorin (ZnPh), and bacteriochlorin (ZnBPh) are anchored with ligated maquette His residues. Metal-free, open chain tetrapyrroles phycoerythrobilin (PEB), phycocyanobilin (PCB) and biliverdin (BV) are anchored by coupling of the bilin vinyl to a maquette Cys. Carotenoid β-carotene is non-covalently bound. Spectra are normalized to absorption maxima near 400 nm.

Assembling tetrapyrroles with different spectral properties within the maquettes leads to intraprotein excitation energy transfer (EET) and emission from the lowest energy chromophore.[51,67] Biliverdin, PEB, PCB and other fluorescing bilin chromophores bind to maquettes to support light-harvesting-antennae-like energy transfer to circular tetrapyrroles. In a design intended to couple maquette EET with natural phycobilisomes, the phycobilisome protein subunit CpcA has been fused to a maquette binding two different tetrapyrroles for three-step energy transfer (Figure 1.5).[67]

1.8.3 Interprotein Electron Transfer

Maquettes interact with a number of different natural proteins while being expressed *in vivo*, notably with heme and bilin lyase as well as membrane transporters that facilitate heme C formation. Maquettes also participate in productive electron transfer with natural proteins *in vivo*. For example, single-electron reduction of a maquette heme A by NADPH is moderated by the two-electron flavo-protein ferredoxin-NADPH reductase,[48] the natural redox partner of PSI. Transmembrane heme B maquette[43] interacts with decyl-ubiquinol and water soluble cytochrome *c*

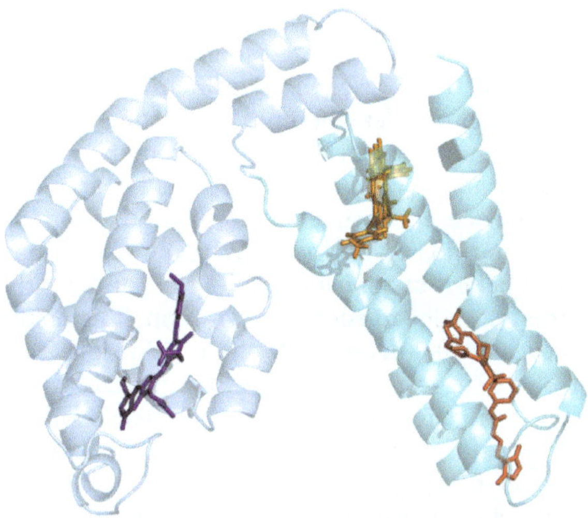

Figure 1.5 Molecular model illustrating the fusion of a natural phycobilisome protein subunit (gray) with bound bilin PEB (purple) that supports EET with chlorin (orange) and bacteriochlorin (red) bound in a maquette four-helix bundle frame (cyan). This construct supports light-harvesting antennae-like multi-step excitation energy transfer from bilin to bacteriochlorin. Such designs are aimed towards engineering the ability to tap into light-harvesting systems *in vivo* to direct energy towards customized electron-transfer catalysis.

to reproduce quinol-cytochrome *c* oxidoreductase activity analogous to natural cytochrome bc_1.[90]

Stopped flow studies show that interprotein electron-transfer between water soluble heme maquettes and natural protein cytochrome *c* is electrostatically modulated by ionic strength and the charge pattern of surface residues,[91] analogous to the modulation observed in natural electron-transfer proteins. Details of faster intraprotein electron transfer are resolved by CO-photolysis. Just as O_2 binds to some heme maquettes, when ferrous bis–His ligated maquette hemes are exposed to sufficient concentrations of CO, one histidine is typically displaced by CO. A laser pulse releases CO reforming the bis–His state with a relatively low redox midpoint potential. This bis–His heme is now competent for rapid reduction of higher potential cofactors such as heme C in natural cytochrome *c*.

1.8.4 Intraprotein Electron Transfer

In multi-cofactor maquettes, both Zn tetrapyrroles and flavins engage in kinetically resolvable light-activated intraprotein electron transfer.

1.8.4.1 Light Activated Zn Tetrapyrroles

Zn tetrapyrroles are attractive chromophores for initiating electron transfer because intersystem crossing from excited singlet to triplet states is relatively efficient (30–70%) with triplet excited state lifetimes nearly six orders of magnitude longer than singlet lifetimes. According to eqn (1.2) and Figure 1.3, this means redox partners can be placed up to 10 Å further away and still achieve comparable yield of light-activated charge separation. The typical ~7 ms lifetime of excited triplet Zn porphyrin is shortened by placing a heme within electron-transfer distance.[92] With mono-His and bis–His sites on different pairs of helices at opposite ends of the bundle with a separation of four helical turns, the edge-to-edge electron-tunnelling distance is ~20 Å. Light-activation reveals ~0.8 ms heme B photo-reduction at a driving force of 0.52 eV. Eqn (1.2) suggests a plausible reorganization energy for this intraprotein electron transfer of ~0.9 eV, which would place the ~1.2 eV charge recombination to ground state in the Marcus inverted region but closer to the estimated reorganization energy. Thus heme is transiently photo-reduced, with charge recombination about three times faster than the charge separation.

A transmembrane maquette has been engineered with three tetrapyrrole cofactor binding sites.[43] The ligating histidines of the middle site lie three and four helical turns from the two terminal binding sites. Binding both Zn porphyrin and heme B cofactors leads to 20 μs light induced electron transfer from the pigment to the heme acceptor, a rate consistent with the shorter and faster three helical turn tetrapyrrole spacing.[43]

1.8.4.2 Light Activated Flavins

In natural bioenergetic proteins, flavins typically interface $n=1$ and $n=2$ electron redox centres in natural electron-transfer chains such as in respiratory Complex I and Complex II, exploiting the relative stability of the half-reduced flavin semiquinone state as an intermediate. However, flavins are light-activated as photochemical sensitizers in photo-lyase or blue light and magnetic field sensors as in cryptochromes. While most natural flavoproteins bind flavin non-covalently, in nearly 10% of cases, flavins are covalently linked. We use the covalent linkage strategy in flavo-maquette construction to hold flavins at appropriate distances for intraprotein electron transfer.[63,64] Light activation of the maquette flavin initiates photo-oxidation of sufficiently close tryptophan residues, analogous to the action of natural cryptochromes.[64] Similar to natural cryptochromes, reduced flavin-oxidized Trp radical pairs in these maquettes with lifetimes in the microsecond domain are conspicuously sensitive to applied magnetic fields. Alternatively, in the presence of another electron donor, photo-reduced flavin can reduce nearby heme, analogous to the action of flavo-cytochrome b_2.[64]

1.9 Photosynthetic Charge Separation Engineering

1.9.1 Photochemical Dyad and Triad Dynamics

It is difficult to achieve light-induced charge separation lifetimes long enough to support millisecond or slower redox catalysis in two-cofactor photochemical dyads. As Figure 1.3 shows, competition between productive charge separation by electron tunnelling and the typically few nanoseconds or faster decay of the chromophore excited singlet state, leads to charge separation lifetimes of hundreds of nanseconds at best. Speeding charge separation by increasing the driving force tends to speed charge recombination with the loss by P+A- short-circuiting. Lengthening excited-state lifetime by intersystem crossing to a triplet state enables longer distance electron transfer and longer charge separated lifetimes, albeit with some loss of electrochemical energy in singlet to triplet conversion.

Stabilization of charge separation for timescales long enough to support millisecond or slower catalysis requires the addition of at least one more cofactor to permit multistep, long-distance charge separation. The number of cofactors required will depend upon the energetics of the system, the required lifetime for catalysis and whether charge separation is initiated from the excited singlet or triplet state. An example of the expected performance of triplet-borne charge separating cofactor triad is shown in Figure 1.7. As in the cofactor dyads of Figure 1.3, the expected charge-separation yield is shown as dependent on two parameters that can be adjusted by design: the driving force for P to A electron-tunnelling, and the P to A edge-to-edge distance. For this figure, we included a triad donor D at

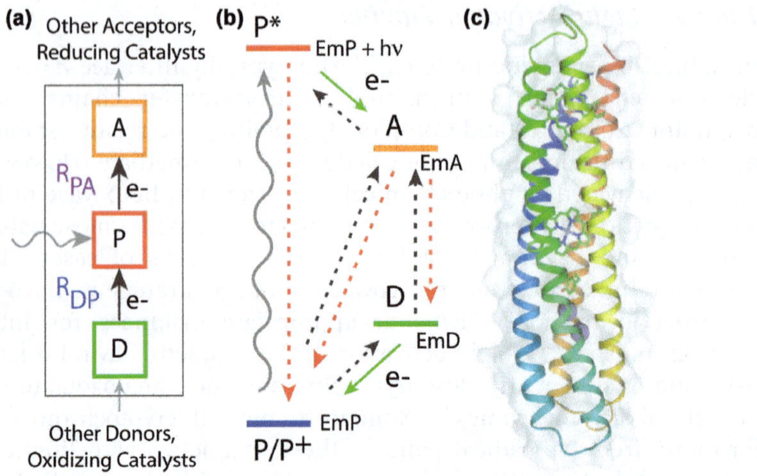

Figure 1.6 (a) Light-induced charge separation engineering is dramatically improved by using three cofactors instead of two to assemble a charge-separation triad, with an electron acceptor (A) and donor (D) on either side of the excitable pigment (P). A nearly linear arrangement of cofactors supports longer charge separation lifetimes. (b) Electron-tunnelling rates between cofactors (dashed lines) depend on inter-cofactor distances (R) and driving forces for electron transfer, estimated from the difference in relevant redox midpoint potentials (E_m). (c) An X-ray crystal structure of a multi-cofactor, light-activated charge separating reaction centre maquette that stabilizes charge separation for hundreds of milliseconds.

4.6 Å from P with a driving force of 0.19 eV, in the range expected for a tyrosinate donor to a metallated tetrapyrrole cation. The charge separation yield at 0.1 ms in the dyad (Figure 1.3B) is much smaller than in the triad (Figure 1.7A). Indeed, electron-tunnelling analysis indicates that charge separation should persist in cofactor triads out to tens of milliseconds (Figure 1.7B). In our experience, maquette triads can achieve charge separation lifetimes of hundreds of milliseconds.

Singlet-borne charge separation is more demanding because the initial charge separation must compete with typically nanosecond or shorter excited state decay lifetimes, which requires a shorter distance for the first step of charge separation. Under these circumstances, a cofactor tetrad will be required for millisecond or longer charge-separation lifetimes.

1.9.2 Photochemical Tyrosine Oxidation

Not all biological electron-transfer reactions in native electron-transfer chains are rate limited by electron tunnelling. A classic example is proton-coupled electron transfer during the oxidation of tyrosine, an essential step in water oxidation catalysed by PSII. The redox potential of deprotonated

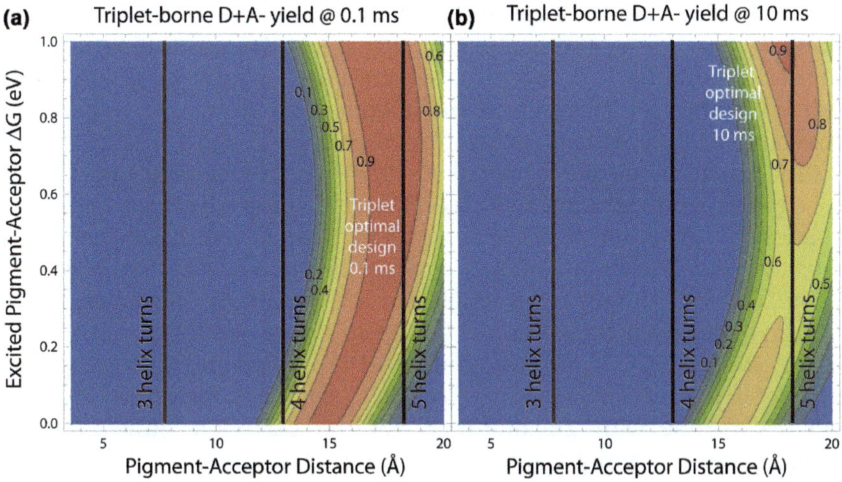

Figure 1.7 Solving a system of differential equations using electron-tunnelling rates for all possible electron transfers in a cofactor triad for the yield of charge separation at a given time after excitation predicts performance for given designs. (a) The yield of triplet-borne charge-separated triad at 0.1 ms is considerably greater than the triplet-borne dyad in Figure 3B. Indeed, charge separation yields remain high at 10 ms (b), long enough for certain catalytic reactions.

tyrosinate in maquettes is close to 0.74 V,[79] which provides a favourable driving force for oxidation by a photo-oxidized Zn porphyrin cation with an E_m value near 0.91 V. At pH 9.5, approaching the reported pK_a of tyrosine of 10 to 11,[93] we observe tyrosinate oxidation in maquettes using a Zn porphyrin excitable pigment with an E_m of about 0.91 V together with a photo-reduced heme acceptor. The charge separation lifetime is more than 100 ms.

At pH 7.5, where the majority of Tyr is protonated, the appropriate E_m value for estimating the relevant driving force of this reaction is not the equilibrium value near 0.87 V,[79] but that of the protonated Tyr.[94] This E_m is likely closer to 1.1 to 1.4 V, based on reported operating values for Yz of PSII[95] and the pH dependence of the formal redox midpoint potential in a helical bundle.[79] Thus photo-oxidation of Tyr near neutral pH is likely endergonic in this triad. Photochemical triad maquettes do photo-oxidize Tyr at pH 7.5, albeit with smaller yield. Charge-separation lifetimes are modulated by inserting nearby protonatable residues such as a His; crystal structures show this His hydrogen bonds to Tyr much as in natural PSII.

1.9.3 Photochemical Metal Cluster Oxidation

The oxygen evolving Mn_4CaO_4 cluster in natural PSII is ligated by a cluster of aspartates, glutamates and histidines. This cluster assembles by photo-oxidizing Mn(II) in a series of light and dark steps.[96,97] Photo-assembly is required each time PSII is replaced after photo-damage, which can be as short as 30 minutes.[98] To better understand the engineering of Mn cluster photo-assembly, two groups have modified natural proteins to couple light-activatable centres with metal-ligating amino acids. Allen and co-workers modified natural purple bacterial reaction centres by inserting ligating carboxylates near the light-activated bacteriochlorophyll dimer pigment.[99] Additionally, the E_m of the bacteriochlorophyll dimer was raised from 0.51 V to 0.80 V by adjacent mutations, enough to photo-oxidize Mn(II) with an E_m of 0.63 V.[100] Although this single Mn site cannot oxidize water, it does perform light-driven oxygen production from superoxide.[101] In collaboration with T. Wydrzynski,[102] we re-engineered the natural iron storage protein bacterioferritin for light activated electron transfer by substituting the heme iron tetrapyrrole with Zn-chlorin-e6. Natural bacterioferritin employs a cluster of two His and 4 Glu to ligate Fe(II) and a nearby Tyr. Exposing the metal binding site to Mn(II) instead of Fe(II) created an artificial reaction centre proto-type that photo-oxidized tyrosine only in the presence of di-nuclear Mn(II).

When we installed a bacterioferritin-like carboxylate/histidine ligation motif in photochemical triad maquettes adjacent to the photo-oxidizable Tyr we created a photo-oxidizable metal binding site analogous to PSII. Crystal structures reveal that metals bind in the site as designed (Figure 1.6c) while light activation shows that metal oxidation is effective. As in PSII, Mn(II) oxidation in maquettes is a relatively slow, adiabatic electron-transfer reaction compared to a simple electron tunnelling reaction. Also as in natural photosystems, a redox pool supports repeated light-activated turnover to accumulate oxidizing equivalents. We have observed interprotein electron transfer from photo-reduced heme in maquettes to added cyto-chrome *c*. Under continuous illumination in the absence of Mn(II) little photo-reduction of cyt *c* is seen. In the presence of Mn(II), cyt *c* acts as an oxidizing redox pool that permits quantitation of the turnover number of reaction centre maquettes. Under continuous illumination, reduction levels off after up to four oxidizing equivalents are accumulated, indicating oxidation of multiple Mn(II) to Mn(III) and possibly Mn(IV).

1.10 Conclusion

The maquette approach allows an iterative, modular approach to reproduction of bioenergetic function that clarifies what is and is not important in engineering redox protein function. Such an approach is a valuable complement to mutagenesis of natural proteins, which often shows intolerance to progressive changes of multiple amino acids due to marginally stable folding.[103] Natural proteins are generally only as stable as they need to be under the forces of natural selection.[104] The broad tolerance of maquettes to insertion of a wide range of natural and unnatural redox cofactors, combined with the ability to control cofactor placement, enables ready design of a wide range of electron-tunnelling cofactor chains for long-distance electron transfer and charge separation. The most critical factors are setting appropriate edge-to-edge distances between cofactors and driving forces guided by eqn (1.2). Such designs support hundreds of millisecond charge separated lifetimes that are longer than presently reported lifetimes in other natural or synthetic cofactor triads. The design challenge now for maquettes and artificial proteins in general, is to exploit long-lived charge separated states to engineer the bond making and breaking reactions to support efficient catalysis.[105] In this way we can hope not only to match, but to go beyond natural design, as predicted by Emil Fisher in his Nobel lecture in 1902.[106] The photochemical maquettes described here are aiming for designs capable of splitting water into hydrogen and oxygen using earth abundant and economical materials as a means of addressing human sustainable energy needs.

Acknowledgements

This research was carried out as part of the Photosynthetic Antenna Research Center (PARC), an Energy Frontier Research Center funded by the U.S. Department of Energy, Office of Science, Office of Basic Energy Sciences, under Award DESC0001035 supporting, CCM, NME, JAM and PLD in maquette construction and characterization.

References

1. J. Deisenhofer, O. Epp, I. Sinning and H. Michel, *J. Mol. Biol.*, 1995, **246**, 429–457.
2. K. N. Ferreira, T. M. Iverson, K. Maghlaoui, J. Barber, J. Barber and S. Iwata, *Science*, 2004, **303**, 1831–1838.
3. Y. Umena, K. Kawakami, J. R. Shen and N. Kamiya, *Nature*, 2011, **473**, 55–65.
4. A. Ben-Shem, F. Frolow and N. Nelson, *Nature*, 2003, **426**, 630–635.
5. R. Baradaran, J. M. Berrisford, G. S. Minhas and L. A. Sazanov, *Nature*, 2017, **494**, 443–448.
6. J. Zhu, K. R. Vinothkumar and J. Hirst, *Nature*, 2016, **536**, 354–358.
7. F. Sun, X. Huo, Y. Zhai, A. Wang, J. Xu, D. Su, M. Bartlam and Z. Rao, *Cell*, 2005, **121**, 1043–1057.
8. X. Gao, X. Wen, L. Esser, B. Quinn, L. Yu, C.-A. Yu and D. Xia, *Biochemistry*, 2003, **42**, 9067–9080.
9. S. Iwata, C. Ostermeier, B. Ludwig and H. Michel, *Nature*, 1995, **376**, 660–669.
10. T. Tsukihara, H. Aoyama, E. Yamashita, T. Tomizaki, H. Yamaguchi, K. Shinzawa-Itoh, R. Nakashima, R. Yaono and S. Yoshikawa, *Science*, 1995, **269**, 1069–1074.
11. G. Kurisu, H. Zhang, J. L. Smith and W. A. Cramer, *Science*, 2003, **302**, 1009–1014.

12. G. McDermott, S. M. Prince, A. A. Freer, A. M. Hawthornthwaite-Lawless, M. Z. Papiz, R. J. Cogdell and N. W. Isaacs, *Nature*, 1995, **374**, 517–521.
13. C. C. Moser, J. M. Keske, K. Warncke, R. S. Farid and P. L. Dutton, *Nature*, 1992, **355**, 796–802.
14. C. C. Page, C. C. Moser, X. Chen and P. L. Dutton, *Nature*, 1999, **402**, 47–52.
15. C. C. Moser, T. A. Farid, S. E. Chobot and P. L. Dutton, *Biochim. Biophys. Acta*, 2006, **1757**, 1096–1109.
16. C. C. Moser and P. L. Dutton, ed. J. H. Golbeck, Springer, Dordrecht (Academic Press), Dordrecht, 2006, vol. 24, pp. 583–594.
17. C. C. Moser, C. C. Page and P. L. Dutton, *Photochem. Photobiol. Sci.*, 2005, **4**, 933–939.
18. C. C. Moser, J. L. R. Anderson and P. L. Dutton, *Biochim. Biophys. Acta*, 2010, **1797**, 1573–1586.
19. J. T. Bridgham, E. A. Ortlund and J. W. Thornton, *Nature*, 2009, **461**, 515–519.
20. G. C. Finnigan, V. Hanson-Smith, T. H. Stevens and J. W. Thornton, *Nature*, 2012, **481**, 394–398.
21. H. J. Muller, *Genetics*, 1918, **3**, 422–499.
22. H. J. Muller, *Mutat. Res.*, 1964, **1**, 2–9.
23. C. Darwin, *The Origin of Species by Means of Natural Selection*, Earlton House, New York, 1872.
24. C. C. Moser, C. C. Page and P. L. Dutton, *Philos. Trans. R. Soc., B*, 2006, **361**, 1295–1305.
25. J. M. Brisendine and R. L. Koder, *Biochim. Biophys. Acta, Bioenerg.*, 2016, **1857**, 485–492.
26. R. A. Jensen, *Annu. Rev. Microbiol.*, 1976, **30**, 409–425.
27. L. C. Wheeler, S. A. Lim, S. Marqusee and M. J. Harms, *Cur. Opin. Struct. Biol.*, 2016, **38**, 37–43.
28. S. C. Patel, L. H. Bradley, S. P. Jinadasa and M. H. Hecht, *Protein Sci.*, 2009, **18**, 1388–1400.
29. R. Paz, *Phys. Today*, 1989, **42**, 88.
30. M. Fahnenschmidt, R. Bittl, E. Schlodder, W. Haehnel and W. Lubitz, *Phys. Chem. Chem. Phys.*, 2001, **3**, 4082–4090.
31. H. K. Rau, N. DeJonge and W. Haehnel, *Proc. Natl. Acad. Sci. U. S. A.*, 1998, **95**, 11526–11531.
32. H. K. Rau and W. Haehnel, *J. Am. Chem. Soc.*, 1998, **120**, 468–476.
33. L. Regan and W. F. DeGrado, *Science*, 1988, **241**, 976–978.
34. S. Kamtekar, J. M. Schiffer and H. Xiong, *Science*, 1993, **262**, 1680–1685.
35. A. D. McLachlan and M. Stewart, *J. Mol. Biol.*, 1975, **98**, 293–304.
36. F. H. C. Crick, *Acta Crystal*, 1953, **6**, 689–697.
37. P. S. Huang, G. Oberdorfer, C. Xu, X. Y. Pei, B. L. Nannenga, J. M. Rogers, F. DiMaio, T. Gonen, Ben Luisi and D. Baker, *Science*, 2014, **346**, 481–485.
38. S. A. Marshall and S. L. Mayo, *J. Mol. Biol.*, 2001, **305**, 619–631.

39. E. Marsh and W. F. DeGrado, *Proc. Natl. Acad. Sci. U. S. A.*, 2002, **99**, 5150–5154.
40. C. M. Summa, M. M. Rosenblatt, J. K. Hong, J. D. Lear and W. F. DeGrado, *J. Mol. Biol.*, 2002, **321**, 923–938.
41. C. C. Moser, M. M. Sheehan, N. M. Ennist, G. Kodali, C. Bialas, M. T. Englander, B. M. Discher and P. L. Dutton, *Methods Enzymol.*, 2016, **580**, 365–388.
42. B. M. Discher, D. Noy, J. Strzalka, S. Ye, C. C. Moser, J. D. Lear, J. K. Blasie and P. L. Dutton, *Biochemistry*, 2005, **44**, 12329–12343.
43. G. Goparaju, B. A. Fry, S. E. Chobot, G. Wiedman, C. C. Moser, P. L. Dutton and B. M. Discher, *Biochim. Biophys. Acta, Bioenerg.*, 2016, **1857**, 503–512.
44. D. E. Robertson, R. S. Farid, C. C. Moser, S. E. Mulholland, J. L. Urbauer, R. Pidikiti, J. D. Lear, A. J. Wand, W. F. DeGrado and P. L. Dutton, *Nature*, 1994, **368**, 425–431.
45. S. S. Huang, R. L. Koder, M. Lewis, A. J. Wand and P. L. Dutton, *Proc. Natl. Acad. Sci.*, 2004, **101**, 5536–5541.
46. L. A. Solomon, G. Kodali, C. C. Moser and P. L. Dutton, *J. Am. Chem. Soc.*, 2014, **136**, 3192–3199.
47. J. M. Shifman, B. R. Gibney, R. E. Sharp and P. L. Dutton, *Biochemistry*, 2000, **39**, 14813–14821.
48. B. R. Gibney, Y. Isogai, F. Rabanal, K. S. Reddy, A. M. Grosset, C. C. Moser and P. L. Dutton, *Biochemistry*, 2000, **39**, 11041–11049.
49. J. L. R. Anderson, C. T. Armstrong, G. Kodali, B. R. Lichtenstein, D. W. Watkins, J. A. Mancini, A. L. Boyle, T. A. Farid, M. P. Crump, C. C. Moser and P. L. Dutton, *Chem. Sci.*, 2014, **5**, 507–514.
50. S. S. Huang, B. R. Gibney, S. E. Stayrook, P. L. Dutton and M. Lewis, *J. Mol. Biol.*, 2003, **326**, 1219–1225.
51. G. Kodali, J. A. Mancini, L. A. Solomon, T. V. Episova, N. Roach, C. J. Hobbs, P. Wagner, O. A. Mass, K. Aravindu, J. E. Barnsley, K. C. Gordon, D. L. Officer, P. L. Dutton and C. C. Moser, *Chem. Sci.*, 2017, **8**, 316–324.
52. I. Cohen-Ofri, M. van Gastel, J. Grzyb, A. Brandis, I. Pinkas, W. Lubitz and D. Noy, *J. Am. Chem. Soc.*, 2011, **133**, 9526–9535.
53. P. Braun, E. Goldberg, C. Negron, M. von Jan, F. Xu, V. Nanda, R. L. Koder and D. Noy, *Proteins*, 2011, **79**, 463–476.
54. D. J. Sommer, M. D. Vaughn, B. C. Clark, J. Tomlin, A. Roy and G. Ghirlanda, *Biochim. Biophys. Acta*, 2016, **1857**, 598–603.
55. D. W. Watkins, C. T. Armstrong, J. L. R. Anderson and J. R. Anderson, *Curr. Opin. Chem. Biol.*, 2014, **19**, 90–98.
56. D. W. Watkins, C. T. Armstrong, J. L. Beesley, J. E. Marsh, J. M. X. Jenkins, R. B. Sessions, S. Mann and J. L. Ross Anderson, *Biochim. Biophys. Acta*, 2016, **1857**, 493–502.
57. V. Nanda, S. Senn, D. H. Pike, A. Rodriguez-Granillo, W. A. Hansen, S. D. Khare and D. Noy, *Biochim. Biophys. Acta, Bioenerg.*, 2016, **1857**, 531–538.

58. S. E. Mulholland, B. R. Gibney, F. Rabanal and P. L. Dutton, *J. Am. Chem. Soc.*, 1998, **120**, 10296–10302.
59. B. R. Gibney, S. E. Mulholland, F. Rabanal and P. L. Dutton, *Proc. Natl. Acad. Sci. U. S. A.*, 1996, **93**, 15041–15046.
60. K. B. Musgrave, C. E. Laplaza, R. H. Holm, B. Hedman and K. O. Hodgson, *J. Am. Chem. Soc.*, 2002, **124**, 3083–3092.
61. J. Grzyb, F. Xu, L. Weiner, E. J. Reijerse, W. Lubitz, V. Nanda and D. Noy, *Biochim. Biophys. Acta*, 2010, **1797**, 406–413.
62. V. Nanda, M. M. Rosenblatt, A. Osyczka, H. Kono, Z. Getahun, P. L. Dutton, J. G. Saven and W. F. DeGrado, *J. Am. Chem. Soc.*, 2005, **127**, 5804–5805.
63. R. E. Sharp, C. C. Moser, F. Rabanal and P. L. Dutton, *Proc. Natl. Acad. Sci. U. S. A.*, 1998, **95**, 10465–10470.
64. C. Bialas, L. E. Jarocha and K. B. Henbest, *J. Am. Chem. Soc.*, 2016, **138**, 16584–16587.
65. S. Hay, K. Westerlund and C. Tommos, *J. Phys. Chem. B*, 2007, **111**, 3488–3495.
66. Z. Zhao, PhD Thesis, University of Pennsylvania, 2016.
67. J. A. Mancini, G. Kodali, J. Jiang, K. R. Reddy, J. S. Lindsey, D. A. Bryant, P. L. Dutton and C. C. Moser, *J. R. Soc., Interface*, 2017, **14**, 20160896.
68. G. R. Dieckmann, D. K. McRorie, D. L. Tierney, L. M. Utschig, C. P. Singer, T. V. O'Halloran, J. E. Penner-Hahn, W. F. DeGrado and V. L. Pecoraro, *J. Am. Chem. Soc.*, 1997, **119**, 6195–6196.
69. B. T. Farrer, C. P. McClure, J. P. Penner-Hahn and V. L. Pecoraro, *Inorg. Chem.*, 2000, **39**, 5422–5423.
70. D. S. Touw, C. E. Nordman, J. A. Stuckey and V. L. Pecoraro, *Proc. Natl. Acad. Sci. U. S. A.*, 2007, **104**, 11969–11974.
71. M. Tegoni, F. Yu, M. Bersellini, J. E. Penner-Hahn and V. L. Pecoraro, *Proc. Natl. Acad. Sci.*, 2012, **109**, 21234–21239.
72. M. L. Zastrow, A. F. A. Peacock, J. A. Stuckey and V. L. Pecoraro, *Nat. Chem*, 2012, **4**, 118–123.
73. J. R. Calhoun, F. Nastri, O. Maglio, V. Pavone, A. Lombardi and W. F. DeGrado, *Biopolymers*, 2005, **80**, 264–278.
74. S. Geremia, L. Di Costanzo, L. Randaccio, D. E. Engel, A. Lombardi, F. Nastri and W. F. DeGrado, *J. Am. Chem. Soc.*, 2005, **127**, 17266–17276.
75. J. R. Calhoun, W. Liu, K. Spiegel, M. Dal Peraro, M. L. Klein, K. G. Valentine, A. J. Wand and W. F. DeGrado, *Structure*, 2008, **16**, 210–215.
76. H. Irving and R. J. P. Williams, *J. Chem. Soc.*, 1953, 3192–3219.
77. C. Tommos, J. J. Skalicky, D. L. Pilloud, A. J. Wand and P. L. Dutton, *Biochemistry*, 1999, **38**, 9495–9507.
78. S. D. Glover, C. Jorge, L. Liang, K. G. Valentine, L. Hammarstrom and C. Tommos, *J. Am. Chem. Soc.*, 2014, **136**, 14039–14051.
79. B. W. Berry, M. C. Martínez-Rivera and C. Tommos, *Proc. Natl. Acad. Sci.*, 2012, **109**, 9739–9743.

80. Q.-H. Dai, C. Tommos, E. J. Fuentes, M. R. A. Blomberg, P. L. Dutton and A. J. Wand, *J. Am. Chem. Soc.*, 2002, **124**, 10952–10953.
81. S. Hay, K. Westerlund and C. Tommos, *Biochemistry*, 2005, **44**, 11891–11902.
82. K. Westerlund, S. D. Moran, H. K. Privett, S. Hay, J. Jarvet, B. R. Gibney and C. Tommos, *Prot. Eng., Des. Sel.*, 2008, **21**, 645–652.
83. B. R. Lichtenstein, V. R. Moorman, J. F. Cerda, A. J. Wand and P. L. Dutton, *Chem. Commun.*, 2012, **48**, 1997–1999.
84. B. R. Lichtenstein, C. Bialas, J. F. Cerda, B. A. Fry, P. L. Dutton and C. C. Moser, *Angew. Chem., Int. Ed.*, 2015, **127**, 13830–13833.
85. R. A. Marcus and N. Sutin, *Biochim. Biophys. Acta, Bioenerg.*, 1985, **811**, 265–322.
86. M. R. Gunner, D. E. Robertson and P. L. Dutton, *J. Phys. Chem.*, 1986, **90**, 3783–3795.
87. N. W. Woodbury, W. W. Parson, M. R. Gunner, R. C. Prince and P. L. Dutton, *Biochim. Biophys. Acta*, 1986, **851**, 6–22.
88. A. Pesce, D. De Sanctis, M. Nardini, S. Dewilde, L. Moens, T. Hankeln, T. Burmester, P. Ascenzi and M. Bolognesi, *TBMB*, 2004, **56**, 657–664.
89. R. L. Koder, J. L. R. Anderson, L. A. Solomon, K. S. Reddy, C. C. Moser and P. L. Dutton, *Nature*, 2009, **458**, 305–309.
90. S. C. Hokanson, PhD Thesis, University of Pennsylvania, 2010.
91. B. A. Fry, L. A. Solomon, P. L. Dutton and C. C. Moser, *Biochim. Biophys. Acta*, 2016, **1857**, 513–521.
92. T. A. Farid, G. Kodali, L. A. Solomon, B. R. Lichtenstein, M. M. Sheehan, B. A. Fry, C. Bialas, N. M. Ennist, J. A. Siedlecki, Z. Zhao, M. A. Stetz, K. G. Valentine, J. L. R. Anderson, A. J. Wand, B. M. Discher, C. C. Moser and P. L. Dutton, *Nat. Chem. Biol.*, 2013, **9**, 826–833.
93. M. M. Martinez-Rivera, B. W. Berry, K. G. Valentine, K. Westerlund, S. Hay and C. Tommos, *J. Am. Chem. Soc.*, 2011, **133**, 17786–17795.
94. J. Bonin, C. Costentin, M. Robert, M. Routier and J.-M. Savéant, *J. Am. Chem. Soc.*, 2013, **135**, 14359–14366.
95. F. Rappaport and B. Diner, *Coord. Chem. Rev.*, 2008, **252**, 259–272.
96. R. L. Burnap, *Phys. Chem. Chem. Phys.*, 2004, **6**, 4803–4807.
97. H. Bao and R. L. Burnap, *Front. Plant Sci.*, 2016, **7**, 578.
98. N. Keren, A. Berg, P. VanKan, H. Levanon and I. Ohad, *Proc. Natl. Acad. Sci. U. S. A.*, 1997, **94**, 1579–1584.
99. M. Thielges, G. Uyeda, A. Cámara-Artigas, L. Kálmán, J. C. Williams and J. P. Allen, *Biochemistry*, 2005, **44**, 7389–7394.
100. L. Kálmán, J. C. Williams and J. P. Allen, *Biochemistry*, 2011, **50**, 3310–3320.
101. J. P. Allen, T. L. Olson, P. Oyala, W.-J. Lee, A. A. Tufts and J. C. Williams, *Proc. Natl. Acad. Sci.*, 2012, **109**, 2314–2318.
102. B. Conlan, N. Cox, J. H. Su, W. Hillier, J. Messinger, W. Lubitz, P. L. Dutton and T. Wydrzynski, *Biochim. Biophys. Acta, Bioenerg.*, 2009, **1787**, 1112–1121.

103. D. M. Taverna and R. A. Goldstein, *Proteins*, 2001, **46**, 105–109.
104. G. I. Makhatadze and P. L. Privalov, *Adv. Protein Chem.*, 1995, **47**, 307–425.
105. Y. Maeda, N. Javid, K. Duncan, L. Birchall, K. F. Gibson, D. Cannon, Y. Kanetsuki, C. Knapp, T. Tuttle, R. V. Ulijn and H. Matsui, *J. Am. Chem. Soc.*, 2014, **136**, 15893–15896.
106. E. Fischer, *Nobel Lectures, Chemistry 1901–1921*, Elsevier Publ. Co., Amsterdam, 1966.

CHAPTER 2

Structure of Respiratory Complex I: "Minimal" Bacterial and "De luxe" Mammalian Versions

LEONID A. SAZANOV

Institute of Science and Technology Austria (IST Austria), Am Campus 1, 3400 Klosterneuburg, Austria
Email: sazanov@ist.ac.at

2.1 Introduction

Complex I catalyses the transfer of two electrons from NADH to quinone, coupled to the translocation of protons across the membrane, contributing to the proton-motive force required for the synthesis of ATP,[1–5] according to eqn (1):

$$\mathrm{NADH + Q + H^+ + nH^+_{in} \rightarrow NAD^+ + QH_2 + nH^+_{out}} \quad (1)$$

Here "in" refers to the mitochondrial matrix or bacterial cytoplasm (N-side of the membrane), and "out" to the intermembrane space or bacterial periplasm (P-side). The reaction is fully reversible – in the presence of high proton-motive force complex I can reduce NAD^+ using quinol as a source of electrons.[6] The number of protons translocated per NADH oxidized ("n") is currently considered to be four, although this stoichiometry

Chemical Biology No. 5
Mechanisms of Primary Energy Transduction in Biology
Edited by Mårten Wikström

Published by the Royal Society of Chemistry, www.rsc.org

was studied mostly for mitochondrial enzyme,[7–9] and less so for bacterial counterpart.[10] Recent studies with both systems (bovine and *Paracoccus denitrificans*) confirm this number.[11] Different quinones are used as electron acceptors by complex I from different species. The mitochondrial enzyme uses ubiquinone, complex I from *Thermus thermophilus* uses menaquinone and *Escherichia coli* complex I uses either ubiquinone or menaquinone, depending on the growth conditions encountered by the bacteria.[2,3]

Complex I is one of the largest macromolecular assemblies known and is the largest component of the respiratory chain. The mammalian mitochondrial enzyme consists of 45 different subunits of about 980 kDa in total.[12–14] The best studied examples are complex I from bovine heart,[1,13] ovine heart,[14,15] the obligate aerobic yeast *Yarrowia lipolytica*[16–18] and the fungus *Neurospora crassa*.[19] The prokaryotic enzyme is simpler and generally consists of 14 conserved "core" subunits with a combined molecular mass of about 550 kDa.[2,20] *Escherichia coli* and few other bacterial enzymes consist of 13 subunits because two subunits are fused (NuoC and NuoD).[21] *Thermus thermophilus* and its closest relatives contain additional frataxin-like and chaperone-like subunits and so have 16 subunits in the entire complex.[20,22] The best studied examples of bacterial complex I are from *E. coli*,[23] *T. thermophilus*,[20,24] *Paracoccus denitrificans*[25,26] and *Aquifex aeolicus*.[27] Prokaryotic complex I is also called NDH-1 (NADH dehydrogenase-1), in contrast to the simpler non-proton pumping NDH-2 enzyme.[28] In bacteria, such as *E. coli*, expression of complex I depends on the growth conditions and can be induced at low oxygen concentrations, when higher effectiveness of the respiratory chain is required.[29,30] At high oxygen concentrations complex I may be replaced by faster, non energy-conserving NDH-2 enzyme.[30]

Apart from canonical complex I, its less well characterized analogues are also found in chloroplasts, cyanobacteria and archaea. In these cases, the three-subunit NADH dehydrogenase module of complex I is replaced by $F_{420}H_2$ dehydrogenase in archaea and by an as-yet unknown electron input module (interacting with ferredoxin) in chloroplasts and cyanobacteria.[31–33]

Complex I contributes about 40% of the proton flux across inner mitochondrial membranes, used for ATP synthesis, and so is central to energy production in eukaryotes. Since neuronal tissues rely heavily on respiration for their energy needs, even a relatively small drop in complex I activity due to mutations can lead to severe human neurodegenerative diseases.[34] Complex I has also been suggested to be a major source of reactive oxygen species (ROS) in mitochondria, which can damage mitochondrial DNA (mtDNA) and may be one of the causes of aging.[35] Parkinson's disease, at least in its sporadic form, may be caused by increased ROS production from malfunctioning complex I.[36] ROS production also increases dramatically during the reverse electron transport (RET).[37] The physiological role of RET is not yet established. It may contribute to increased ROS levels when macrophages repurpose mitochondria from ATP synthesis to ROS production in order to promote a pro-inflammatory state.[38] RET has also been implicated in metabolic signalling.[39]

Both mitochondrial and bacterial enzymes have a characteristic L-shaped structure, with the hydrophobic arm embedded in the membrane and the hydrophilic peripheral arm protruding into the mitochondrial matrix or the bacterial cytoplasm.[27,40,41] Nomenclature of complex I subunits historically differs for different species, complicating comparisons (Table 2.1). Using harsh detergents or unfavourable pH, the enzyme can be split into three main domains, likely reflecting its evolutionary origins.[23,31,42] The hydrophilic peripheral arm consists of the dehydrogenase domain and connecting domain, providing a link to the hydrophobic membrane domain. The peripheral arm consists of subunits Nqo1–6 and 9 (*T. thermophilus* nomenclature) and a hydrophobic membrane arm consists of Nqo7–8 and Nqo10–14.[31] The order of genes in bacterial operons encoding complex I subunits generally reflects these domains, with the exception that subunits Nqo7 and Nqo6 have swapped places.[43] All known cofactors, the primary electron acceptor flavin mononucleotide (FMN) and eight to nine iron–sulphur (Fe–S) clusters,[2,44,45] are found in the hydrophilic arm, comprising the catalytic core of the enzyme. The proton-pumping machinery must reside in the membrane arm, which contains 63–64 transmembrane (TM) helices in bacteria[22,46] and about 80 in mitochondria.[15,17,18,47] The mechanism of coupling between electron transfer and proton translocation is not fully established. Two main models were discussed previously: indirect with some contributions from direct coupling (redox-driven through chemical intermediates, usually employing modifications of the Q cycle)[48,49] and purely indirect or conformation-driven coupling.[2–4,22,50,51] Recently consensus has moved towards the latter.[15,18,22,52]

Analogues of all 14 conserved subunits of bacterial complex I are found in the mitochondrial enzyme,[1] and they contain equivalent redox components.[2] Thus, the mechanism is likely to be conserved throughout the species and the bacterial enzyme represents a useful 'minimal' model of mitochondrial complex I. In addition to the core subunits, mitochondrial complex I contains many (31 in bovine enzyme[13]) additional subunits, termed "supernumerary" or "accessory". These subunits are found in both peripheral and membrane domains of the complex and most are smaller than 20 kDa.[4,13,53] Many of them have similarities to the proteins of known function, such as acyl-carrier protein or a cell death regulatory protein.[54] We have purified and characterized complex I from *Paracoccus denitrificans*, a close relative of mitochondria's ancestors. Surprisingly, this enzyme was found to contain three "accessory" subunits common to all eukaryotes,[26] suggesting that the elaboration of complex I started before the original symbiotic event. The physiological role of the "accessory" subunits is not fully established. Some of them provide structural scaffold around the core, stabilizing the complex, and some may have regulatory roles.[15,18,52]

Table 2.1 Complex I core subunits.

Module	*Escherichia coli*	*Thermus thermophilus*	*Yarrowia lipolytica*	*Bos taurus* (Bovine)	*Homo sapiens*[a]	Cofactors[b]
Peripheral arm						
Dehydrogenase (N)	NuoF 445[c] 49.3[d]	Nqo1 438 48.6	NUBM 470 51.6	51 kDa 444 48.5	NDUFV1	FMN N3 (4Fe[51])
	NuoE 166 18.6	Nqo2 180 20.2	NUHM 215 24.1	24 kDa 217 23.8	NDUFV2	N1a (2Fe[24])
	NuoG 907 100.2	Nqo3 783 86.5	NUAM 694 75.2	75 kDa 704 77.0	NDUFS1	N1b (2Fe[75]) N4 (4Fe[75]C) N5 (4Fe[75]H) (N7)[e]
Connecting (Q)	NuoD (NuoCD)[f] 595 68.1	Nqo4 409 46.4	NUCM 444 49.9	49 kDa 430 49.1	NDUFS2	No cofactor
	NuoC see above	Nqo5 207 23.4	NUGM 251 29.2	30 kDa 228 26.4	NDUFS3	No cofactor
	NuoI 180 20.5	Nqo9 182 20.1	NUIM 198 22.3	TYKY 176 20.2	NDUFS8	N6a (4Fe[TY]1) N6b (4Fe[TY]2)
	NuoB 220 25.1	Nqo6 180 20.1	NUKM 183 20.4	PSST 179 20.1	NDUFS7	N2 (4Fe[PS])

Membrane arm														TMH[g]
—	NuoH	325	36.2	Nqo8	365	41.0	NU1M	341	38.3	ND1	318	35.7	ND1	8–9
Pumping (P)	NuoA	147	16.5	Nqo7	119	13.1	NU3M	128	14.5	ND3	115	13.1	ND3	3
	NuoJ	184	19.9	Nqo10	176	18.6	NU6M	185	20.8	ND6	175	19.1	ND6	5
	NuoK	100	10.8	Nqo11	95	10.0	NULM	89	9.8	ND4L	98	10.8	ND4L	3
	NuoN	485	52.0	Nqo14	427	44.4	NU2M	469	53.3	ND2	347	39.3	ND2	11–14
	NuoM	509	56.5	Nqo13	469	49.2	NU4M	486	54.5	ND4	459	52.1	ND4	14
	NuoL	613	66.5	Nqo12	606	65.1	NU5M	655	73.7	ND5	606	68.3	ND5	16–17
Total core		4876	540		4636	507		4808	538		4496	503.5		

[a] *Homo sapiens* core subunit names are shown for reference, no structures of human Complex I have been elucidated.
[b] The traditional nomenclature for Fe–S clusters (Nx, derived from initially described electron paramagnetic resonance (EPR) signatures [50]), as well as the nomenclature proposed recently [51] on the basis of re-assignment of EPR signals to structurally observed clusters, is shown. In the new nomenclature, clusters are named according to their nuclearity (2Fe or 4Fe), their subunit location (using bovine nomenclature) and when necessary, as ligated by four Cys (C) or three Cys and one His (H).
[c] Number of amino acids in the mature subunit.
[d] MW of the mature subunit in kilo-Daltons.
[e] Cluster N7 is present only in some bacteria (for example, *E. coli* and *T. thermophilus*).
[f] Subunits NuoC and NuoD are fused in *E. coli* and some other bacteria.
[g] Number of transmembrane helices.

2.2 Overview of the Bacterial Enzyme Structure

The first X-ray crystal structure for complex I was that of the hydrophilic arm of the enzyme from *T. thermophilus*, determined to 3.1 Å resolution, in both the apo and NADH bound forms.[55,56] These structures allowed identification of the subunits and prosthetic groups within this sub-complex. The X-ray crystal structure of the membrane arm from *E. coli* to 3.9 Å resolution allowed identification and determination of the organization of subunits in the membrane arm.[51] Later the resolution was improved to 3.0 Å resolution and the atomic model of the membrane domain was described, revealing many unusual features of the protein fold.[46] Later X-ray crystal structure of the entire complex from *T. thermophilus* to 3.3 Å resolution finally revealed how the hydrophilic and membrane arms are put together.[22]

In the overview of this structure, shown in Figure 2.1, the peripheral arm of complex I is a Y-shaped assembly about 140 Å high and the membrane arm is bean-shaped, about 180 Å long. The peripheral arm from *T. thermophilus* contains nine subunits: Nqo1–6, Nqo9 and unexpectedly, two additional subunits not part of the *nqo* operon, Nqo15[56] and loosely-bound

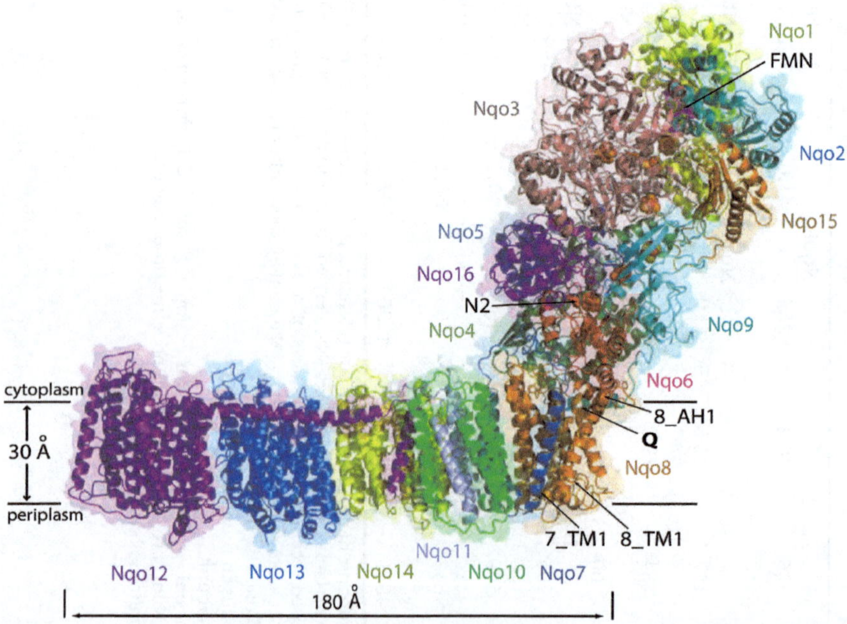

Figure 2.1 Overview of the entire complex I from *T. thermophilus*. Subunits are coloured differently and labelled (PDB code 4HEA). FMN and Fe–S clusters are shown as magenta and red-orange spheres, respectively, with cluster N2 labelled. Key helices around the entry point (Q) into the quinone reaction chamber, and approximate membrane position are indicated. Reprinted by permission from Macmillan Publishers Ltd: *Nature* (ref. 22), copyright (2013).

Nqo16.[22] One uppermost tip of the peripheral arm is formed by the subunits Nqo1 and Nqo2, and the other by the C-terminal domain of Nqo3. The main stem is formed by the N-terminal domain of Nqo3 and the connecting domain subunits. Its lower part consists of subunits Nqo4 and Nqo6 (the latter coordinates the terminal Fe–S cluster N2), and it forms an interface with the membrane domain, sitting on top of subunit Nqo8.

The membrane arm comprises seven subunits: Nqo7–8 and Nqo10–14, with 64 TM helices, most of them lying normal to the membrane.[22,46,51] Sixteen TM helices are present in subunit Nqo12, 14 in Nqo13, 14 in Nqo14, 5 in Nqo10, 3 in Nqo11, 3 in Nqo7 and 9 in Nqo8. Subunits Nqo12, Nqo13 and Nqo14 are arranged, like carriages in the train, towards the distal end of the membrane arm, with Nqo12 furthest away from the hydrophilic arm. These three subunits share sequence similarity with Na^+/H^+ Mrp (Multiple resistance and pH adaptation) antiporter complex subunits (MrpA, MrpD and MrpD, respectively[57,58]), contain 14 conserved TM helices each and are likely responsible for pumping one proton each across the membrane. A fourth proton channel is formed by smaller subunits Nqo7, 8, 10 and 11,[22] near the interface with the hydrophilic arm. Unexpected at the time, the C-terminus of Nqo12 was found to form an extended 110 Å alpha helix, which runs along the membrane surface and terminates adjacent to Nqo14.[22,46,51] This amphipathic helix, termed helix HL, links together antiporter-like subunits as a putative coupling/stabilizing element. Another such element (βH motif) is formed from series of connected β-hairpins and helices on the opposite side of the domain.[46] Recent mutagenesis studies suggest that helix HL plays a mostly stabilizing role, strapping antiporter-like subunits together,[59] although the absence of any direct coupling role remains to be confirmed.

Most subunits of the complex have structural homology to other proteins, which apparently served as smaller "building blocks" during the evolution of the enzyme. These "blocks", containing different redox centres, fit together in the peripheral arm in such a way that a continuous electron transfer pathway through the enzyme is formed. The evolutionary origins of complex I can thus be traced to different types of ferredoxins (subunits Nqo2 and Nqo9), FeFe-hydrogenases (N-terminus of subunit Nqo3), molybdopterin-containing enzymes (C-terminus of subunit Nqo3) and NiFe-hydrogenases (subunits Nqo4 and Nqo6). Such similarities were noted also from sequence comparisons.[1,4,31,60]

Many protein complexes with an as yet unknown structure seem to share bigger "building blocks" with complex I. Several NAD^+-reducing enzymes, for example cytoplasmic NiFe-hydrogenase[1] and formate dehydrogenase[61] from *Ralstonia eutropha* contain analogues of subunits Nqo1–3, the "dehydrogenase domain" of complex I. Membrane-bound NiFe-hydrogenases contain analogues of the "connecting domain" subunits (Nqo4–6, and Nqo9) and of most membrane domain subunits, which may be involved in proton pumping.[60] It is likely that complex I originated from the unification of two pre-evolved complexes, a soluble hydrogenase and an Mrp-like antiporter.[62]

2.3 Electron Transfer Pathway

2.3.1 NADH/FMN Binding Site

The substrate NADH binds in subunit Nqo1 adjacent to the FMN moiety near the tip of the hydrophilic arm. The two electrons that are extracted from NADH are first transferred to a FMN molecule as a hydride ion before being passed down a series of iron–sulphur centres one at a time to a quinone bound at the interface with the membrane arm.

Complex I contains a non-traditional Rossmann fold, which binds both FMN and NADH through the incorporation of an additional glycine rich loop (Figure 2.2b).[55,56] The residues involved in interactions with both of the cofactors are very well conserved suggesting this binding pocket is consistent between species as varied as bacteria and humans. Within the binding pocket the nicotinamide ring of NADH stacks against the exposed face of the isoalloxazine ring of the bound FMN, in a manner similar to that seen in other nucleotide binding flavoenzymes.[55] This positions the B-face of the nicotinamide ring against the *re* face of the isoalloxazine ring to allow the 4B hydrogen of NADH to transfer as a hydride to the N5 atom of FMN (Figure 2.2b). The transfer of the hydride is unusually fast and the distance between these two atoms is shorter than usual, at 3.2 Å. The short distance between the two nucleotides can be explained by the close proximity of the Cβ atom of Nqo1 Glu97 which is within van der Waals contact distance with the C4N atom of the bound NADH, forcing the two nucleotides together.

The binding of the adenine ring of the NADH moiety causes a shift of around 1.5 Å of the loop formed by Nqo1 residues 202 to 207 towards the bound NADH. This positions the side chain of Nqo1 Phe205 1.7 Å closer to the adenine ring promoting a stacking interaction and a hydrogen bond forms between the side chain of Nqo1 Lys202 and the phosphate moiety of the NADH molecule.[55] Complex I is unusual in its ability to use deamino-NADH as a substrate. This is due to NADH binding, unusually, in an extended conformation, so that N6A of the adenine ring (which is an oxygen in deamino-NADH) forms no interactions with the protein.

When the structure of the hydrophilic domain with NADH bound and Fe–S clusters reduced is compared to the oxidised domain structure, apart from the changes listed above around the NADH binding site, a reproducible shift of around 1 Å can be seen in a 4-helix bundle in Nqo4 and helices H1 and H2 in Nqo6.[55] These conformational changes are likely driven by the reduction of nearby cluster N2, and may be propagated into the membrane arm as part of the coupling mechanism, since these helices interact with adjacent TM helices.

2.3.2 Fe–S Clusters

Seven iron–sulphur clusters form a redox chain between the FMN and the quinone-binding site, spanning a distance of some 95 Å (Figure 2.2a,

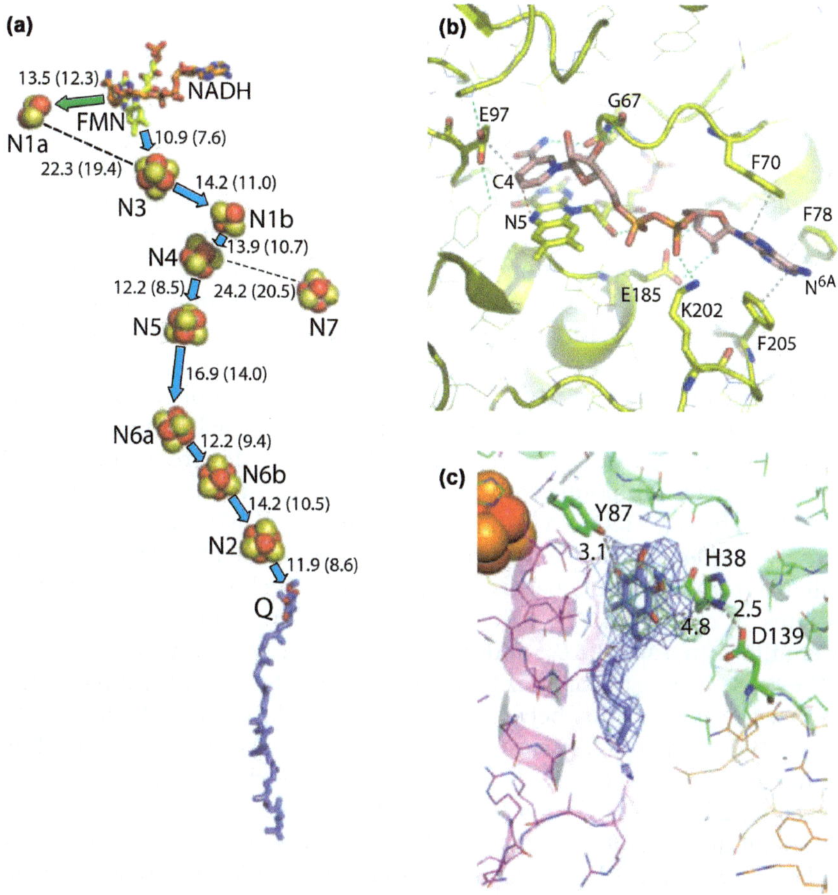

Figure 2.2 Redox centres and substrate binding sites in complex I. (a) Arrangement of redox centres in *T. thermophilus* complex I. The main pathway of electron transfer is indicated by blue arrows, and a diversion to cluster N1a by a green arrow. The Fe–S clusters are labelled according to the traditional nomenclature (see also Table 2.1). The distances between the centres given in Å were calculated both centre-to-centre and edge-to-edge (shown in parentheses). Clusters N3 and N4 are separated by 17.6 Å (13.8 Å edge-to-edge), and clusters N1b and N5 by 19.2 Å (16.7 Å edge-to-edge). Positions of NADH[55] and quinone[22] are based on experimental data, except that the entire ubiquinone tail was modelled into the Q-binding cavity. (b) The NADH binding site,[55] viewed from the solvent-exposed side (PDB code 3IAM). FMN and residues involved in NADH binding are shown as sticks with carbon in yellow and NADH with carbon in salmon. Potential interactions with Nqo1 residues are indicated. Hydrogen bonds are represented as dotted green lines, hydrophobic stacking interactions in grey, the hydride (H^-) transfer path in red and van der Waals contact between Glu97 and C4 of NADH in grey. (c) Bound decyl-ubiquinone, shown with experimental electron density.[22] Nqo4 residues interacting with the headgroup are indicated. Potential polar interactions are shown labelled with distances in Å. Reprinted by permission from Macmillan Publishers Ltd: *Nature Reviews Molecular Cell Biology*, (ref. 77), copyright (2015).

Table 2.1).[45] After the reduction of FMN by NADH, one electron is transferred from FMN to a tetranuclear cluster N3 that also resides in subunit Nqo1. From cluster N3 the electron is then passed to the binuclear cluster N1b and then on to tetranuclear clusters N4 and then N5 within Nqo3. The cluster N7 is also present in Nqo3 in some bacterial species but with 20 Å between this cluster and its closest cluster, N4, it is too removed from the redox chain to be part of the electron transfer pathway.[45] As cluster N7 is also only conserved in a small number of bacterial species it may represent an evolutionary remnant in complex I, required for stability in certain species.[63] From N5 the electrons are transferred to the two tetranuclear clusters in Nqo9; N6a and N6b, before being passed to the final tetranuclear cluster, N2, within Nqo6.

The electron transfer pathway can be traced unambiguously through the structure, with all redox centres within 14 Å from each other, a maximal distance for electron transfer relevant in biology [64] (Figure 2.2a). At pH 7, the two-electron midpoint redox potential (E_m) of NADH is about −320 mV, of FMN about −340 mV and of ubiquinone (UQ) about +110 mV (*T. thermophilus* utilizes menaquinone (MQ), about −80 mV). The free energy available from NADH/UQ pair is about 430 mV, which means that complex I operates at close to 100% efficiency, achieving stoichiometry of $4H^+/2e^-$ (at equilibrium $4\Delta p = 2\Delta E_h$, and Δp is about 200 mV in mitochondria).

The Fe–S clusters of complex I have a "rollercoaster" profile (in part due to electrostatic interactions between neighbouring clusters), with clusters N3, N4 and N6a being roughly equipotential at ∼−250 mV, while cluster N2 has the highest redox potential of ∼−100 mV.[65] These clusters (as well as N1b) are reduced by NADH during turnover while clusters in between have very low potential (therefore not observed by EPR) and serve as "stepping stones" in the redox chain. Cluster N2, due to its high potential, draws electrons from the chain and passes them onto the quinone substrate, ubiquinone or menaquinone depending on the organism.[2] N2 lies only 12 Å from quinone head-group binding in a channel formed between Nqo6 and the adjacent Nqo4 subunit, reaching into the hydrophobic domain.[22] Nqo4 and Nqo6 comprise the base of the hydrophilic arm and form the main interaction with the hydrophobic arm.

Most clusters in complex I are [2Fe–2S] or [4Fe–4S] in standard all Cys coordination, except for clusters N5 and N2. [4Fe–4S] cluster N5 is 1His-3Cys coordinated, and clusters N5 and N6a are separated by the longest distance in the chain (Figure 2.2a), so they probably comprise a rate-limiting step in the N3 to N2 pathway. It remains to be established whether the conserved His ligation of N5 may be somehow related to its position at this "bottleneck", allowing regulation of the overall electron transfer rate.

The [4Fe–4S] cluster N2 is all Cys coordinated, but two of the cysteines are consecutive, forming an unusual tandem cysteine motif consisting of Nqo6 Cys45 and Cys46, with Cys111 and Cys140 completing the ligation.[55] The tandem cysteine motif is fully conserved in complex I and the *T. thermophilus* structure revealed strained geometry in these residues. This leads to disconnection of one of the tandem cysteines in the reduced hydrophilic arm structures and associated shifts of nearby helices.[55] Such disconnection may not happen in the more stable entire complex, but in any case the preservation of such unfavourable geometry during evolution indicates that conformational flexibility around cluster N2 is important for the mechanism.

Modelling shows that an electron can be transferred from FMN to Q within 50 μs, much faster than about 5 ms required for one catalytic turnover.[66] Real-time EPR measurements confirmed that first electron is transferred from NADH to cluster N2 within 90 μs.[67] This is consistent with the fact that most EPR-visible clusters of complex I are reduced under steady-state NADH oxidation, so that it is likely that N2 oxidation and quinone binding/release are rate-limiting.[68] Thus, all electron transfer steps up to and including cluster N2 are fast and probably not linked to large conformational re-arrangements, indicating that most redox energy is released (and utilised for proton pumping) during quinone redox reactions.

2.3.3 Quinone Binding Site

The quinone-binding site, formed between subunits Nqo4, Nqo6, Nqo7 and Nqo8, is unusually elongated and enclosed from the solvent. This is in stark contrast to other membrane proteins where quinone sites are usually quite open and accommodate mostly the headgroup only. The hydrophilic head of the quinone molecule binds in the deep end of a cavity, about 15 Å away from the membrane surface. It is hydrogen bonded to Tyr87 and His38 from Nqo4; both residues are invariant and essential for activity (Figure 2.2c).[22] The entry point for the quinone head-group is very narrow (approximately 2–3×4–5 Å) and so it may not allow any solvent into the cavity when quinone is bound, as its tail will block the entrance (Figure 2.3b). Surprisingly, the chamber is lined mostly by hydrophilic residues, which may guide quinone headgroup deep into the cavity (Figure 2.3a).[22] Only one side of the chamber is hydrophobic and so can accommodate the hydrophobic tail of the quinone. In contrast to some recent proposals,[69] the structure provides no evidence for any second quinone-binding site.

Why does complex I have such an unusual enclosed Q-site? One possible reason is that this allows complex I to fully control the protonation of quinone, since the headgroup is tightly surrounded by protein side-chains. In this case double negative charge from electrons delivered to this area (and stored probably not on quinone itself but on coordinating side-chains) can be used to drive conformational changes in the membrane arm, leading to proton pumping. Once the work is done, the coordinating residues may be protonated and the quinol released. It is amazing that in nature during the catalytic turnover the very long ubiquinone molecule slips all the way in and out of the reaction chamber 200 times per second. The intricacies of this tightly coordinated process are being studied by molecular dynamics modelling (unpublished data).

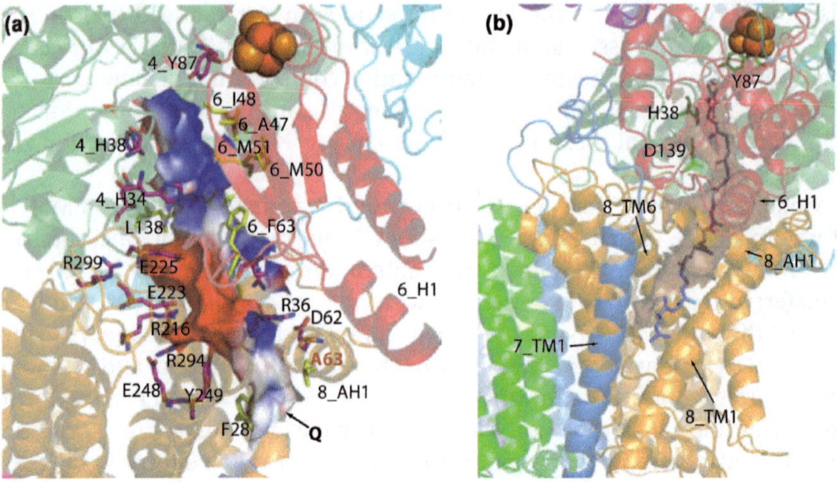

Figure 2.3 The quinone-binding cavity. Hydrophilic arm subunit Nqo4 is shown in green, Nqo6 in red and hydrophobic arm subunit Nqo8 in orange. Iron–sulfur cluster N2 is shown as red–orange spheres. (a) The internal solvent-accessible surface of the quinone chamber is shown, coloured red for negative, white for neutral and blue for positive surface charges. Charged residues lining the cavity are shown with carbon in magenta and hydrophobic residues in yellow. Residues are labelled with prefix indicating subunit (omitted for Nqo8). Ala63, the site of the primary LHON disease mutation,[110] is labelled in red. (b) Theoretical model of bound ubiquinone-10. The headgroup is positioned according to experimental observations with decyl-ubiquinone[22] and the structure of the hydrophobic tail was modelled to fit within the cavity. Carbon atom in cyan indicates the 8th isoprenoid unit. The quinone chamber is shown with surface in transparent brown and helices framing its entry point are indicated. Movable helix 6_H1,[55] interacting with 8_AH1, is also labelled. Reprinted by permission from Macmillan Publishers Ltd: *Nature* (ref. 22), copyright (2013).

2.3.4 Cluster N1a and ROS Production

Adjacent to FMN, within subunit Nqo2, lies the binuclear cluster N1a. It does not form part of the main redox chain, representing a diversion (Figure 2.2a). However, this cluster is fully conserved and so must be important for function. Its potential is usually lower (*E. coli* is an exception) than that of isopotential clusters. We have suggested that N1a may play the role of an antioxidant, preventing excessive generation of ROS by complex I.[45] Flavin is now generally regarded as a main source of ROS in complex I,[3,70,71] although cluster N2 and (semi)quinone[72] are also being discussed. FMN accepts two electrons simultaneously (as a hydride) from NADH and transfers them one at a time to one-electron carriers Fe–S clusters. The one-electron redox potential of N1a (∼−380 mV in bovine) is too low for accepting the first electron from reduced $FMNH_2$, but it is suitable for accepting the second electron, from flavosemiquinone (midpoint potentials are about −300 mV for $FMNH_2$–flavosemiquinone and about −390 mV for flavosemiquinone–oxidized flavin[73]). Thus, two electrons from flavin can be donated nearly simultaneously to two nearby clusters, N3 (first) and N1a (second). This mechanism will prevent any significant accumulation of the flavosemiquinone intermediate, which could otherwise react with oxygen, leading to ROS production. The flavin is exposed to the solvent at the deep end of the NADH-binding cavity (Figure 2.2b), whereas cluster N1a is shielded and so it is suitable for such a temporary storage of electrons. As oxidation of cluster N2 by quinone is a likely rate-limiting step, electrons can move from N1a, *via* FMN, towards cluster N3 only when, and as soon as, N2 is re-oxidized. The flavosemiquinone formed during this transfer will be very short-lived, as the electron transfer between redox centers is several orders of magnitude faster than quinone binding/release and the flavosemiquinone–FMN pair represents a small, but unfavourable redox potential barrier in this pathway. This is in broad agreement with studies by Ransac *et al.* who have suggested that N1a is mainly reduced by the flavosemiquinone species and can be oxidized by a different flavosemiquinone species.[74] In their modelling study they showed that this could lead to a decrease in the lifetime of the flavosemiquinone and thus reduce ROS production. Complex II appears to possess a similar mechanism (by having an off-pathway heme) for efficient coupling of two- and one- electron transfers, avoiding radical formation.[75] Studies with site-directed mutants indicated that changes to N1a environment do influence enzyme activity and ROS production, although the difference was deemed to be small.[76] We have observed larger differences with different set of mutants (unpublished data), therefore further work is required to fully establish the role of N1a.

2.4 Proton-translocating Channels

Proton pumping occurs in the membrane arm, most likely through the four channels identified in the structure, as mentioned above—three in antiporter-like subunits and one at the interface with the hydrophilic domain (Figure 2.4). As the four proton-pumping units are up to 120 Å from the quinone binding site this leads to the question of how reduction of the quinone is linked to proton pumping. The most likely answer, supported by the structural features, is that long-range conformational coupling is at play.

One of the most fascinating features of the structure is the presence of conserved charged and polar residues in the middle of the membrane, which extend from the quinone-binding site at the junction between the two main arms to the tip of the hydrophobic arm subunit Nqo12 (Figure 2.4).[22,46] These residues are found on the breaks in discontinuous TM helices and are likely surrounded by water molecules, spanning the entire length of the hydrophobic arm.[22] The whole arrangement thus represents a flexible hydrophilic central axis of the membrane domain.

Many residues from the central axis represent key residues of the putative proton-translocating channels. In the antiporter-like subunits, two sets of

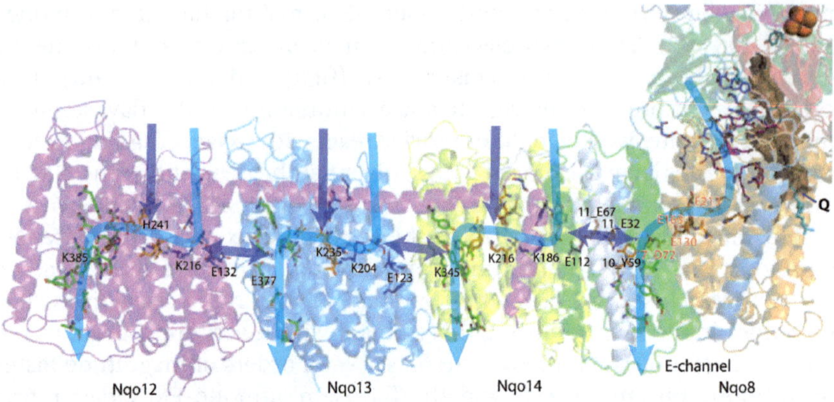

Figure 2.4 Proton translocation channels and the central hydrophilic axis. Illustrated using *T. thermophilus* structure (PDB 4HEA). Polar residues lining the channels are shown as sticks with carbon in dark blue for the first (N-terminal) half-channel, in green for the second (C-terminal) half-channel and in orange for connecting residues. Key residues for antiporter-like subunits, GluTM5 and LysTM7 from the first half-channel, Lys/HisTM8 from the connection and Lys/GluTM12 from the second half-channel, are labelled. Residues playing similar roles in the E-channel are also labelled (Glu/Asp quartet in red), with prefix indicating subunit (omitted for Nqo8). All key residues sit on the central hydrophilic axis of the membrane domain. The quinone binding cavity is shown in brown, with modelled ubiquinone molecule in cyan and residues connecting cavity to the E-channel in magenta. Originally[22] suggested proton translocation pathways are indicated by blue arrows, and additional proposed[77] paths (new entry sites and inter-subunit transfer) by violet arrows. Subunits are coloured as in Figure 2.1.

five conserved helices each (TMs 4–8 and TMs 9–13), are related to each other by internal symmetry. Crucially, the symmetry-related helices TM7 and TM12 are interrupted in the middle of the bilayer by an extended loop. Such discontinuous helices are functionally important for ion transport, introducing flexibility and charge to the middle of the membrane.[52,53] TM7 contains a central lysine (termed LysTM7), which is in close proximity to conserved key glutamate on TM5 (GluTM5), which can modulate pK_a of LysTM7. Jointly, TMs 4–8 form the cytoplasm-linked half-channel for proton translocation. The symmetry-related half-channel formed by TMs 9–13 is linked to the periplasm and also contains a central lysine (exceptionally a glutamate in Nqo13) within TM12 (Lys/GluTM12). Lysine from another discontinuous helix TM8 (histidine in Nqo12) and other conserved charged residues form the connection in the middle of the membrane.[22] Thus, each antiporter-like subunit contains a single proton channel formed from two connected half-channels (Figure 2.4). More detailed analysis[77] suggests that additional proton input pathways into the central parts of the subunits are possible: one from the cytoplasm roughly central along TM8 and another as a 'side-entry' from the interface between subunits, *via* GluTM5 (Figure 2.4, violet arrows). Such multiple input pathways would allow for effective capture of protons present in very low concentrations in the cytoplasm (high pH). Conversely, exit pathways into the periplasm are not extensive, consistent with the necessity for the protein to tightly control ejection of protons against the gradient into the low pH periplasm.

The fourth proton channel is also formed from two connected half-channels, cytoplasm-linked in subunit Nqo8 and periplasm-linked *via* subunits Nqo10 and Nqo11.[22] The arrangement of the TM helices in Nqo8 is unusual, as nearly all helices are dramatically tilted relative to membrane normal (core TM helices 2–6 are tilted by up to 45°). Surprisingly, Nqo8 core presents antiporter-like half-channel fold, with Glu130 and Glu163 in the GluTM5 position and Glu213 and Glu248 near the LysTM7 position. The C-terminal half of Nqo8 TM5 is completely unwound into the loop forming a large part of the quinone cavity. Overall, there are many more charged residues in the membrane part of Nqo8 compared to the antiporter-like subunits, with many of these residues being conserved in complex I. In the second half-channel, Glu32 in subunit Nqo11 mimics the GluTM12, which interacts with the conserved essential Tyr59 in subunit Nqo10. A Glu/Asp quartet and putative water molecules in the center of the membrane link the two half-channels into a single channel[22] (Figure 2.4). Thus, this channel is referred to as the "E-channel" due to this abundance of glutamate residues in its centre.

It is expensive for the protein to fold and assemble with such a high concentration of charged residues in the middle of the membrane. The overall architecture thus suggests that the central hydrophilic axis probably plays a dominating role in the propagation and coordination of conformational changes during the catalytic cycle.

2.5 Additional Subunits in Bacterial Enzyme

An unexpected find in the structure of the hydrophilic arm of *T. thermophilus* complex I was the presence of an additional subunit, Nqo15.[20,56] This protein is not part of the *nqo* operon and is only present in a few extremophilic prokaryotic species. Nqo15 has a fold that is highly homologous to frataxin,[78,79] a protein which is involved in the biogenesis of iron–sulphur centres and has no sequence homologue in *T. thermophilus*. Although the exact role of frataxin is still under some debate, it has been shown to form spherical oligomers, which can bind iron, suggesting a potential role in iron storage akin to ferritin.[80] Although the structures of frataxin and Nqo15 are homologous, sequentially these two proteins are very different. Frataxin contains several fully conserved acidic residues on the edge of the first β-stand and α-helix, which are involved in iron binding.[81] These residues are not present within Nqo15, which instead has a series of four histidine residues along a different β-strand on the face of its β-sheet. Within *T. thermophilus* complex I these histidines face into a channel formed between subunits Nqo1, 2, 3 and 15. Side chains of histidines from Nqo1 and Nqo3 also face into this channel. Divalent cations, including iron mimic manganese, have been shown to bind in this channel, in close proximity to the iron–sulphur clusters N3 and N1a.[55] This leads to the intriguing possibility that this subunit may be involved in iron binding and may have a role in iron–sulphur cluster regeneration. Nqo15 tightly interacts with its neighbours and can be removed from complex I only upon fragmentation of the hydrophilic domain (unpublished data).

In the structure of the entire *T. thermophilus* complex, another novel subunit was identified, which is not necessary for oxidoreductase or proton-pumping activity of the enzyme, but is essential for crystallization as it is involved in crystal contacts.[22] This 14.2 kDa subunit, dubbed Nqo16, has a fold consisting of five-stranded β-sheet flanked by two α-helices on each side, and belongs to the superfamily DUF3197 (NCBI) of proteins with unknown function, only found in thermophiles. Nqo16 might play a role as an assembly factor, as the proportion of *T. thermophilus* complex I containing Nqo16 varies depending on the cell growth.[22]

2.6 Coupling Between Electron Transfer and Proton Translocation

The total lack of redox groups in the hydrophobic domain begs the question of how does the electron transport drive proton pumping? The structure of bacterial complex I provides clear implications for the mechanism of coupling between electron transfer and proton translocation. Most of the redox energy is released during the quinone chemistry, which must drive conformational changes.

The most plausible scenario of the reaction mechanism is as follows: the transfer of two electrons from NADH to the quinone *via* the FMN and the Fe–S clusters results in charged species in the Q chamber, which interact electrostatically with the Glu/Asp quartet *via* a funnel of charged residues in between, driving conformational changes first in the E-channel. Shifts of helices observed upon reduction of cluster N2[55] probably help with these changes, allowing for full redox energy to be used. The conformational changes in the E-channel then propagate to the neighbouring antiporter-like subunit Nqo14, and on to distal Nqo13 and Nqo12, all through the flexible central hydrophilic axis (Figure 2.7).[22] The result of these concerted conformational changes are the changes in exposure to solvent and in pK_a of key residues in the half-channels, resulting in proton translocation. However, the exact extent of such movements is currently unclear. How exactly other coupling (or connecting) elements, helix HL and βH motif, are involved, also remains to be established by the determination of structures of different redox states of the complex.

2.7 Recent Structural Studies on Mammalian Complex I

Due to the high conservation of the core subunits across the species bacterial enzyme structure is a good basic model for the eukaryotic enzyme, as the catalytic mechanisms must be the same. However, the understanding of the role of about 30 "supernumerary" subunits, doubling the molecular weight of the enzyme, of course requires the solution of the structure of the mitochondrial complex. Recently, X-ray crystallography brought us the structure of the mitochondrial enzyme from aerobic yeast *Yarrowia lipolytica* at 3.8 Å resolution, with some supernumerary subunits modelled as poly-alanine and partial atomic models of core subunits. Overall, the complex was modelled at the atomic level for about 25% of the protein.[18] Meanwhile new developments in cryo-electron microscopy[82] gave the ability to solve structures of biological complexes in atomic detail. With the new direct electron detectors and better software it became possible to obtain high-resolution maps of difficult to crystalize biological assemblies like complex I. The first cryo-EM study of mammalian enzyme (*Bos taurus*) at 5 Å resolution resulted in poly-alanine models for the core and 14 supernumerary subunits,[83] and later a further eight supernumerary subunits were tentatively assigned.[84]

In 2016, two cryo-EM structures of mammalian complex I were published independently and nearly simultaneously. A 4.2 Å resolution model[15] for the bovine complex I had all subunits assigned but due to relatively low resolution the completeness of the atomic model was low for supernumerary subunits (27% atomic) and also for the core subunits (65% atomic), especially for 51, 24 and 75 kDa subunits (see extended data Tables 1 and 2 in ref. 15). We have published a 3.9 Å resolution structure of the ovine (*Ovis aries*) enzyme which is nearly complete (88% atomic).[15]

Later in 2016, the structure of porcine "respirasome" (supercomplex I.III_2.IV) was published, describing the supercomplex at 4.0 Å resolution and complex I at 3.6 Å.[85] This structure is nominally the highest resolution model of mammalian complex I and is relatively complete (84% atomic), although it suffers from several drawbacks. Most EM maps describing the intermediate states of processing are of the wrong hand (mirror image).[85,86] Furthermore, even though the paper was submitted one month after the publication of the complete complex I structure,[15] it purports to present the first structure of mammalian enzyme, describing as novel all the features already reported. In the porcine complex I model (PDB 5GUP) the B-factors are set to random values and not refined, the Fe–S clusters geometry and environment are incorrect, the FMN isoalloxazine ring is flipped, and the statistics of model refinement or model completeness were not reported. Finally, the mechanism of electron transfer between CI and $CIII_2$, proposed by the authors,[85] is inconsistent with all the existing extensive knowledge on $CIII_2$, dispensing with the Q-cycle. Therefore, on the balance, we use mainly the ovine structure for discussion in this review, as it is currently the most complete refined model of mammalian complex I.

Sequence comparisons indicate that all species studied (bovine, ovine and porcine)[14] are equally good models of human enzyme (∼84% identical). We chose ovine enzyme for our studies as it was found to be very stable and highly active after purification, fully retaining 42 kDa subunit (easily lost from bovine complex).[14] Extensive trials with different detergents revealed Brij-35 as the detergent giving the most homogeneous spread of particles on cryo-EM grids and the lowest background in images. Therefore this preparation was used for data collection, even though Brij-35 inactivates the complex and destabilizes the attachment of supernumerary 4TM subunit B14.7 to the complex (we use the established bovine subunit nomenclature, shown in Table 2.1, for the ovine enzyme). This subunit and area immediately next to it is disordered in our maps and was modeled as poly-ALA based on our "respirasome" EM maps,[87] where these areas are well defined.

All three studies reveal some degree of flexibility in the overall structure of the complex, with different 3D classes in each study showing somewhat different angles between the two main arms (from "closed" to "open") and different degrees of rotation of matrix arm versus membrane arm. The differences are not huge (within ∼3–4 degrees) but significant. Ovine structures were resolved to 4.6 Å resolution for the "closed" class and 3.9 Å for the "open" class.[15] The bovine study presents three forms of the enzyme; classes 1 and 2 similar to the ovine "closed" conformation and class 3 analogous to the "open" state.[52] In the porcine study the highest resolution structure is similar to the "closed" state.[85] However, none of these classes are identical – additionally to the hinge like movement, the peripheral arm can rotate with regard to the membrane domain (MD) to a different extent. Additionally, in the bovine study most of the particles were assigned to classes 1 and 2 corresponding to the ovine "closed" state. In the ovine it was opposite, the "open" state (corresponding to bovine class 3) gave the highest resolution map, grouping most of the particles. Also in both cases, ovine "open" and bovine class 3, B14.7 is disordered, suggesting that this subunit is rather weakly attached to the complex. Different observed conformations of the enzyme suggest that it does not occur only in two discrete forms (at least in the detergent solubilized sample). Presumably more intermediate conformations can be obtained by better particles separation and classification during image processing.

Overall, there is an intriguing correlation between the angle between arms and the conformation of loops lining the quinone-binding cavity at the interface between the two arms. In the "closed" classes (bovine class 2 and porcine) these loops are better ordered and the access path for quinone is unhindered, while in the ovine "open" class loops are partly disordered and the quinone path is blocked at the deep end of the cavity. As discussed below, the latter conformation may resemble the "deactive" state of the enzyme and/or all (or most) of the observed conformations may be related to the conformational changes which protein undergoes during the catalytic cycle.

2.8 Structure of the Ovine Mitochondrial Complex I

The cryo-EM density (map combined from several refinements focused on different domains of the complex) was of excellent quality, allowing for *de novo* atomic model building.[15] Modelling of the core subunits was helped by the fact that their folds are preserved from bacteria to mammals. However, the assignment of the 31 supernumerary subunits to the remaining density was challenging, as most of the subunits do not have structural homologues in the PDB. Therefore, to provide experimental verification for previous assignments and to locate unassigned subunits, we performed extensive cross-linking/mass-spectrometry mapping experiments. They allowed us not only to locate subunits relative to each other, but also provided restrains on the fold of individual subunits. The model was improved by density-guided iterative re-building in Rosetta.[88] The final model is of excellent quality, with a Molprobity score of 2.5 and EMringer score 1.96, significantly better than average score of ~1.0 for this resolution.[89]

2.8.1 Core Subunits

The fold of all fourteen core subunits (Figure 2.5a, see Table 2.1 for nomenclature) is well conserved from the bacterial version, with only slight shifts of subunits as rigid-bodies and minor shifts of helices and β-sheets in most subunits. Fe–S clusters are arranged in the redox chain with distances similar to bovine[83] and *T. thermophilus*,[56] with only slight deviations in positions from the bacterial enzyme.[56] The environment of the Fe–S clusters, including the charged residues nearby, and NADH binding site are well conserved from bacteria, including all residues interacting with FMN and NADH. Thus, a clear path for electron transfer from NADH *via* FMN and the series of Fe–S clusters towards quinone is conserved between species. Cluster N2 terminates the Fe–S chain and donates electrons to the hydrophobic quinone but is about 25 Å away from the membrane surface, as in bacteria.

The quinone-binding site lies at the interface of the hydrophilic 49 kDa/PSST and membrane ND1/ND3 subunits. The unique structure of this site is preserved with one difference—a conserved loop connecting the first and second strands of the N-terminal β-sheet from the 49 kDa subunit ($\beta1$–$\beta2^{49\text{-kDa}}$ loop) extends much further into the cavity, clashing with the position of the bound quinone from the bacterial structure. Conserved residues His59 and Tyr108 from the 49 kDa subunit in bacteria interact with the quinone headgroup (Figure 2.2c). In the ovine structure His59 instead occupies the position of the headgroup, with the rest of the loop clashing with the isoprenoid tail. Such a conformation will prevent quinone access closer than 20 Å to cluster N2, blocking electron transfer. A similar shift of this loop was observed also in the yeast mitochondrial enzyme, resulting in the proposal that it represents the "deactive" (D) state of the complex.[18] In the absence of substrates mitochondrial complex I exists in the D state and only upon turnover converts into the "active" (A) state.[90] This transition is

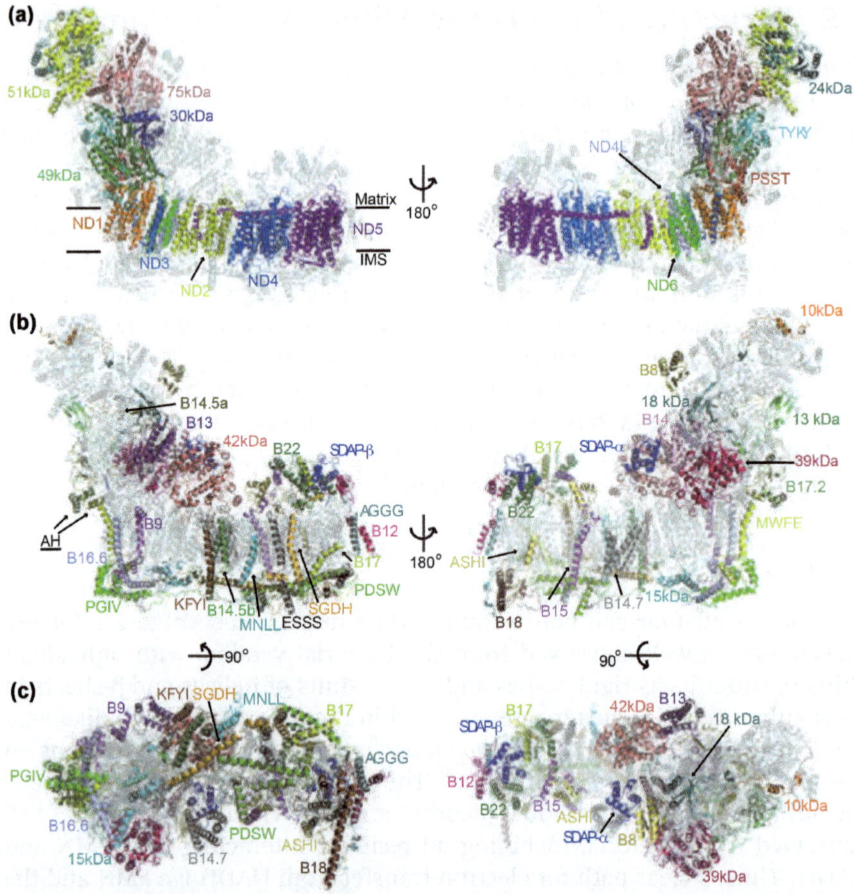

Figure 2.5 Structure of ovine complex I. (a) Structure depicted as a cartoon, with core subunits coloured and labelled, and supernumerary in grey and transparent. Approximate lipid bilayer boundaries are indicated. (b,c) Structure depicted with core subunits in grey and supernumerary subunits coloured and labelled (left–right, matrix–IMS views). Amphipathic helices at the "back" of the complex, likely attached to the lipid bilayer, are indicated as AH.
Reprinted by permission from Macmillan Publishers Ltd: *Nature*, (ref. 15), copyright (2016).

linked to a change in the conformation of the TMH1-2^{ND3} loop, as conserved Cys39^{ND3} becomes accessible to chemical modification only in the D state.[91] This loop, as well as the TMH5–6^{ND1} loop (unwound C-terminal half of TMH5), both contact the β1–β2$^{49\text{-kDa}}$ loop, and all line the quinone cavity. These three loops are better ordered in bovine class 2 and porcine structures, representing "closed" conformation of the complex, where the β1–β2$^{49\text{-kDa}}$ loop is retracted from the cavity to the position similar to bacterial structure, allowing full quinone access. These observations suggest that overall conformational changes in the complex may be linked to (and driven by) the re-arrangements in the Q cavity, with concerted loop movements possibly helping to eject reduced quinol at the end of catalytic cycle. Such a state, when quinone access is prevented, may also be "frozen" (possibly by some chemical modification) in the D state and requires turnover to regain the flexibility.

The largest core subunits in the membrane domain are three antiporter-like subunits, ND2, ND4 and ND5. Uniquely in metazoans, the ND2 subunit lost three N-terminal TM helices and their space is occupied by lipid molecules in our structure. The architecture of the four putative proton translocation channels is well conserved from bacteria, including all discontinuous helices and the positions of all key residues. These residues are surrounded by conserved charged and polar residues, with cavities sufficient to hold many water molecules, so that a continuous central hydrophilic flexible axis extends from the Q site all the way to the tip of the membrane domain and likely plays a prominent role in the mechanism.

2.8.2 Supernumerary Subunits

Supernumerary subunits form a shell around the core subunits, as noted previously,[18,83] especially around the membrane domain and its interface with the peripheral arm (Figure 2.5b, c). With few exceptions, most supernumerary subunits are not globular structures but rather form extended coils, with some secondary structure elements, mostly α-helices. These extended structures allow supernumerary subunits to make numerous interactions at interfaces with other subunits, forming an extensive network of close contacts and polar bonds, stabilising the complex from the outer side of the core. In contrast, core subunits each form mostly a compact globular structure, so that the bacterial enzyme is stabilised rather by interactions at compact surfaces buried between two to three core subunits. Supernumerary subunits interweave extensively between each other and with the core subunits, making the whole mitochondrial complex assembly much more interlinked, with very large total buried surface area and a large number of inter-subunit bonds. Such intertwined structures of subunits suggest that they can be added to the complex only in a certain order, so the intricate assembly process of the seven hydrophobic mtDNA-encoded subunits and 38 imported nuclear-encoded subunits has to be tightly controlled.[92] Overall patterns of conservation show that interiors of core subunits are the most conserved, and some interfaces between core and supernumerary subunits are also well conserved.

Supernumerary subunits associated with the membrane domain can be categorized into single TM-helix domain (STMD)[93] subunits scattered around the entire domain, disulphide-rich subunits forming a layer of long α-helices traversing the IMS side of the domain, like the "structural keel" of a ship,[94] an acyl-carrier protein (ACP)/LYR motif subunit pair forming the bulk of the large matrix "bulge" at the tip of the domain and the globular 42 kDa subunit at the interface with the peripheral arm. Subunits associated with the peripheral arm include another ACP/LYR pair and B13 jointly "embracing" the 42 kDa subunit, the NADPH-containing 39 kDa subunit and a range of smaller subunits mostly forming a shell around the core with their intertwined extended structures.

There are 12 STMD subunits in mammalian complex I, defined by a single hydrophobic α-helical region separating hydrophilic N- and C-terminal domains. The N-termini of all STMD subunits are located in the mitochondrial matrix and TM helices are spread around the membrane domain. The matrix "bulge" on the tip of the MD is formed from the highly intertwined N-terminal domains of STMD subunits B15, ASHI, B12, AGGG, B17 and SGDH, which surround the tip with their TM helices. The ACP/LYR pair (SDAP-β/B22) also contributes to the "bulge". B12 forms the very tip of MD, contributing the most distal TM helix and interacting with SDAP-β/B22 pair in the matrix. B12 interacts extensively with complex IV in the supercomplex.[87] The only two supernumerary subunits with more than one TM helix are the 4-TM helix B14.7, clamping on the C-terminus of ND5 subunit, and 2-TM helix B14.5b, attached to ND2 on the opposite side of the domain.

The STMD subunit SGDH has a large C-terminal IMS domain that forms a long coil wrapped around a coil from subunit PDSW. These two subunits are thus "locked" *via* their backbone and they together account for three long α-helices traversing nearly the entire intermembrane surface, stabilizing this side of the membrane domain (Figure 2.5c, left). PDSW contains four globally conserved cysteines, which are positioned to form two intra-subunit disulphide bonds, stabilizing the fold in the oxidizing environment of IMS. Three subunits facing the IMS (PGIV, 15 kDa and B18) contain double Cx_9C domains (CHCH domain), which stabilise helix-turn-helix structures with two disulphides.[95] The rigid disulphide-rich interlocked helices of these subunits appear to replace extended β-hairpin/helix motif (βH), which stabilizes this side of the domain in bacteria.

In subunit B18 one of the helices from the CHCH motif is very long and protrudes far out of the complex. This helix and the C-terminal PDSW helix project towards complexes III and IV in the supercomplex.[96] Subunit B14.7 intimately interacts with complex III[85,87] and overall, complex I interacts with its partners in the supercomplexes mainly through the supernumerary subunits, consistent with the general absence of supercomplexes in bacteria, which lack supernumerary subunits.

The 42 kDa subunit is a large globular protein, which preserves a nucleoside kinase family fold and is attached to the matrix side of ND2. Subunit B13 (bound to the core 30 kDa subunit) and the SDAP-α/B14 pair form matrix arm extensions that approach the 42 kDa subunit, providing possible additional stabilizing contact between the peripheral and membrane arms. In the "closed" class conformation B13 moves about 3 Å closer to the 42 kDa subunit, allowing for direct contacts.

At the "heel" of the complex, subunit MWFE forms extensive interactions in the membrane with ND1 as a helix anti-parallel to $TM1^{ND1}$ and in the IMS with PGIV, B16.6 and ND6. Subunit B16.6 is identical to the cell death regulator GRIM19[54] and it contains a long N-terminal coil interacting extensively with the matrix arm, followed by an exceptionally long (65 residues) α-helix traversing the membrane and ending with a loop and short helix interacting with several subunits on the IMS side. Other subunits stabilizing the interaction between two main domains are B17.2 and B14.5a, both containing N-terminal amphipathic α-helices, bound at the membrane interface, with the rest of the polypeptides wrapping around the hydrophilic arm. Subunits PSST, TYKY and B9 also contain such amphipathic helices, all located at the "back" of the complex (Figure 2.5b), probably helping to ensure proper position of the hydrophilic arm over the lipid bilayer. The 18 kDa subunit also plays a stabilizing role running nearly the entire length of the matrix arm with a long C-terminal coil.

In the matrix arm, the small globular subunit B8 adopts a thioredoxin fold[97] and binds at the interface of two large sub-domains of the 75 kDa subunit. B8 is the only matrix subunit that contains a possible disulphide bond (Cys23–Cys57). The cysteines are likely to be reduced in the mitochondrial matrix and they are exposed in the complex, adjacent to interface with 75 kDa subunit. B8 may play a role as a sensor of oxidative damage, as in patients with Parkinson's disease B8 is lost from complex I and this loss is correlated with the accumulation of oxidative damage to the enzyme.[98] At the tip of matrix arm, the 10 kDa subunit binds along the surface of 24 kDa and 51 kDa core subunits mostly as an extended coil. At its C-terminus it reaches the 75 kDa core subunit and the supernumerary 18 kDa subunit. Given its position at the interface of the three core subunits and its extended structure the 10 kDa subunit is probably involved in the stabilization of this distal region of the complex.

2.8.3 Cofactors in Supernumerary Subunits

Several additional enzyme cofactors present in supernumerary subunits are well resolved in the structure (Figure 2.6). The 39 kDa subunit belongs to the NADPH-binding short-chain dehydrogenase/reductase family and the fold is well preserved, including the NADPH-binging site. The subunit is wedged into the side of the peripheral arm, interacting with many subunits (75 kDa, 30 kDa, 13 kDa, TYKY and PSST) near the interface with membrane. The density for bound NADPH is very clear (Figure 2.6a). In mitochondrial complex I, NADPH is tightly bound to the 39 kDa subunit and does not seem to play a catalytic role, mainly stabilizing the fold of this subunit.[99] In the structure the nicotinamide ring is ~45 Å away from the nearest Fe–S cluster N2, arguing against the direct redox function of NADPH. The nucleotide is buried deep in the subunit, except for the extra phosphate, which points out and forms strong hydrogen bonds with the guanidinium of conserved $Arg178^{PSST}$, which sits on the C-terminal helix directly connected to

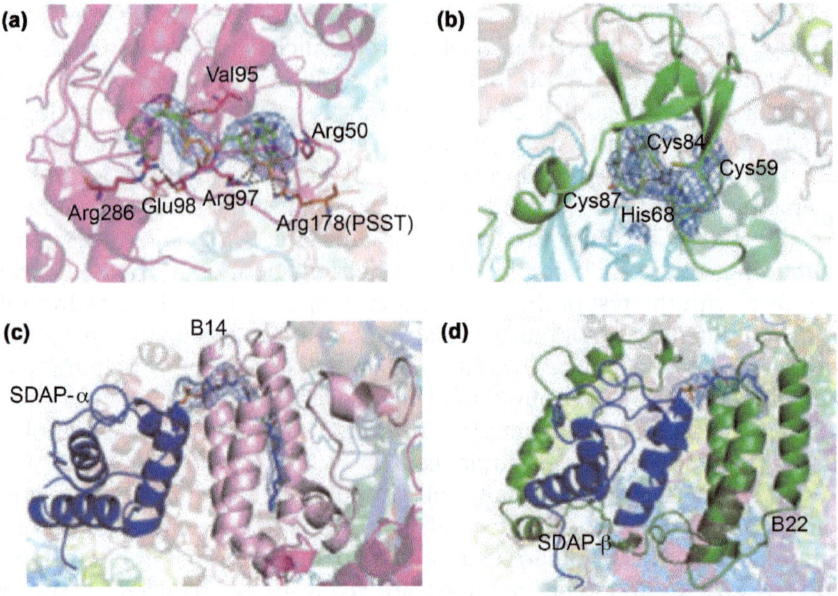

Figure 2.6 Additional cofactors identified in the ovine structure. (a) NADPH in the 39 kDa subunit. Interacting residues are shown. (b) Zn^{2+} ion in the 13 kDa subunit, with coordinating residues. (c) Phosphopantetheine in SDAP-α. (d) Phosphopantetheine in SDAP-β.
Reprinted by permission from Macmillan Publishers Ltd: *Nature,* (ref. 15), copyright (2016).

cluster N2. This helix along with another PSST helix next to it (H1) undergoes conformational change upon cluster reduction.[55] It is conceivable that such a structure allows the complex to sense the redox poise of the cell *via* the presence of NADP(H) and communicate it to the "heart" of its machinery at cluster N2.

There are two copies of the acyl-carrier protein (ACP) in mammalian complex I, SDAP-α and -β. ACPs play an essential role in fatty acid synthesis by carrying intermediates between enzymes participating in the synthesis of a growing acyl chain, attached to protein *via* a phosphopantetheine moiety.[100] The acyl chain is provided to ACP partners by the entire chain "flipping" from the hydrophobic cavity between ACP helices into hydrophobic cavities in partner proteins.[100] We observe clear density for phosphopantetheine moieties in both copies of SDAP. In SDAP-α this is covalently attached to conserved $Ser44^{SDAP}$ and extends in the "flipped out" conformation far into the hydrophobic crevice between three helices of the SDAP-α partner protein, LYR motif containing subunit B14.[101] The density is clear for a 12-carbon acyl chain extension on the phosphopantetheine moiety (Figure 2.6c). Therefore SDAP-α is strongly bound to B14 due to this large buried hydrophobic surface. SDAP-α only interacts with complex I *via* B14, and both subunits form a subdomain of the hydrophilic arm pointing towards the 42 kDa subunit.

Subunits B14 and B22 belong to the LYR (leucine/tyrosine/arginine) motif family of proteins, mostly found in mitochondria, acting as assembly factors for OXPHOS complexes and involved in Fe–S cluster biogenesis, often in partnership with ACPs.[101,102] Subunits B22 and SDAP-β form part of the matrix "bulge" at the tip of the membrane domain. Subunit SDAP-β also interacts with B22 through a "flipped out" phosphopantetheine chain buried within the three-helix bundle of B22, although here there is no clear density for an acyl chain extension past the thiol (Figure 2.6d). Thus, two pairs of ACP–LYR proteins are tightly linked *via* phosphopantetheine-acyl chain cofactors, with one pair bound to the hydrophilic arm and another to the membrane arm. In both cases, residues of the LYR motif are directly involved in interactions with SDAP.

The B14 subunit interacts extensively and intimately with the TMH1-2^{ND3} loop, which wraps around the Q site and changes conformation upon the A to D transition,[91] as noted above. This may explain why, when B14 is deleted, the complex becomes inactive even though still assembled,[102] as the deletion may lock the TMH1-2^{ND3} loop in the D conformation. This suggests that complex I will not become active until fully assembled with the SDAP-α/B14 pair containing an extended acyl chain, possibly providing a regulatory link between fatty acid synthesis and OXPHOS activity in mitochondria. Nearby subunit B13 is also involved in A/D transitions.[103] On the other side of complex, SDAP-β/B22 pair makes numerous close interactions with antiporter-like subunit ND5 and so could in principle regulate proton-pumping activity.

The 13 kDa subunit, bound to the side of the matrix arm, is one of the three accessory subunits already found in α-proteobacteria[26] and harbours a conserved Zn-binding motif, confirmed to bind Zn^{2+} in *Y. lipolytica.*[104] We observe a clear density for a Zn^{2+} coordinated by three conserved cysteines and a histidine (Figure 2.6b). The subunit has an extended structure interacting mostly with TYKY but also with the 49 kDa, 75 kDa, B17.2 and 39 kDa subunits. Zn-containing proteins are sensitive to oxidative stress,[24] and loss of the 13 kDa subunit leads to loss of cluster N6a,[25] as it becomes exposed. In this way CI may be equipped with another oxidative stress "sensor", in addition to bound NADPH.

Twelve tightly bound lipids were identified in density at crevices between hydrophobic subunits. Four lipid molecules clearly have four acyl chains each and so were assigned as cariolipins, known to be essential for activity of the complex.[105] Others have been tentatively assigned as phosphatidylcholines and phosphatidylethanolamines, known to co-purify with the complex along with cardiolipin.[105,106] Strikingly, a cardiolipin and three other lipid molecules fill the void left by the missing (in metazoans) three ND2 N-terminal helices. This void is encircled by TM helices from supernumerary subunits B14.5b and KFYI, hinting that the ND2 helices may have been lost in evolution to accommodate a specific binding site for lipids. Two cardiolipins fill a large gap between the antiporter-like ND4 and ND5 subunits, preventing potential proton leaks and instability. Another cardiolipin stabilizes amphipathic helices at the "back" of the complex. The structure thus shows the basis for the essential role of cardiolipin and other lipids in the activity and integrity of the complex.

2.9 Mitochondrial Complex I Mechanism: Similar to Bacterial but with Extra "Stabilisers" and "Sensors"

Complex I deficiency is a cause of many severe mitochondria related diseases, ranging from lethal neonatal disease to adult-onset neurodegenerative disorders, and known mutations involve all core and many of supernumerary subunits.[107] Mutations can affect function either directly by interfering with redox reactions or proton pumping, or indirectly by affecting complex stability and assembly. The ovine structure provided the basis for understanding most harmful mutations at an atomic level (manuscript in preparation).

The mechanism of coupling between two processes separated by large distances, electron transfer occurring in the peripheral arm and proton translocation in the membrane arm, is one of the key outstanding questions in the field. Positions of all the key redox cofactors and residues involved in proton pumping are conserved from bacteria to mammals, suggesting that the basic mechanism is likely the same. The central axis of charged residues in the membrane domain probably plays a key role (Figure 2.7). In each catalytic cycle, the unusual enclosed Q site may allow control of the protonation of quinone and use of the negative charge from the two electrons (stored either on Q or more likely on coordinating Tyr/His residues) to drive conformational changes in ND1, which consists of highly tilted TM helices, full of polar residues and thus is likely flexible. These changes, starting from ND1/E-channel, can be propagated *via* the central axis to the antiporter-like subunits, including distal ND5, resulting in changes of pKa and accessibility of key residues. The net result would be pumping of four protons per cycle, one per each channel (Figure 2.7). A notable observation is the more closed conformation of loops in the Q site in the "open" class ovine structure (and in *Y. lipolytica* enzyme[18]), likely reflecting the D state of the complex. This conformation might also occur during normal function when quinol is ejected from the site into the lipid bilayer, if A/D transitions are related to conformations encountered during the catalytic cycle.[18] If such a link exists, supernumerary subunits implicated in A/D transitions (39 kDa, B13 and SDAP-α/B14 pair) could also participate in catalytic conformational changes, by interacting with the key ND3 loop flanking the Q site or additionally through interactions with the 42 kDa subunit (Figure 2.7). Different orientations of the peripheral arm relative to the membrane domain are observed in the "closed" class, bringing B13 and SDAP-α closer to 42 kDa, hinting at such a possibility. The role of the 42 kDa subunit is probably more in "fine tuning" any movements, as this subunit is metazoan-specific.[4]

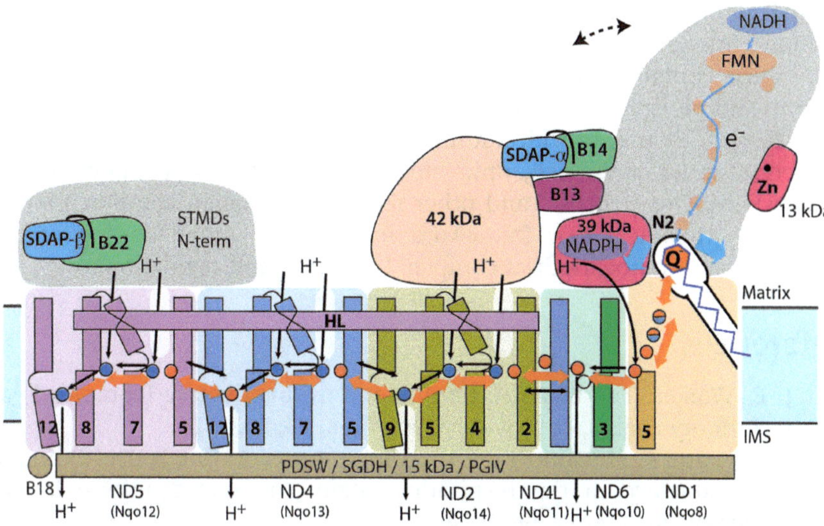

Figure 2.7 Mechanism of complex I. Core and some putatively regulatory supernumerary subunits in mitochondrial enzyme are shown (*T. thermophilus* nomenclature for core membrane subunits is in brackets). Basic coupling mechanism involving core subunits is expected to be conserved from bacteria to mitochondria, but the mitochondrial enzyme is equipped with additional "stabilisers" and "sensors". Upon transfer of two electrons from cluster N2, negative charge around quinone initiates a cascade of conformational changes, propagating from the E-channel (ND1/ND6/ND4L or Nqo8/10/11) to the antiporters *via* the central hydrophilic axis (red arrows). Cluster N2-driven shifts of 49 kDa/PSST or Nqo4/6 helices[55] (blue arrows) likely assist overall conformational changes. ND5/Nqo12 helix HL and traverse helices from four supernumerary subunits on the IMS side (βH motif in bacteria) may serve as "stators". Dashed line indicates the shift of peripheral arm between "closed" and "open" conformations, which may happen during the catalytic cycle as a consequence of re-arrangements of loops around the Q site. NADPH-containing 39 kDa subunit and Zn-containing 13 kDa subunit are essential for activity and may serve as redox "sensors". Both SDAP subunits interact with their LYR partners *via* "flipped out" phosphopantetheine (black line) and may provide a "sensor" for lipid biosynthesis. The 42 kDa subunit can interact with SDAP-α/B14 pair and B13 subunit, possibly "fine tuning" conformational changes. The net result of one conformational cycle, driven by NADH:ubiquinone oxidoreduction, is the translocation of four protons across the membrane (black lines indicate possible pathways).
Reprinted by permission from Macmillan Publishers Ltd: *Nature*, (ref. 15), copyright (2016).

Traverse helix HL from ND5 appears to play mainly a stabilizing "stator" role[59] rather than being a moving element.[51] Rigid disulphide-rich supernumerary subunits traversing the IMS side of the membrane domain may represent another "stator" element unique to the mitochondrial enzyme (Figure 2.7).

Supernumerary subunits clearly stabilize the complex through numerous extended inter-subunit contacts over large distances. Some of them, especially those containing additional cofactors (39 kDa, SDAP's, B14, B22 and 13 kDa) and phosphorylated residues (42 kDa, ESSS, MWFE, B14.5a, B14.5b and B16.6),[108,109] may provide additional regulatory links as the "sensors" of the redox status of the cell, lipid biosynthesis and overall mitochondrial homeostasis. Our structure and other recent mammalian complex I structures now provide a new framework for detailed understanding of the mechanisms of mitochondrial complex I function and regulation.

References

1. J. E. Walker, The NADH – ubiquinone oxidoreductase (complex I) of respiratory chains, *Q. Rev. Biophys.*, 1992, **25**, 253–324.
2. T. Yagi and A. Matsuno-Yagi, The proton-translocating NADH-Quinone oxidoreductase in the respiratory chain: the secret unlocked, *Biochemistry*, 2003, **42**, 2266–2274.
3. L. A. Sazanov, Respiratory complex I: mechanistic and structural insights provided by the crystal structure of the hydrophilic domain, *Biochemistry*, 2007, **46**, 2275–2288.
4. U. Brandt, Energy converting NADH: quinone oxidoreductase (complex I), *Annu. Rev. Biochem.*, 2006, **75**, 69–92.
5. M. Wikström, V. Sharma, V. R. Kaila, J. P. Hosler and G. Hummer, New perspectives on proton pumping in cellular respiration, *Chem. Rev.*, 2015, **115**, 2196–2221.
6. A. D. Vinogradov, Catalytic properties of the mitochondrial NADH-ubiquinone oxidoreductase (complex I) and the pseudo-reversible active/inactive enzyme transition, *Biochim. Biophys. Acta*, 1998, **1364**, 169–185.
7. M. Wikström, Two protons are pumped from the mitochondrial matrix per electron transferred between NADH and ubiquinone, *FEBS Lett.*, 1984, **169**, 300–304.
8. A. S. Galkin, V. G. Grivennikova and A. D. Vinogradov, $H^+/2e^-$ stoichiometry in NADH-quinone reductase reactions catalyzed by bovine heart submitochondrial particles, *FEBS Lett.*, 1999, **451**, 157–161.
9. A. Galkin, S. Drose and U. Brandt, The proton pumping stoichiometry of purified mitochondrial complex I reconstituted into proteoliposomes, *Biochim. Biophys. Acta*, 2006, **1757**, 1575–1581.
10. A. V. Bogachev, R. A. Murtazina and V. P. Skulachev, H^+/e^- stoichiometry for NADH dehydrogenase I and dimethyl sulfoxide reductase in anaerobically grown *Escherichia coli* cells, *J. Bacteriol.*, 1996, **178**, 6233–6237.
11. A. J. Jones, J. N. Blaza, F. Varghese and J. Hirst, Respiratory Complex I in Bos taurus and Paracoccus denitrificans Pumps Four Protons across the Membrane for Every NADH Oxidized, *J. Biol. Chem.*, 2017, **292**, 4987–4995.

12. J. Carroll, I. M. Fearnley, R. J. Shannon, J. Hirst and J. E. Walker, Analysis of the subunit composition of complex I from bovine heart mitochondria, *Mol. Cell. Proteomics*, 2003, **2**, 117–126.
13. J. Carroll *et al.*, Bovine complex I is a complex of 45 different subunits, *J. Biol. Chem.*, 2006, **281**, 32724–32727.
14. J. A. Letts, G. Degliesposti, K. Fiedorczuk, M. Skehel and L. A. Sazanov, Purification of Ovine Respiratory Complex I Results in a Highly Active and Stable Preparation, *J. Biol. Chem.*, 2016, **291**, 24657–24675.
15. K. Fiedorczuk *et al.*, Atomic structure of the entire mammalian mitochondrial complex I, *Nature*, 2016, **538**, 406–410.
16. S. Kerscher, L. Grgic, A. Garofano and U. Brandt, Application of the yeast *Yarrowia lipolytica* as a model to analyse human pathogenic mutations in mitochondrial complex I (NADH:ubiquinone oxidoreductase), *Biochim. Biophys. Acta*, 2004, **1659**, 197–205.
17. C. Hunte, V. Zickermann and U. Brandt, Functional modules and structural basis of conformational coupling in mitochondrial complex I, *Science*, 2010, **329**, 448–451.
18. V. Zickermann *et al.*, Structural biology. Mechanistic insight from the crystal structure of mitochondrial complex I, *Science*, 2015, **347**, 44–49.
19. A. Videira, Complex I from the fungus *Neurospora crassa*, *Biochim. Biophys. Acta*, 1998, **1364**, 89–100.
20. P. Hinchliffe, J. Carroll and L. A. Sazanov, Identification of a novel subunit of respiratory complex I from *Thermus thermophilus*, *Biochemistry*, 2006, **45**, 4413–4420.
21. D. Schneider *et al.*, Assembly of the *Escherichia coli* NADH:ubiquinone oxidoreductase (complex I), *Biochim. Biophys. Acta*, 2008, **1777**, 735–739.
22. R. Baradaran, J. M. Berrisford, G. S. Minhas and L. A. Sazanov, Crystal structure of the entire respiratory complex I, *Nature*, 2013, **494**, 443–448.
23. H. Leif, V. D. Sled, T. Ohnishi, H. Weiss and T. Friedrich, Isolation and characterization of the proton-translocating NADH: ubiquinone oxidoreductase from *Escherichia coli*, *Eur. J. Biochem.*, 1995, **230**, 538–548.
24. S. W. Meinhardt *et al.*, Studies on the NADH-menaquinone oxidoreductase segment of the respiratory chain in *Thermus thermophilus* HB-8, *J. Biol. Chem.*, 1990, **265**, 1360–1368.
25. T. Yano *et al.*, Characterization of cluster N5 as a fast-relaxing [4Fe-4S] cluster in the Nqo3 subunit of the proton-translocating NADH-ubiquinone oxidoreductase from *Paracoccus denitrificans*, *J. Biol. Chem.*, 2003, **278**, 15514–15522.
26. C. Y. Yip, M. E. Harbour, K. Jayawardena, I. M. Fearnley and L. A. Sazanov, Evolution of respiratory complex I: "supernumerary" subunits are present in the alpha-proteobacterial enzyme, *J. Biol. Chem.*, 2011, **286**, 5023–5033.
27. G. Peng *et al.*, Isolation, characterization and electron microscopic single particle analysis of the NADH:ubiquinone oxidoreductase

(complex I) from the hyperthermophilic eubacterium *Aquifex aeolicus*, *Biochemistry*, 2003, **42**, 3032–3039.
28. T. Yagi, T. Yano, S. DiBernardo and A. MatsunoYagi, Procaryotic complex I (NDH-1), an overview, *Biochim. Biophys. Acta*, 1998, **1364**, 125–133.
29. L. A. Sazanov, J. Carroll, P. Holt, L. Toime and I. M. Fearnley, A role for native lipids in the stabilization and two-dimensional crystallization of the Escherichia coli NADH-ubiquinone oxidoreductase (Complex I), *J. Biol. Chem.*, 2003, **278**, 19483–19491.
30. Q. H. Tran, J. Bongaerts, D. Vlad and G. Unden, Requirement for the proton-pumping NADH dehydrogenase I of *Escherichia coli* in respiration of NADH to fumarate and its bioenergetic implications, *Eur. J. Biochem.*, 1997, **244**, 155–160.
31. T. Friedrich and D. Scheide, The respiratory complex I of bacteria, archaea and eukarya and its module common with membrane-bound multisubunit hydrogenases, *FEBS Lett.*, 2000, **479**, 1–5.
32. L. A. Sazanov, P. A. Burrows and P. J. Nixon, The plastid *ndh* genes code for an NADH-specific dehydrogenase: isolation of a complex I analogue from pea thylakoid membranes, *Proc. Natl. Acad. Sci. USA*, 1998, **95**, 1319–1324.
33. T. Shikanai, Chloroplast NDH: A different enzyme with a structure similar to that of respiratory NADH dehydrogenase, *Biochim. Biophys. Acta*, 2016, **1857**, 1015–1022.
34. A. H. Schapira, Human complex I defects in neurodegenerative diseases, *Biochim. Biophys. Acta*, 1998, **1364**, 261–270.
35. R. S. Balaban, S. Nemoto and T. Finkel, Mitochondria, oxidants, and aging, *Cell*, 2005, **120**, 483–495.
36. T. M. Dawson and V. L. Dawson, Molecular pathways of neurodegeneration in Parkinson's disease, *Science*, 2003, **302**, 819–822.
37. E. T. Chouchani *et al.*, Ischaemic accumulation of succinate controls reperfusion injury through mitochondrial ROS, *Nature*, 2014, **5105**, 431–435.
38. E. L. Mills *et al.*, Succinate Dehydrogenase Supports Metabolic Repurposing of Mitochondria to Drive Inflammatory Macrophages, *Cell*, 2016, **167**, 457–470 e413.
39. F. Scialo *et al.*, Mitochondrial ROS Produced via Reverse Electron Transport Extend Animal Lifespan, *Cell Metab.*, 2016, **23**, 725–734.
40. D. J. Morgan and L. A. Sazanov, Three-dimensional structure of respiratory complex I from *Escherichia coli* in ice in the presence of nucleotides, *Biochim. Biophys. Acta*, 2008, **1777**, 711–718.
41. V. Guenebaut, A. Schlitt, H. Weiss, K. Leonard and T. Friedrich, Consistent structure between bacterial and mitochondrial NADH:ubiquinone oxidoreductase (complex I), *J. Mol. Biol.*, 1998, **276**, 105–112.
42. P. J. Holt, D. J. Morgan and L. A. Sazanov, The location of NuoL and NuoM subunits in the membrane domain of the *Escherichia coli* complex I: implications for the mechanism of proton pumping, *J. Biol. Chem.*, 2003, **278**, 43114–43120.

43. T. Yano, S. S. Chu, V. D. Sled, T. Ohnishi and T. Yagi, The proton-translocating NADH-quinone oxidoreductase (NDH-1) of thermophilic bacterium *Thermus thermophilus* HB-8. Complete DNA sequence of the gene cluster and thermostable properties of the expressed NQO2 subunit, *J. Biol. Chem.*, 1997, **272**, 4201–4211.
44. T. Ohnishi, Iron-sulfur clusters/semiquinones in complex I, *Biochim. Biophys. Acta*, 1998, **1364**, 186–206.
45. P. Hinchliffe and L. A. Sazanov, Organization of iron-sulfur clusters in respiratory complex I, *Science*, 2005, **309**, 771–774.
46. R. G. Efremov and L. A. Sazanov, Structure of the membrane domain of respiratory complex I, *Nature*, 2011, **476**, 414–420.
47. R. G. Efremov and L. A. Sazanov, Respiratory complex I: 'steam engine' of the cell? *Curr. Opin. Struct. Biol.*, 2011, **21**, 532–540.
48. P. L. Dutton, C. C. Moser, V. D. Sled, F. Daldal and T. Ohnishi, A reductant-induced oxidation mechanism for complex I, *Biochim. Biophys. Acta*, 1998, **1364**, 245–257.
49. S. T. Ohnishi, J. C. Salerno and T. Ohnishi, Possible roles of two quinone molecules in direct and indirect proton pumps of bovine heart NADH-quinone oxidoreductase (complex I), *Biochim. Biophys. Acta*, 2010, **1797**, 1891–1893.
50. T. Friedrich, Complex I: a chimaera of a redox and conformation-driven proton pump? *J. Bioenerg. Biomembr.*, 2001, **33**, 169–177.
51. R. G. Efremov, R. Baradaran and L. A. Sazanov, The architecture of respiratory complex I, *Nature*, 2010, **465**, 441–445.
52. J. Zhu, K. R. Vinothkumar and J. Hirst, Structure of mammalian respiratory complex I, *Nature*, 2016, **536**, 354–358.
53. L. A. Sazanov, S. Y. Peak-Chew, I. M. Fearnley and J. E. Walker, Resolution of the membrane domain of bovine complex I into subcomplexes: implications for the structural organization of the enzyme, *Biochemistry*, 2000, **39**, 7229–7235.
54. I. M. Fearnley *et al.*, Grim-19, a cell death regulatory gene product, is a subunit of bovine mitochondrial nadh:ubiquinone oxidoreductase (complex I), *J. Biol. Chem.*, 2001, **276**, 38345–38348.
55. J. M. Berrisford and L. A. Sazanov, Structural basis for the mechanism of respiratory complex I, *J. Biol. Chem.*, 2009, **284**, 29773–29783.
56. L. A. Sazanov and P. Hinchliffe, Structure of the hydrophilic domain of respiratory complex I from *Thermus thermophilus*, *Science*, 2006, **311**, 1430–1436.
57. I. M. Fearnley and J. E. Walker, Conservation of sequences of subunits of mitochondrial complex I and their relationships with other proteins, *Biochim. Biophys. Acta*, 1992, **1140**, 105–134.
58. C. Mathiesen and C. Hagerhall, Transmembrane topology of the NuoL, M and N subunits of NADH:quinone oxidoreductase and their homologues among membrane-bound hydrogenases and bona fide antiporters, *Biochim. Biophys. Acta*, 2002, **1556**, 121–132.

59. G. Belevich, J. Knuuti, M. I. Verkhovsky, M. Wikström and M. Verkhovskaya, Probing the mechanistic role of the long alpha-helix in subunit L of respiratory Complex I from *Escherichia coli* by site-directed mutagenesis, *Mol. Microbiol.*, 2011, **82**, 1086–1095.
60. P. M. Vignais, B. Billoud and J. Meyer, Classification and phylogeny of hydrogenases, *FEMS Microbiol. Rev.*, 2001, **25**, 455–501.
61. J. I. Oh and B. Bowien, Structural analysis of the fds operon encoding the NAD+-linked formate dehydrogenase of Ralstonia eutropha, *J. Biol. Chem.*, 1998, **273**, 26349–26360.
62. R. G. Efremov and L. A. Sazanov, The coupling mechanism of respiratory complex I – A structural and evolutionary perspective, *Biochim. Biophys. Acta*, 2012, **1817**, 1785–1795.
63. T. Pohl *et al.*, Iron-sulfur cluster N7 of the NADH:ubiquinone oxidoreductase (complex I) is essential for stability but not involved in electron transfer, *Biochemistry*, 2007, **46**, 6588–6596.
64. C. C. Page, C. C. Moser, X. Chen and P. L. Dutton, Natural engineering principles of electron tunnelling in biological oxidation-reduction, *Nature*, 1999, **402**, 47–52.
65. H. R. Bridges, E. Bill and J. Hirst, Mossbauer spectroscopy on respiratory complex I: the iron-sulfur cluster ensemble in the NADH-reduced enzyme is partially oxidized, *Biochemistry*, 2012, **51**, 149–158.
66. C. C. Moser, T. A. Farid, S. E. Chobot and P. L. Dutton, Electron tunneling chains of mitochondria, *Biochim. Biophys. Acta*, 2006, **1757**, 1096–1109.
67. M. L. Verkhovskaya, N. Belevich, L. Euro, M. Wikström and M. I. Verkhovsky, Real-time electron transfer in respiratory complex I, *Proc. Natl. Acad. Sci. U. S. A.*, 2008, **105**, 3763–3767.
68. A. B. Kotlyar, V. D. Sled, D. S. Burbaev, I. A. Moroz and A. D. Vinogradov, Coupling site I and the rotenone-sensitive ubisemiquinone in tightly coupled submitochondrial particles, *FEBS Lett.*, 1990, **264**, 17–20.
69. T. Ohnishi, E. Nakamaru-Ogiso and S. T. Ohnishi, A new hypothesis on the simultaneous direct and indirect proton pump mechanisms in NADH-quinone oxidoreductase (complex I), *FEBS Lett.*, 2010, **584**, 4131–4137.
70. L. Kussmaul and J. Hirst, The mechanism of superoxide production by NADH:ubiquinone oxidoreductase (complex I) from bovine heart mitochondria, *Proc. Natl. Acad. Sci. U. S. A.*, 2006, **103**, 7607–7612.
71. A. P. Kudin, N. Y. Bimpong-Buta, S. Vielhaber, C. E. Elger and W. S. Kunz, Characterization of superoxide-producing sites in isolated brain mitochondria, *J. Biol. Chem.*, 2004, **279**, 4127–4135.
72. A. J. Lambert and M. D. Brand, Inhibitors of the quinone-binding site allow rapid superoxide production from mitochondrial NADH:ubiquinone oxidoreductase (complex I), *J. Biol. Chem.*, 2004, **279**, 39414–39420.
73. V. D. Sled, N. I. Rudnitzky, Y. Hatefi and T. Ohnishi, Thermodynamic analysis of flavin in mitochondrial NADH:ubiquinone oxidoreductase (complex I), *Biochemistry*, 1994, **33**, 10069–10075.

74. S. Ransac, C. Arnarez and J. P. Mazat, The flitting of electrons in complex I: A stochastic approach, *Biochim. Biophys. Acta*, 2010, **1797**, 641–648.
75. R. F. Anderson, R. Hille, S. S. Shinde and G. Cecchini, Electron transfer within complex II. Succinate:ubiquinone oxidoreductase of *Escherichia coli*, *J. Biol. Chem.*, 2005, **280**, 33331–33337.
76. J. A. Birrell, K. Morina, H. R. Bridges, T. Friedrich and J. Hirst, Investigating the function of [2Fe-2S] cluster N1a, the off-pathway cluster in complex I, by manipulating its reduction potential, *Biochem. J.*, 2013, **456**, 139–146.
77. L. A. Sazanov, A giant molecular proton pump: structure and mechanism of respiratory complex I, *Natl. Rev. Mol. Cell Biol.*, 2015, **16**, 375–388.
78. S. J. Cho *et al.*, Crystal structure of *Escherichia coli* CyaY protein reveals a previously unidentified fold for the evolutionarily conserved frataxin family, *Proc. Natl. Acad. Sci. U. S. A.*, 2000, **97**, 8932–8937.
79. S. Dhe-Paganon, R. Shigeta, Y. I. Chi, M. Ristow and S. E. Shoelson, Crystal structure of human frataxin, *J. Biol. Chem.*, 2000, **275**, 30753–30756.
80. O. Gakh *et al.*, Physical evidence that yeast frataxin is an iron storage protein, *Biochemistry*, 2002, **41**, 6798–6804.
81. Y. He *et al.*, Yeast frataxin solution structure, iron binding, and ferrochelatase interaction, *Biochemistry*, 2004, **43**, 16254–16262.
82. W. Kuhlbrandt, Biochemistry. The resolution revolution, *Science*, 2014, **343**, 1443–1444.
83. K. R. Vinothkumar, J. Zhu and J. Hirst, Architecture of mammalian respiratory complex I, *Nature*, 2014, **515**, 80–84.
84. J. Zhu *et al.*, Structure of subcomplex Ibeta of mammalian respiratory complex I leads to new supernumerary subunit assignments, *Proc. Natl. Acad. Sci. U. S. A.*, 2015, **112**, 12087–12092.
85. M. Wu, J. Gu, R. Guo, Y. Huang and M. Yang, Structure of Mammalian Respiratory Supercomplex I1III2IV1, *Cell*, 2016, **167**, 1598–1609 e1510.
86. J. Gu *et al.*, The architecture of the mammalian respirasome, *Nature*, 2016, **537**, 639–643.
87. J. A. Letts, K. Fiedorczuk and L. A. Sazanov, The architecture of respiratory supercomplexes, *Nature*, 2016, **537**, 644–648.
88. F. DiMaio *et al.*, Atomic-accuracy models from 4.5-A cryo-electron microscopy data with density-guided iterative local refinement, *Nat. Methods*, 2015, **12**, 361–365.
89. B. A. Barad *et al.*, EMRinger: side chain-directed model and map validation for 3D cryo-electron microscopy, *Nat. Methods*, 2015, **12**, 943–946.
90. A. D. Vinogradov and V. G. Grivennikova, The mitochondrial complex I: progress in understanding of catalytic properties, *IUBMB Life*, 2001, **52**, 129–134.
91. A. Galkin *et al.*, Identification of the mitochondrial ND3 subunit as a structural component involved in the active/deactive enzyme transition of respiratory complex I, *J. Biol. Chem.*, 2008, **283**, 20907–20913.

92. L. Sanchez-Caballero, S. Guerrero-Castillo and L. Nijtmans, Unraveling the complexity of mitochondrial complex I assembly: A dynamic process, *Biochim. Biophys. Acta*, 2016, **1857**, 980–990.
93. V. Zickermann, H. Angerer, M. G. Ding, E. Nubel and U. Brandt, Small single transmembrane domain (STMD) proteins organize the hydrophobic subunits of large membrane protein complexes, *FEBS Lett.*, 2010, **584**, 2516–2525.
94. C. Wirth, U. Brandt, C. Hunte and V. Zickermann, Structure and function of mitochondrial complex I, *Biochim. Biophys. Acta*, 2016, **1857**, 902–914.
95. R. Szklarczyk *et al.*, NDUFB7 and NDUFA8 are located at the intermembrane surface of complex I, *FEBS Lett.*, 2011, **585**, 737–743.
96. J. A. Letts, K. Fiedorczuk & L. A. Sazanov Architecture of Respiratory Supercomplexes. *Submitted*, (2016).
97. C. Brockmann *et al.*, The oxidized subunit B8 from human complex I adopts a thioredoxin fold, *Structure*, 2004, **12**, 1645–1654.
98. P. M. Keeney, J. Xie, R. A. Capaldi and J. P. Bennett Jr., Parkinson's disease brain mitochondrial complex I has oxidatively damaged subunits and is functionally impaired and misassembled, *J. Neurosci.*, 2006, **26**, 5256–5264.
99. A. Abdrakhmanova, K. Zwicker, S. Kerscher, V. Zickermann and U. Brandt, Tight binding of NADPH to the 39-kDa subunit of complex I is not required for catalytic activity but stabilizes the multiprotein complex, *Biochim. Biophys. Acta*, 2006, **1757**, 1676–1682.
100. J. E. Cronan, The chain-flipping mechanism of ACP (acyl carrier protein)-dependent enzymes appears universal, *Biochem. J.*, 2014, **460**, 157–163.
101. H. Angerer *et al.*, The LYR protein subunit NB4M/NDUFA6 of mitochondrial complex I anchors an acyl carrier protein and is essential for catalytic activity, *Proc. Natl. Acad. Sci. U. S. A.*, 2014, **111**, 5207–5212.
102. H. Angerer, Eukaryotic LYR Proteins Interact with Mitochondrial Protein Complexes, *Biology*, 2015, **4**, 133–150.
103. A. V. Ushakova, M. Duarte, A. D. Vinogradov and A. Videira, The 29.9 kDa subunit of mitochondrial complex I is involved in the enzyme active/de-active transitions, *J. Mol. Biol.*, 2005, **351**, 327–333.
104. K. Kmita *et al.*, Accessory NUMM (NDUFS6) subunit harbors a Zn-binding site and is essential for biogenesis of mitochondrial complex I, *Proc. Natl. Acad. Sci. U. S. A.*, 2015, **112**, 5685–5690.
105. M. S. Sharpley, R. J. Shannon, F. Draghi and J. Hirst, Interactions between phospholipids and NADH:ubiquinone oxidoreductase (complex I) from bovine mitochondria, *Biochemistry*, 2006, **45**, 241–248.
106. K. Shinzawa-Itoh *et al.*, Bovine heart NADH-ubiquinone oxidoreductase contains one molecule of ubiquinone with ten isoprene units as one of the cofactors, *Biochemistry*, 2010, **49**, 487–492.
107. E. Fassone and S. Rahman, Complex I deficiency: clinical features, biochemistry and molecular genetics, *J. Med. Genet.*, 2012, **49**, 578–590.

108. V. A. Morais *et al.*, PINK1 loss-of-function mutations affect mitochondrial complex I activity via NdufA10 ubiquinone uncoupling, *Science*, 2014, **344**, 203–207.
109. G. Palmisano, A. M. Sardanelli, A. Signorile, S. Papa and M. R. Larsen, The phosphorylation pattern of bovine heart complex I subunits, *Proteomics*, 2007, **7**, 1575–1583.
110. K. Huoponen, J. Vilkki, P. Aula, E. K. Nikoskelainen and M. L. Savontaus, A new mtDNA mutation associated with Leber hereditary optic neuroretinopathy, *Am. J. Hum. Genet.*, 1991, **48**, 1147–1153.

CHAPTER 3

Structure and Function of Respiratory Complex I

VOLKER ZICKERMANN[a,b]

[a] Structural Bioenergetics Group, Institute of Biochemistry II, Medical School, Goethe-University, Frankfurt am Main, Germany; [b] Cluster of Excellence Frankfurt "Macromolecular Complexes", Goethe-University, Frankfurt am Main, Germany
Email: Zickermann@med.uni-frankfurt.de

3.1 Introduction

Respiratory complex I is a very large membrane protein complex that couples electron transfer from NADH to ubiquinone with the translocation of protons across the inner membrane of mitochondria or the cell membrane of bacteria.[1,2] The resulting proton motive force is utilized to drive ATP synthase. Complex I dysfunction is implicated in a number of hereditary and degenerative disorders.[3] The structures of partial and complete bacterial and mitochondrial complex I were determined by X-ray crystallography (Figure 3.1).[4–9] The recent revolution in cryo electron microscopy opened a new avenue for structure determination of mammalian complex I (Figure 3.1) and supercomplexes consisting of complexes I, III and IV.[10–17] The minimal form of complex I comprises 14 central subunits that are arranged in an L-shaped architecture of a membrane arm and a peripheral arm. The central subunits are conserved throughout species from bacteria to humans. In eukaryotes some 30 accessory subunits are distributed around the core of central subunits.[18,19] A limited number of accessory subunits are present in selected bacterial species.[20] Unfortunately, there is no uniform

Chemical Biology No. 5
Mechanisms of Primary Energy Transduction in Biology
Edited by Mårten Wikström

Published by the Royal Society of Chemistry, www.rsc.org

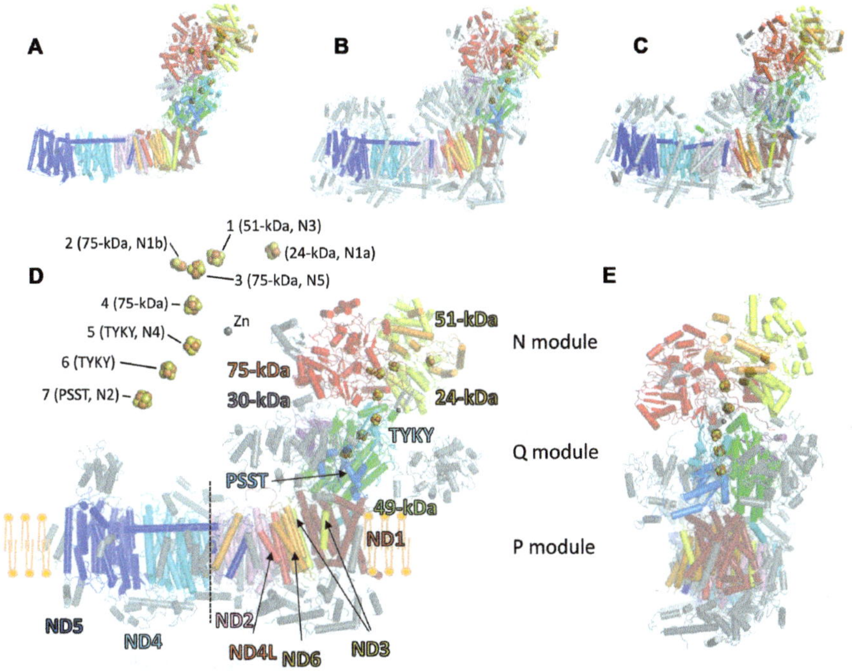

Figure 3.1 Structures of bacterial, yeast and mammalian complex I. The structures of complex I from (A) *T. thermophilus* (PDB ID: 4HEA), (B) *O. aries* (PDB ID: 5LNK), (C) *B. taurus* (PDB ID: 5LC5) and (D,E) *Y. lipolytica* (PDB ID: 4WZ7 with addition of recent cryoEM data;[13] D, side view; E, view from the back) are shown in cartoon representation with central subunits in color and accessory subunits in grey. Complex I consists of functional modules for NADH oxidation (N module), ubiquinone reduction (Q module) and proton pumping (P module); the dashed line in (D) indicates the boundary between proximal and distal P modules. The inset in (D) shows the relative arrangement of iron sulfur clusters and zinc, enlarged and slightly turned.

nomenclature for complex I subunits. Table 3.1 provides a translation for subunit designations. The evolution of complex I and its relation to Mrp type cation/proton antiporters and [NiFe] hydrogenases has been reviewed recently.[21] The assembly of complex I is an intricate process that is aided by assembly factors. A complete model for complex I assembly in mitochondria was recently reported.[22]

The canonical set of redox active groups comprises eight iron–sulphur centres[23] and one FMN, the immediate acceptor for electrons from NADH. In some bacterial species, additional iron–sulfur clusters were identified but their function is unclear.[24,25] Unexpectedly, initial electron microscopy studies showed that the reduction site for the highly hydrophobic substrate ubiquinone is not membrane intrinsic, indicating that electron transfer in the peripheral arm and proton pumping by membrane arm subunits are spatially separated.[26,27] Tight coupling between the two processes is

Table 3.1 Subunits of complex I.

Homo sapiens	*Bos taurus*	*Yarrowia lipolytica*	*Thermus thermophilus*	Comment
NDUFS1	75 kDa	NUAM	Nqo3	2 Fe_4S_4; 1 Fe_2S_2
NDUFV1	51 kDa	NUBM	Nqo1	FMN, NADH, Fe_4S_4
NDUFS2	49 kDa	NUCM	Nqo4	Q-binding
NDUFS3	30 kDa	NUGM	Nqo5	
NDUFV2	24 kDa	NUHM	Nqo2	Fe_2S_2
NDUFS8	TYKY	NUIM	Nqo9	2 Fe_4S_4
NDUFS7	PSST	NUKM	Nqo6	Q-binding, Fe_4S_4
NU1M	ND1	NU1M	Nqo8	mtDNA
NU2M	ND2	NU2M	Nqo14	mtDNA
NU3M	ND3	NU3M	Nqo7	mtDNA
NU4M	ND4	NU4M	Nqo13	mtDNA
NU5M	ND5	NU5M	Nqo12	mtDNA
NU6M	ND6	NU6M	Nqo10	mtDNA
NULM	ND4L	NULM	Nqo11	mtDNA
NDUFA2	B8	NI8M		Thioredoxin fold
NDUFS4	AQDQ/18 kDa	NUYM		
NDUFS6	13 kDa	NUMM		Zn^{2+}
NDUFA12	B17.2	N7BM		
NDUFA7	B14.5a	NUZM		
NDUFA5	B13	NUFM		
NDUFA9	39 kDa	NUEM		NADPH, short chain dehydrogenase fold
NDUFA6	B14	NB4M		LYR motif
NDUFAB1	SDAP	ACPM1		Phosphopantetheine
		ACPM2		Phosphopantetheine
NDUFA1	MWFE	NIMM		STMD
NDUFA3	B9	NI9M		STMD
NDUFA8	PGIV	NUPM		Mia-Cx_9C
NDUFA11	B14.7	NUJM		4 TMH
NDUFA13	B16.6	NB6M		STMD/GRIM-19
NDUFS5	PFFD/15 kDa	NIPM		Mia-Cx_9C
NDUFB3	B12	NB2M		STMD
NDUFB4	B15	NB5M		STMD
NDUFB7	B18	NB8M		Mia-Cx_9C
NDUFB8	ASHI	NIAM		STMD
NDUFB9	B22	NI2M		LYR motif
NDUFB10	PDSW	NIDM		Cx_yC
NDUFB11	ESSS	NESM		STMD
NDUFV3	9 kDa			
NDUFC1	KFYI			STMD
NDUFC2	B14.5b			STMD
NDUFA10	42 kDa			
NDUFB1	MNLL			STMD
NDUFB2	AGGG			STMD
NDUFB5	SGDH			STMD
NDUFB6	B17			STMD

Table 3.1 (*Continued*)

Homo sapiens	*Bos taurus*	*Yarrowia lipolytica*	*Thermus thermophilus*	Comment
		NUXM		2 TMH
		NEBM		STMD
		NUNM		STMD
		NUUM		STMD
		ST1		Sulphotransferase
			NQO15	Frataxin family
			NQO16	DUF3197 family

illustrated by the fact that mutation of key residues even in the most distant part of the membrane arm were shown to abolish ubiquinone reductase activity.[28] Moreover, under specific conditions, *i.e.* in the presence of a membrane potential and a highly reduced ubiquinone pool, complex I can operate in reverse and reduce NAD^+. The electron transfer activity of complex I is inhibited by a plethora of chemically diverse compounds.[29] Under conditions of reverse electron transfer or after inhibition, when electrons back up in the electron transfer chain, complex I releases increased amounts of deleterious oxygen radicals.[30–32] Complex I from many eukaryotic species can undergo a reversible active/deactive transition that was first described by Kotlyar and Vinogradov.[33] In the absence of substrate a conformational switch[34] converts the active A form in the deactive or dormant D form. The D form still has a residual catalytic capacity and transition to the A form is induced by turnover. The relevance of the A/D transition for protection against ROS induced ischemia/reperfusion injury was demonstrated.[35]

3.2 Overall Structure of Complex I

The X-ray structure of complete complex I from the bacterium *Thermus thermophilus* was solved by Sazanov and colleagues at 3.3 Å resolution.[8] The X-ray structure of complete mitochondrial complex I from the aerobic yeast *Yarrowia lipolytica* was refined at 3.6 to 3.9 Å resolution.[9] Recently, cryoEM structures of bovine, ovine and porcine complex I were reported at 4.2, 3.9, and 3.6 Å resolution.[11,12,17] Despite the larger mass of mitochondrial complex I (1 MDa) as compared to the bacterial enzyme (550 kDa), the L-shaped overall architecture of bacterial, yeast and mammalian complex I is rather similar (Figure 3.1).[36]

The membrane arm of *Y. lipolytica* complex I is about 210 Å long, 100 Å high and 75 Å wide while the peripheral arm has a size of about 130×110×100 Å (Figure 3.1). The membrane arm can be divided into a distal and proximal proton pumping module (P_P and P_D modules). The peripheral arm comprises the NADH oxidation module (N module) and the ubiquinone reduction module (Q module).[1,7] The NADH oxidation site with the primary electron acceptor FMN is connected with the ubiquinone reduction site by a chain of seven iron–sulphur clusters. N2 is the last cluster in the chain and the electron donor for ubiquinone. The mass difference between bacterial and mitochondrial complex I reflects the addition of accessory subunits that will be briefly described in the next paragraph.

3.3 Accessory Subunits

The number of accessory subunits for complex I from *Y. lipolytica* is 28 and 31 for mammalian complex I. Most accessory subunits occur in all eukaryotes but some are specific for either fungal, mammalian or plant complex I.[37,38] It is interesting to note that subunit NDUFV3 exists in two isoforms that have a tissue specific expression profile.[39–41] Several different types of accessory subunits can be distinguished. In the lipid bilayer, the core of the membrane arm is surrounded by single transmembrane domain (STMD) subunits.[42] STMDs are also found in other respiratory chain complexes. Accessory subunit NUJM/B14.7 has four transmembrane helices (TMHs) and has an important structural role for joining P_P and P_D modules and for supercomplex formation.[10,15,43,44] A layer of subunits carrying characteristic $CysX_9Cys$ patterns that indicate processing by the Mia40/Erv1 relay pathway[45] are located on the intermembrane space side of the membrane arm. Together with the transmembrane accessory subunits they form a frame or cage that appears to have an important structural role for the membrane arm in eukaryotic complex I. Deletion of the gene for the Mia40 substrate subunit NB8M caused separation of the distal part of the membrane arm.[46] In contrast, the peripheral arm is stabilized by several accessory subunits that are arranged along the interfaces of subunits and domains.[6] The NUMM/B13 subunit belongs to this group and is noteworthy because it carries a zinc binding site. Deletion of the corresponding gene in *Y. lipolytica* or mutation of Zn binding residues blocked a late step in complex I assembly.[47] Deletion of the gene for subunit NUYM/18 kDa was shown to disturb two iron–sulphur clusters of the N module.[48] NUYM is the ortholog of NDUFS4 that is a hot spot for pathogenic mutations in humans.[49] It has been known for a long time that complex I harbours an acyl carrier protein (ACPM).[50,51] It was shown later that two different ACPMs, ACPM1 and ACPM2,[52] are associated with *Y. lipolytica* complex I while mammalian complex I comprises two identical copies (SDAP-α and SDAP-β).[10] Both ACPMs are anchored to complex I *via* proteins of the LYRM family.[53,54] Mass spectrometry analysis suggested that bovine SDAP carries 3-hydroxy-tetradecanoate attached to its phosphopantethein cofactor[55] and the structure of ovine complex I revealed that the acyl-group is inserted into the interior of the associated LYR protein.[12] SDAP was also detected as a free matrix protein[56] and it was shown that the essential function of

ACPM1 for *Y. lipolytica* is independent of complex I.[52,54] Most likely the free ACPM is required for the synthesis of the octanoic acid precursor of lipoic acid,[57] the prosthetic group of α-ketoacid dehydrogenases and it remains to be elucidated why acyl carrier proteins associate with complex I. The NUEM/39 kDa subunit has the typical fold of short chain dehydrogenases and tightly binds NADPH.[58] The function of this subunit is unclear but it is interesting to note that it might play a role in controlling the A/D transition.[59] The 42 kDa subunit is specific for selected branches of higher eukaryotes and contains a nucleoside kinase fold. So far no activity was reported but the subunit was found to be phosphorylated by PINK1 establishing a possible link with Parkinson's disease.[60] Finally, complex I from *Y. lipolytica* comprises a sulphur transferase subunit (ST1).[13,61] Subunit ST1 is a mercaptopyruvate sulfur transferase and the subunit was shown to release H_2S in the presence of suitable substrates.

3.4 Central Subunits of the Membrane Arm

The seven central subunits of the membrane arm, ND1–ND6 and ND4L (Figure 3.2), are typically encoded by mitochondrial DNA in eukaryotes. The three largest subunits ND2, ND4 and ND5 share a common core fold[5] and are related to polypeptides of multi-subunit Mrp type cation/proton antiporters (Figure 3.3).[62] ND2 and ND4 correspond to MrpD while ND5 corresponds to MrpA. Each subunit comprises two inverted repeats of TMHs 4–8 and TMHs 9–13.[5] TMHs 7 and TMHs 12 are discontinuous helices with loop regions within the membrane phase. Discontinuous helices were frequently shown or suggested to play a critical role in transmembrane transport processes and the intersecting loop regions typically harbour highly conserved charged residues.[63,64] In ND2, ND4 and ND5 the loop region of TMH7 comprises a strictly conserved lysine that is close to a strictly conserved glutamate residue in the neighbouring TMH5 (Figure 3.2). The loop region of TMH12 comprises a lysine in ND2, a glutamate in ND4, and an aspartate/lysine pair in ND5. Subunit ND5 is distinguished by a substantial C-terminal extension including the long lateral helix that lines a large part of the membrane arm. The lateral helix terminates with TMH16 that is followed in a V shaped arrangement by TMH17 in fungal complex I. This transmembrane anchor of the lateral helix is clamped to the membrane arm by accessory subunit NUJM/B14.7. The lateral helix was originally suggested to be a piston-like energy transmission device[7,65] but this view was not supported by functional data.[66–68] For as-yet unknown reasons TMH1–3 are missing in ND2 of selected branches of higher metazoans.[69] Towards the peripheral arm, the linear arrangement of ND5, ND4 and ND2 is followed by subunits ND4L and ND6. ND4L is homologous to subunit C of Mrp type antiporters, while ND6 has been discussed to be related to the C terminal domain of MrpA (Figure 3.3).[70,71] Two of the three TMHs in ND4L harbour strictly conserved glutamate residues. The major part of the interface between membrane arm and peripheral arm is formed by subunit ND1.

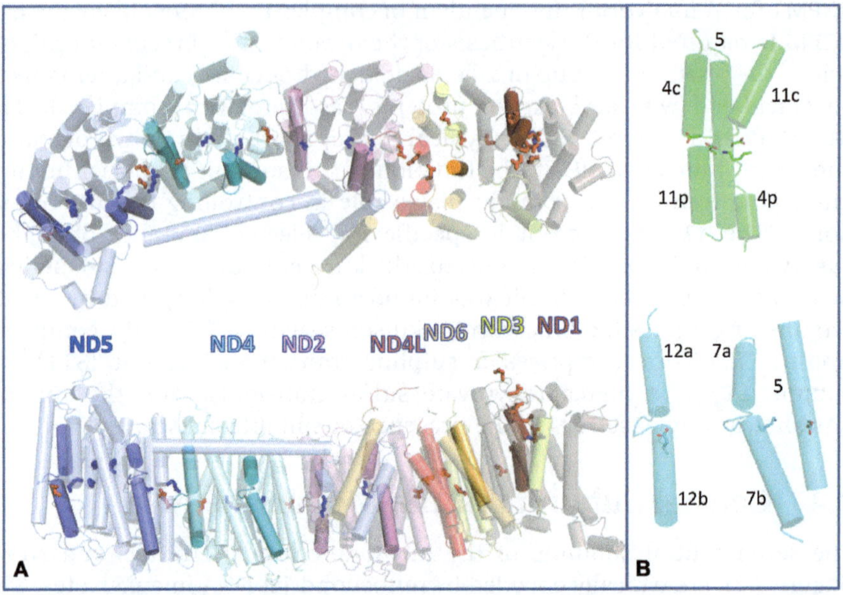

Figure 3.2 Central subunits of the membrane arm and residues forming the hydrophilic axis. (A) The central subunits of the membrane arm of *Y. lipolytica* complex I (PDB ID: 4WZ7) are shown in cartoon representation, the discontinuous helices of ND5, ND4, ND2, and ND1 and the π bulge helix of ND6 are accentuated, selected residues of the hydrophilic axis are shown in stick representation (basic residues, blue; acidic residues, red). (B) The arrangement of discontinuous helices 4 and 11 and TMH5 of NhaA (PDB ID: 1ZCD) (green, selected residues in stick representation, including those critical for ion binding) is compared with the arrangement of discontinuous helices 7 and 12 and TMH5 of subunit ND4 (blue, selected residues in stick representation, see text); TMH5 carries a glutamate residue that is strictly conserved in complex I and Mrp antiporters.

The subunit has typically eight TMHs but in *T. thermophilus* a ninth helix is present. This helix is replaced by accessory subunit NB6M/B16.6 in eukaryotes. Several of the TMH in ND1 are highly tilted relative to the membrane normal. The connection between TMH1 and 2 is formed by a long loop and amphipathic surface helix α1. The segment of TMHs 2–6 is structurally similar to one of the repeats in the large antiporter like subunits.[72] TMH5 is a discontinuous helix in *Y. lipolytica* complex I,[9] however, the fold of the shorter segment oriented towards the peripheral arm is not helical in other complex I structures.[8,11,12] Remarkably, this sequence stretch and the following connection to TMH6 comprise a series of highly conserved acidic residues. A conserved arginine residue in the intersecting loop region of discontinuous TMH5 is reminiscent of the conserved lysine in TMH7 of the antiporter-like subunits. Conserved acidic residues in transmembrane

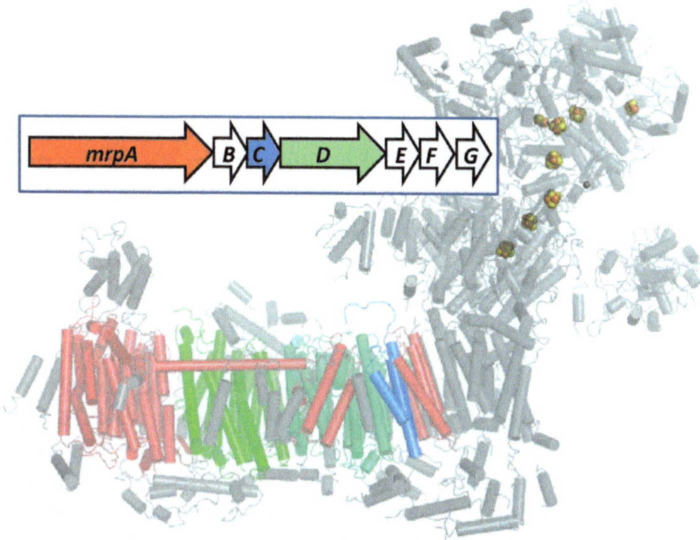

Figure 3.3 Relation of complex I with Mrp type cation/proton antiporters. Three of the seven gene products of a group 1 Mrp operon are related to complex I subunits. MrpA corresponds to ND5 and ND6 (red), MrpC corresponds to ND4L (blue) and MrpD corresponds to ND4 and ND2 (different shades of green).

segments of ND1 are found in TMH4 and 6. Subunit ND3 has a remarkable topology with a spatial separation of the first from the second and third TMH. TMH2 of ND3 comprises a pair of conserved acidic residues. The long loop connecting the first two helices lines the interface region of membrane and peripheral arm and harbours a cysteine residue that was used as reporter group for the A/D transition because it is only accessible in the D form.[34,73,74]

3.5 Proton Translocation Pathways and the Central Hydrophilic Axis

A proton pump stoichiometry of $4H^+/2e^-$ was experimentally determined for eukaryotic complex I.[75–77] This number has been debated[78] and a lower stoichiometry has been suggested for those bacterial species that use menaquinone instead of ubiquinone. Nevertheless, four distinct proton translocation pathways are expected to operate in complex I. There is a wide consensus that H^+ is the coupling ion of complex I but Na^+/H^+ antiporter activity was reported for the enzyme from selected bacterial species and suggested to have an auxiliary function.[79,80] Antiporter activity was also found for mitochondrial complex I in the D form.[81] The relevance of Na^+ for complex I linked transport processes has been reviewed recently.[82]

The prime candidates for proton translocation activity are subunits ND2, ND4, and ND5 because they are related to cation/proton antiporters and harbour discontinuous helices that are hallmarks of membrane transport proteins. However, it should be noted that discontinuous helices 7 and 12 in the antiporter like subunits of complex I are spatially separated in contrast to the situation found in members of a different Na^+/H^+ antiporter family, NhaA or NapA, where two discontinuous helices cross each other close to the proposed ion binding site (Figure 3.2).[83,84] As a first step to locate putative proton channels, chains of protonable and hydrophilic residues were defined in the structures of complex I from *T. thermophilus* and *Y. lipolytica*.[5,8,9] Possible proton exit pathways were consistently found along discontinuous TMH12 and TMH13 in each of the three antiporter-like subunits. The proton input pathways in bacterial complex I were suggested to target the arrangement of TMH5 and discontinuous TMH7. This resulted in a symmetrical system of two half channels that are connected within the membrane phase by suitable residues including a conserved basic residue in TMH8 and water molecules. In contrast, a different proton uptake pathway was suggested for the antiporter like subunits of complex I from *Y. lipolytica*. The likely entry point was found between TMH7 and 8 close to the lateral helix. From here, the H^+ pathway was suggested to follow a chain of residues leading to the intersecting loop of discontinuous TMH12. A combination of both types of proton uptake routes was suggested later.[85] In the model developed based on the complex I structure of *Y. lipolytica*, the TMH5/TMH7 element is apparently bypassed. However, a remarkable feature of the membrane arm subunits ensures that none of the putative pump elements is isolated. From the intersecting loop of TMH12 of ND5 a wire of protonable and hydrophilic residues runs in the centre of the antiporter like subunits and forms a continuous connection of all TMH12 and TMH5/7 putative pump elements. This hydrophilic axis is continued by the two acidic residues in ND4L, a π bulge helix in ND6, the two acidic residues in ND3, and the two acidic residues in TMH4 and TMH6 of ND1. The charged residues in the sequence stretch that form the intersecting loop region of TMH5, its short helical segment and the loop connecting to TMH6, link the hydrophilic axis with the peripheral arm interface. Strikingly, the cluster of acidic residues in subunit ND1 is located right below the catalytic site for ubiquinone reduction. The continuation of the hydrophilic axis towards the peripheral arm was coined E channel by Sazanov and colleagues and suggested as a proton translocation pathway.[8] However, in contrast to the assignment of three pump units to the antiporter-like subunits, there is less agreement on the supposed forth pump site. A possible channel was initially identified at the interface of subunits NuoN(ND2) and NuoK(ND4L) of *E. coli* complex I.[5] In MD simulations hydration of this putative proton entry path was observed in response to the protonation state of the acidic residues in subunit NuoK (ND4L).[86] Later a similar proton translocation pathway was suggested in complex I from *Y. lipolytica*.[9] However, it should be noted that the critical threonine residue in the equivalent channel of *E. coli* complex I ($Thr160_N$) is not conserved in *Y. lipolytica* complex I.

3.6 Peripheral Arm and Ubiquinone Binding Site

The peripheral arm comprises seven central subunits that can be assigned to the functional modules for NADH oxidation and ubiquinone reduction. With the exception of the 30 kDa and 49 kDa subunits all central subunits of the N and Q module harbour one or several iron–sulphur clusters.[87] Their designations and the assignment of EPR signals are inconsistently used in the literature. I will here adhere to the scheme proposed by Roessler and Hirst (Figure 3.1).[88] The NADH binding site with the primary electron acceptor FMN is located in the 51 kDa subunit. Together with the 24 kDa subunit, it forms the so-called flavoprotein fragment of complex I. From FMN electrons are transferred to the 4Fe4S iron–sulphur cluster in the 51 kDa subunit (EPR signal N3) and then further on *via* a chain of iron–sulphur clusters to the ubiquinone reduction site. The 24 kDa subunit harbours a binuclear iron–sulphur cluster (signal N1a) in electron transfer distance to FMN. However, this cluster is not part of the extended electron transfer wire and its function is under debate.[32] The 75 kDa subunit harbours a binuclear cluster (signal N1b) and two tetranuclear clusters. One of the latter has an unusual ligation by three cysteines and one histidine and is EPR silent. The other cluster gives rise to signal N5. The TYKY subunit has a ferredoxin fold and binds two tetranuclear clusters. Signal N4 was assigned to the cluster oriented towards the 75 kDa subunit, the cluster oriented towards the PSST subunit is EPR silent. The last cluster in the chain is liganded by subunit PSST. The tetranuclear cluster (signal N2) with a remarkable ligation involving two adjacent cysteines is the electron donor for ubiquinone. The redox potential of the N2 cluster is pH dependent and the associated redox-Bohr group was identified to be a histidine residue in the 49 kDa subunit.[89] Removal of the redox-Bohr effect by site-directed mutation of the histidine did not affect proton pumping. The PSST subunit and the 49 kDa subunit are both evolutionary related to Ni–Fe hydrogenases.[1] Before structural information became available, a large body of mutagenesis data and inhibitor labelling studies had indicated that these two subunits form the major part of the ubiquinone binding and reduction site.[90–92] The two subunits enclose a central cavity that opens towards subunit ND1 in the membrane arm (Figure 3.4). At the deepest end of the cavity in close vicinity of cluster N2, a conserved tyrosine residue (Tyr144 in *Y. lipolytica* corresponding to Tyr87 in *T. thermophilus*) was identified to bind the ubiquinone headgroup.[93] Remarkably, ubiquinone has to penetrate a considerable distance into the peripheral arm to be reduced. A hot spot of functionally important residues was found in the loop connecting the first and the second β strands of the N-terminal β sheet of the 49 kDa subunit (Figures 3.4 and 3.5).[94] This loop points into the cavity and

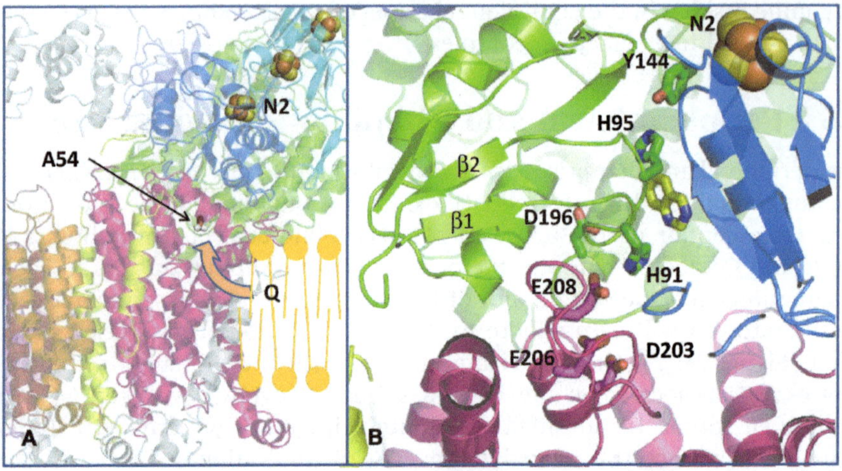

Figure 3.4 The ubiquinone access path and binding site. (A) Interface of membrane arm and peripheral arm of *Y. lipolytica* complex I (PDB ID: 4WZ7, colors compare to Figure 3.1, accessory subunit NUEM removed for clarity), ubiquinone (Q) enters the catalytic site *via* a portal formed by subunit ND1. Mutation of an alanine residue in the α1–2 surface helix (A54 in *Y. lipolytica* corresponding to A52 in humans) is known to interfere with ubiquinone reduction and to cause LHON. (B) Close up view of the ubiquinone reduction site (colors see Figure 3.1, parts of PSST removed for clarity). The loop connecting the first two β strands of the N-terminal β sheet of the 49 kDa subunit points into the ubiquinone cavity and comprises two highly conserved histidine residues (His91 and His95). The toxophore of quinazoline type inhibitors (yellow stick representation) was modeled based on anomalous Fourier analysis. Details, see text.

harbours two strictly conserved histidine residues (His91 and His95 in *Y. lipolytica* corresponding to His34 and His38 in *T. thermophilus*). A sector of the cavity is lined by residues of the prominent four-helix bundle of the 49 kDa subunit. Several lines of evidence indicate that a conserved aspartate residue (Asp196 in *Y. lipolytica* corresponding to Asp139 in *T. thermophilus*) of this structural element is functionally important.[95,96] An access path to the cavity is provided by membrane arm subunit ND1. An opening is formed by TMH1 and 6 and the surface helix α1 (Figure 3.4). Mutation of an alanine residue of this amphipathic helix pointing into the entrance interferes with ubiquinone reduction and is one of the most frequent causes for LHON disease.[97]

Different strategies were followed to obtain structural data on the binding position of ubiquinone in the catalytic site. Baradaran *et al.* co-crystallized complex I with a short chain ubiquinone analogue or the complex I inhibitor piericidine.[8] Ubiquinone and inhibitor were found to bind to Tyr87 close to cluster N2 and to His38 of the β1 β2 loop (Figure 3.5). This configuration is in excellent agreement with mutagenesis data

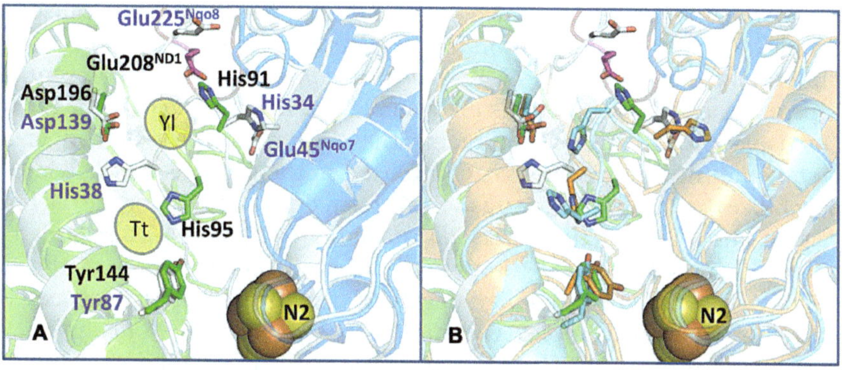

Figure 3.5 Conformational changes in the ubiquinone reduction site. View from cluster N2 towards the membrane arm, helix α2 of subunit PSST removed for clarity. (A) Overlay of complex I from *Y. lipolytica* (PDB ID: 4WZ7, cartoon and stick representation of selected residues in color – compare to Figure 3.1; residue numbering in black, residues without superscript 49 kDa subunit) and *T. thermophilus* (PDB ID: 4HEA, cartoon and stick representation in grey; residue numbering in violet, residues without superscript Nqo4). The different approximate positions of inhibitor/ ubiquinone are indicated and labeled Yl for *Y. lipolytica* (compare to Figure 3.4) and Tt for *T. thermophilus* (see text). (B) The 49 kDa and PSST subunits of bovine (class II, PDB ID: 5LC5, orange) and ovine (open conformation, PDB ID: 5LNK, cyan) complex I were added to the overlay shown in (A), the positions of corresponding histidine residues in the β1 β2 loop are clearly different.

obtained with *Y. lipolytica* complex I.[91] The position of His38 was found to be stabilized by Asp139 of the four-helix bundle. We co-crystallized complex I with two brominated derivatives of the complex I ubiquinone antagonist inhibitor decyl-quinazolineamine (DQA) and determined the bromine positions by anomalous Fourier analysis.[9] The inhibitor toxophore was modelled in a position sandwiched between the tip of the β1 β2 loop of the 49 kDa subunit and the second helix of subunit PSST (Figure 3.5). The clear difference in the results of the two studies (Figure 3.5) was suggested to reflect an important functional difference between bacterial complex I and complex I from many eukaryotes that will be discussed below.

3.7 Structural Basis of the A/D transition

Complex I from many eukaryotic species including *Y. lipolytica* can undergo a reversible transition between an active (A) form and a deactive (D) form.[98] The D form dominates under conditions of substrate limitation, *e.g.* during an ischaemic period, and is converted back to the A form by turnover. A conformational switch between A and D form was demonstrated based on differential accessibility of a cysteine residue in the loop connecting the first two TMHs of subunit ND3.[34] Slowing reactivation by reversible nitrosation of this cysteine residue was shown to alleviate the tissue damage caused by complex I linked ROS formation in the reperfusion phase after myocardial infarction.[35] The structure of *Y. lipolytica* complex I must represent the D form because (i) the A form is meta stable and the enzyme is in the D form after isolation, (ii) turnover inducing the A form can be excluded in the crystal, and (iii) the enzyme was crystallized in the presence of Ca^{2+} known to stabilize the D form. In contrast, the D form does not exist for the bacterial enzyme and the structure of *T. thermophilus* complex I was regarded to represent the A form in a comparison between the structures of the two enzymes.[9] Overall, there is a high degree of structural similarity between the central subunits of complex I from *T. thermophilus* and *Y. lipolytica*. Remarkably, prominent differences are observed in the ubiquinone binding site and in the binding position of ubiquinone and inhibitors (Figure 3.5). The conformation of the β1–β2 loop in *Y. lipolytica* complex I blocks access deep into the central cavity and His95 contacts Tyr144. In contrast, ubiquinone was bound close to iron–sulphur cluster N2 by the two corresponding residues Tyr87 and His38 in *T. thermophilus* complex I.[8] In bacterial complex I His34 of the β1–β2 loop binds the highly conserved Glu45 of the long Nqo7 (ND3) loop. This aspartate is followed by a serine in *T. thermophilus* and by a cysteine residue in many other species. Intriguingly, this is the cysteine residue used as the reporter group for the D form in bovine complex I.[34] In *Y. lipolytica* complex I, His91 of the β1–β2 loop is flipped to a different position and binds Glu208 of the ND1 TMH 5–6 loop, which corresponds to Glu225 in Nqo8 of *T. thermophilus* complex I (Figure 3.5). Releasing the binding interaction with the ND3 loop might induce a more open conformation in this structural domain, which would be in perfect agreement with the observed selective accessibility of the ND3 cysteine in the D form and a largely disordered conformation of the loop in the *Y. lipolytica* complex I structure. Taken together the comparison of the two structures provided a viable model for the structural basis of the A/D transition. In the D form access to the Q site is blocked and transition to the A from requires a concerted rearrangement of the ubiquinone reduction site including the β1–β2 loop of the 49 kDa subunit, the TMH1–2 loop of ND3 and the TMH5–6 loop of ND1.[9] In the recent cryoEM study on ovine complex I, an "open" and a "closed" conformation were observed.[12,15] The "open" conformation was tentatively assigned to the D form because the conformation of the β1–β2 loop would cause blockage of the ubiquinone cavity similar to the situation observed for complex I from *Y. lipolytica*. However, the model for the structural basis of the A/D transition presented above was challenged by the cryoEM structure of bovine complex I.[11] Two different classes were discussed to represent A and D form that were biochemically quantified to be present in an approximate 1 : 1 relation in the sample. Class 1 was tentatively assigned to the D form because the ND3 loop was disordered, class 2 was assigned to the A form. At this stage evaluation of bovine complex I class 1 is difficult because the resolution is lower and the β1–β2 loop is disordered. Inhibitor binding was suggested as a decisive parameter for better resolution of the loop in the X-ray structure. Class 2 does not closely match either of the two X-ray structures (Figure 3.5). For a thorough evaluation a clear assignment of A and D form to individual classes/structures with sufficient resolution will be needed. Nevertheless, it is intriguing that all of the available structures of complete complex I show conformational differences in the β1–β2 loop of the 49 kDa site (Figure 3.5).

3.8 Catalytic Mechanism of Redox Linked Proton Translocation

The topology of complex I is fundamentally different from that of the other redox driven proton pumps of the respiratory chain. Electron transfer occurs exclusively in the peripheral arm. Thus, it is spatially separated from proton translocation in the membrane arm (Figure 3.6). Electron transfer from NADH to cluster N2 is fast compared to the turnover of complex I.[99] Although recent time resolved EPR studies suggest that electron transfer reactions in complex I are more complex,[100] currently there is no compelling

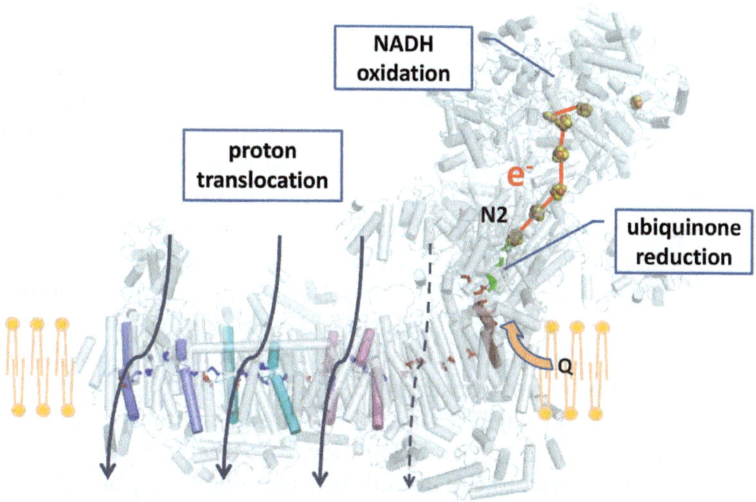

Figure 3.6 Redox-linked proton translocation by complex I. Electrons are transferred to the ubiquinone reduction site near cluster N2; selected critical residues of the catalytic site are shown in green stick representation (compare to Figure 3.4). Ubiquinone enters the site *via* subunit ND1 (discontinuous helix of ND1, red). The three antiporter like subunits of the membrane arm comprise discontinuous helices (ND5, blue; ND4, cyan; ND2, pink) and putative proton translocation pathways (blue arrows). A putative forth channel was identified at the interface of subunits ND2 and ND4L (dashed arrow). A series of protonable residues (basic, blue; acidic, red) extends from subunit ND5 to subunit ND1 and terminates below the ubiquinone reduction site. Conformational changes linked with the chemistry of ubiquinone reduction trigger proton translocation *via* the hydrophilic axis of the membrane arm.

evidence that any of the inter-cluster electron transfer steps are linked to proton pumping. Several lines of evidence indicate that the redox energy driving proton translocation is largely released during the reduction of ubiquinone.[101,102] Full reduction of ubiquinone requires transfer of two electrons and two protons. The observation of semiquinone species by EPR[102–104] indicated that ubiquinone reduction is a stepwise process that proceeds *via* defined intermediates with sufficiently long lifetimes to be observable. However, it should be noted that a consistent picture about the presence, number, properties and roles of semiquinone intermediates is still lacking.[32]

It was initially proposed by Euro *et al.* that formation of charged semiquinone species might trigger the pump sites by transmission of the negative charge *via* an electrostatic coupling mechanism involving conserved charged residues in the membrane domain.[105]

The two-state stabilization change mechanism proposed by Brandt[106] assumes that electron and proton transfer events are strictly separated and are joined *via* two distinct conformational states of the catalytic site. In the E state ubiquinone is situated in electron transfer distance to N2 and $Q^{\bullet -}$ is generated by transfer of one electron from cluster N2. The semiquinone anion is stabilized by rearrangement of the site and is now situated in a position that permits protonation to $QH^{\bullet}$, but not uptake of the second electron. This conformational state is called the P state. The conformational change associated with stabilization of the semiquinone species drives the first of two pump strokes. The same principle is repeated, *i.e.* E state reduction of $QH^{\bullet}$ to QH^{-}, followed by its stabilization driving the second pump stroke. In the subsequent P state QH^{-} is protonated to QH2 and released to the bulk. The proposed cycling of the Q site between E and P state was found to fit intriguingly well to the observed structural differences for the A and D form of complex I discussed above.[9] The A form with ubiquinone bound close to N2 is reminiscent of the hypothetical E state while the D form with a more remote position of Q might correspond to the P state. This perspective would suggest that the D form essentially results from a controlled arrest of the P state. Intriguingly, the differences between A and D form are not restricted to the ubiquinone reduction site, but also involve conformational changes of the first loop of subunit ND3 and the TMH5–6 loop of ND1, establishing a connection to the hydrophilic axis. Assuming that these conformational changes occur in the catalytic cycle of complex I it was hypothesized that redox-linked conformational changes of the ubiquinone reduction site are converted into an electrostatic pulse that is propagated along the hydrophilic axis to trigger the proton pumps in the membrane arm.[9] In contrast to other proposed mechanistic schemes, the two state stabilization change mechanism explicitly implies two pump strokes per cycle. A two stroke mechanism at the level of the ubiquinone chemistry could still be translated into a one stroke mechanism at the level of the proton pumps. However, the full reversibility of complex I supports a mechanism that subdivides the entire energy difference of the reaction to two separate steps.

Baradaran *et al.* modelled ubiquinone between Tyr87 and the His38/Asp139 ion pair (Figure 3.5) of subunit Nqo4 (49 kDa subunit) and in their mechanistic scheme ubiquinone is reduced to the doubly charged Q^{2-} in the catalytic site.[8] This intermediate was suggested to couple electrostatically with acidic residues in the so-called E channel in subunit ND1 to drive conformational changes along the hydrophilic axis that ultimately lead to proton pumping. A detailed model for redox linked conformational changes in the Q site was reported by Sharma *et al.* based on QM/MM simulations performed with the structure of *T. thermophilus* complex I.[95] The quinone molecule was initially placed in the crystallographic site (Figure 3.5) and the dynamics were compared for oxidized ubiquinone, semiquinone ($Q^{\bullet -}$) and the doubly charged Q^{2-}. In the case of Q^{2-} *i.e.* after two electron reduction, a massive rearrangement of the site was observed. Proton transfer from the Tyr87 and the His38 led to formation of QH2 and disruption of the His38/Asp139 ion pair. This was followed by a 10–12 Å movement of Asp139 towards the membrane phase and a movement of His38 towards Tyr87. Interestingly, the new position of the histidine was similar to the position of the corresponding residue in the *Y. lipolytica* structure. The *in silico* movement of Asp139 disturbed a network of charged residues in subunit Nqo8 (ND1). Consequently, side chain movements of E channel residues towards the N side of the membrane were observed in MD simulations. Interestingly, electrostatic calculations suggested a higher affinity for protons of two acidic residues in the N side position and a return to the original position after protonation. This computational approach thus offers some clues to how events in the ubiquinone reduction site might trigger proton uptake by acidic residues of ND1.

For secondary transporters variations of alternating access models are discussed.[64,107] Transmembrane transport of the substrate bound to a central binding site is essentially executed by the relative movement of two domains of the protein. There is no indication that a "classical" alternating access mechanism is operating in the antiporter like subunits of complex I. It was rather suggested that the transient formation of water chains for loading the pump site with protons from the N side and subsequent release to the P side might implement an alternative alternating access mechanism in complex I.[108] In MD simulations the formation of a water chain and the uptake of a proton from the N side at the NuoN/NuoK (ND2/ND4L) interface was found to be coupled to the protonation state of the two conserved acidic residues in NuoK (ND4L).[86] These two acidic residues are connected to the E channel and might thus be triggered by the events in the charged cluster of subunit Nqo8 (ND1) described above. Towards the distal part of the membrane arm, the hydrophilic axis constitutes a continuous connection with the putative pump elements in the antiporter like subunits. Hummer and Wikström suggested that an electrostatic domino effect initiated by formation of Q^{2-} might trigger a concerted flipping of side chains that essentially transfers part of the negative charge into the membrane arm.[108] Loading the pump sites from the N side through transiently formed water channels would permit charge compensation by protonation. Formation of QH_2 would trigger reversal of the charging process and thus promote proton release. Gating would be achieved by moving the protons in the hydrophilic axis and changing accessibility to the P side. It remains to be established how such a gating mechanism could work in detail. Moreover, a charging mechanism would also require strict control of proton accessibility and seems to exclude proton uptake in the proximal segment of the hydrophilic axis, because it would counteract the charging process.

Conformational changes of the ubiquinone reduction are a common key principle of the mechanistic schemes addressed in this review. The pronounced flexibility of key residues in the β1–β2 loop of the 49 kDa subunit, as shown by a comparison of four complex I structures (Figure 3.5), lends strong support to the suggestion that the site can adopt different conformations that are linked with function. However, at present, the molecular details of any of the proposed mechanisms are unknown and more work is needed to reach a comprehensive understanding of redox-linked proton translocation by complex I.

Acknowledgements

This work was supported by the German Research Foundation (ZI 552/4-1) and the Excellence Initiative of the German Federal and State Governments (EXC 115). I thank Stefan Dröse and Ulrich Brandt for critically reading the manuscript.

References

1. U. Brandt, *Annu. Rev. Biochem.*, 2006, **75**, 69.
2. J. Hirst, *Annu. Rev. Biochem.*, 2013, **82**, 551.
3. R. J. Rodenburg, *Biochim. Biophys. Acta*, 2016, **1857**, 938.
4. L. A. Sazanov and P. Hinchliffe, *Science*, 2006, **311**, 1430.
5. R. G. Efremov and L. A. Sazanov, *Nature*, 2011, **476**, 414.
6. J. Zhu, M. S. King, M. Yu, L. Klipcan, A. G. Leslie and J. Hirst, *Proc. Natl. Acad. Sci. U. S. A.*, 2015, **112**, 12087.
7. C. Hunte, V. Zickermann and U. Brandt, *Science*, 2010, **329**, 448.
8. R. Baradaran, J. M. Berrisford, G. S. Minhas and L. A. Sazanov, *Nature*, 2013, **494**, 443.
9. V. Zickermann, C. Wirth, H. Nasiri, K. Siegmund, H. Schwalbe, C. Hunte and U. Brandt, *Science*, 2015, **347**, 44.
10. K. R. Vinothkumar, J. Zhu and J. Hirst, *Nature*, 2014, **515**, 80.
11. J. Zhu, K. R. Vinothkumar and J. Hirst, *Nature*, 2016, **536**, 354.

12. K. Fiedorczuk, J. A. Letts, G. Degliesposti, K. Kaszuba, M. Skehel and L. A. Sazanov, *Nature*, 2016, **538**, 406.
13. E. D'Imprima, D. J. Mills, K. Parey, U. Brandt, W. Kuhlbrandt, V. Zickermann and J. Vonck, *Biochim. Biophys. Acta*, 2016, **1857**, 1935.
14. J. Gu, M. Wu, R. Guo, K. Yan, J. Lei, N. Gao and M. Yang, *Nature*, 2016, **537**, 639.
15. J. A. Letts, K. Fiedorczuk and L. A. Sazanov, *Nature*, 2016, **537**, 644.
16. J. S. Sousa, D. J. Mills, J. Vonck and W. Kuhlbrandt, *Elife*, 2016, **5**, e21290.
17. M. Wu, J. Gu, R. Guo, Y. Huang and M. Yang, *Cell*, 2016, **167**, 1598.
18. J. Hirst, *Biochem. J.*, 2011, **437**, e1.
19. K. Kmita and V. Zickermann, *Biochem. Soc. Trans.*, 2013, **41**, 1272.
20. C. Y. Yip, M. E. Harbour, K. Jayawardena, I. M. Fearnley and L. A. Sazanov, *J. Biol. Chem.*, 2011, **286**, 5023.
21. G. J. Schut, O. Zadvornyy, C. H. Wu, J. W. Peters, E. S. Boyd and M. W. Adams, *Biochim. Biophys. Acta*, 2016, **1857**, 958.
22. L. Sanchez-Caballero, S. Guerrero-Castillo and L. Nijtmans, *Biochim. Biophys. Acta*, 2016, **1857**, 980.
23. T. Ohnishi, *Biochim. Biophys. Acta*, 1998, **1364**, 186.
24. E. Nakamaru-Ogiso, T. Yano and T. Ohnishi, *J. Biol. Chem.*, 2005, **280**, 301.
25. T. Pohl, T. Bauer, K. Dorner, S. Stolpe, P. Sell, G. Zocher and T. Friedrich, *Biochemistry*, 2007, **46**, 6588.
26. V. Zickermann, M. Bostina, C. Hunte, T. Ruiz, M. Radermacher and U. Brandt, *J. Bio.l Chem.*, 2003, **278**, 29072.
27. T. Clason, V. Zickermann, T. Ruiz, U. Brandt and M. Radermacher, *J. Struct. Biol.*, 2007, **159**, 433.
28. M. Sato, P. K. Sinha, J. Torres-Bacete, A. Matsuno-Yagi and T. Yagi, *J. Biol. Chem.*, 2013, **288**, 24705.
29. M. Murai and H. Miyoshi, *Biochim. Biophys. Acta*, 2016, **1857**, 884.
30. L. Kussmaul and J. Hirst, *Proc. Natl. Acad. Sci. U. S. A.*, 2006, **103**, 7607.
31. A. Galkin and U. Brandt, *J. Biol. Chem.*, 2005, **280**, 30129.
32. J. Hirst and M. M. Roessler, *Biochim. Biophys. Acta*, 2016, **1857**, 872.
33. A. B. Kotlyar and A. D. Vinogradov, *Biochim. Biophys. Acta*, 1990, **1019**, 151.
34. A. Galkin, B. Meyer, I. Wittig, M. Karas, H. Schagger, A. Vinogradov and U. Brandt, *J. Biol. Chem.*, 2008, **283**, 20907.
35. E. T. Chouchani, C. Methner, S. M. Nadtochiy, A. Logan, V. R. Pell, S. Ding, A. M. James, H. M. Cocheme, J. Reinhold, K. S. Lilley, L. Partridge, I. M. Fearnley, A. J. Robinson, R. C. Hartley, R. A. Smith, T. Krieg, P. S. Brookes and M. P. Murphy, *Nat. Med.*, 2013, **19**, 753.
36. T. Clason, T. Ruiz, H. Schagger, G. Peng, V. Zickermann, U. Brandt, H. Michel and M. Radermacher, *J. Struct. Biol.*, 2010, **169**, 81.
37. C. Wirth, U. Brandt, C. Hunte and V. Zickermann, *Biochim. Biophys. Acta*, 2016, **1857**, 902.
38. N. Subrahmanian, C. Remacle and P. P. Hamel, *Biochim. Biophys. Acta*, 2016, **1857**, 1001.

39. S. Guerrero-Castillo, A. Cabrera-Orefice, M. A. Huynen and S. Arnold, *Biochim. Biophys. Acta*, 2016, **1858**, 208.
40. H. R. Bridges, K. Mohammed, M. E. Harbour and J. Hirst, *Biochim. Biophys. Acta*, 2016, **1858**, 197.
41. M. G. Dibley, M. T. Ryan and D. A. Stroud, *FEBS Lett.*, 2017, **591**, 109.
42. V. Zickermann, H. Angerer, M. G. Ding, E. Nubel and U. Brandt, *FEBS Lett.*, 2010, **584**, 2516.
43. B. Andrews, J. Carroll, S. Ding, I. M. Fearnley and J. E. Walker, *Proc. Natl. Acad. Sci. U. S. A.*, 2013, **110**, 18934.
44. S. Guerrero-Castillo, F. Baertling, D. Kownatzki, H. J. Wessels, S. Arnold, U. Brandt and L. Nijtmans, *Cell Metab.*, 2016, **25**, 128.
45. J. M. Herrmann and J. Riemer, *J. Biol. Chem.*, 2012, **287**, 4426.
46. S. Dröse, S. Krack, L. Sokolova, K. Zwicker, H. D. Barth, N. Morgner, H. Heide, M. Steger, E. Nübel, V. Zickermann, S. Kerscher, B. Brutschy, M. Radermacher and U. Brandt, *PLoS Biol.*, 2011, 9.
47. K. Kmita, C. Wirth, J. Warnau, S. Guerrero-Castillo, C. Hunte, G. Hummer, V. R. Kaila, K. Zwicker, U. Brandt and V. Zickermann, *Proc. Natl. Acad. Sci. U. S. A.*, 2015, **112**, 5685.
48. F. Kahlhofer, K. Kmita, I. Wittig, K. Zwicker and V. Zickermann, *Biochim. Biophys. Acta*, 2016, **1858**, 175.
49. M. E. Breuer, P. H. Willems, J. A. Smeitink, W. J. Koopman and M. Nooteboom, *IUBMB. Life*, 2013, **65**, 202.
50. U. Sackmann, R. Zensen, D. Roehlen, U. Jahnke and H. Weiss, *Eur. J.Biochem.*, 1991, **200**, 463.
51. M. J. Runswick, I. M. Fearnley, J. M. Skehel and J. E. Walker, *Febs Lett.*, 1991, **286**, 121.
52. K. Dobrynin, A. Abdrakhmanova, S. Richers, C. Hunte, S. Kerscher and U. Brandt, *Biochim. Biophys. Acta*, 2010, **1797**, 152.
53. H. Angerer, *Biochem. Soc. Trans.*, 2013, **41**, 1335.
54. H. Angerer, M. Radermacher, M. Malkowska, M. Steger, K. Zwicker, H. Heide, I. Wittig, U. Brandt and V. Zickermann, *Proc. Natl. Acad. Sci. U. S. A.*, 2014, **111**, 5207.
55. J. Carroll, I. M. Fearnley, R. J. Shannon, J. Hirst and J. E. Walker, *Mol. Cell Proteomics.*, 2003, **2**, 117.
56. J. E. Cronan, I. M. Fearnley and J. E. Walker, *FEBS Lett.*, 2005, **579**, 4892.
57. J. K. Hiltunen, Z. Chen, A. M. Haapalainen, R. K. Wierenga and A. J. Kastaniotis, *Prog. Lipid Res.*, 2010, **49**, 27.
58. A. Abdrakhmanova, K. Zwicker, S. Kerscher, V. Zickermann and U. Brandt, *Biochim. Biophys. Acta*, 2006, **1757**, 1676.
59. M. Ciano, M. Fuszard, H. Heide, C. H. Botting and A. Galkin, *FEBS Lett.*, 2013, **587**, 867.
60. V. A. Morais, D. Haddad, K. Craessaerts, P. J. De Bock, J. Swerts, S. Vilain, L. Aerts, L. Overbergh, A. Grunewald, P. Seibler, C. Klein, K. Gevaert, P. Verstreken and S. B. De Strooper, *Science*, 2014, **344**, 203.
61. A. Abdrakhmanova, K. Dobrynin, K. Zwicker, S. Kerscher and U. Brandt, *FEBS Lett.*, 2005, **579**, 6781.

62. C. Mathiesen and C. Hägerhäll, *Biochim. Biophys. Acta*, 2002, **1556**, 121.
63. E. Screpanti and C. Hunte, *J. Struct. Biol.*, 2007, **159**, 261.
64. D. Drew and O. Boudker, *Annu. Rev. Biochem.*, 2016, **85**, 543.
65. R. G. Efremov and L. A. Sazanov, *Curr. Opin. Struct. Biol.*, 2011, **21**, 532.
66. G. Belevich, J. Knuuti, M. I. Verkhovsky, M. Wikström and M. Verkhovskaya, *Mol. Microbiol.*, 2011, **82**, 1086.
67. S. Steimle, C. Schnick, E. M. Burger, F. Nuber, D. Kramer, H. Dawitz, S. Brander, B. Matlosz, J. Schafer, K. Maurer, U. Glessner and T. Friedrich, *Mol. Microbiol.*, 2015, **98**, 151.
68. S. Zhu and S. B. Vik, *J. Biol. Chem.*, 2015, **290**, 20761.
69. J. A. Birrell and J. Hirst, *FEBS Lett.*, 2010, **584**, 4247.
70. C. Mathiesen and C. Hagerhall, *FEBS Lett.*, 2003, **549**, 7.
71. E. Virzintiene, V. K. Moparthi, Y. Al-Eryani, L. Shumbe, K. Gorecki and C. Hagerhall, *FEBS Lett.*, 2013, **587**, 3341.
72. B. C. Marreiros, A. P. Batista, A. M. Duarte and M. M. Pereira, *Biochim. Biophys. Acta*, 2013, **1827**, 198.
73. E. V. Gavrikova and A. D. Vinogradov, *FEBS Lett.*, 1999, **455**, 36.
74. A. Galkin and S. Moncada, *J. Biol. Chem.*, 2007, **282**, 37448.
75. M. K. F. Wikström, *FEBS Lett.*, 1984, **169**, 300.
76. A. S. Galkin, V. G. Grivennikova and A. D. Vinogradov, *FEBS Lett.*, 1999, **451**, 157.
77. A. Galkin, S. Dröse and U. Brandt, *Biochimica et Biophysica Acta*, 2006, **1757**, 1575.
78. M. Wikström and G. Hummer, *Proc. Natl. Acad. Sci. U. S. A.*, 2012, **109**, 4431.
79. S. Stolpe and T. Friedrich, *J. Biol. Chem.*, 2004, **279**, 18377.
80. A. P. Batista, A. S. Fernandes, R. O. Louro, J. Steuber and M. M. Pereira, *Biochim. Biophys. Acta*, 2010, **1797**, 509.
81. P. G. Roberts and J. Hirst, *J. Biol. Chem.*, 2012, **287**, 34743.
82. P. J. Castro, A. F. Silva, B. C. Marreiros, A. P. Batista and M. M. Pereira, *Biochim. Biophys. Acta*, 2016, **1857**, 928.
83. C. Hunte, E. Screpanti, M. Venturi, A. Rimon, E. Padan and H. Michel, *Nature*, 2005, **435**, 1197.
84. M. Coincon, P. Uzdavinys, E. Nji, D. L. Dotson, I. Winkelmann, S. Abdul-Hussein, A. D. Cameron, O. Beckstein and D. Drew, *Nat. Struct. Mol. Biol.*, 2016, **23**, 248.
85. L. A. Sazanov, *Nat. Rev. Mol. Cell Biol.*, 2015, **16**, 375.
86. V. R. Kaila, M. Wikström and G. Hummer, *Proc. Natl. Acad. Sci. U. S. A.*, 2014, **111**, 6988.
87. T. Ohnishi, *Biochim. Biophys. Acta*, 1998, **1364**, 186.
88. M. M. Roessler, M. S. King, A. J. Robinson, F. A. Armstrong, J. Harmer and J. Hirst, *Proc. Natl. Acad. Sci. U. S. A.*, 2010, **107**, 1930.
89. K. Zwicker, A. Galkin, S. Dröse, L. Grgic, S. Kerscher and U. Brandt, *J. Biol. Chem.*, 2006, **281**, 23013.
90. M. Murai, Y. Mashimo, J. Hirst and H. Miyoshi, *Biochemistry*, 2011, **50**, 6901.

91. M. A. Tocilescu, V. Zickermann, K. Zwicker and U. Brandt, *Biochim. Biophys. Acta*, 2010, **1797**, 1883.
92. H. Angerer, H. R. Nasiri, V. Niedergesass, S. Kerscher, H. Schwalbe and U. Brandt, *Biochim. Biophys. Acta*, 2012, **1817**, 1776.
93. M. A. Tocilescu, U. Fendel, K. Zwicker, S. Dröse, S. Kerscher and U. Brandt, *Biochim. Biophysic. Acta*, 2010, **1797**, 625.
94. M. A. Tocilescu, U. Fendel, K. Zwicker, S. Kerscher and U. Brandt, *J. Biol. Chem.*, 2007, **282**, 29514.
95. V. Sharma, G. Belevich, A. P. Gamiz-Hernandez, T. Rog, I. Vattulainen, M. L. Verkhovskaya, M. Wikström, G. Hummer and V. R. Kaila, *Proc. Natl. Acad. Sci. U. S. A.*, 2015, **112**, 11571.
96. M. Murai, H. Inaoka, T. Masuya, S. Aburaya, W. Aoki and H. Miyoshi, *Biochemistry*, 2016, **55**, 3189.
97. V. Zickermann, B. Barquera, M. K. F. Wikström and M. Finel, *Biochemistry*, 1998, **37**, 11792.
98. A. B. Kotlyar and A. D. Vinogradov, *Biochim. Biophys. Acta*, 1990, **1019**, 151.
99. M. L. Verkhovskaya, N. Belevich, L. Euro, M. Wikström and M. I. Verkhovsky, *Proc. Natl. Acad. Sci. U. S. A.*, 2008, **105**, 3763.
100. V. S. de, K. Dorner, M. J. Strampraad and T. Friedrich, *Angew. Chem. Int. Ed. Engl.*, 2015, **54**, 2844.
101. M. Wikström, V. Sharma, V. R. Kaila, J. P. Hosler and G. Hummer, *Chem. Rev.*, 2015, **115**, 2196.
102. S. Magnitsky, L. Toulokhonova, T. Yano, V. D. Sled, C. Hagerhall, V. G. Grivennikova, D. S. Burbaev, A. D. Vinogradov and T. Ohnishi, *J. Bioenerg. Biomembr.*, 2002, **34**, 193.
103. T. Ohnishi, S. Magnitsky, L. Toulokhonova, T. Yano, T. Yagi, D. S. Burbaev and A. D. Vinogradov, *Biochem. Soc. Trans.*, 1999, **27**, 586.
104. E. Nakamaru-Ogiso, M. Narayanan and J. A. Sakyiama, *J. Bioenerg. Biomembr.*, 2014, **46**, 269.
105. L. Euro, G. Belevich, M. I. Verkhovsky, M. Wikström and M. Verkhovskaya, *Biochim. Biophys. Acta*, 2008, **1777**, 1166.
106. U. Brandt, *Biochim. Biophys. Acta*, 2011, **1807**, 1364.
107. E. Padan and M. Landau, *Met. Ions. Life Sci.*, 2016, **16**, 391.
108. G. Hummer and M. Wikström, *Biochim. Biophys. Acta*, 2016, **1857**, 915.

CHAPTER 4

Multi-scale Molecular Simulations on Respiratory Complex I

VILLE R. I. KAILA

Department Chemie, Technische Universität München,
Lichtenbergstraße 4, 85747 Garching, Germany
Email: ville.kaila@ch.tum.de

4.1 Introduction to Structure and Function of Complex I

Complex I is one of the largest (0.5–1 MDa) and most intricate respiratory enzymes. It catalyzes electron transfer (eT) from nicotine amide dinucleotide (NADH) to quinone (Q) in its hydrophilic domain along a *ca.* 100 Å wire composed of flavin mononucleotide (FMN) and 8–9 FeS centers.[1-5] This reduces Q to quinol (QH_2) in a process that is coupled to pumping of four protons across the membrane domain of complex I,[6,7] up to *ca.* 200 Å away from the site of Q reduction (Figure 4.1). Despite recently resolved X-ray[8-12] and cryo-EM[13-15] structures from several species, data from labeling-,[16-18] crosslinking-,[19] and site-directed mutagenesis studies,[20-26] as well as biophysical experiments,[27,28] the mechanism by which complex I catalyzes this remarkable long-range proton-coupled electron transfer (PCET) process still remains unclear. Elucidating the molecular mechanism of complex I is not

Chemical Biology No. 5
Mechanisms of Primary Energy Transduction in Biology
Edited by Mårten Wikström

Published by the Royal Society of Chemistry, www.rsc.org

only crucial for understanding primary energy transduction in biology, but it is also of great biomedical relevance, since almost half of all known mitochondrial disorders are linked to mutations in complex I.[1,2,29]

The membrane domain of complex I comprises three antiporter-like subunits, NuoN (*E. coli* nomenclature, Nqo14 in *T. thermophilus*, ND2 in human/*Bos taurus*), NuoM (Nqo13/ND4), and NuoL (Nqo12/ND5), each of which contains two pseudo-symmetric trans-membrane (TM) segments, TM4-8 and TM9-13 (Figure 4.1, *inset*).[9,10,12,14,15] Helices TM7 and TM12 are broken by short loops, and may participate in the pumping process, similar to the structurally related carrier-type transporters, which employ such motifs to transport ions across the membrane.[30] The X-ray structures also

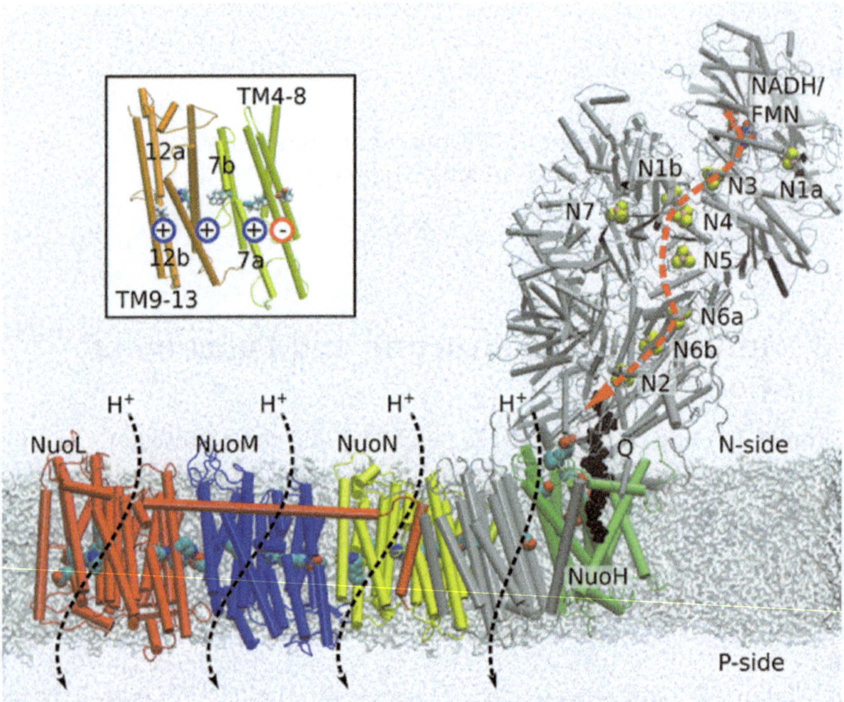

Figure 4.1 The structure and function of complex I. The figure shows an MD simulation setup of complex I from *Thermus thermophilus* (PDB ID:4HEA) in a water-membrane-ion environment, comprising *ca.* 840 000 atoms. Q (in black) has been modeled into the structure. Electron transfer from NADH/FMN *via* the FeS centers (red dotted line) to Q activates the proton pumping in the antiporter-like subunits NuoL (in red), NuoM (in blue), NuoN (in yellow), NuoH (in green). Conserved titratable residues along the membrane domain are also shown in van-der-Waals representation. *Inset*: the structure of TM4-8 and TM9-13 of NuoN. Each antiporter-like subunit (NuoN/M/L) comprises a conserved Lys (indicated with "+")/Glu (indicated with "−") ion-pair, a central Lys (+) and a terminal Lys (+, or Glu in NuoM).

show that despite a low sequence similarity, TM2-6 of subunit NuoH (Nqo8/ND1) is structurally similar to the antiporter-like NuoN/M/L subunits.[10] The membrane domain comprises a chain of buried titratable residues (Figure 4.1), which are central for the proton pumping process.[20–26] The membrane domain is connected by an unusual long transverse HL-helix, originally suggested to provide a piston function that is involved in establishing proton pumping across the membrane.[9] Recent work, however, indicates that the HL-helix may function as a molecular clamp that connects the antiporter-subunits together,[31] but *cf.* also ref. 32 and 33.

The proton pumping in complex I is highly efficient, employing up to 97% of the redox potential gap between NADH (−320 mV) and Q (+90 mV in the membranes, but see below).[34] The pumping machinery is fully reversible and electrons can also be extracted from QH_2 to drive the reverse eT along the FeS chain to NAD^+ by using an external pH gradient across the membrane.[1–5] Such reverse eT conditions are also of physiological relevance, since this increases the production of reactive oxygen species (ROS).[1,2,29] The Q reduction and proton pumping are strongly coupled, and mutations of residues in the terminal NuoL subunit therefore also inhibit the Q-reduction activity,[25,26] as expected based on the principles of microscopic reversibility. Nevertheless, the molecular principles of this coupling remain poorly understood.

Elucidation of molecular structures of complex I in recent years has opened up mechanistic studies of the enzyme for molecular simulations.[32,35–43] These techniques can provide a powerful methodology to study the structure, function, and dynamics of complex biological systems on a wide range of timescales and spatial resolutions. Molecular simulations have played an important role in bioenergetics, providing mechanistic insight into the function of, cytochrome *c* oxidase,[44–55] photosystem II,[56–62] cytochrome bc_1,[63,64] F_oF_1-ATPase,[65–68] as well as light-driven ion-pumps.[69,70] Here, how methods of computational biochemistry can be used to study the function of complex I, and central information that such simulations may provide is described. In section 2, the basic theory of the methods, which have been employed in computational studies of complex I, as well as techniques that may provide important input for future work are briefly reviewed, followed by discussion on computational results on complex I in sections 3–8.

4.2 Computational Models and Methods

4.2.1 Classical Molecular Dynamics Simulations

The goal of classical molecular dynamics (MD) simulations is to model molecular interactions to describe the microscopic dynamics of the system of interest. To obtain a computationally efficient evaluation of inter- and intra-molecular interactions, a pre-parameterized force field potential is employed,[71–73]

$$V = \sum_{\text{bonds}} \frac{1}{2} k_b (r - r_0)^2 + \sum_{\text{angles}} \frac{1}{2} k_\theta (\theta - \theta_0)^2 + \sum_{\text{improper}} \frac{1}{2} k_\chi (\chi - \chi_0)^2 + \sum_{\text{dihedrals}} k_\phi [1 + \cos(n\phi - \delta)] + \sum_{i>j} (q_i q_j / 4\pi\varepsilon_0 r_{ij}) + \sum_{i>j} \varepsilon_{ij} [(\sigma_{ij}/r_{ij})^{12} - 2(\sigma_{ij}/r_{ij})^6], \tag{4.1}$$

where bond-, angle-, improper-, and dihedral-terms describe bonded (covalent) interactions within the molecule, and allow for fluctuation around some equilibrium reference values (r_0, θ_0, χ_0, ϕ, δ). The non-bonded interactions within and between molecules are modeled using a Coulombic electrostatic potential (q_i – point charges; ε_0 vacuum permittivity) and a Lennard-Jones 12-6 term (ε_{ij} – potential depth; σ_{ij} – combined van der Waals radii for atoms i and j) to describe dispersive interactions. The functional form of the force field expression allows the molecule to undergo conformational changes, but covalent bonds cannot form or break due to the employed harmonic approximation, as the energy increases parabolically upon displacement of the atoms from their equilibrium values. Biomolecular force fields, such as CHARMM,[71] GROMOS,[72] and AMBER,[73] are parameterized using quantum mechanical calculations and experimental data in order to reproduce, for example, structural data, diffusion properties, and solvation free energies. Force field parameters are available for common biomolecules, such as amino acids, lipids, nucleic acids, and sugars.[71–73] Complex I, however, comprises several co-factors, including 8–9 iron–sulfur centers, FMN, NADH, and Q that can reside in different redox and protonation states during the catalytic cycle. For these co-factors, standard force field parameterizations are not available, and they thus require parameterization based on quantum chemical calculations (see below). For modeling transient catalytic states of complex I, the cofactors are parameterized in different charge states, which can be used for probing the dynamics of the systems, for example, prior and after reduction of the co-factors. In addition to so-called type-I force fields (eqn (4.1)), polarizable models,[74] where the point charge distribution can fluctuate, and reactive force fields,[75,76] where the bonding topology is parameterized to change during the dynamics, have also been developed. Such simulation methods have not yet been employed in studies of complex I, but they may provide important mechanistic insight in the future work. For example, empirical valence bond (EVB) simulations[50,51] have contributed to our understanding of the proton pumping in cytochrome *c* oxidase. In addition to the atomistic MD simulations discussed here, coarse-grained (CG) force fields[77] that model residue interactions on a bead-level instead of considering explicit atomic interactions, are also being developed, and provide access to longer simulation time-scales, *i.e.*, milliseconds rather than microseconds.

In MD simulations, the Newtonian equations of motion (EOM),

$$\boldsymbol{F} = -\nabla V = m\boldsymbol{a} = m\frac{\mathrm{d}^2\boldsymbol{r}}{\mathrm{d}t^2} \tag{4.2}$$

are discretized and integrated numerically. The Verlet algorithm, for example, can be used to calculate the atomic positions in the next time step, $\boldsymbol{r}(t+\Delta t)$, based on their current, $\boldsymbol{r}(t)$, and previous positions, $\boldsymbol{r}(t-\Delta t)$,

$$\boldsymbol{r}(t+\Delta t) = 2\boldsymbol{r}(t) + \boldsymbol{r}(t-\Delta t) + (\boldsymbol{f}(t)/m)\Delta t^2 \tag{4.3}$$

with forces, $\boldsymbol{f}(t)$, and masses, m, of the particles. The forces between atoms are obtained from the gradient of eqn (4.1), and initial velocities at $t = 0$ are assigned from a Maxwell–Boltzmann velocity distribution, for a given simulation temperature. In order to properly model molecular collisions, the EOMs are integrated using a short time step, Δt, of 1–2 fs, which captures the fastest bond-vibrations within the system. The time-ordered atomic positions obtained by integrating the EOMs give the MD trajectory, which contains central dynamic information of the system.

The bottleneck in the classical MD simulations are estimations of the long-range electrostatic interactions, since the interaction of each N atom needs to be evaluated with all N-1 atoms in the system. As an MD simulation setup of the bacterial complex I comprises *ca.* 10^6 atoms (see below), this would result in evaluation on the order of 10^{12} interactions, which is not possible to achieve even with the fastest computers. The N^2-scaling of such computation can, however, be reduced by using the Particle Mesh Ewald (PME) algorithm, in which interactions between atoms far in space are calculated in reciprocal (Fourier)-space and added to their direct space contributions, using a fast-Fourier transformation (FFT) algorithm. This procedure lowers the computational scaling to $N\log(N)$-scaling for a system with N atoms. In classical molecular simulations, the temperature (T) and pressure (p) are also modeled by using thermostats and barostats and are normally set to $T = 310$ K and $p = 1$ bar in biomolecular modeling. To this end, eqn (4.2–4.3) are modified to the Langevin equations, that also take into account stochastic Brownian forces within the system.

Due to the tremendous increase in computational power in recent years, MD simulations today allow access to microsecond timescales. Special computers, such as the ANTON supercomputer, have pushed this limit even further, allowing simulations to be carried out on the millisecond timescales, at least for small proteins.[78] Information about processes taking place beyond the microseconds timescale can also be obtained from shorter simulations. To this end, non-equilibrium or transient catalytic states and their relaxation can be studied or data is collected from several independent simulations using Markov-State models (MSM)[79] that project out molecular motion based on different timescales. Moreover, instead of following direct brute-force dynamics of an individual system, free-energy simulation techniques[80] can be employed to perturb the system to systematically undergo rare conformational changes, followed by reconstruction of the underlying free energy landscape using methods of statistical mechanics (see below). Reaction rates can this way be related to the free energy barriers using transition state theory.

To study the dynamics of complex I, a realistic computational model of the enzyme in its biological surrounding must be built. This starts by modeling all hydrogen atoms that are commonly not resolved in the protein X-rays structures. To this end, protonation states for all titratable residues are assigned by performing Poisson–Boltzmann (PB) continuum electrostatic calculations with Monte-Carlo sampling (see below).[81,82] After all atoms have been explicitly modeled, complex I is inserted in a lipid bilayer, and solvated with water molecules and ions. Either single component lipid models, such as 1-palmitoyl-2-oleoyl-*sn*-glycero-3-phosphocholine (POPC), or multi-component lipids that mimic the composition of the inner mitochondrial membrane, are commonly used in membrane protein simulations.[35–37] The molecular simulation setup of complex I from *Thermus thermophilus* results in a system with *ca.* 840 000 atoms, shown in Figure 4.1. Although the eukaryotic complex I contains 46 subunits and has a molecular mass of *ca.* 1 MDa, the MD model for the *Bos taurus* enzyme results in only a slightly larger system of *ca.* 1.1 million atoms, since a larger part of the simulation box comprises water molecules. Unfortunately, *ca.* 5% of the residues in the eukaryotic complex I structures still remain unresolved. Although these missing parts can be modeled using protein prediction methods,[83,84] such simulations may have larger uncertainties in comparison to simulations performed based on fully experimentally resolved structures. After the structure of complex I has been inserted in the membrane–water–ion environment, the model is energy minimized and equilibrated followed by production simulations. Up to date, microsecond-timescale MD simulations for several independent trajectories have been reached.[35–37] Simulation of such timescales, requires *ca.* 5×10^8 integration steps ($\Delta t = 2$ fs, eqn (4.3)) for each of the *ca.* 10^6 atoms, and therefore access to high-performance supercomputers is necessary.

4.2.2 Free-energies and Electrostatic Poisson–Boltzmann Calculations

It can be difficult to observe rare events with high-energy barriers in classical MD simulations. Sampling of such events can, however, be achieved by applying external potentials on a reaction–coordinate of interest that flattens its free-energy landscape. The unbiased free energy profile, *i.e.*, the probability of observing the event of interest without introducing external forces, is obtained from the probability distribution of the reaction coordinate observed in the biased (restrained) simulations by re-weighting the statistics with the employed biasing potential. Commonly used computational

free-energy approaches are; umbrella sampling, metadynamics, and free energy perturbation methods (for further discussion see, ref. 80). Generating converged free energy profiles requires a careful choice of the reaction-coordinate(s) as well as a significant statistical overlap in these coordinates. Application of free energy simulation techniques can thus be very challenging for large systems such as complex I.

The electrostatic free-energy for protonation (pK_a) and reduction (E_m) processes are particularly relevant for understanding the function of complex I. These properties can also be estimated by solving the (linearized) PB equation, that relates an electrostatic potential (ψ) to a charge distribution (ρ),[81,82]

$$\nabla\varepsilon(r)\cdot\nabla\psi(r)-\varepsilon(r)\lambda(r)\left[\frac{8\pi Iq^2}{\varepsilon(r)k_BT}\right]\psi(r)=-4\pi\rho(r), \tag{4.4}$$

where $\varepsilon(r)$ is the dielectric constant, λ is a switching function, I is the ionic strength, k_BT is the thermal energy, and q are the point charges in the system. Integration of the obtained potential at the point charges gives the electrostatic free energy,

$$\Delta G=\frac{1}{2}\sum_i q_i\psi_i. \tag{4.5}$$

To calculate pK_a and/or E_m values, a thermodynamic cycle is employed, where the electrostatic free energies resulting from transfer of the protonated (AH) and deprotonated (A^-) (or reduced:oxidized) species from aqueous ($\varepsilon=80$) phase to the protein interior ($\varepsilon=4$) are estimated. In addition to this so-called Born desolvation energy, the interaction of AH and A^- is computed with all protein background charges. Inserting AH/A^- into the protein might also result in changes of the 2^N possible protonation states of the protein with N titratable residues. Due to the high dimensionality of these possible protonation states, the last effect is often calculated by Monte Carlo sampling. For example, the membrane domain complex I, contains *ca.* 350 titratable residues, giving 2^{350} possible protonation states. The electrostatics shifts are commonly calculated relative to an experimentally measured pK_a or E_m values of the model compound, for example, in water where the values are known to high accuracy.

4.2.3 Quantum Chemical Density Functional Theory Models

In order to capture the energetics of a chemical process, for example, Q reduction to QH_2 or proton transfer across the membrane domain, the motion of electrons must be rigorously described based on the Schrödinger equation. Although the exact solution of this equation is not possible for large molecules, density functional theory (DFT) calculations provide an accurate methodology to approximate the many particle Schrödinger

equation, with a good balance between computational cost and accuracy. DFT is in principle exact in the non-relativistic limit, thus representing a rigorous reformulation of the Schrödinger equation, and awarded its developer Walter Kohn a Nobel Prize in 1997. The exact dependence of how the energy is related to the electron density is, however, still unknown, and it was not until the early 1990s when accurate approximations of the so-called exchange-correlation term were developed. One commonly employed density functional is Becke's three-parameter hybrid functional, B3LYP,[85,86] which has become important in computational biochemistry.[38,44,47,56–58,61,62] In this functional, an empirical amount (20%) of Pauli electron-exchange is introduced from the mean-field Hartree–Fock theory. Different density functionals have a benchmarked error of *ca.* 1–5 kcal mol^{-1} in reaction energies, whereas geometries are usually predicted within an accuracy of 0.05 Å (in bond distances) relative to experimental geometries on model compounds.[87] Despite these and other systematic errors (see ref. 87), DFT nevertheless remains one of the most powerful and accurate techniques to treat complex biochemical systems at a QM level.

In order to perform DFT calculations, the electrons are modeled by finite basis sets, constructed from a linear-combination of atomic orbitals to give molecular orbitals (LCAO-MO). Basis sets are often composed of a sum of Gaussian functions, and benchmarking studies suggest that basis-set convergence within DFT is reached when the electrons are modeled using a triple set of functions (triple-zeta quality basis set, *e.g.* def2-TZVP or 6-311G**).[88] The basis sets are used to solve the Kohn–Sham equations in DFT, until self-consistent solutions are obtained. This procedure gives molecular orbitals with optimized weight of each orbital contribution, as well as a total energy for the system that can be compared between different states for geometry-optimized structures. DFT calculations now allow treatment of systems with *ca.* 100–500 QM atoms and can be used for structure or reaction pathway optimization, first-principles dynamic simulations, or molecular property calculations.

Care must be taken when treating certain properties at DFT level that result from electron correlation effects.[87] DFT is not able to rigorously capture dispersion interactions, but this central interaction is modeled in dispersion-corrected density functionals (DFT-D),[89] by introduction of an empirical r^{-6} dispersion term to the functional, and is commonly used in computational biochemistry. Another challenging problem central to biochemical systems, and particularly for the FeS centers in complex I, is the accurate treatment of spin energetics. It is known that the B3LYP functional has a tendency to systematically overestimate the stability of high-spin configurations,[90] as the high spin energy becomes favored by increasing the amount of introduced exchange, an empirical parameter within the hybrid density functionals. It can therefore be important to benchmark the performance of different functionals on the property of interest. The spin-energetics of the FeS centers in complex I have a particularly challenging electronic structure, as each of the irons have 4–5 unpaired electrons, which

are anti-ferromagnetically coupled together to yield a $S=1/2$ or 0 system. For the tetranuclear FeS centers (N2, N7, N6a, N6b, N5, N4, N3), there are six unique anti-ferromagnetically coupled spin-states that may each have different energies. In order to achieve proper spin states, the broken-symmetry spin-flip approach can provide a particularly useful approximation in estimating spin-coupling parameters within the framework of DFT.[91,92]

In order to study a protein function using DFT methods, a QM cluster model, capturing the central chemical environment can be constructed.[93] In such models, the active system of interest together with central first- and second sphere residue interactions, are cut out from the biological environment. The terminal atoms are saturated with hydrogen atoms, and fixed during structure optimization to mimic strain that arises from the protein environment. Moreover, the electric response of the environment can be modeled using implicit dielectric polarizable medium models with a dielectric constant usually set at 4–20. The total energy of the QM cluster models in different states can be systematically compared, as proper structure optimizations can be performed. When the system size becomes very large, however, it is increasingly difficult to find proper energy minima, leading to uncertainties in the total energy of the system. Such problems may also arise when explicit dynamics of the system is considered.

4.2.4 Hybrid Quantum Mechanics/Classical Mechanics Models

In hybrid quantum mechanics/classical mechanics (QM/MM) calculations the QM system of interest is embedded and polarized by a protein surroundings, which is described at the classical force field level.[94] In so-called additive QM/MM schemes, the total energy and forces are calculated from the sum of the QM and MM subsystems and the interactions between the QM and MM parts ($E_{QM/MM}=E^{MM}+E^{QM}+E_{QM\text{-}MM}$).[95] There are also subtractive QM/MM schemes, for example, the ONIOM method ("our own *N*-layered integrated molecular orbital and molecular mechanics"),[96] where the total classical energy is corrected by the energy differences between the MM and QM parts for the central region of interest ($E_{QM/MM}=E_{large}{}^{MM}-E_{small}{}^{MM}+E_{small}{}^{QM}$; where "*large*" refers to the complete system and "*small*" is the central region of interest). In QM/MM calculations, the QM region is defined and separated from its classical surroundings by cutting between Cα and Cβ atoms for each residue involved, and introducing link atoms between these regions that chemically saturate the QM system. The link atoms are made invisible for the MM region, and charge distributions at the boundary regions are spread out to prevent over-polarization effects. In QM/MM simulations, the energy and forces for the central QM region are calculated *on-the-fly* using DFT, which replaces the pre-parameterized force field expression (eqn (4.1)) for this region. It can be confusing for a non-expert to distinguish between different QM/MM simulations as the QM treatment may refer to DFT, *ab initio* theory, or semi-empirical methods. For example, the self-consistent charge-tight binding density functional theory (SCC-DFTB) has provided mechanistic insight into the function of bacteriorhodopsin[69] and cytochrome *c* oxidase.[54] In SCC-DFTB, the electron density is replaced by point-charges instead of the electron density as in DFT. Although it is difficult to benchmark the exact computational errors of different QM/MM methods, DFT-based QM/MM simulations normally allow accessing some tens of picoseconds timescale, whereas semi-empirical QM/MM can normally be extended to the 100 ps – 1 ns timescales.

4.3 Dynamics of Electron Transfer

While NADH is a two electron carrier, the FeS centers in the hydrophilic domain of complex I undergo one-electron oxidoreduction, switching between their reduced ($2Fe^{2+}2Fe^{3+}$ or $Fe^{2+}Fe^{2+}$) and oxidized ($3Fe^{2+}Fe^{3+}$ or $Fe^{3+}Fe^{2+}$) forms.[97] The hydrophilic domain of complex I thus functions as a "two-to-one" electron converter, that bifurcates the eT from NADH to Q. The first electron from NADH is transferred by PCET *via* FMN to the binuclear N1a center. Flavosemiquinone species ($FMN^{-/\bullet}$ or $FMNH^{\bullet}$) have not yet been observed, and the mechanism for this putative hydride (H^-) transfer also remains unclear.[2,5] Experiments, however, show that after reduction of N1a, the second electron rapidly moves from NADH/FMN along the *ca.* 100 Å chain of FeS clusters to the high potential N2 center in *ca.* 20 µs,[27,28] while the resulting NAD^+ is kinetically trapped,[98] possibly to avoid leaking the electron from N1a to oxygen in bulk solvent. All FeS clusters except N7, which resides *ca.* 20 Å from the main eT pathway,[27] participate in the transfer process, and the FeS centers range in redox mid-point potentials (E_m) from *ca.* −330 mV to *ca.* −200 mV (N2). Experiments further show that reduction of N2 results in a slower (millisecond) redistribution of the electron from N1a to the other FeS centers.[27]

The rate for this eT process can be estimated from the empirical Moser–Dutton ruler,[99]

$$\log k_{eT} = 13 - (1.2 - 0.8\rho)(R[\text{Å}] - 3.6) - 3.1[\text{eV}^{-1}]\frac{(\Delta G - \lambda)^2}{\lambda} \tag{4.6}$$

with free energies of the inter-FeS center eT processes (ΔG) derived from electrochemical experiments, edge-to-edge distances (R) from X-ray structures, and by employing typical reorganization energies ($\lambda = 0.5$–0.7 eV), and protein packing densities ($\rho = 0.76$). This gives an eT rate in the milliseconds timescale,[2,5,100] which is somewhat slower or comparable to the overall complex I turnover of *ca.* 2 ms. It is, however, challenging to assign the ΔGs based on experiments, since the electrostatic couplings between the centers are not directly obtained from the measured redox potentials. To this end, Couch *et al.*[40] and Medvedev *et al.*[39] calculated electrostatic couplings between the FeS centers based on PB electrostatics calculations and experimental constraints from the measured E_m values. These "coupling" corrected E_m values obtained using different dielectric environments ($\varepsilon = 4$–20) upshift some FeS redox potentials, and give eT rates of *ca.* 0.4–4 ms based on a Moser–Dutton treatment.[5] Hayashi *et al.*[41] also addressed whether tunneling pathways could increase the overall transfer rate, by explicitly calculating the electronic overlaps along different pathways. By performing semi-empirical calculations at the broken-symmetry ZINDO/S level on the FeS centers and their nearby surroundings, they found that many of the pair-wise transfer processes are indeed faster than the overall turnover. However, these calculations suggested that the transfer between N5 → N6a (10 s^{-1}) and N3 → N1b (10^3 s^{-1}) form bottlenecks in the process. They further found that internal water molecules, which have not yet been resolved in the experimental structures of complex I, may increase the eT rate by bridging unfavorable gaps between the FeS centers and increase the electronic coupling along the pathways. MD simulations on the microseconds timescale suggest that some water molecules can indeed enter the hydrophilic domain, *e.g.*, the region between N5 and N6a (Figure 4.2A). The simulations, however, also show that there are relatively large fluctuations in many of the inter-FeS distances of *ca.* 2–3 Å (Figure 4.2B). In addition to the

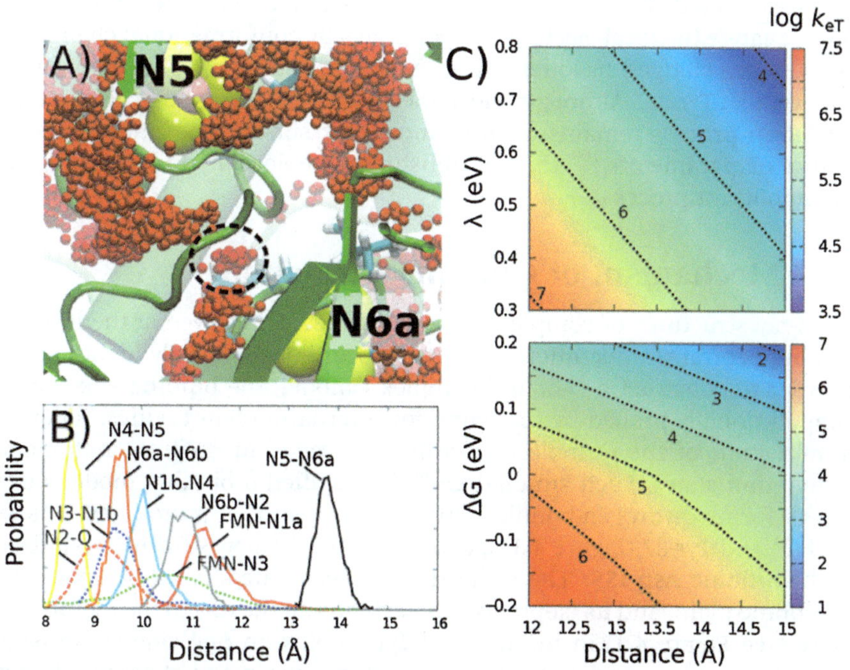

Figure 4.2 (A) Averaged dynamics of water molecules (in red) near N5 and N6a suggest that water molecules may bridge empty gaps in the X-ray structure of complex I. (B) *Edge-to-edge* distances between the eT centers during a microsecond MD trajectory. (C) Sensitivity of the rate-limiting eT rate (log k_{eT}) between N5-N6a on the eT parameters, R, λ, and ΔG.

uncertainties in ΔGs and λ, fluctuations in R may also modulate the eT rate by several orders of magnitude, as shown for the rate-limiting eT step between N5 and N6a in Figure 4.2C. Moreover, although many of the predicted tunneling rates[41] are indeed consistent with the experimental turnover-constraints, a potential error source in these pathway calculations could also arise from the treatment of the highly challenging FeS spin-energetics (see section 2.3). A DFT treatment could offer a more accurate, yet computationally expensive option to probe the spin-energetics relative to semi-empirical methods, whereas development of novel quantum chemical multi-reference methods, such as the density-matrix renormalization group (DMRG) theory,[101] might in future open up an accurate correlated *ab initio* treatment of such challenging electronic structure problems.

Interestingly, de Vries *et al.*[28] recently found that the eT between N5 → N6a (N4Fe[75]H → N4 in *E. coli*) becomes sixfold slower upon reduction of N2, suggesting that complex I might utilize a feedback regulation mechanism to modulate the rate of Q reduction, possibly in order to time it with the pumping cycle. A direct Coulombic-interaction between N2 and N6a, would be expected to tune the ΔG for this eT by *ca.* 60 mV, decreasing the eT rate by *ca.* 20%, which may account for this effect in part. PB calculations[38] and experiments,[102] suggest that His-169 (*T. thermophilus* numbering, if not otherwise stated) becomes protonated upon reduction of N2. Such effect might change the local electrostatic couplings by conformational changes in surrounding charged residues near the terminal FeS centers, and modulate the overall eT rate. Although the molecular mechanism for this putative regulation process remains unclear, molecular simulations can be used to predict eT-parameters,[103] thus establishing a molecular understanding of this important process.

4.4 Mechanism of Quinone Reduction

The X-ray structures of complex I lack a Q molecule resolved at the binding site, but central residue interactions have been biochemically identified.[104] In order to construct a Q-bound complex I model, the binding site can be computationally probed by searching for internal protein cavities, followed by relaxation of the Q molecule within the cavity in different redox and protonation states. Such simulations[36,38] identified a binding mode where the Q forms hydrogen-bonding interactions with a protonated His-38 ($HisH^+$) and Tyr-87 (Figure 4.3A), and where the former is further stabilized by the anionic Asp-139. The Q head group is located *ca.* 20 Å above the membrane plain, and its isoprenoid tail extends all the way to the membrane phase (see Figure 4.1) in the unusual Q-tunnel, with one side comprising non-polar residues and the other side comprising many charged Glu/Arg ion-pairs.[10,12] In addition to this hydrogen-bonded binding mode, the simulations suggest that Q can also bind in an alternative conformation, where the Q head group forms a stacking interaction with His-38, while retaining its hydrogen bond with Tyr-87 (Figure 4.3B).[38] DFT calculations

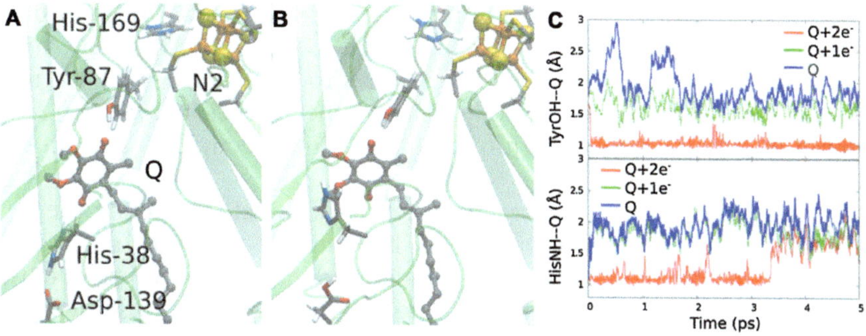

Figure 4.3 Q-binding site in complex I. (A) Hydrogen-bonded and (B) stacked binding modes of Q, forming contacts with Tyr-87 and His-38. (C) QM/MM MD simulations of Q in oxidized state (in blue), one-electron reduced state ($Q+1e^-$, in green), and two-electron reduced state ($Q+2e^-$, in red). The data suggest that two-electron reduction of Q leads to formation of QH_2 by proton transfer from Tyr-87 and His-38. Data in C is obtained from ref. 36.

further suggest that when Q is oxidized, the stacked conformation is energetically slightly preferred over the hydrogen-bonded conformation. The energetic preference for this stacked conformation becomes somewhat more pronounced upon N2 reduction, suggesting that the redox state of the terminal FeS center might regulate the binding energetics of Q.

QM/MM simulations can be used to probe the eT dynamics between N2 and Q, when both groups and their nearby surroundings are included in the same QM region. Such simulations also require "diabatization" of the initial electron transfer state so that the dynamics are initiated from a state with a reduced electron donor (N2) and an oxidized electron acceptor (Q). In such simulations,[38] the eT between N2 and Q takes place on picoseconds timescales when the latter resides in its relaxed hydrogen-bonded binding mode, whereas no eT was observed in stacked Q-conformation on accessible simulation timescales. Electrostatic PB calculations suggest that the E_m of the Q/SQ redox pair is around −260 mV and −380 mV in the hydrogen-bonded and stacked-binding modes, respectively, indicating that the eT is exergonic in the hydrogen-bonded binding mode, and endergonic in the stacked-binding mode. Importantly, these redox potential calculations support results from earlier electrochemical measurements,[105] which indicate that the E_m of Q/SQ is <-300 mV in the Q-binding site, since no semiquinone radicals were observed. Mechanistically, these findings imply that there is no significant redox-drop between NADH (−320 mV) and Q (−300 mV) when the latter is bound near N2. However, since Q in membranes have an E_m of *ca.* +90 mV, movement of Q towards the membrane would be expected to couple to the release of *ca.* 400 mV (*ca.* 9 kcal mol^{-1}) redox-energy, which in turn could thermodynamically drive pumping of two protons across an electrochemical proton gradient of *ca.* 200 mV. The Q tunnel contains several aromatic and charged conserved residues, that could form transient cation-π and/or π–π interactions with Q. If Q indeed would have a second binding-site in this tunnel, it could explain findings from earlier EPR studies, which suggest that SQ resides <15 Å and/or *ca.* 30 Å from the N2 center.[106] These findings might favor a mechanistic model where the piston-like dynamics of Q within its tunnel could couple to the proton pumping process.[4,34]

After formation of SQ, the second electron is transferred along the FeS chain to re-reduce N2, followed by eT between N2 and SQ. In QM/MM simulation of the first eT step, SQ remains deprotonated (Figure 4.3C). However, the second eT step from N2 to SQ, leads to a stepwise deprotonation of Tyr-87 and His-38, in a process that results in the formation of QH_2. Interestingly, electrostatic calculations suggest that N2 reduction is coupled to the protonation of His-169 near the N2 cluster, consistent with earlier experiments on the pH-dependence of the N2 centers, and lack of this pH-dependence in the H169M mutant.[102] Simulations suggest that the second eT step is strongly favored by the deprotonation of His-169, a process that could kinetically control the rate of Q reduction.

4.5 Redox-linked Conformational Changes in the Membrane Domain

The DFT-based QM/MM MD simulations on the QH_2 formation process, described above, can be used to probe the dynamics on some tens of picosecond timescales, whereas classical MD simulations are necessary for probing the dynamics on longer (ns–μs) timescales. Classical MD simulations show that formation of QH_2 by proton transfer from Tyr-87 and His-38 increases the dissociation probability of the Asp-139/His-38 pair. This in turn induces conformational changes in carboxylates and arginines along the E-channel,[10] that propagate to Glu-213/Glu-163, located on a flexible loop in the subunit NuoH (Nqo8/ND1).[36] Interestingly, Glu-213 has been refined in a different conformation in complex I from *Y. lipolytica*,[12] supporting that the residue might indeed undergo conformational changes. Electrostatic calculations further suggest that the conformational changes of these glutamates lead to an increase in their protonation probability, which may lead to uptake of protons from the N-side of the membrane. In order to probe the involvement of Asp-139 in triggering this force propagation from the Q-site, the residue was mutated to asparagine *in silico* and *in vitro*. Pumping experiments on the D139N mutant show that the activity of complex I is inhibited by 75% relative to the wild type, supporting the important putative function of this residue.[36] Moreover, QM/MM simulations suggest that two-electron reduction leads to formation of QH^- instead of QH_2, since His-38 is likely to be neutral in D139N. Moreover, the simulations show that the resulting QH^- species moves *ca.* 10 Å downwards in the Q-cavity, mimicking in part the conformational changes around the Asp-139 residue, possibly due to repulsion from the anionic Tyr-87 ($TyrO^-$). Although not yet extensively studied in MD simulations, SQ ($Q^{\bullet/-}$) electrostatically resembles such QH^- species and could also trigger similar coupled conformational and electrostatic changes within the E-channel. More experimental data is, however, currently required to determine whether the one- or two-electron reduced Q species are involved in triggering proton pumping.[107–110]

4.6 Function of the Proton Pump

Classical MD simulations suggest that the membrane domain of complex I undergoes significant hydration changes on *ca.* 200–400 ns timescales.[35,37] These hydration changes are triggered by the protonation states of buried charged residues, to which quasi-one dimensional water chains provide hydrogen-bonded connectivity from the N-side of the membrane. These residues are located in the antiporter-like subunits, and provide also further connectivity along terminal Lys/Glu residues within each subunit to the P-side of the membrane (Figure 4.4). QM/MM simulations can be further used to probe the proton conduction properties of such classically formed water wires. To this end, the water molecules within the wire and their

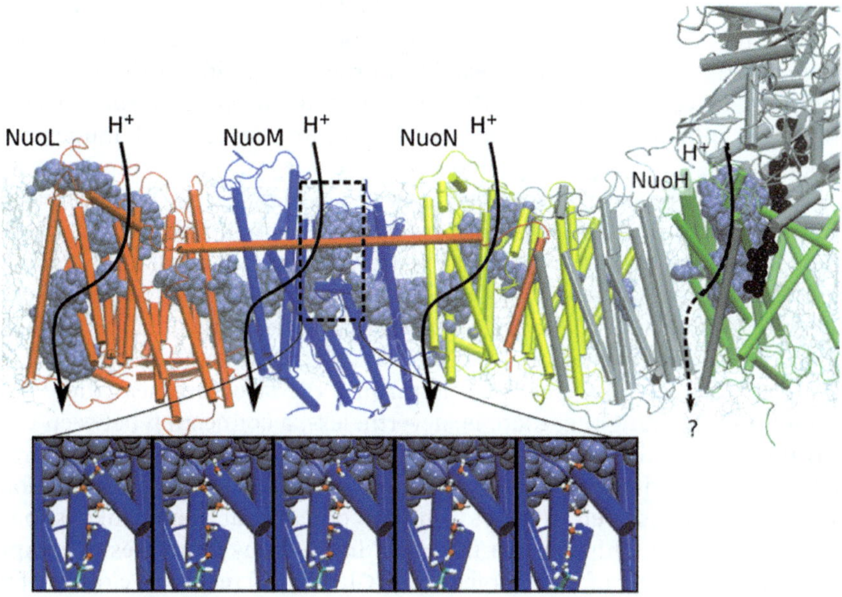

Figure 4.4 Formation of hydrogen-bonded water arrays that establish protonic connectivity across the membrane domain based on microsecond MD simulations of complex I from *Thermus thermophilus*. Proton-channels are formed at four symmetry-related locations in NuoL, NuoM, NuoN along TM7a and TM12b, and along the E-channel region in NuoH (see text). *Inset*: QM/MM MD simulations on Grotthuss-type proton transfer along the classically formed water chains. The proton transfer process takes place on picosecond timescales in water wires formed in the μs-MD trajectories.

nearby surroundings are moved into a QM region and the water molecule near the bulk surface modified into a H_3O^+ species, while modeling the remaining protein surroundings at the classical force field (MM) level. QM/MM MD simulations of such states further support that the water chains provide effective catalyst for Grotthuss-type proton transfer reactions (Figure 4.4, *inset*), and also that lateral proton transfer along the antiporter-like subunits is indeed possible.[37]

MD simulations suggest that the proton channels are established at four symmetry-related locations,[37] with an input site near TM7b and output near TM12b in NuoN, NuoM, and NuoL. A "fourth" input channel from the N-side is also observed in the NuoH subunits, forming a hydrated structure near TM6. This hydrated region forms contacts with the glutamate region of the E-channel, which undergoes conformational and protonation changes as a result of the Q-reduction process (see above). This region is also close to the Q-tunnel (see above), suggesting that movement of the Q within its tunnel might be strongly coupled to the proton-pumping machinery.

Importantly, the simulations show that the continuous connectivity between the N- and P-sides is broken by bulky phenylalanine and leucine residues along the hydrophilic axis in each antiporter-like subunit, suggesting that complex I strictly regulates the hydrated connectivity across the membrane. These bulky residues might provide important gating points that prevent the protons from leaking to the wrong side of the membrane and help in establishing an efficient pumping machinery. Such residues thus also provide important targets for future site-directed mutagenesis experiments.

Many carrier-type transporters operate by large-scale conformational changes that provide alternate access across the membrane.[111,112] Despite their homology to Mrp-type (multi-resistance and pH-regulated) transporters,[10,113] the antiporter-like subunits in complex I seem to undergo subtle conformational changes in their broken helices and surrounding elements upon the hydration changes.[35,37] Interestingly, the hydration state in complex I can also be modulated by perturbing the structure near these helix elements, suggesting that there is, nevertheless, a connection between these two processes. The channel hydration state is also sensitive to the protonation states of the central lysine residue, as well as to the conformational state of the conserved Glu–Lys ion-pair of each antiporter-like subunit (Figure 4.1, *inset*). This ion-pair is broken in the X-ray structures of complex I, whereas modeling of the charged state (Glu^-/Lys^+) results in closure of the ion-pair, with occasional inter-subunit contacts between the NuoN–NuoM and NuoL–NuoM subunits.

Mechanistically it is important to understand how complex I employs direct (electrostatic) and/or indirect (conformational) coupling principles to drive the proton pumping across the membrane.[1–5,34,107–110] Electrostatic calculations suggest that while the interaction between charged residue pairs is very strong, up to 20 kcal mol^{-1}, coupling beyond residue pairs may not allow for large enough pK_a modulations that would effectively release the proton across the membrane.[35,37] However, inter-subunit contacts that form between glutamates and lysines in neighboring (NuoN–NuoM and NuoM–NuoL) subunits might help in releasing the proton to the P-side. Due to the relatively weak coupling beyond residue pairs, this may imply that certain pumping models, for example, the wave-spring model,[5,21] where protons are released in synchronized steps form NuoL/NuoN and NuoM/NuoH, may require indirect (conformational) coupling. Simulations also show that the dissociation of the Lys^{+}/Glu^{-} ion pair decreases the proton affinity of the central lysine residue, whereas further deprotonation of this residue leads to a decreased water access from the N-side. These findings may provide an important clue into the function of the pumping machinery (see below).

4.7 Putative Model for Redox-driven Proton-pumping

Results from the molecular simulations provide important mechanistic insights that can be used to derive a minimal mechanistic model for the function of the proton pump, schematically shown in Figure 4.5.[35–38] In this model, reduction of Q triggers conformational changes in NuoH that lead to uptake of a proton from the N-side by water chains, as suggested by the MD simulations.[37] Although the exact details are still unclear, this could in turn increase the dissociation of the Glu/Lys ion-pair in NuoN, which would further lead to proton transfer towards the P-side due to destabilization of the protonated "middle" Lys. Redistribution of the charge in NuoN would in turn dissociate the Glu/Lys ion-pair in NuoM, which would trigger a similar electrostatic perturbation, and lead to an internal proton transfer along the

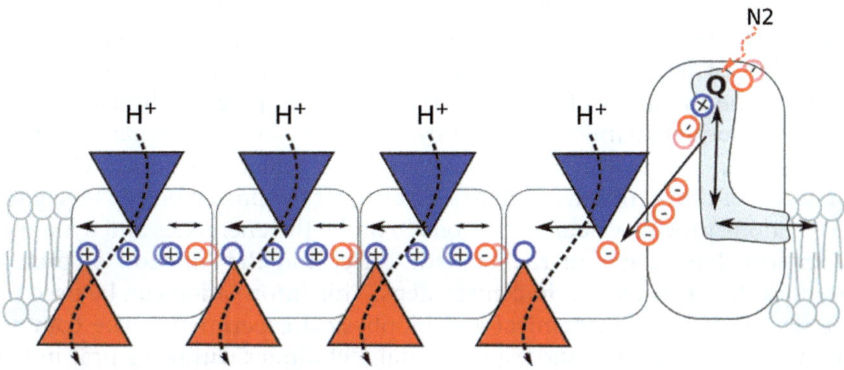

Figure 4.5 Putative schematic pumping model in complex I. Q reduction triggers local electrostatic changes in the active site that propagate to the NuoH subunit, which leads to proton uptake by water wires (blue triangle). Intrinsic proton transfer reactions induce conformational changes in the Lys–Glu ion pair of each antiporter-like subunit, and opens up proton uptake from the N-side (blue triangle) by sequential propagation along the membrane domain. Movement of Q along its tunnel (in gray) couples to release of redox energy that is employed to push the proton(s) across the membrane.

subunit. This charge redistribution would then propagate to NuoL, induce a Glu/Lys dissociation, followed by a proton release from the middle Lys. The MD simulations indicate that deprotonation of the central lysine might close the contact with the N-side, which is further expected to prevent the "released" proton from leaking towards its thermodynamically favored N-side of the membrane. Based on thermodynamic arguments discussed above, movement of the reduced quinone species towards the membrane domain is expected to lead to an energy transduction event, which could push out the proton(s) to the P-side and re-establish the initial state of the catalytic machinery. The sequential propagation of such conformational and electrostatic changes along each antiporter-like subunit could, at least in part, explain why mutations of conserved residues in NuoL inhibit the Q-reduction activity. In this putative model, the conformation of the Lys–Glu ion-pair in TM4–8 is strongly coupled with the proton transfer along the middle Lys → terminal Lys (Glu) in TM9-13, and *vice versa*. Introduction of residues that disturb either of these processes would therefore be expected to affect the global pumping energetics, as indeed observed in certain NuoL mutants.[7,25,26]

4.8 Conclusions

The respiratory complex I is a redox-driven proton pump that couples a 100 Å electron transfer along its hydrophilic domain to proton transfer across its membrane domain, up to 200 Å away from its active site. Multi-scale molecular simulations can provide important insight into the energetics and dynamics of this remarkable long-range PCET process. Simulations show that complex I employs coupled conformational and electrostatic changes that trigger pK_a shifts in buried conserved residues. This in turn controls the formation of proton-conducting water wires that provide alternate access between the two sides of the membrane. The pumping machinery in complex I is thermodynamically driven by the redox state and dynamics of the quinone, and it is therefore important to elucidate the exact chemical character of transient intermediates that trigger the proton pump. Molecular simulations provide valuable methods to probe the energetics and dynamics of intermediate states of the catalytic cycle, and the function of protein residues that are involved in central steps. This information can be used for the design of new biochemical and biophysical experiments. The combination of computational and experimental techniques will be central in elucidating the function of the intricate pumping machinery in complex I.

Abbreviations

B3LYP	Becke's three-parameter functional – a density functional
DFT	density functional theory
E_m	midpoint redox potential

EOM	equations-of-motion
eT	electron transfer
FeS	iron sulfur
λ	reorganization energy
MD	molecular dynamics
MM	molecular mechanics
PB	(electrostatic) Poisson–Boltzmann calculations
PCET	proton-coupled electron transfer
pK_a	acid dissociation constant
pT	proton transfer
QM	quantum mechanics
QM/MM	hybrid quantum mechanics/molecular mechanics

Acknowledgements

I acknowledge insightful discussion on complex I with Prof. Mårten Wikström, Dr Gerhard Hummer, Dr Vivek Sharma, Dr Ana P. Gamiz-Hernandez, and Andrea Di Luca. Our work on biological energy conversion machineries has been supported by grants from the German Research Foundation (DFG), the Jane and Aatos Erkko Foundation, the German Academic Exchange Service (DAAD), and the European Research Foundation (ERC). Computing time for our research on complex I is provided by the Leibniz Rechenzentrum/ SuperMUC.

References

1. U. Brandt, *Annu. Rev. Biochem.*, 2006, **75**, 69.
2. J. Hirst, *Annu. Rev. Biochem.*, 2013, **82**, 551.
3. L. A. Sazanov, *Nat. Rev. Mol. Cell Biol.*, 2015, **16**(6), 375.
4. M. Wikström, V. Sharma, V. R. I. Kaila, J. P. Hosler and G. Hummer, *Chem. Rev.*, 2015, **115**(5), 2196.
5. M. L. Verkhovskaya and D. A. Bloch, *Int. J. Biochem. Cell Biol.*, 2013, **45**(2), 491.
6. M. Wikström, *FEBS Lett*, 1984, **169**(2), 300.
7. S. Dröse, S. Krack, L. Sokolova, K. Zwicker, H.-D. Barth, N. Morgner, H. Heide, M. Steger, E. Nübel, V. Zickermann, S. Kerscher, B. Brutschy, M. Radermacher and U. Brandt, *PLoS Biol.*, 2011, **9**(8), e1001128.
8. L. A. Sazanov and P. Hinchliffe, *Science*, 2006, **311**(5766), 1430.
9. R. G. Efremov and L. A. Sazanov, *Nature*, 2011, **465**(7297), 414.
10. R. Baradaran, J. M. Berrisford, G. S. Minhas and L. A. Sazanov, *Nature*, 2013, **494**(7438), 443.
11. C. Hunte, V. Zickermann and U. Brandt, *Science*, 2010, **329**(5990), 448.
12. V. Zickermann, C. Wirth, H. Nasiri, K. Siegmund, H. Schwalbe, C. Hunte and U. Brandt, *Science*, 2015, **347**(6217), 44.
13. K. R. Vinothkumar, J. Zhu and J. Hirst, *Nature*, 2014, **515**(7525), 80.

14. J. Zhu, K. R. Vinothkumar and J. Hirst, *Nature*, 2016, **536**(7616), 354.
15. K. Fiedorczuk, J. A. Letts, G. Degliesposti, K. Kaszuba, M. Skehel and L. A. Sazanov, *Nature*, 2016, **538**(7625), 406.
16. M. Murai, Y. Mashimo, J. Hirst and H. Miyoshi, *Biochemistry*, 2011, **50**(32), 6901.
17. S. Nakanishi, M. Abe, S. Yamamoto, M. Murai and H. Miyoshi, *Biochim. Biophys. Acta*, 2011, **1807**(9), 1170.
18. M. Murai, S. Murakami, T. Ito and G. Miyoshi, *Biochemistry*, 2015, **54**(17), 2739.
19. S. Zhu and S. B. Vik, *J. Biol. Chem.*, 2015, **290**(34), 20761.
20. B. Amarneh and S. B. Vik, *Biochemistry*, 2003, **42**(17), 4800.
21. L. Euro, G. Belevich, M. I. Verkhovsky, M. Wikström and M. Verkhovskaya, *Biochim. Biophys. Acta*, 2008, **1777**(9), 1166.
22. J. Torres-Bacete, P. K. Sinha, A. Matsuno-Yagi and T. Yagi, *J. Biol. Chem.*, 2011, **286**(39), 34007.
23. M. C. Kao, E. Nakamaru-Ogiso, A. Matsuno-Yagi and T. Yagi, *Biochemistry*, 2005, **44**(27), 9545.
24. M. Kervinen, J. Pätsi, M. Finel and I. E. Hassinen, *Biochemistry*, 2004, **43**(3), 773.
25. E. Nakamaru-Ogiso, M. C. Kao, H. Chen, S. C. Sinha, T. Yagi and T. Ohnishi, *J. Biol. Chem.*, 2010, **285**(50), 39070.
26. J. Michel, J. DeLeon-Rangel, S. Zhu, K. Van Ree and S. B. Vik, *PLoS One*, 2011, **6**(2), e17420.
27. M. L. Verkhovskaya, N. Belevich, L. Euro and M. Wikström, *Proc. Natl. Acad. Sci. U. S. A.*, 2008, **105**(10), 3763.
28. S. de Vries, K. Dörner, M. J. F. Straampraad and T. Friedrich, *Angew. Chem., Int. Ed.*, 2015, **54**(9), 2844.
29. M. Mimaki, X. Wang, M. McKenzie, D. R. Thorburn and M. T. Ryan, *Biochim. Biophys. Acta*, 2012, **1817**(6), 851.
30. E. Screpanti and C. Hunte, *J. Struct. Biol.*, 2007, **159**(2), 261.
31. G. Belevich, J. Knuuti, M. I. Verkhovsky, M. Wikström and M. Verkhovskaya, *Mol. Microbiol.*, 2011, **82**(5), 1086.
32. P. Tan, Z. Feng, L. Zhang, T. Hou and Y. Li, *J. Recept. Signal Transduction Res.*, 2015, **35**(2), 170.
33. S. Steimle, M. Willistein, P. Hegger, M. Janoschke, H. Erhardt and T. Friedrich, *FEBS Lett.*, 2012, **586**(6), 699.
34. M. Wikström and G. Hummer, *Proc. Natl. Acad. Sci. U. S. A.*, 2012, **109**(12), 4431.
35. V. R. I. Kaila, M. Wikström and G. Hummer, *Proc. Natl. Acad. Sci. U. S. A.*, 2014, **111**(19), 6988.
36. V. Sharma, G. Belevich, A. P. Gamiz-Hernandez, T. Róg, I. Vattulainen, M. L. Verkhovskaya, M. Wikström, G. Hummer and V. R. I. Kaila, *Proc. Natl. Acad. Sci. U. S. A.*, 2015, **112**(37), 11571.
37. A. Di Luca, A. P. Gamiz-Hernandez and V. R. I. Kaila, *Proc. Natl. Acad. Sci. U. S. A.*, 2017, **114**(31), E6314.

38. A. P. Gamiz-Hernadez, A. Jussupow, M. P. Johansson and V. R. I. Kaila, *submitted*.
39. E. S. Medvedev, V. A. Couch and A. A. Stuchebrukhov, *Biochim. Biophys. Acta*, 2010, **1797**(9), 1665.
40. V. A. Couch, E. S. Medvedev and A. A. Stuchebrukhov, *Biochim. Biophys. Acta*, 2009, **1787**(10), 1266.
41. T. Hayashi and A. A. Stuchebrukhov, *Proc. Natl. Acad. Sci. U. S. A.*, 2010, **107**(45), 19157.
42. T. Hayashi and A. A. Stuchebrukhov, *J. Phys. Chem. B*, 2011, **115**(18), 5354.
43. G. Hummer and M. Wikström, *Biochim. Biophys. Acta*, 2016, **1857**(7), 915.
44. M. R. A. Blomberg and P. E. M. Siegbahn, *Biochim. Biophys. Acta*, 2012, **1817**, 495.
45. M. Wikström, M. I. Verkhovsky and G. Hummer, *Biochim. Biophys. Acta*, 2003, **1604**(2), 61.
46. V. R. I. Kaila, M. I. Verkhovsky, G. Hummer and M. Wikström, *Proc. Natl. Acad. Sci. U. S. A.*, 2008, **105**(17), 6255.
47. S. Superkar, A. P. Gamiz-Hernadez and V. R. I. Kaila, *Angew. Chem., Int. Ed.*, 2016, **55**(39), 11940.
48. E. Fadda, C. H. Yu and R. Pomès, *Biochim. Biophys. Acta*, 2008, **1777**(3), 277.
49. R. Sugitani and A. A. Stuchebrukhov, *Biochim. Biophys. Acta*, 2009, **1787**(9), 1140.
50. A. V. Pisliakov, P. K. Sharma, Z. T. Chu, M. Haranczyk and A. Warshel, *Proc. Natl. Acad. Sci. U. S. A.*, 2008, **105**(22), 7726.
51. R. Liang, J. M. Swanson, Y. Peng, M. Wikström and G. A. Voth, *Proc. Natl. Acad. Sci. U. S. A.*, 2016, **113**(27), 7420.
52. V. Sharma, G. Enkavi, I. Vattulainen, T. Róg and M. Wikström, *Proc. Natl. Acad. Sci. U. S. A.*, 2015, **112**(7), 2040.
53. A. L. Woelke, G. Galstyan and E. W. Knapp, *Biochim. Biophys. Acta*, 2014, **1837**(12), 1998.
54. P. Goyal, S. Yang and Q. Cui, *Chem. Sci.*, 2015, **6**(1), 826.
55. J. Lu and M. R. Gunner, *Proc. Natl. Acad. Sci. U. S. A.*, 2014, **111**(34), 12414.
56. P. E. M. Siegbahn, *Biochim. Biophys. Acta*, 2013, **1827**(8–9), 1003.
57. E. M. Sproviero, J. A. Gascón, J. P. McEvoy, G. W. Brudvig and V. S. Batista, *J. Am. Chem. Soc.*, 2008, **130**(11), 3428.
58. I. Ugur, A. W. Rutherford and V. R. I. Kaila, *Biochim. Biophys. Acta*, 2016, **1857**(6), 740.
59. D. Narzi, D. Bovi and L. Guidoni, *Proc. Natl. Acad. Sci. U. S. A.*, 2014, **111**(24), 8723.
60. K. Saito, A. W. Rutherford and H. Ishikita, *Nat. Commun.*, 2015, **6**, 8488.
61. D. A. Pantazis, W. Ames, N. Cox, W. Lubitz and F. Neese, *Angew. Chem., Int. Ed.*, 2012, **51**(39), 9935.
62. M. Shoji, H. Isobe and K. Yamaguchi, *Chem. Phys. Lett.*, 2015, **636**, 172.

63. S. Izrailev, A. R. Crofts, E. A. Berry and K. Schulten, *Biophys. J.*, 1999, 77(4), 1753.
64. P. A. Postila, K. Kaszuba, P. Kuleta, I. Vattulainen, M. Sarewicz, A. Osyczka and T. Róg, *Sci. Rep.*, 2016, 33607.
65. M. Dittrich, S. Hayashi and K. Schulten, *Biophys. J.*, 2003, **85**(4), 2253.
66. S. Mukherjee and A. Warshel, *Proc. Natl. Acad. Sci. U. S. A.*, 2012, **109**(37), 14876.
67. K. Nam, J. Pu and M. Karplus, *Proc. Natl. Acad. Sci. U. S. A.*, 2014, **111**(50), 17851.
68. K. Okazaki and G. Hummer, *Proc. Natl. Acad. Sci. U. S. A.*, 2015, **112**(34), 10720.
69. P. Goyal, N. Ghosh, P. Phatak, M. Clemen, M. Gaus, M. Elstner and Q. Cui, *J. Am. Chem. Soc.*, 2011, **133**(38), 14981.
70. C. M. Suomivuori, A. P. Gamiz-Hernandez, D. Sundholm and V. R. I. Kaila, *Proc. Natl. Acad. Sci. U. S. A.*, 2017, **114**(276), 7043.
71. A. D. MacKerell, D. Bashford, M. Bellott, R. L. Dunbrack, J. D. Evanseck, M. J. Field, S. Fischer, J. Gao, H. Guo, S. Ha, D. Joseph-McCarthy, L. Kuchnir, K. Kuczera, F. T. K. Lau, C. Mattos, S. Michnick, T. Ngo, D. T. Nguyen, B. Prodhom, W. E. Reiher, B. Roux, M. Schlenkrich, J. C. Smith, R. Stote, J. Straub, M. Watanabe, J. Wiórkiewicz-Kuczera, D. Yin and M. Karplus, *J. Phys. Chem. B*, 1998, **102**(18), 3586.
72. C. Oostenbrink, A. Villa, A. E. Mark and W. F. Van Gunsteren, *J. Comput. Chem.*, 2004, **25**(13), 1656.
73. J. Wang, R. M. Wolf, J. W. Caldwell, P. A. Kollman and D. A. Case, *J. Comput. Chem.*, 2004, **25**(9), 1157.
74. G. Lamoureux and B. Roux, *J. Chem. Phys.*, 2003, **119**(6), 3025.
75. A. Warshel and R. M. Weiss, *J. Am. Chem. Soc.*, 1980, **102**(20), 6218.
76. S. C. L. Kamerlin and A. Warshel, *Wiley Interdiscip. Rev.: Comput. Mol. Sci.*, 2011, **1**(1), 30.
77. L. Monticelli, S. K. Kandasamy, X. Periole, R. G. Larson, D. P. Tieleman and S. J. Marrink, *J. Chem. Theory Comput.*, 2008, **4**(5), 819.
78. K. Lindorff-Larsen, S. Piana, R. O. Dror and D. E. Shaw, *Science*, 2011, **334**(6055), 517.
79. J. D. Chodera and F. Noé, *Curr. Opin. Struct. Biol.*, 2014, **25**, 135.
80. *Free-energy Methods Review, Free Energy Calculations: Theory and Applications in Chemistry and Biology*, ed. C. Chipot and A. Pohorille, Springer Series in Chemical Physics, 2007.
81. M. R. Gunner and N. A. Baker, *Methods Enzymol.*, 2016, **578**, 1.
82. G. M. Ullmann and E. W. Knapp, *Eur. Biophys. J.*, 1999, **28**(7), 533.
83. C. A. Rohl, C. E. M. Strauss, K. M. S. Misura and D. Baker, *Methods Enzymol.*, 2004, **383**, 66.
84. A. Sali and T. L. Blundell, *J. Mol. Biol.*, 1993, **234**(3), 779.
85. A. D. Becke, *J. Chem. Phys.*, 1993, **98**, 5648.
86. C. T. Lee, W. T. Yang and R. G. Parr, *Phys. Rev. B*, 1988, **37**, 785.
87. A. J. Cohen, P. Mori-Sánchez and W. Yang, *Chem. Rev.*, 2012, **112**(1), 289.

88. F. Jensen, *J. Chem. Phys.*, 2002, **116**(17), 7372.
89. S. Grimme, J. Antony, S. Ehrlich and H. Krieg, *J. Chem. Phys.*, 2010, **132**(15), 154104.
90. M. Reiher, O. Salomon and B. A. Hess, *Theory Chem. Acc.*, 2001, **107**, 48.
91. L. Noodleman, *J. Chem. Phys.*, 1981, **74**(10), 5737.
92. E. Ruiz, J. Cano, S. Alvarez and P. Alemany, *J. Comput. Chem.*, 1999, **20**(13), 1391.
93. P. E. M. Siegbahn and F. Himo, *J. Biol. Inorg. Chem.*, 2009, **14**(5), 643.
94. A. Warshel and M. Levitt, *J. Mol. Biol.*, 1976, **103**(2), 227.
95. H. M. Senn and W. Thiel, *Angew. Chem., Int. Ed.*, 2009, **48**(7), 1198.
96. L. W. Chung, W. M. C. Sameera, R. Ramozzi, A. J. Page, M. Hatanaka, G. P. Petrova, T. V. Harris, X. Li, Z. Ke, F. Liu, H.-B. Li, L. Ding and K. Morokuma, *Chem. Rev.*, 2015, **115**(12), 5678.
97. T. Ohnishi, *Biochim. Biophys. Acta*, 1998, **1364**(2), 186.
98. L. Euro, G. Belevich, D. A. Bloch, M. I. Verkhovsky, M. Wikström and M. Verkhovskaya, *Biochim. Biophys. Acta*, 2009, **1787**, 68.
99. C. C. Page, C. C. Moser, X. Chen and P. L. Dutton, *Nature*, 1999, **402**(6757), 47.
100. C. C. Moser, T. A. Farid, S. E. Chobot and P. L. Dutton, *Biochim. Biophys. Acta*, 2006, **1757**(9–10), 1096.
101. G. K.-L. Chan and S. Sharma, *Ann. Rev. Phys. Chem.*, 2011, **62**(1–668), 465.
102. K. Zwicker, A. Galkin, S. Drose, L. Grgic, S. Kerscher and U. Brandt, *J. Biol. Chem.*, 2006, **281**(32), 23013.
103. J. Blumberger, *Chem. Rev.*, 2015, **115**(20), 11191.
104. M. A. Tocilescu, U. Fendel, K. Zwicker, S. Dröse, S. Kerscher and U. Brandt, *Biochim. Biophys. Acta*, 2010, **1797**(6–7), 625.
105. M. Verkhovskaya and M. Wikström, *Biochim. Biophys. Acta*, 2014, **1837**(2), 246.
106. T. Yano, W. R. Dunham and T. Ohnishi, *Biochemistry*, 2005, **44**, 1744.
107. T. Ohnishi, E. Nakamaru-Ogiso and S. T. Ohnishi, *FEBS Lett.*, 2010, **584**(19), 4131.
108. J. R. Treberg and M. D. Brand, *J. Biol. Chem.*, 2011, **286**(20), 17579.
109. U. Brandt, *Biochim. Biophys. Acta*, 2011, **1807**(10), 1364.
110. M. Verkhovsky, D. A. Bloch and M. Verkhovskaya, *Biochim. Biophys. Acta*, 2012, **1817**(9), 1550.
111. O. Jardetzky, *Nature*, 1966, **211**(5052), 969.
112. L. R. Forrest, Y. W. Zhang, M. T. Jacobs, J. Gesmonde, L. Xie, B. H. Honig and G. Rudnick, *Proc. Natl. Acad. Sci. U. S. A.*, 2008, **105**(30), 10338.
113. B. C. Marreiros, A. P. Batista, A. M. Duarte and M. M. Pereira, *Biochim. Biophys. Acta*, 2013, **1827**(2), 198.

CHAPTER 5

Coupling Hydride Transfer to Proton Pumping: the Swiveling Mechanism of Transhydrogenase

SANGJIN HONG,[a] JOSEPHINE H. LEUNG,[b] CHANG SUN,[a] PAWEEN MAHINTHICHICHAN,[a] LICI SCHURIG-BRICCIO,[a] PIUS S. PADYATTI[b] AND ROBERT B. GENNIS*[a]

[a] Department of Biochemistry, MC-712, University of Illinois, Urbana, IL 61801, USA; [b] Department of Integrative Structural and Computational Biology, The Scripps Research Institute, 10550 North Torrey Pines Rd., La Jolla, CA 92037, USA
*Email: r-gennis@illinois.edu

5.1 Introduction

The generation and subsequent utilization of a proton (or sodium) motive force is central to virtually all life. Making a steady state proton motive force (PMF or $\Delta\mu_{H^+}$) requires coupling a strongly favorable biochemical reaction (*e.g.*, respiration) to moving charges and protons across the membrane bilayer (inside→outside). Enzymes that utilize the PMF couple the flux of protons in the reverse direction (outside→inside) to drive what would otherwise be an unfavorable reaction. This chapter is focused on the nicotinamide nucleotide transhydrogenase (Nnt), a consumer of the PMF that maintains the highly reduced state of the NADPH/$NADP^+$ "redox buffer".

Chemical Biology No. 5
Mechanisms of Primary Energy Transduction in Biology
Edited by Mårten Wikström

Published by the Royal Society of Chemistry, www.rsc.org

The enzyme catalyzes hydride transfer (two-electron transfer plus one proton) between two metabolically critical dinucleotides (shown below) and couples this reaction to the translocation of one proton across the membrane.[1]

$$\text{NADP} + \text{NADP}^+ + \text{H}^+_{\text{out}} \leftrightarrow \text{NAD}^+ + \text{NADPH} + \text{H}^+_{\text{in}} \qquad \text{(Reaction 5.1)}$$

There are relatively few enzymes that have evolved to perform primary energy transduction, *i.e.*, to couple the PMF to biochemical reactions. Examples covered in this book, besides the transhydrogenase, include Complex I (NADH : quinone oxidoreductase); Complex III (bc_1 complex; quinol : cyt c oxidoreductase); Complex IV (cytochrome oxidase; cyt c : O_2 oxidoreductase); and the F_1F_O-ATP synthase. For each of these four central enzymes of mitochondrial bioenergetics, X-ray structures have been determined and, with the exception of Complex I, the coupling mechanism is reasonably well understood. Each of these enzymes has a unique way to couple biochemical function to the PMF. There is no universal coupling mechanism. As the material covered in this book makes clear, Nature has evolved a relatively small number of very distinct mechanisms to couple the free energy of the chemiosmotic gradient to biochemical reactions at the heart of metabolism.

5.2 Location and Physiological Roles of Transhydrogenase

The nicotinamide nucleotide transhydrogenase is present in the cytoplasmic membrane of most prokaryotes and in the inner mitochondrial membrane of most eukaryotes, including human. The reaction catalyzed (Reaction 5.1) couples the PMF-driven flux of protons through the enzyme to favor the formation of NADPH at the expense of NADH. In the absence of a PMF, the equilibrium of Reaction 5.1 would maintain the redox poise of the [NADH/NAD^+] and [NADPH/$NADP^+$] couples equal since the midpoint potentials ($E_m^{o'}$) of these redox couples are virtually equal, −315 mV and −320 mV, respectively. In the presence of a PMF (positive outside), the thermodynamic driving force will favor a higher [NADPH/$NADP^+$] ratio than [NADH/NAD^+] due to the favorable free energy change from transferring a proton from the outside to the inside (Reaction 5.1).

NADPH is required for: (1) the reduction of H_2O_2 *via* either theoredoxin or glutathione; and (2) anabolic reactions such as the biosynthesis of amino acids, lipids and nucleotides. Figure 5.1 summarizes the connection between the PMF generated by respiration, the major source of H_2O_2, and the transhydrogenase, generating NADPH which is used ultimately to reduce H_2O_2.[2–4] A mutant of *Caenorhabditis elegans* lacking transhydrogenase, for example, has an increased sensitivity to oxidative stress.[5] The parasitic pathogen of oysters, *Perkinsus marinus*,[6,7] encodes a record number of seven distinct transhydrogenases, possibly to detoxify H_2O_2 from the oysters' defense mechanisms.[8]

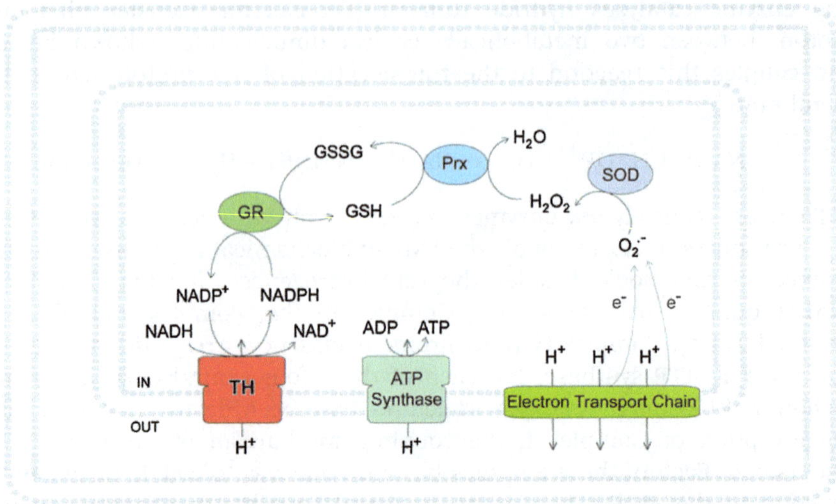

Figure 5.1 Schematic illustration of a mitochondrion or a prokaryote (represented as a Gram negative bacterium) showing the generation of hydrogen peroxide by superoxide dismutase (SOD) and the electron transport chain, and the role of the transhydrogenase (TH) in the reduction of peroxide. NADPH is required by glutathione reductase (GR) to reduce glutathione (GSSG/GSH) which is used by peroxidases (Prx) to reduce hydrogen peroxide to water.

The mitochondrial transhydrogenase is not an essential enzyme because it is not the only way in which NADPH can be generated within the mitochondrion.[9,10] Although present in mammalian mitochondria, the transhydrogenase is not universal. For example, within fungi, the enzyme is found in *Neurospora* and *Aspergillus* but is not present in *Saccharomyces* species. Nevertheless, a deficiency of the mammalian mitochondrial transhydrogenase is consequential and detrimental in animals as well as in humans (see next section), demonstrating that it is responsible for a significant fraction of the NADPH supply within the mammalian mitochondrion under relevant conditions.

Similarly, not all prokaryotes express an energy-coupled transhydrogenase, but rely on alternative pathways to generate NADPH.[11–13] Some prokaryotes have a water-soluble transhydrogenase which is not energy-coupled and, when expressed, will drive the [NADH/NAD^+] and [NADPH/$NADP^+$] ratios towards equality. *E. coli* has both the energy-coupled and the soluble transhydrogenase,[11] though they are under different transcriptional regulation.[14] The *E. coli* energy-coupled transhydrogenase is present under conditions of high growth rate where the demand for NADPH is greatest,[14] though the enzyme is not essential for survival.[12] In *Pseudomonas putida*, the energy-coupled transhydrogenase is important for maintaining redox homeostasis particularly during the biodegradation of aromatic compounds.[15] Changes in the expression of the energy-coupled transhydrogenase are important for the adaptation of *Methylobacterium extorquens* to engineered changes to metabolic pathways.[16] The importance of the supply of NADPH for optimal operation of numerous biosynthetic pathways is reflected in the purposeful manipulation of the energy-coupled transhydrogenase in bioengineered bacterial strains for commercial production of chemicals.[17–20]

It has been established that each hydride transfer is coupled to the electrogenic translocation of one proton across the bilayer.[1,21] In mitochondria and in bacteria, the PMF is used to generate NADPH, but *in vitro* the reverse reaction will pump protons across the membrane in the opposite direction.[22,23] Indeed, transhydrogenase-catalyzed hydride transfer from NADPH to NAD^+ to form NADH has been shown to generate a PMF in the mitochondria of the parasitic helminth *Hymenolepsis diminuta*, driving ATP synthesis under anaerobic conditions.[24] The transhydrogenase is a true proton pump and it is reversible.

5.3 Significance of the Transhydrogenase to Human Health

Recent research has resulted in an increased appreciation of the important role of the transhydrogenase to maintain mitochondrial and cellular redox balance. The transhydrogenase is responsible for about 50% of the NADPH generated in the mitochondrion[25] and is critical for the detoxification of H_2O_2.[2–4] In addition, NADPH is required for anabolic pathways for nucleotide, lipid and amino acid biosynthesis. Diminished transhydrogenase activity, due to low gene expression (*e.g.*, ageing) or mutations, results in numerous clinical manifestations depending on the severity of the deficiency. People with the most severe mitochondrial transhydrogenase mutations display familial glucocorticoid deficiency.[26–28] Reduction of transhydrogenase activity to 60% is sufficient to result in defects in mitochondrial DNA copy number. At 30% of normal function, defects in oxidative phosphorylation are observed.[27] There is a correlation between the expression of the gene encoding the transhydrogenase (*Nnt*) with human obesity and visceral fat distribution.[29] A three-dimensional model of the human mitochondrial transhydrogenase, based on the X-ray structure of the enzyme from the hyperthermophilic bacterium *T. thermophilus*,[30] has been made and the locations of disease-causing mutations have been mapped on this structure to assist in correlating structure and the nature of the mitochondrial disorder.[31]

Animal studies show that a deficiency of transhydrogenase results in mitochondrial redox imbalance and hypertension,[32] glucose intolerance, low insulin secretion, low oxidative phosphorylation,[33] ROS toxicity leading to neurodegeneration[4] and altered brain function.[2,34,35] Increasing transhydrogenase function has been suggested as a possible therapeutic strategy.[2] On the other hand, NADPH generated by the transhydrogenase is important for the metabolic adaptation of cancer cells to utilize glutamine to supply the tricarboxylic cycle to support accelerated cell growth and proliferation.[36–38] A mouse model for heart failure suggested that under a large cardiac workload, the rapid depletion of NADH required for energy could result in driving the transhydrogenase backwards (Reaction 5.1), depleting the mitochondrial matrix of NADPH resulting in oxidative damage.[39,40] Interestingly, a widely used mouse laboratory strain C57BL/6J has a spontaneous mutation in the *Nnt* gene encoding the transhydrogenase and as a consequence manifests mitochondrial redox abnormalities.[33,41,42]

Another point relating transhydrogenase to human health is that *Acinetobacter baumannii*, a human pathogen responsible for a significant and growing cause of bloodstream infections, has been shown to exhibit a fitness defect under conditions encountered during infection when the bacterial transhydrogenase is genetically inactivated.[43]

5.4 Domains, Subunits and Sequence Conservation

The transhydrogenase contains no prosthetic groups, *e.g.*, no heme, flavin, quinone or metals. The enzyme is remarkably conserved in all three domains of life[12,44–46] and appears to always be a homodimer. Each protomer (two protomers per dimer) consists of three domains: dI, dII and dIII, each with distinct functions (Figure 5.2).

Domains I and III bind NAD(H) and NADP(H), respectively, and hydride transfer takes place across the dI/dIII interface. (Note: the notation

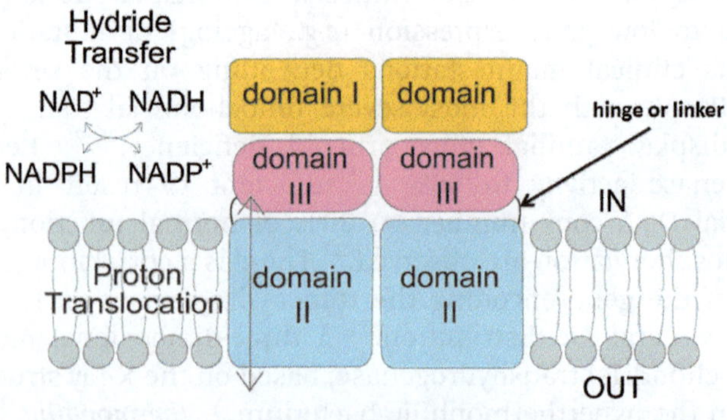

Figure 5.2 Schematic of the domain organization of the transhydrogenase, shown as a dimer in which each protomer contains one copy of each of domain I, domain II and domain III. Hydride transfer takes place at the interface between domain I and domain III and proton translocation is through domain II. Domains I and III contain, respectively the NAD(H) and NADP(H) binding sites. The linker between domains II and III is also shown and is always present. The notations "IN" and "OUT" denote, respectively, the mitochondrial matrix (or bacterial cytoplasm) and the mitochondrial intermembrane space (or bacterial periplasm or external solution).
Modified from ref. 30. Reprinted with permission from AAAS.

"NAD(H)" implies that the statement refers to either NADH or NAD^+, *etc.*). Domain II is embedded in the membrane and contains 12–14 transmembrane spans (depending on species). Although the three domains are conserved, the number of subunits and their N- and C-termini vary between species. This is illustrated in Figure 5.3. There are four different arrangements of how the three domains of each protomer are encoded within one, two or three subunits. One feature that is conserved in all the variants is that dIII is always linked by a "hinge" or "linker" region to the last transmembrane helix of dII. Although the protomer unit (dI/dII/dIII) is encoded by 1, 2 or 3 genes depending on the organism, there are no data to suggest that the properties of the assembled enzymes are altered by, whether there are 1, 2 or 3 subunits.

5.4.1 Single-subunit Transhydrogenase, Variant 1

Found in animal mitochondria as well as in fungi (Figure 5.3C). The single gene is nucleus-encoded and imported into the mitochondrion despite having 14 transmembrane spans. The N-terminus is at the start of dI, which is followed by dII, the hinge region and finally dIII. The 14 transmembrane spans are denoted TM1 to TM14 and the numbering is maintained for homologous transmembrane spans in all species regardless of which subunit they are in.

5.4.2 Single-subunit Transhydrogenase, Variant 2

Found uniquely in protozoa such as *Entamoeba histolytica*[47,48] (Figure 5.3D). The N-terminus is on the matrix side of TM5 within dII and the subunit continues through the hinge region, dIII, dI and ends on the matrix side of TM4 of dII. The most notable aspect of this arrangement is that dIII is tethered at both the N-terminal side (to dII) and also at the C-terminal side (to dI). This must constrain the dynamics of dIII, which is proposed to move through very different orientations during catalysis.[30,46]

5.4.3 Two-subunit Transhydrogenase

Found in many prokaryotes, including *E. coli* on which most of the biochemical studies have been done, reviewed in Ref. 44–46. The α subunit consists of dI along with the first 4 transmembrane helices of dII (Figure 5.3B). The β subunit starts on the periplasmic side of TM6 and includes the hinge region and dIII. Sequence alignments with the single-subunit transhydrogenases predict that dII contains 13 transmembrane helices and is missing the equivalent of TM5 in the mitochondrial transhydrogenases. Indeed, the α and β subunits of the *E. coli* transhydrogenase have been linked by a peptide that mimics the mitochondrial TM5 to yield a functional single-subunit enzyme.[49]

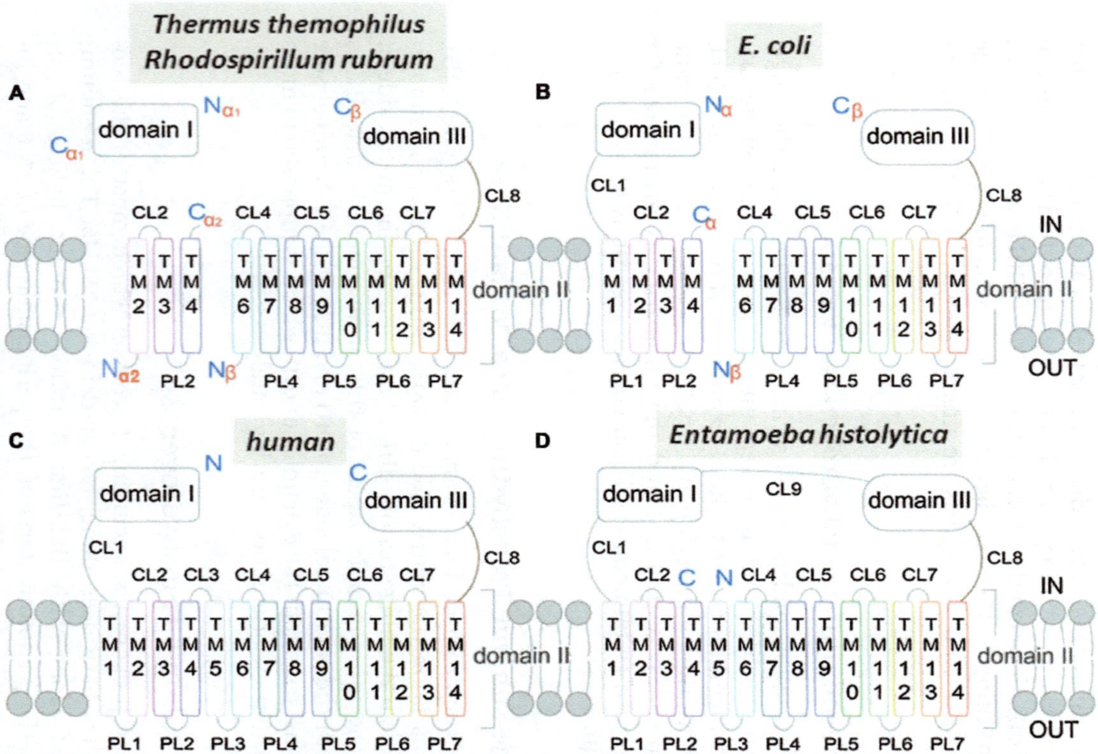

Figure 5.3 Four distinct variations of the subunit organization of transhydrogenases. (A) Three-subunit variants. (B) Two-subunit variants. (C) Single-subunit variant in which the N-terminus is in domain I. (D) Single-subunit variant present in protozoa in which the N-terminus is on the cytoplasmic side of TM5 in the membrane domain. The N- and C-termini of each subunit are shown along with the standard notation of the transmembrane helices (TM1 to TM14), the cytoplasmic "loops" (CL1 to CL9) and the exterior or periplasmic "loops" (PL1 to PL7).
Modified from ref. 30. Reprinted with permission from AAAS.

5.4.4 Three-subunit Transhydrogenase

Found in prokaryotes, including *T. thermophilus* and *Rhodospirillum rubrum*. The only X-ray structures of the entire enzyme and of the membrane domain II are of the transhydrogenase from *T. thermophilus*.[30] In these enzymes, dI is present as a separate hydrophilic subunit, α_1, that is not covalently tethered to the transmembrane portion of the enzyme (Figure 5.3A). In the *T. thermophilus* enzyme, α_2 subunit consists of first three transmembrane helices of dII, corresponding to TM2, TM3 and TM4 of the mitochondrial enzyme, and the β subunit is the same as in the two-subunit transhydrogenases, starting on the periplasmic side of TM6 (Figure 5.3B).[30] The membrane domain in the three-subunit transhydrogenases contains 12 transmembrane domains, including 3 from the α_2 subunit and 9 from the β subunit. The transmembrane helix corresponding to TM5 of the mitochondrial enzymes is not present in any of the three-subunit transhydrogenases.

Despite the differences in subunit arrangements, the amino acid sequence conservation is considerable. For example, the human transhydrogenase protomer is a single polypeptide (Nnt) with 1086 amino acids. The two subunits of the *E. coli* enzyme, α (PntA) and β (PntB), have 510 and 462 amino acids, respectively, for a total of 972 residues. Aligning the combined two *E. coli* sequences with the human sequence shows 47% identity, and the 56 amino acid stretch covering TM13 and TM14, which are part of the proton pump apparatus, shows 75% identity. Another indication of the extent of conservation is the fact that functional hybrid complexes can be formed between constructs of domains I and III from different species, including *R. rubrum*/*E. coli*[50–52] and *R. rubrum*/human.[53]

5.5 Steady State Assays of the Transhydrogenase

Since the interconversion of NADH/NADP$^+$ to NAD$^+$/NADPH does not result in any change in the UV/vis spectrum, the reaction cannot be conveniently monitored spectrophotometrically. For convenience, analogues of the nucleotides are used that have different spectral changes upon reduction/oxidation than the natural substrates. The assay of the "reverse reaction" (Figure 5.4) substitutes acetylpyridine adenine dinucleotide (AcPdAD$^+$ or AcPyAD$^+$) for NAD$^+$. The reaction can be monitored by the absorption changes at 375 nm. Importantly, this "reverse reaction" is coupled to proton translocation.

$$\text{AcPyAD}^+ + \text{NADPH} + \text{H}^+_{\text{in}} \rightarrow \text{AcPyADH} + \text{NADP}^+ + \text{H}^+_{\text{out}} \quad \text{(Reaction 5.2)}$$

The midpoint potential of the AcPyAD$^+$/AcPyADH couple ($E_m^{o'} = -250$ mV) is more positive than for NADP$^+$/NADPH ($E_m^{o'} = -320$ mV) so if the nucleotides are all at the same concentration the reaction runs in the forward direction, favoring the reduction of AcPyAD$^+$. Furthermore, AcPyAD(H) interacts specifically with the NAD(H) site in dI of the transhydrogenase and not with the NADP(H) site in dIII.[54,55]

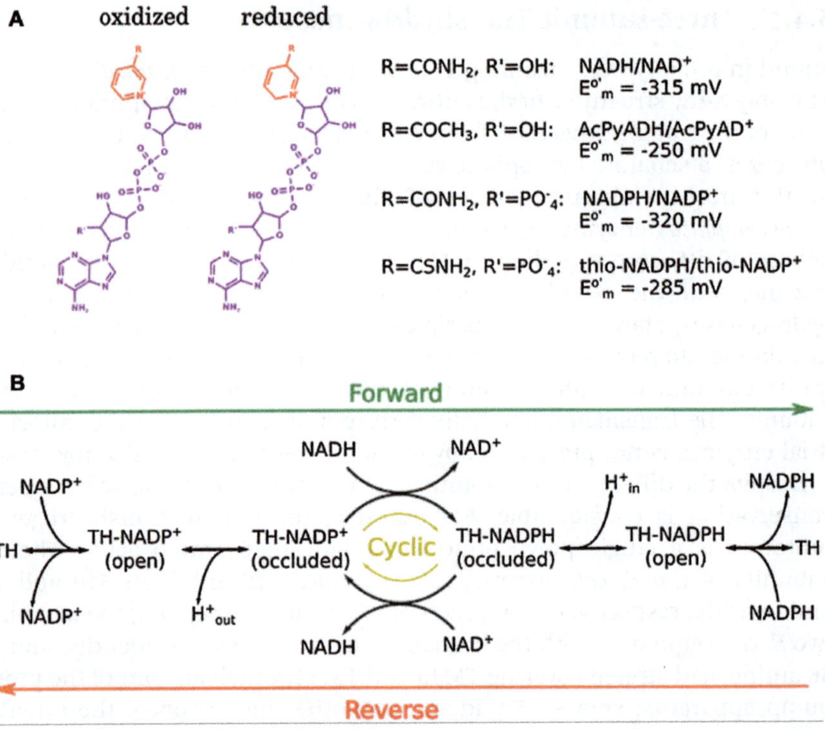

Figure 5.4 (A) Structures and electrochemical properties of the substrates used to assay the transhydrogenase. (B) Schematic showing the sequence of species in the forward, reverse and cyclic steady state assays of the transhydrogenase. The forward assay is the favored direction of the reaction in the most organisms.

In a similar way, thio-$NADP^{+}$[56] is used as a substitute for $NADP^{+}$ to monitor the rate of the forward reaction (Figure 5.4), which is also coupled to proton translocation.

$$\text{NADH} + \text{thio-NADP}^{+} + \text{H}^{+}_{\text{out}} \rightarrow \text{NAD}^{+} + \text{thio-NADPH} + \text{H}^{+}_{\text{in}} \quad \text{(Reaction 5.3)}$$

The forward reaction can also be measured by monitoring the reduction of $NADP^{+}$ by NADH, but this requires stopping the reaction and measuring the amount of NADPH made using the enzyme glutathione reductase.[57,58] A third assay that is often used to monitor transhydrogenase activity is the "cyclic assay", illustrated in Figure 5.4. In this assay, NADH is used to reduce $AcPyAD^{+}$ though the intermediary of NADP(H). Typically a small concentration of NADP(H) is used to bind to dIII, and a large concentration of NADH and $AcPyAD^{+}$ are present. NADH binds to the nucleotide binding site on dI and reduces $NADP^{+}$ in dIII. The generated NADPH then serves to reduce $AcPyAD^{+}$ which exchanges into the site in dI vacated by NAD^{+}, also a product of the first reaction. This reaction can proceed even if NADP(H) does not dissociate from dIII. This reaction is not coupled to proton translocation, does not require the membrane domain, and works well just with the complex of the hydrophilic domains dI and dIII (see below). It is worth noting that the K_m values for the various substrates for the forward, reverse and cyclic reactions with the *E. coli* transhydrogenase are in the range 1 μM to 25 μM.[59,60] The turnover number for the *E. coli* transhydrogenase measured by the reverse assay is about 20 s^{-1} and the cyclic reaction can be faster.[57] Not surprisingly, the steady state kinetics parameters are very dependent on the source of the enzyme (species) and the assay conditions, including, buffer, pH, temperature and the type of detergent for purified protein.

Inhibitors of the enzyme include Zn^{2+}, which apparently blocks proton translocation and interacts with a histidine that has not been identified,[57,61] and palmitoyl CoA, which binds with high affinity to the NADP(H) site.[62] The high affinity for palmitoyl CoA has been used for the basis of affinity chromatography of the *E. coli* transhydrogenase.[63]

Proton pumping can be measured using either inverted membrane vesicles prepared from native membranes containing the transhydrogenase, such as *E. coli* membranes[64] or *R. rubrum* chromatophores,[1] or by reconstituting the purified enzyme into liposomes.[1,23,57] In either case, only transhydrogenase molecules with the binding sites for NAD(H) and NADP(H) on the outside of the vesicles/liposomes are available to substrates added to solution. Hence, the reverse reaction (Reaction 2) pumps protons into the vesicles/liposomes, which corresponds to outward proton pumping in the native system. The rate of acidification of the vesicle/liposome interior is qualitatively measured[65,66] by the quenching of fluorescent dyes such as ACMA.[67]

5.6 Structures of the Isolated Domains: Divide and Conquer

5.6.1 Domain I

The crystal structures of isolated recombinant dI have been determined for the transhydrogenases from *E. coli*,[68] *R. rubrum*[69,70] and *T. thermophilus*.[30] Domain I is a stable dimer,[71] and each protomer itself contains two domains, dI-1 and dI-2. An estimate of the dissociation constant between the two dI protomers to form the dimer is $K_d \approx 10^{-10}$ M,[72] showing a very strong interaction. Figure 5.5A shows the structure of the dI dimer from the *R. rubrum* transhydrogenase with NADH bound to one protomer. The interface between the two dimers contains α-helices (helix 6 and helix 11). There is an interesting β-hairpin that extends from helix 6 of one protomer and interacts with the second protomer. Deletion of the β-hairpin in the *E. coli* enzyme reduces the rate of hydride transfer to about 30%.[68] The $(dI)_2$

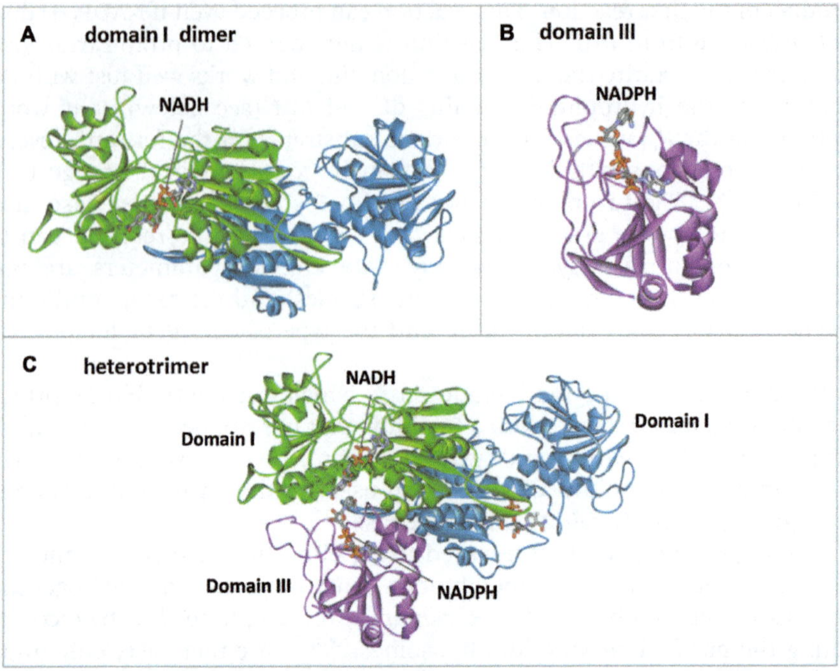

Figure 5.5 Structures of the hydrophilic domains of the transhydrogenase from *Rhodospirillum rubrum*. (A) Domain I is a dimer that binds to either NADH or NAD^+. The structure is shown with NADH bound. (B) Domain III crystallizes as a monomer and can bind to either NADPH or $NADP^+$. The structure is shown with NADPH bound. (C) Heterotrimer, consisting of the dimer of domain I bound to one equivalent of domain III. In the structure, domain I is bound to NADH and domain III is bound to NADPH. (pdb 1L7E, pdb 1PNQ, pdb 1U2D).

structure has been determined in the apo-form, *i.e.*, no bound NAD(H), and with either NADH or NAD^+ bound. Each protomer binds to NAD(H) and there is no cooperativity between the two NAD(H) binding sites (Table 5.1). The NAD(H) binds to a Rossman fold in a cleft between domains dI-1 and dI-2. The preference for NAD(H) over NADP(H) is due to hydrogen bond interactions between an aspartate side chain (*Rr*-α_1D202) with the 2′- and 3′-hydroxyls of the adenosine ribose, which is prevented by the phosphorylation of the 2′-hydroxyl in NADP(H). The pyrophosphate interacts with a GXGXXG motif (*Rr*-α_1G179-V182) which is located adjacent to the β-hairpin. There are several structures of the $(dI)_2$-NAD^+ complex which vary in terms of the positioning and apparent mobility of protein loops adjacent to the dinucleotide (TAP loop, RQD loop and mobile loop), referred to as "open", "intermediate" and "closed" conformations (Figure 5.6). The structure of the $(dI)_2$-NADH complex resembles the "intermediate" form. Both the mobile loop and RQD loop directly interact with the bound NAD(H).

Table 5.1 Dissociation constants of dinucleotides to transhydrogenase holo-enzyme and domains.

	Apparent dissociation constant				
	NAD^+	NADH	H_2NADH[a]	$NADP^+$	NADPH
E. coli holo-enzyme	100–500 μM[121]	50 μM[121]		16 μM[121]	0.87 μM 1.7 sites[121]
E. coli holo-enzyme pre-bound to NADH					1.6 μM[121]
E. coli holo-enzyme pre-bound to NAD^+				214 μM[121]	
E. coli holo-enzyme pre-bound to NADPH		> 500 μM[121]			
E. coli dI		45 μM[122]			
R. rubrum $(dI)_2dIII$ pre-bound to NADPH		Site 1: 20 μM Site 2: 300 μM[71]			
R. rubrum $(dI)_2dIII$ pre-bound to $NADP^+$			44 μM 2 identical sites[96]		
R. rubrum $(dI)_2dIII$ pre-bound to NADPH			Site 1: 24 μM Site 2: 176 μM[96]		
R. rubrum Diii				$<10^{-9}$ M[53,58,85]	$<10^{-9}$ M[53,58,85]
E. coli dIII					
Human dIII					
R. rubrum $(dI)_2$	300–500 μM[122,123]	20 μM 2 identical sites[71,122,124]			
Bovine $(dI)_2$		9.7 μM[125]			

[a]H_2NADH: 1,4,5,6-tetrahydro-nicotinamide adenine dinucleotide. An analogue of NADH in which the C5–C6 double bond of the physiological 1,4-dihydronicotinamide has been reduced. This analogue binds in the place of NADH but is not redox active.[96]

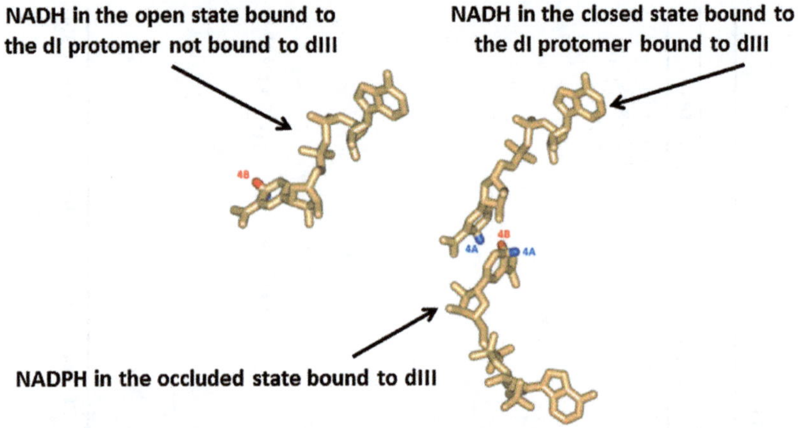

Figure 5.6 The dinucleotides bound to the heterotrimer from *R. rubrum* pictured in Figure 5.5C. Note the two different conformers (open, closed) of NADH in the two copies of domain I. The NADP(H) bound to domain III and the NAD(H) in the closed state bound to domain I are poised for steospecific hydride transfer between C4A of NAD(H) and C4B of NADP(H). (pdb 1U2D).

The mobile loop, which is observable by solution NMR,[73] closes down on the protein surface upon NAD(H) binding. Mutations in the mobile loop of the *R. rubrum* transhydrogenase result in a decrease in binding affinity as well as slower rate of hydride transfer.[74] Mutations in the RQD loop have resulted in modest changes in binding affinity but have a dramatic effect on the rate of hydride transfer.[75,76] The RQD loop appears to position the nicotinamide for optimal hydride transfer.[76] The nicotinamide ring is rotated differently in the NAD^+ and NADH structures,[77] stabilized by interactions with *Rr*-α_1D135 for NADH and *Rr*-α_1Q132 for NAD^+ which are both in the RQD loop. Besides the changes in the immediate vicinity of the bound dinucleotides, there are no significant global changes induced by the binding of either the reduced or oxidized ligand.

5.6.2 Domain III

X-ray structures have been determined for dIII from the transhydrogenases from bovine heart mitochondria,[78] human heart mitochondria[79] and *R. rubrum*.[80] Structural information is also available from solution NMR studies of dIII from *R. rubrum*[81,82] and *E. coli*.[83,84] Whereas the isolated domain I and the isolated membrane domain (dII) are dimers, recombinant dIII is a monomer.[53,58,82,85] Within the enzyme, dIII is always part of a subunit that contains at least a portion of the membrane domain and is connected by a linker or hinge (Figure 5.3). Hence, all the structures of "isolated" dIII are from recombinant protein. The dIII fold is a classic dinucleotide binding Rossmann fold, containing 6 parallel β strands flanked by 6 α helices. Domain III is always isolated with tightly bound NADP(H), and the structures all contain either bound $NADP^+$ or NADPH. Figure 5.5B shows the structure of dIII from the *R. rubrum* transhydrogenase bound to NADPH. The apparent K_d of NADP(H) to isolated dIII is less than 10^{-9} M,[58,85,86] and this very high affinity is the basis for referring to this as the "occluded" conformation of dIII (Table 5.1). The structure of the apo-form of dIII ("apo" denotes no NADP(H) bound) has not been determined, nor has dIII been crystallized in any other conformation.

One unusual feature is that the bound $NADP^+$ is oriented in a flipped orientation in the Rossmann fold compared to most other dinucleotide binding sites.[78,79] The diphosphate interacts with a GXGXXA/G motif that is located in loop E, and the specificity of NADP(H) over NAD(H) is provided by interactions with the 2′-phosphate of the adenine-associated ribose with a conserved "KRS" motif in loop E (bovine-K991-R992-S993).[78] The very high affinity of NADP(H) is at least partially due to the interactions with residues in loop E, which folds over the NADP(H) like a lid. Site-directed mutagenesis of the intact *E. coli* transhydrogenase has demonstrated the importance to catalytic activity of residues in loop E (*Ec*-βK424, *Ec*-βR425, *Ec*-βY431) as well as *Ec*-βD392 in loop D.[87–90] The effects of mutations in loop E and loop D have also been examined in isolated dIII from *E. coli*.[84,91] The aspartate in loop D (*Ec*-βD392) is of particular interest insofar that mutagenesis eliminates binding of NADP(H) to isolated dIII.[84] This residue forms hydrogen bonds to the pyrophosphate and to one of the ribose hydroxyl groups of the bound dinucleotide,[78,79] and has an unusually high pK_a (≈10.1) that is moderately sensitive to whether $NADP^+$ or NADPH is bound to dIII.[82,92,93] At pH 7, *Ec*-βD392 is protonated whether $NADP^+$ or NADPH is bound to dIII, and no pH-dependent conformational change is observed in the physiological range.[93]

Some of the crystal structures of dIII from the *R. rubrum* transhydrogenase contain two non-symmetry related forms of the NADP(H)-dIII complex.[80] The main difference between the two forms is the conformation of loop D. In one *R. rubrum* conformation, loop D is in an "open" state, as it is in the other structures of dIII.[78,79] The other *R. rubrum* conformation has loop D in a "closed state" which is not seen in other structures and is also not observed in the NMR data of isolated dIII.[81,93] The NMR data do, however, show that loop D is conformationally mobile,[80] and site-directed mutagenesis suggests a possible role of loop D in modulating the interaction with NADP(H).[91] An important caveat to all the studies with the isolated domains of transhydrogenase is that the important interactions between the domains, critical for the function of the enzyme, are absent.

5.6.3 Heterotrimer, $(dI)_2(dIII)$

When isolated dIII is mixed in solution with the isolated dI dimer, a heterotrimeric complex is formed, indicating a profound asymmetry in the way in which the dimeric dI binds to the monomeric dIII . The crystal structure of the heterotrimer from *R. rubrum* has also been determined[77,94–96] (Figure 5.5C), and the heterotrimeric stoichiometry of the $(dI)_2(dIII)$ complex has been confirmed in solution studies.[71] The *R. rubrum* heterotrimer is isolated with NADP(H) bound to dIII, primarily as $NADP^+$, which dissociates very slowly, showing that the dIII in the heterotrimer is in the occluded form, as in isolated dIII. The rate constant for dissociation of $NADP^+$ is $4\times10^{-3}s^{-1}$, which is about 5-fold slower than the dissociation rate constant of $NADP^+$ from isolate dIII (0.022 s^{-1}).[71] The X-ray structure confirms that loop E is closed over the bound NADP(H) in the heterotrimer, just as it is in the structure of isolated dIII, consistent with the very high affinity and very slow rate of dissociation. Loop D is in the open form, and modeling studies show that if loop D were in the closed conformation it would clash with residues in dI at the $(dI)_2$/dIII interface.[95] Both loop D and loop E in dIII are concluded to be important both for NADP(H) binding and for the interaction between dIII and the domain I dimer, which is the interface where hydride transfer occurs.[91,95] Importantly, aside from the interface between dIII and $(dI)_2$ there are no significant changes in the protein structure compared to isolated dIII and isolated $(dI)_2$.

The structure of the heterotrimer shows that dIII interacts significantly with both protomers of dI and the asymmetry of the interaction prevents a second equivalent of dIII to bind.[95] NMR studies show evidence of a very weak interaction of the $(dI)_2(dIII)$ heterotrimer with a second dIII but the $(dI)_2(dIII)_2$ tetramer has never been observed. The affinity between isolated dIII and the dI dimer has been measured to have $K_d \approx 6\times10^{-8}$ M,[82] and the affinity of a second equivalent of dIII binding must be at least 1000-fold weaker.[82] This asymmetry is not an artifact from combining separately prepared recombinant domains. The transhydrogenase from the protozoan *Entamoeba histolytica* is a single subunit (variant 2, Figure 5.3D), allowing the entire hydrophilic part of the enzyme to be cloned and expressed as a single tethered dIII-dI unit.[47,48] The monomeric form of the protein is inactive, but the dimeric *E. histolytica* $(dIII\text{-}dI)_2$ has biochemical properties very much like the *R. rubrum* heterotrimer and appears to have one copy of dIII tightly bound to domain I and one copy of dIII that is not complexed to domain I.[47] This same asymmetric interaction between the dI dimer and dIII is also apparent in the structure of the holoenzyme from *T. thermophilus.*[30]

The structure of the $(dI)_2$/(dIII) heterotrimer has been determined with different combinations of dinucleotides as well as redox-inactive dinucleotide analogues.[77,94–96] Different crystal structures reveal a number of variants in the nucleotide-bound structures which suggest protein conformational changes required to bring the two reactive species (e.g., NADH and $NADP^+$) together in a way to allow hydride transfer. It is likely that once the occluded form of dIII interacts with $(dI)_2$ there are changes in the orientations of residues at the interface that seal the site from water and position the nicotinamide rings for hydride transfer.[77] Figure 5.6 shows the NADH and NADPH dinucleotides as they appear in the structure of the *R. rubrum* heterotrimer (Figure 5.5C) with the protein removed for clarity. The NADH is in a closed state in the copy of dI which is bound to dIII and in the open state in the other copy of dI. The nicotinamide moieties at the dI/dIII interface are poised for direct, stereospecific hydride transfer.

5.6.4 The Membrane Domain (dII)

The structure of the isolated membrane domain of the transhydrogenase from *T. thermophilus* has been determined (Figure 5.7).[30] The *T. thermophilus* transhydrogenase is a 3-subunit enzyme and the construct used to obtain dII consists of subunits α_2 plus a modified version of the β subunit that is truncated after TM14 (see Figure 5.3A) to remove dIII. The protein is a dimer

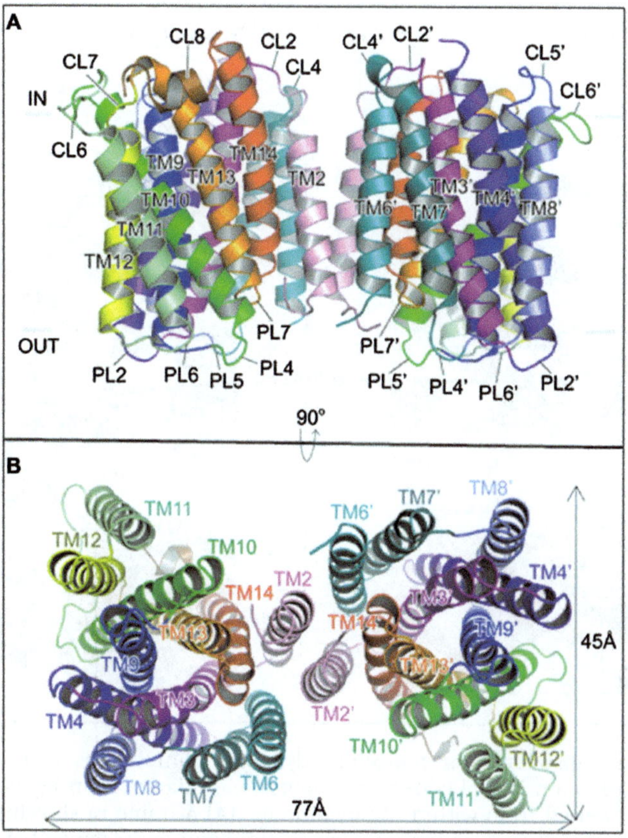

Figure 5.7 Atomic model of the dimer of domain II of the transhydrogenase from *T. thermophilus*. (A) View from the membrane. (B) View from the periplasm. Each of the transmembrane helices (TM), cytoplasmic loops (CL) and periplasmic loops (PL) is indicated.
From ref. 30. Reprinted with permission from AAAS. (pdb 4O93).

and each protomer contains 12 transmembrane helices, 3 in α_2 and 9 in β. The dII/dII interface holding the dimer together consists of contacts only along one transmembrane helix (TM2).

5.6.4.1 *The Proton Channel*

The most important feature is that each protomer contains what appears to be a proton-conducting channel. The central part of the channel is comprised of a 3-helix bundle of TM3, TM9 and TM13. Around this central unit are TM4, TM10 and TM14. Molecular dynamics studies[97] show water-filled cavities on both the cytoplasmic and periplasmic sides and a region in the middle which is largely dry but where a continuous water wire can be observed sporadically. Figure 5.8A shows a snapshot showing water molecules penetrating from the bulk solutions on both sides of the membrane, separated by a dry area. The snapshot shown in Figure 5.8B shows that water can penetrate the dry area transiently to connect both sides.

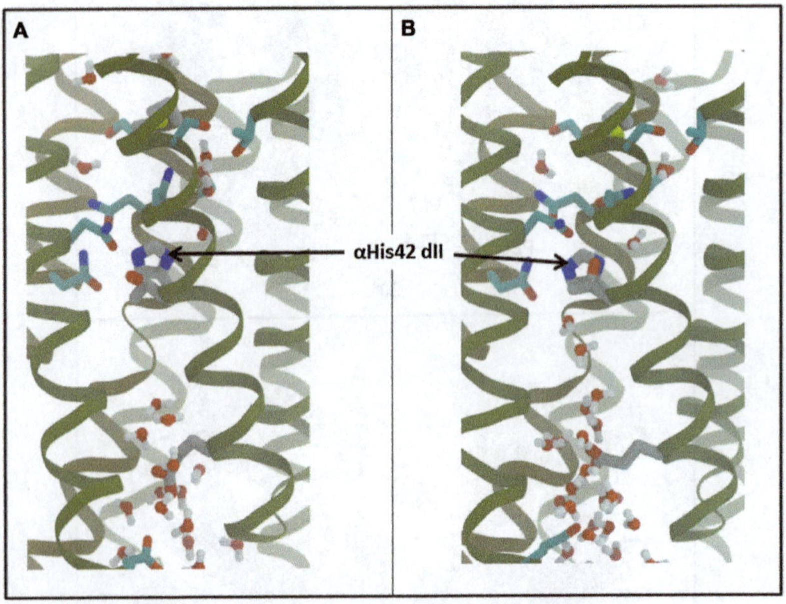

Figure 5.8 Snapshots from molecular dynamics simulation of *T. thermophilus* domain II embedded in a membrane bilayer in an aqueous medium, viewed from within the membrane. (A) A frame in showing water-filled cavities on both the cytoplasmic (top) and periplasmic (bottom) sides of the membrane. There is a dry region in between where water is not present. (B) A frame showing the transient penetration of the dry region by water, connecting the two water-filled cavities within the protein. *Tt*-α_2H42, the single protonatable residue in the proton channel, is shown. See ref. 97 for details.

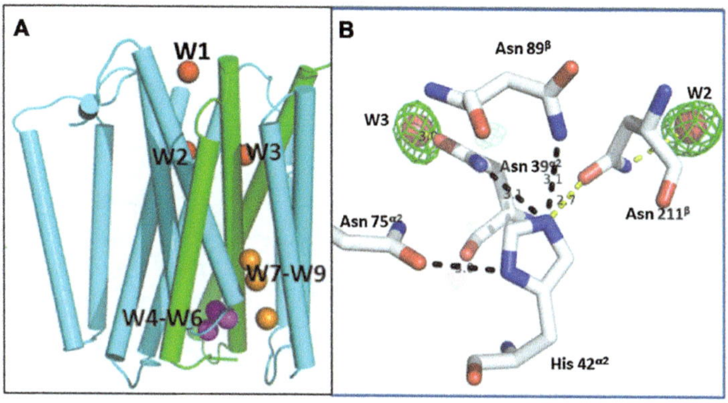

Figure 5.9 (A) Schematic of one protomer of domain II from the *T. thermophilus* transhydrogenase showing the location of nine water molecules resolved by X-ray crystallography within the protein.[97] (B) Hydrogen bond network within the proton channel centered on *Tt*-α_2H42 with four asparagine residues and two water molecules.
Modified from *Structure*, Critical role of water molecules in proton translocation by the membrane-bound transhydrogenase, P. S. Padayatti, J. H. Leung, P. Mahinthichaichan, A. Ishchenko, V. Cherezov, M. Soltis, J. B. Jackson, E. Tajkhorshid, C. D. Stout, R. B. Gennis, and Q. Zhang, Copyright 2017, with permission from Elsevier.[97] (pdb 5UNI).

The X-ray structure of *T. thermophilus* domain II at 2.2 Å resolution[97] shows 9 water molecules buried within the protein at positions consistent with the MD simulated water (Figure 5.9A). Two of these water molecules are part of a hydrogen bond network that includes residues *Tt*-α_2N39 (TM3), *Tt*-βN89 (TM9), *Tt*-βS132 (TM10) and *Tt*-βN211 (TM13) (Figure 5.9B).[97]

Cysteine labeling experiments[98] demonstrate a solvent-accessible cavity open to the cytoplasmic side of the membrane in the *E. coli* transhydrogenase, allowing access to residues *Ec*-βN222 (*Tt*-βN211), *Ec*-βH91 (*Tt*-βN89) and *Ec*-βV87 (*Tt*-βV85), shown in Figure 5.10. The locations of these residues are consistent with the water-filled cavity shown in Figure 5.8.

The only protonatable residue in the middle of the proton channel of the *E. coli* transhydrogenase is *Ec*-βH91(TM9) (Figure 5.8) and this residue is also critical for proton translocation.[22,64,99] This residue is absent in the *T. thermophilus* transhydrogenase and is replaced by an asparagine, *Tt*-βN89 (TM9). There are many examples where the same replacement is found, but in most cases, a "replacement" histidine is present in TM3, *Tt*-α_2H42 in the *T. thermophilus* transhydrogenase, which replaces a serine in the *E. coli* transhydrogenase (*Ec*-αS461).

5.6.4.2 *Three Groups of Sequence Variants Within the Proton Channel*

The *E. coli* and *T. thermophilus* transhydrogenases are representatives of two of the three groups of sequence variants of residues in the middle of the

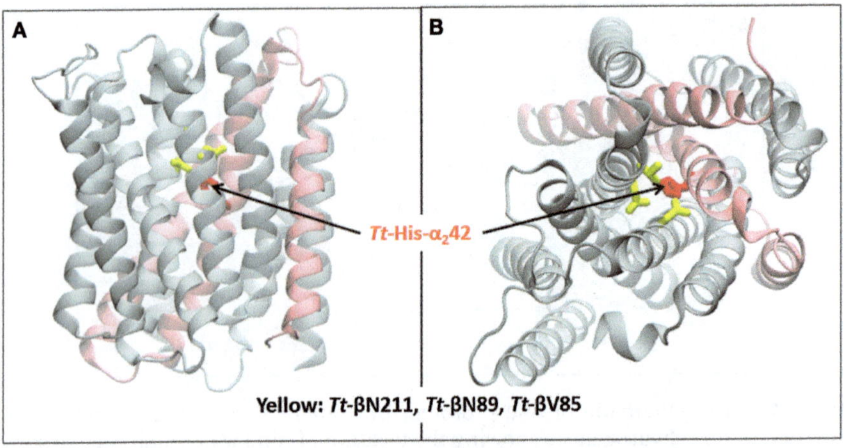

Figure 5.10 Structure of one protomer of domain II of the transhydrogenase from *T. thermophilus* showing the location of three residues (*Tt*-βN89, *Tt*-βN211 and *Tt*-βV85) at the equivalent positions of those identified in the *E. coli* transhydrogenase as being accessible to the solvent on the cytoplasmic side of the membrane.[98] The residues are shown in yellow. *Tt*-α_2H42 is shown in red. (A) View from within the membrane. (B) View from the cytoplasmic side. (pdb 4O9U).

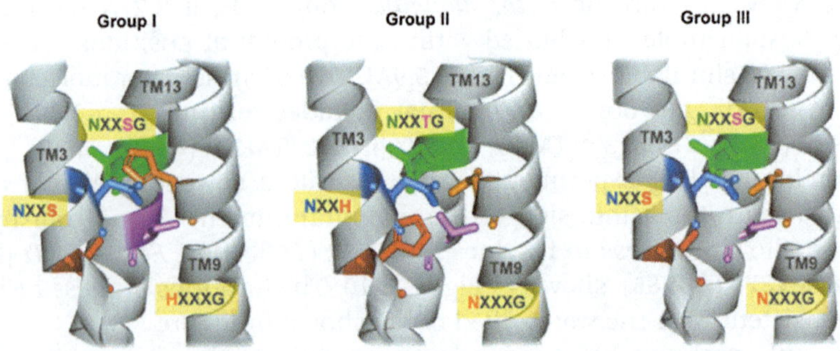

Figure 5.11 The three sequence variants within TM3, TM9 and TM13 that line the proton channel of domain II of the transhydrogenase.[30]

proton channel within TM3, TM9 and TM13 (Figure 5.11).[30] Whereas the enzymes of Groups I and II have one histidine near the center of the channel, those in Group 3 lack any protonatable group in this region. Many of the organisms containing Group III transhydrogenases are found in hot springs or in marine habitats, raising the possibility that perhaps these enzymes couple hydride transfer to sodium translocation instead of proton translocation across the membrane. *P. marinus*,[6,7] the oyster protozoan pathogen with seven transhydrogenases encoded in its genome, has representatives of both Group I and Group III transhydrogenases. Interestingly, replacing *Ec*-βH91 by asparagine, which switches a Group I enzyme to one mimicking

Group III lacking a protonatable group in the middle of the channel, yields an enzyme that retains nearly full reverse activity but no longer coupled to the proton motive force.[22]

5.6.4.3 *An Important Salt Bridge and the Hinge Region*

Another notable structural feature of the membrane domain is a salt bridge between *Tt*-βD202, in the cytoplasmic loop connecting TM12 and TM13, and *Tt*-βR254, within the linker connecting dII and dIII (Figure 5.12). These residues correspond to *Ec*-βD213 and *Ec*-βR265 which were predicted to form a salt bridge based on cysteine crosslinking experiments.[66] The susceptibility of the peptide bond following *Ec*-βR265 to trypsin cleavage is substantially enhanced in the presence of NADP(H) (either reduced or oxidized form) bound to dIII,[100] and the rate of proteolysis is modulated by mutations of *Ec*-βH91 in the membrane domain II.[22] Mutations in either *Ec*-βR265 or *Ec*-βD213 result in strong inhibition (over 10-fold) of the reverse reaction and increasing the apparent K_M of NADPH.[66] The hinge region of the protein and the *Ec*-βR265 – *Ec*-βD213 salt bridge is clearly implicated in coupling the proton translocation through dII to the binding of NADP(H) in dIII. Unfortunately, only a portion of the linker is retained and can be observed in the crystal structure of dII, since the remainder of the linker has been truncated along with dIII.

5.7 Structure of the Holo-enzyme

The crystal structure of the intact, complete transhydrogenase (holo-enzyme) from *T. thermophilus* has been solved to 6.9 Å resolution.[30] The holo-enzyme is an asymmetric dimer in which two copies of dIII are sandwiched between the dI dimer and the dII dimer (Figure 5.13).

5.7.1 Two Orientations of Domain III

The most remarkable feature of the holo-enzyme structure is that the two copies of dIII have orientations that differ by a rotation of approximately 180° (Figure 5.11). One copy of dIII is "face-up", with the NADP(H) binding site in dIII facing the NAD(H) binding site in dI, as observed in the structure of the $(dI)_2$dIII heterotrimer. The second copy of dIII is flipped so that the NADP(H) binding site on dIII is "face-down", making direct contact with dII near the opening of the proton channel. There is no direct contact between the two copies of dIII in the holo-enzyme structure. The existence of the two orientations of dIII in the detergent-solubilized transhydrogenase is confirmed by crosslinking between cysteine residues introduced in positions predicted by the structure to be in proximity.[30] In separate experiments, the same residue in dIII, *Tt*-βD284C, forms crosslinks with *Tt*-α_1T164C in dI (face-down orientation) and with *Tt*-α_2V28C in dII (face-up orientation). The two different orientations of dIII is not a crystallization artifact.

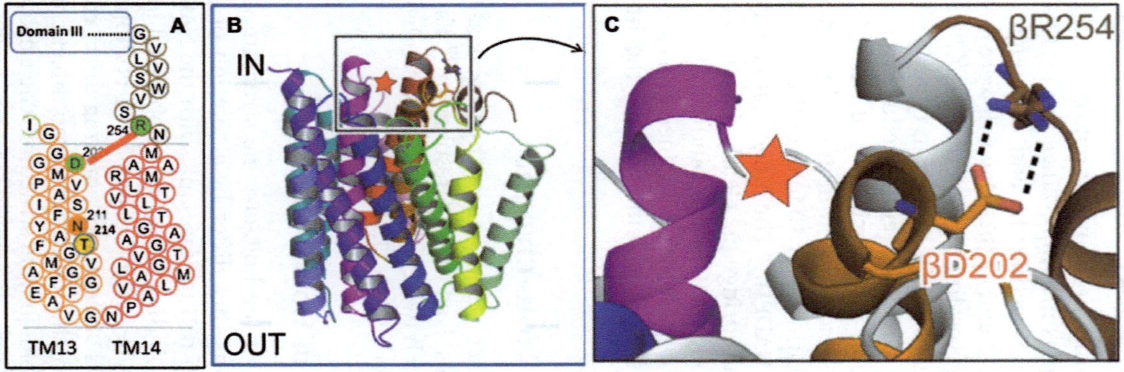

Figure 5.12 The salt bridge between *Tt*-βD202 and *Tt*-βR254 is functionally important for coupling hydride transfer and proton translocation. (A) Schematic of TM13 and TM14 in domain II showing the salt bridge in relation to residues that are part of the proton channel, *Tt*-βN211 and *Tt*-βT214.[97] (B) Structure of domain II showing the location of the salt bridge on the cytoplasmic (IN) surface near the presumed opening to the proton channel (red star). (C) Enlarged region highlighted in part (B) showing the salt bridge and the opening to the proton channel.
Modified from ref. 30. Reprinted with permission from AAAS. (pdb 4O93).

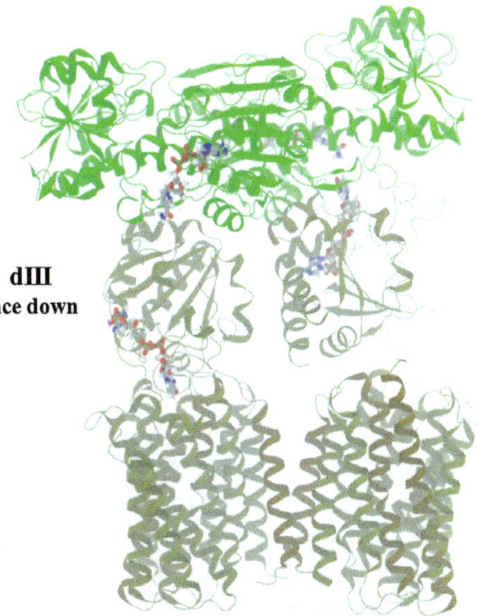

Figure 5.13 Structure of the holo-enzyme from *T. thermophilus* showing domain III in two different orientations within the dimer.[30] (pdb 4O9U).

The crosslinking data do not confirm that the two dIII orientations are simultaneously present in the same molecule, but are consistent with the X-ray model. Single-particle cryoEM was resolved to only 18 Å, and showed that one copy of dIII of the holo-enzyme in solution is more disordered than the other.[30] Rigid body docking of the crystal structure into the cryoEM structure suggests that the disordered copy of dIII is face-up (as in the heterotrimer) and that the face-down dIII interacts tightly with the membrane domain II. As with the heterotrimer, it is not possible to model both copies of dIII in the face-up orientation without steric clashes. These clashes are resolved when the two copies of dIII have different orientations. In addition, in the face-down conformation of dIII, loop D cannot be in the closed conformation without altering the conformations of the membrane domain and linker. This is similar to the conformational constraint of loop D in dIII at the interface with domain I in the $(dI)_2$dIII heterotrimer.[95]

5.7.2 Asymmetry of the Two Protomers of Domain II

Analysis of the domain organization of the 6.9 Å structure of the holo-enzyme reveals an asymmetry between the two dII protomers.[97] Whereas the dII protomer in contact with the face-up dIII fits well with the protomer in the isolated dII dimer, the copy of dII in contact with the face-down oriented dIII is displaced, suggesting that the transmembrane helices do not have identical conformations in the two protomers in the intact enzyme. Furthermore, the electron density that corresponds to a portion of the dII-dIII linker is different in the two dII protomers in the structure of the holo-enzyme. The relatively low resolution is insufficient to provide further detail, but does suggest that the dramatic asymmetry of the two copies of dIII is reflected by structural differences in the proton channels in the dII protomers.

5.7.3 Proton Translocation and the Face-down Orientation of dIII

The face-down dIII brings residues critical to the binding of NADP(H) in contact with or very close to residues involved in proton translocation across domain II. The structure is not sufficiently resolved to provide any detail, but it is clear that loop E and loop D in domain III as well as the nicotinamide ring of bound NADP(H) are very close to the opening of the proton channel in dII as well as the *Tt*-βD202 – *Tt*-βR265 salt bridge in the dII-dIII linker. This is the basis for the current model of how hydride transfer is coupled to proton translocation (see Section 5.11). Before describing the model, it will be useful to briefly review the biochemical observations that underlie the "binding change model" of coupling.

5.8 Key Biochemical Observations

5.8.1 Cooperativity Between the Two Protomers

Isolated protomers with all three domains (*i.e.*, monomers) are not active.[47,72] Titrations with high affinity inhibitors of the intact bovine mitochondrial transhydrogenase show that binding one equivalent of inhibitor per transhydrogenase dimer results in complete inhibition of the enzyme,[101,102] *i.e.*, inhibition of one protomer prevents the second protomer within the dimer from functioning. The current view is that each protomer catalyzes the same reaction. Since one copy of face-up dIII to dI prevents the second copy from binding with this orientation, the two protomers must function out-of-phase with each other. Furthermore, if one copy of dIII were frozen in the face-up conformation, perhaps due to the binding of an inhibitor, the second protomer could not complete the catalytic cycle.

5.8.2 Direct Hydride Transfer Across the dI/dIII Interface

Hydride transfer occurs directly between bound nucleotides with no redox intermediates[60,103] and is stereospecific. In the forward reaction (Figure 5.4), the *pro-R* proton at $C4_N$ of the dihydonicotinamide ring of NADH is transferred to $C4_N$ on the *si* face of the $NADP^+$ nicotinamide ring.[104,105] This is illustrated in Figure 5.6 which shows NADH and NADPH bound to the heterotrimer pictured in Figure 5.5C. The NADH bound to the "free" dI protomer is in a different conformation from the NADH bound at the dI/dIII interface. The nicotinamide ring has rotated to bring it within about 3.5 Å from the nicotinamide ring of the NADPH bound to dIII.

Steady state kinetics analysis of the reverse assay (NADPH → $AcPyAD^+$) is consistent with a random, sequential model in which the substrates bind to the enzyme and can form a ternary complex.[106,107] The ternary complex is also consistent with direct hydride transfer from NADH to $NADP^+$.[60,103] When the substrate concentrations are sufficiently high, the dissociation of NADP(H) from dIII is rate limiting for the proton-coupled reactions.[60,108]

5.8.3 Hydride Transfer Between Enzyme-bound Dinucleotides Favors the Formation of NADPH

Pre-steady state kinetics studies (burst kinetics) with the heterotrimeric $(dI)_2dIII$ complex from *R. rubrum* have been performed using stopped-flow[103,109,110] and continuous flow[111] techniques. Fluorescence from a tryptophan introduced into dIII responds to the redox state of NADP(H) and can be used to monitor the reaction. The addition of a high concentration of NADH to the heterotrimer with pre-bound $NADP^+$ yields a first order rate constant (k_f) of 21 200 s^{-1}, much faster than redox reactions involving NAD(H) or NADP(H) in other soluble enzymes (typically <1000 s^{-1}).[111] The rate constant for the dissociation of NADPH is 3×10^{-4} s^{-1}, so only one turnover is observed in these experiments.[71,110] A similar measurement of the reverse reaction rate (NAD^+ reduction by the heterotrimer complexed to NADPH) yields a much slower apparent first order rate constant, $k_{rev} < 590$ s^{-1}. The ratio k_f/k_{rev} shows that the equilibrium for the direct hydride transfer between the dinucleotides bound to the heterotrimer favors the formation of NADPH by at least 36-fold. The membrane domain is not present so dIII is in the occluded state and there is no effect of the proton motive force.

$$\text{dI-NADH} + \text{dIII-NADP}^+ \leftrightarrow \text{dI-NAD}^+ + \text{dIII-NADPH} \qquad \text{(Reaction 5.4)}$$

In solution, the equilibrium constant for this reaction is 1. The shift in the equilibrium constant when the substrates are bound to the protein in the occluded state is due to the favored binding of the products compared to the reactants to the enzyme. The binding free energies of the substrates/products to the enzyme contributes to the driving force of the forward reaction when the substrates are bound to the enzyme.

An independent measurement determined that the redox midpoint potential for the $NADP^+$/NADPH pair bound to dIII from *R. rubrum* is shifted by +73 mV compared to the value in solution (−320 mV).[86] This means that the affinity of NADPH to the occluded state of dIII is about 270-fold larger compared to that of $NADP^+$. Since the binding of NADH to dI is about 10-fold greater than for NAD^+ (Table 5.1), the equilibrium constant of Reaction 5.4 is about 27, favoring NAD^+/NADPH at the active site for hydride transfer, in the same range estimated by burst kinetics.[111] This favors hydride transfer for the forward reaction, NADH→$NADP^+$.

The shift of the protein conformation from the open state to the occluded state is a central feature of the **"binding change mechanism"**.[44,45,92] Proton translocation across the membrane is coupled to the transitions between the open and occluded states of dIII bound to NADP(H), illustrated below for the forward reaction scheme. The PMF drives the conformational transitions in steps 1 and 3, with hydride transfer in between.

$$\text{Step 1:}\quad H^+_{out} + dII_{open}dIII(NADP^+)_{open} \underset{}{\overset{PMF}{\rightleftharpoons}} dII(H^+)_{occluded}dIII(NADP^+)_{occluded}$$

$$\text{Step 2:}\quad dII(H^+)_{occluded}dIII(NADP^+)_{occluded} \underset{\text{hydride transfer}}{\rightleftharpoons} dII(H^+)_{occluded}dIII(NADPH)_{occluded}$$

$$\text{Step 3:}\quad dII(H^+)_{occluded}dIII(NADPH)_{occluded} \overset{PMF}{\rightleftharpoons} H^+_{in} + dII_{open}dIII(NADP^+)_{open}$$

The dependence of the forward reaction rate on the PMF has been examined with the transhydrogenase in chromatophores from the photosynthetic bacterium *Rhodobacter capsulatus*.[112] Illumination of the chromatophores results in generating a PMF which is used by the endogenous transhydrogenase to drive the forward reaction. This reaction was studied prior to the establishment of the steady state and in steady state. The data support a model in which the PMF accelerates a step in the mechanism following substrate binding but prior to product dissociation, consistent with the sequence of steps shown above. The role of the PMF is analogous to mechanism of driving the conformational changes in the F_1 portion of the F_1F_o-ATP synthase during ATP synthesis.[113]

Moving one proton from the outside to the inside in the presence of a PMF of 150 mV (outside positive) shifts the equilibrium state of Reaction 5.1 to favor the generation of NADPH ("forward" direction) by about 500-fold (25 °C). In addition, the proton motive force also increases the rate of the steady state forward reaction (NADH→$NADP^+$) up to the point where product dissociation becomes rate limiting.[63,114] Under conditions in which hydride transfer is rate limiting, the reaction rate exhibits a deuterium isotope effect ([4A-^{2}H]NADH vs [4A-^{1}H]NADH or [4B-^{2}H]NADPH vs [4B-^{1}H]NADPH) which is not observed when product dissociation becomes rate limiting.[108,111]

5.9 Insights from Binding Studies

A variety of experimental techniques have been used to evaluate the dissociation constants of dinucleotides to both the holo-enzyme and the isolated domains. Results are summarized in Table 5.1. There are several lessons to be learned.

5.9.1 Binding of NAD(H) to Domain I is Not Influenced by the Presence or Absence of the Membrane Domain

Both NADH and NAD^+ are in rapid equilibrium with the binding sites in the dI dimer under all circumstances with the likely exception during the brief period when hydride transfer is occurring.

5.9.2 Negative Cooperativity Prevents the Formation of "Dead End" States of the Enzyme

A "dead-end" state is one in which no redox reaction can occur, such as when NADH binds to dI and NADPH is bound to dIII, or when NAD^+ and $NADP^+$ bind simultaneously. This is prevented by negative cooperativity resulting in a weakened affinity of NADH to dI when NADPH is present on the adjacent dIII or for NAD^+ binding when $NADP^+$ is occupying the dIII site. Binding studies with redox-inactive analogues H_2NADH and H_2NAD^+ in place of NADH and NAD^+, respectively, show that the negative cooperativity is not due to the binding of dIII *per se,* but that it is the redox state of the dinucleotide bound to dIII that is critical to the cooperative coupling between the dI and dIII sites. There is no cooperativity between the two NAD(H) binding sites in the dI dimer so, in the absence of dIII, the two NAD(H) binding sites have the same affinity. In the heterotrimer, however, the presence of NADPH bound to dIII lowers the affinity of the adjacent dI protomer to NADH but has no effect on NADH binding to the second protomer.

5.9.3 The Occluded State of dIII Has a Very High Apparent Affinity for NADP(H)

The dissociation constant is not measured directly but is estimated by the value of the rate constant of dissociation. If the steady state reaction rate is limited by the rate of dissociation of NADP(H), this rate constant can be estimated to be on the order of 10 s^{-1}. In the occluded state, the rate constants for NADP(H) dissociation are 1000 to 10 000-fold slower.[71] Assuming the same second order rate constant for the binding of NADP(H) as for the holo-enzyme, this yields the estimated binding constant in the range of 10^{-9} M or less for the apparent dissociation constant of NADP(H) to the occluded state. It is more likely, however, that the on-rate constant of NADP(H) to the occluded state is also very slow and that this state is not truly in equilibrium with the open state but, rather, a metastable state that is transiently trapped until "released" by contact with the membrane domain.

5.10 Insights from Mutagenesis Studies of the Membrane Domain

Mutagenesis studies of the *E. coli* transhydrogenase, focusing on highly conserved residues, provide a guide to the functions of several residues in the membrane domain (Figure 5.14).

5.10.1 Mutations in the Middle of the Membrane Domain (dII) Influence Enzyme Activity, Proton Translocation and the Binding of NADP(H)

Four residues in transmembrane helices near the middle of the membrane domain have been implicated as important: *Ec*-βH91, *Ec*-βN222, *Ec*-βS139 and *Ec*-βG252.[22,64,65,89,115] These correspond to *Tt*-βN89 (TM9), *Tt*-βN211(TM13), *Tt*-βS132 (TM10) and *Tt*-βG238 (TM14) respectively. *Ec*-βH91, *Ec*-βN222 and *Ec*-βS139 are adjacent to each other within the membrane

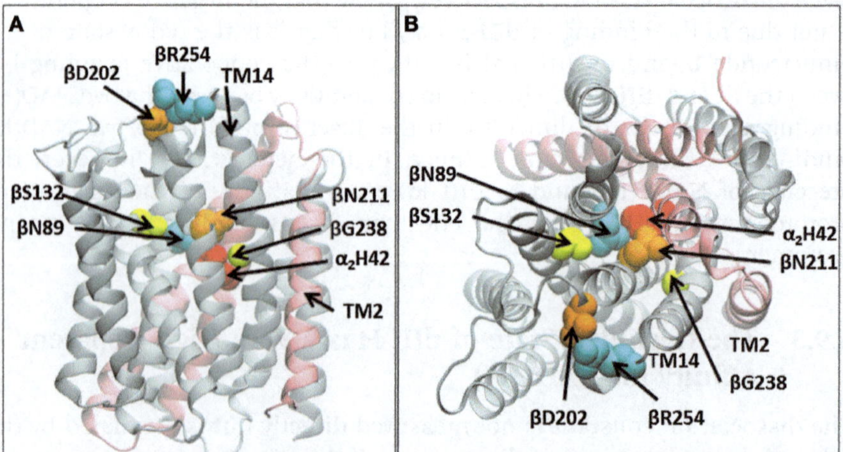

Figure 5.14 The structure of one protomer of domain II of the transhydrogenase from *T. thermophilus* showing the residues implicated in proton translocation and/or coupling between hydride transfer and proton translocation. The backbone of the α_1 subunit is in pink (TM2, TM3 and TM4), and that of the α_2 subunit is gray. TM2 and TM14 are labeled as for the residues referred to in the text. (pdb 4O93).

domain and *Ec*-βG252 is at the interface between TM14 and TM13 (Figure 5.14). Replacing either *Ec*-βH91 or *Ec*-βN222 with a positively charged residue (Lys, Arg) results in drastic reduction (1–3%) of the rate of the proton-coupled reverse reaction (NADPH→AcPyAD$^+$) and locks the holo-enzyme in the occluded state with "permanently" bound NADP(H).[64,115] The presence of bound NADP(H) in the *Ec*-βH91K mutant also results in sensitivity of the peptide bond following *Ec*-βR265 (in the linker region) to proteolysis by trypsin without the need to add exogenous NADP(H).[22] Cyclic activity is substantially retained. These results suggest that the positively charged residue mimics the protonated state of *Ec*-βH91.

In contrast, although the substitution of *Ec*-βH91 by serine, cysteine or threonine also virtually eliminates enzyme activity measured by the reverse assay (NADPH→AcPyAD$^+$), the addition of either NADPH or NADP$^+$ does not result in increased exposure of the peptide bond following *Ec*-βR265 to proteolysis.[22] These mutations to neutral, polar residues appear to lock the enzyme in the open form or, at least, prevents formation of the occluded state.

Of particular interest is the substitution of *Ec*-βH91 by asparagine. The *Ec*-βH91N mutant retains virtually complete reverse enzyme activity but appears to be uncoupled from the proton motive force.[22] The addition of NADPH, but not NADP$^+$, resulted in proteolysis by trypsin. The possibility that a different cation might be transported by the *Ec*-βH91N mutant was not tested. Switching the His/Asn between *Ec*-βH91and *Ec*-βN222 results in an inactive enzyme.

Ec-βG252 is at about the same level in the membrane as *Ec*-βH91and *Ec*-βN222. An alanine substitution, *Ec*-βG252A virtually eliminates reverse transhydrogenase activity (NADPH→AcPyAD$^+$) and locks the enzyme in the occluded form.[65] Doubtless the increased volume of the sidechain perturbs the interactions between TM14 and TM13 and could conceivably perturb *Ec*-βN222 which is on TM13. In addition, TM14 connects to the hinge region and the mutant could alter the position of the helix and perturb this portion of the protein.

5.10.2 Mutations in the Hinge Region (*Ec*-βK261 to *Ec*-βR265) and *Ec*-βD213 Inhibit Transhydrogenase Activity

The structure of domain II of the *T. thermophilus* transhydrogenase shows that the residues corresponding to *Ec*-βR265 and *Ec*-βD213 (*Tt*-βD202 and *Tt*-βR254) form a salt bridge (Figures 5.12 and 5.14). Mutations in each of these two residues in the *E. coli* enzyme result in severe loss of reverse activity and increase the apparent K_M of NADPH.[66] The formation and breaking of this salt bridge, perhaps controlled by the protonation/deprotonation of *Ec*-βD213 have been speculated to be involved in catalysis and in coupling hydride transfer to proton translocation.[44,66,116] Other mutations in the hinge region also result in inhibition of the reverse activity of the transhydrogenase, further supporting the importance of this region of the protein.[66]

5.11 Working Model for Coupling Hydride Transfer and Proton Translocation Across the Membrane

The alternating sites/binding change model that has been developed over several decades of research is supported by considerable biochemical data.[44–46] The model was initially inspired[117] by the mechanism of the F_1F_o-ATP synthase[118–120] and the analogy remains useful. It is proposed that the transition between the face-up and face-down orientations of dIII in each protomer is the key to the coupling mechanism.[46]

5.11.1 The Main Features of the Model are Briefly Summarized

1. NAD(H) binding to domain I is in rapid equilibrium at all times other than the brief period of hydride transfer and is not directly engaged in the mechanism of coupling or gating the proton pump.
2. There are two states defining the binding of NADP(H) to domain III, open and occluded. In the open state, the dissociation constant of NADP(H) is in the range of 10^{-5} to 10^{-6} M, but the open state is not competent for hydride transfer. Presumably, the open state of dIII does not form the appropriate interface with dI required for hydride transfer. In the occluded state, NADP(H) does not dissociate on the timescale of catalytic turnover and for all practical purposes can be considered "locked" into the protein structure. The occluded state of domain III is competent for hydride transfer and one equivalent binds with high affinity ($K_d \approx 60$ nM) to the domain I dimer.[82] There is no direct evidence for the occluded form of dIII in the holo-enzyme, but it is assumed that a similar state forms transiently and is competent for hydride transfer during steady state catalysis. The fact that the addition of NADP(H) to the holoenzyme does not convert the enzyme population to the occluded state[121] demonstrates that this state is not thermodynamically favored in the holoenzyme as it is in isolated dIII or in the $(dI)_2$dIII heterotrimer. Once the $dIII^{open}_{NADP+}dII^{open}_{H+}$ face-down complex forms, the open state and occluded state,$dIII^{open}_{NADP+}dII^{open}_{H+}$, equilibrate.
3. Consistent with the structure of the membrane domain II,[30] it is assumed that the proton channel has one protonatable group in the middle (*Ec*-βH91 or *Tt*-α_2H42) that can connect to the bulk aqueous solutions through either a "half channel" to the cytoplasm or a "half-channel" to the periplasm. Protonic equilibrium to the periplasm occurs only when $NADP^+$ bound to dIII interacts with the membrane domain in the face-down orientation. If NADPH is bound to domain III,

the half-channel to the cytoplasmic side is open. Simultaneous binding of both NADP(H) to $dIII^{open}$ and a proton to dII^{open} are required to shift the holo-enzyme to the occluded state, $dIII^{occluded}_{NADPH}dII^{occluded}_{H^+}$, which allows hydride transfer. These equilibria are shown below.

NADP$^+$ binding/open half-channel to the periplasm

$$dIII^{open}dII^{occluded} \underset{NADP^+}{\rightleftharpoons} dIII^{open}_{NADP^+}dII^{open\ OUT} \underset{H^+_{out}}{\rightleftharpoons} dIII^{open}_{NADP^+}dII^{open}_{H^+}$$

$$\rightleftharpoons dIII^{occluded}_{NADP^+}dII^{occluded}_{H^+}$$

NADPH binding/open half-channel to the cytoplasm

$$dIII^{open}dII^{occluded} \underset{NADPH}{\rightleftharpoons} dIII^{open}_{NADPH}dII^{open/IN} \underset{H^+_{in}}{\rightleftharpoons} dIII^{open}_{NADPH}dII^{open}_{H^+}$$

$$\rightleftharpoons dIII^{occluded}_{NADPH}dII^{occluded}_{H^+}$$

The transition between the open and occluded states is likely to involve the making and breaking of the salt bridge at the surface of the membrane domain (*Ec*-βD213 and *Ec*-βR265) shown in Figure 5.10, perhaps initiated by the protonation/deprotonation of *Ec*-βD213. The proposed sensitivity of the conformation of the open/closed status of the half-channels in the membrane domain to the redox state of NADP(H) is plausible because the nicotinamide ring is at the surface of the membrane domain in the face-down orientation.

Many aspects of the proposed mechanism of the transhydrogenase are analogous to the mechanism of the F_1F_o-ATP synthase,[118–120] including the two half-channels, an open state allowing substrate binding or product dissociation, an occluded state allowing the chemistry to occur, and the use of the PMF to drive the transitions between the open and occluded states.

Figure 5.15 illustrates one version of the catalytic cycle incorporating the features described above as well as the two protomers operating out of phase with each other. The cycle is completely reversible and the direction depends on the proton motive force and the concentrations of each of the four dinucleotides. The probability of protonation of the central histidine in the proton channel depends on the proton motive force.

There are numerous untested assumptions underlying this model, most prominently is the proposed dynamics of domain III as it flips between the face-up and face-down orientations. This swiveling is challenging to accept due to the fact that the dIII is tethered to the membrane domain by a peptide linker in all transhydrogenases, and also simultaneously tethered to domain I in the transhydrogenases present in protozoa (Figure 5.3D). Further experimental tests are clearly needed.

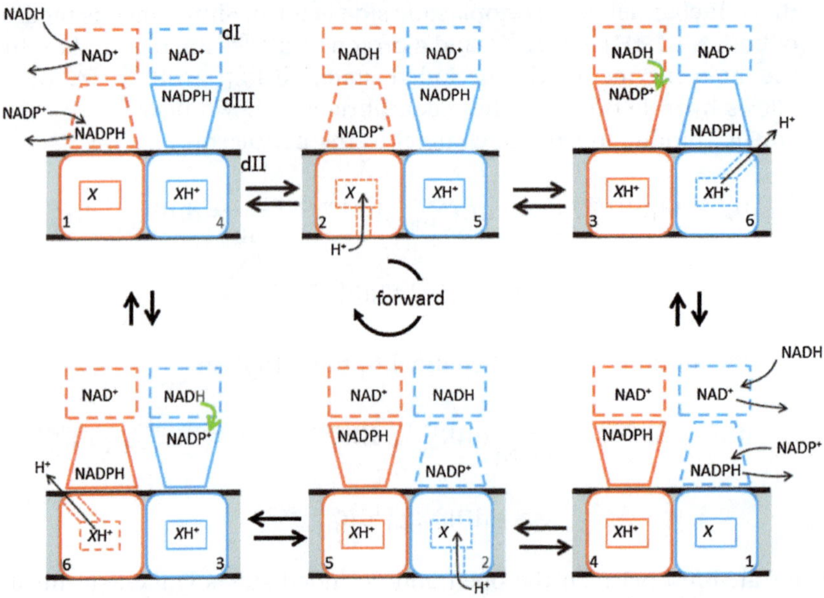

Figure 5.15 A schematic of the proposed mechanism of the transhydrogenase, discussed in the text.
Modified from J. Baz Jackson, J. H. Leung, C. D. Stout, L. A. Schurig-Briccio and R. B. Gennis, *FEBS Lett.*,[46] John Wiley and Sons, © 2015 Federation of European Biochemical Societies.

References

1. T. Bizouarn, L. A. Sazanov, S. Aubourg and J. B. Jackson, *Biochim. Biophys. Acta*, 1996, **1273**, 4–12.
2. P. Lopert and M. Patel, *J. Biol. Chem.*, 2014, **289**, 15611–15620.
3. K. H. Fisher-Wellman, C. T. Lin, T. E. Ryan, L. R. Reese, L. A. Gilliam, B. L. Cathey, D. S. Lark, C. D. Smith, D. M. Muoio and P. D. Neufer, *Biochem. J.*, 2015, **467**, 271–280.
4. D. Ghosh, K. R. Levault and G. J. Brewer, *Aging cell*, 2014, **13**, 631–640.
5. E. L. Arkblad, S. Tuck, N. B. Pestov, R. I. Dmitriev, M. B. Kostina, J. Stenvall, M. Tranberg and J. Rydstrom, *Free Radical Biol. Med.*, 2005, **38**, 1518–1525.
6. H. Sakamoto, S. Suzuki, K. Nagamune, K. Kita and M. Matsuzaki, *J. Eukaryotic Microbiol.*, 2016, **64**, 440–446.
7. A. Villalba, K. S. Reece, M. Camino Ordás, S. M. Casas and A. Figueras, *Aquat. Living Resour.*, 2004, **17**, 411–432.
8. L. Donaghy, E. Kraffe, N. Le Goïc, C. Lambert, A. K. Volety and P. Soudant, *PLoS One*, 2012, **7**, e46594.
9. C. A. Lewis, S. J. Parker, B. P. Fiske, D. McCloskey, D. Y. Gui, C. R. Green, N. I. Vokes, A. M. Feist, M. G. Vander Heiden and C. M. Metallo, *Mol. Cell*, 2014, **55**, 253–263.

10. J. A. Ronchi, A. Francisco, L. A. Passos, T. R. Figueira and R. F. Castilho, *J. Biol. Chem.*, 2016, **291**, 20173–20187.
11. U. Sauer, F. Canonaco, S. Heri, A. Perrenoud and E. Fischer, *J. Biol. Chem.*, 2004, **279**, 6613–6619.
12. H. H. Chou, C. J. Marx and U. Sauer, *PLoS Gen.*, 2015, **11**, e1005007.
13. S. Spaans, R. Weusthuis, J. Van Der Oost and S. Kengen, *Front. Microbiol.*, 2015, 6.
14. B. R. Haverkorn van Rijsewijk, K. Kochanowski, M. Heinemann and U. Sauer, *Microbiology*, 2016, **162**, 1672–1679.
15. P. I. Nikel, D. Perez-Pantoja and V. de Lorenzo, *Environment. Microbiol.*, 2016, **18**, 3565–3582.
16. S. M. Carroll and C. J. Marx, *PLoS Gen.*, 2013, **9**, e1003427.
17. D. Heinrich, M. Raberg and A. Steinbuchel, *FEMS Microbiol. Lett.*, 2015, **362**, fnv038.
18. M. Oldiges, B. J. Eikmanns and B. Blombach, *Applied Microbiol. Biotechnol.*, 2014, **98**, 5859–5870.
19. H. Wu, M. Karanjikar and K. Y. San, *Metab. Eng.*, 2014, **25**, 82–91.
20. H. Ying, S. Tao, J. Wang, W. Ma, K. Chen, X. Wang and P. Ouyang, *Microbial Cell Factories*, 2017, **16**, 52.
21. S. R. Earle and R. R. Fisher, *J. Biol. Chem.*, 1980, **255**, 827–830.
22. N. A. Glavas, C. Hou and P. D. Bragg, *Biochemistry*, 1995, **34**, 7694–7702.
23. M. Yamaguchi, C. D. Stout and Y. Hatefi, *J. Biol. Chem.*, 2002, **277**, 33670–33675.
24. C. F. Fioravanti and K. P. Vandock, *Parasitology*, 2010, **137**, 395–410.
25. J. Rydstrom, *Biochim. Biophys. Acta*, 1757, **2006**, 721–726.
26. O. Jazayeri, X. Liu, C. C. van Diemen, W. M. Bakker-van Waarde, B. Sikkema-Raddatz, R. J. Sinke, J. Zhang and C. M. van Ravenswaaij-Arts, *Eur. J. Med. Gen.*, 2015, **58**, 642–649.
27. Y. Fujisawa, E. Napoli, S. Wong, G. Song, R. Yamaguchi, T. Matsui, K. Nagasaki, T. Ogata and C. Giulivi, *BBA Clin.*, 2015, **3**, 70–78.
28. F. Roucher-Boulez, D. Mallet-Motak, D. Samara-Boustani, H. Jilani, A. Ladjouze, P. F. Souchon, D. Simon, S. Nivot, C. Heinrichs, M. Ronze, X. Bertagna, L. Groisne, B. Leheup, C. Naud-Saudreau, G. Blondin, C. Lefevre, L. Lemarchand and Y. Morel, *Eur. J. Endocrinol.*, 2016, **175**, 73–84.
29. J. T. Heiker, M. Kern, J. Kosacka, G. Flehmig, M. Stumvoll, E. Shang, T. Lohmann, M. Dressler, P. Kovacs, M. Bluher and N. Kloting, *Obesity*, 2013, **21**, 529–534.
30. J. H. Leung, L. A. Schurig-Briccio, M. Yamaguchi, A. Moeller, J. A. Speir, R. B. Gennis and C. D. Stout, *Science*, 2015, **347**, 178–181.
31. L. A. Metherell, J. A. Guerra-Assuncao, M. J. Sternberg and A. David, *Hum. Mutat.*, 2016, **37**, 1074–1084.
32. I. Leskov, A. Neville, X. Shen, S. Pardue, C. G. Kevil, D. N. Granger and D. M. Krzywanski, *J. Am. Soc. Hypertens.*, 2016, **11**, 110–121.
33. J. A. Ronchi, T. R. Figueira, F. G. Ravagnani, H. C. Oliveira, A. E. Vercesi and R. F. Castilho, *Free Radical Biol. Med.*, 2013, **63**, 446–456.

34. M. Picard, M. J. McManus, J. D. Gray, C. Nasca, C. Moffat, P. K. Kopinski, E. L. Seifert, B. S. McEwen and D. C. Wallace, *Proc. Natl. Acad. Sci. U. S. A.*, 2015, **112**, E6614–E6623.
35. M. Picard, M. McManus, J. Gray, C. Nasca, C. Moffat, P. Kopinsky, E. Seifert, B. McEwen and D. Wallace, *Psychoneuroendocrinology*, 2015, **61**, 74.
36. P. A. Gameiro, L. A. Laviolette, J. K. Kelleher, O. Iliopoulos and G. Stephanopoulos, *J. Biol. Chem.*, 2013, **288**, 12967–12977.
37. L. Alberghina and D. Gaglio, *Cell Death Dis.*, 2014, **5**, e1561.
38. A. R. Mullen, Z. Hu, X. Shi, L. Jiang, L. K. Boroughs, Z. Kovacs, R. Boriack, D. Rakheja, L. B. Sullivan, W. M. Linehan, N. S. Chandel and R. J. DeBerardinis, *Cell Rep.*, 2014, **7**, 1679–1690.
39. M. P. Murphy, *Cell Metab.*, 2015, **22**, 363–365.
40. A. G. Nickel, A. von Hardenberg, M. Hohl, J. R. Loffler, M. Kohlhaas, J. Becker, J. C. Reil, A. Kazakov, J. Bonnekoh, M. Stadelmaier, S. L. Puhl, M. Wagner, I. Bogeski, S. Cortassa, R. Kappl, B. Pasieka, M. Lafontaine, C. R. Lancaster, T. S. Blacker, A. R. Hall, M. R. Duchen, L. Kastner, P. Lipp, T. Zeller, C. Muller, A. Knopp, U. Laufs, M. Bohm, M. Hoth and C. Maack, *Cell Metab.*, 2015, **22**, 472–484.
41. A. A. Toye, J. D. Lippiat, P. Proks, K. Shimomura, L. Bentley, A. Hugill, V. Mijat, M. Goldsworthy, L. Moir, A. Haynes, J. Quarterman, H. C. Freeman, F. M. Ashcroft and R. D. Cox, *Diabetologia*, 2005, **48**, 675–686.
42. H. Freeman, K. Shimomura, E. Horner, R. D. Cox and F. M. Ashcroft, *Cell Metab.*, 2006, **3**, 35–45.
43. S. Subashchandrabose, S. Smith, V. DeOrnellas, S. Crepin, M. Kole, C. Zahdeh and H. L. Mobley, *mSphere*, 2016, 1.
44. A. Pedersen, G. B. Karlsson and J. Rydstrom, *J. Bioenerg. Biomembr.*, 2008, **40**, 463–473.
45. J. B. Jackson, *Biochim. Biophys. Acta*, 1817, **2012**, 1839–1846.
46. J. B. Jackson, J. H. Leung, C. D. Stout, L. A. Schurig-Briccio and R. B. Gennis, *FEBS Lett.*, 2015, **589**, 2027–2033.
47. C. J. Weston, J. D. Venning and J. B. Jackson, *J. Biol. Chem.*, 2002, **277**, 26163–26170.
48. C. J. Weston, S. A. White and J. B. Jackson, *FEBS Lett.*, 2001, **488**, 51–54.
49. J. Meuller, K. Mjorn, J. Karlsson, A. Tigerstrom, J. Rydstrom, C. Hou and P. D. Bragg, *Biochim. Biophys. Acta*, 2001, **1506**, 163–171.
50. J. D. Venning, S. J. Peake, P. G. Quirk and J. B. Jackson, *J. Biol. Chem.*, 2000, **275**, 19490–19497.
51. T. Bizouarn, O. Fjellstrom, M. Axelsson, T. V. Korneenko, N. B. Pestov, M. V. Ivanova, M. V. Egorov, M. Shakhparonov and J. Rydstrom, *Eur. J. Biochem.*, 2000, **267**, 3281–3288.
52. O. Fjellstrom, T. Bizouarn, J. W. Zhang, J. Rydstrom, J. D. Venning and J. B. Jackson, *Biochemistry*, 1999, **38**, 415–422.
53. S. J. Peake, J. D. Venning and J. B. Jackson, *Biochim. Biophys. Acta*, 1999, **1411**, 159–169.

54. M. Yamaguchi and Y. Hatefi, *Biochim. Biophys. Acta*, 1997, **1318**, 225–234.
55. A. Pedersen, J. Karlsson, M. Althage and J. Rydstrom, *Biochim. Biophys. Acta*, 1604, **2003**, 55–59.
56. A. Singh, J. D. Venning, P. G. Quirk, G. I. van Boxel, D. J. Rodrigues, S. A. White and J. B. Jackson, *J. Biol. Chem.*, 2003, **278**, 33208–33216.
57. S. J. Whitehead, M. Iwaki, N. P. Cotton, P. R. Rich and J. B. Jackson, *Biochim. Biophys. Acta*, 1787, **2009**, 1276–1288.
58. C. Diggle, T. Bizouarn, N. P. Cotton and J. B. Jackson, *Eur. J. Biochem.*, 1996, **241**, 162–170.
59. J. Meuller, J. Zhang, C. Hou, P. D. Bragg and J. Rydstrom, *Biochem. J.*, 1997, **324**(Pt 2), 681–687.
60. M. Hutton, J. M. Day, T. Bizouarn and J. B. Jackson, *Eur. J. Biochem.*, 1994, **219**, 1041–1051.
61. G. Veronesi, S. J. Whitehead, F. Francia, L. Giachini, F. Boscherini, G. Venturoli, N. P. Cotton and J. B. Jackson, *Biochim. Biophys. Acta*, 1797, **2010**, 494–500.
62. J. Rydstrom, *Eur. J. Biochem.*, 1972, **31**, 496–504.
63. X. Hu, J. W. Zhang, A. Persson and J. Rydstrom, *Biochim. Biophys. Acta*, 1995, **1229**, 64–72.
64. P. D. Bragg and C. Hou, *Arch. Biochem. Biophys.*, 2001, **388**, 299–307.
65. M. Yamaguchi and C. D. Stout, *J. Biol. Chem.*, 2003, **278**, 45333–45339.
66. M. Althage, T. Bizouarn and J. Rydstrom, *Biochemistry*, 2001, **40**, 9968–9976.
67. H. Rottenberg and R. Moreno-Sanchez, *Biochim. Biophys. Acta, Bioenerg.*, 1993, **1183**, 161–170.
68. T. Johansson, C. Oswald, A. Pedersen, S. Tornroth, M. Okvist, B. G. Karlsson, J. Rydstrom and U. Krengel, *J. Mol. Biol.*, 2005, **352**, 299–312.
69. P. A. Buckley, J. Baz Jackson, T. Schneider, S. A. White, D. W. Rice and P. J. Baker, *Structure*, 2000, **8**, 809–815.
70. G. S. Prasad, M. Wahlberg, V. Sridhar, V. Sundaresan, M. Yamaguchi, Y. Hatefi and C. D. Stout, *Biochemistry*, 2002, **41**, 12745–12754.
71. J. D. Venning, D. J. Rodrigues, C. J. Weston, N. P. Cotton, P. G. Quirk, N. Errington, S. Finet, S. A. White and J. B. Jackson, *J. Biol. Chem.*, 2001, **276**, 30678–30685.
72. U. M. Obiozo, T. H. Brondijk, A. J. White, G. van Boxel, T. R. Dafforn, S. A. White and J. B. Jackson, *J. Biol. Chem.*, 2007, **282**, 36434–36443.
73. C. Diggle, N. P. Cotton, R. L. Grimley, P. G. Quirk, C. M. Thomas and J. B. Jackson, *Eur. J. Biochem.*, 1995, **232**, 315–326.
74. S. Gupta, P. G. Quirk, J. D. Venning, J. Slade, T. Bizouarn, R. L. Grimley, N. P. Cotton and J. B. Jackson, *Biochim. Biophys. Acta*, 1998, **1409**, 25–38.
75. T. H. Brondijk, G. I. van Boxel, O. C. Mather, P. G. Quirk, S. A. White and J. B. Jackson, *J. Biol. Chem.*, 2006, **281**, 13345–13354.
76. G. I. van Boxel, P. G. Quirk, N. P. Cotton, S. A. White and J. B. Jackson, *Biochemistry*, 2003, **42**, 1217–1226.

77. O. C. Mather, A. Singh, G. I. van Boxel, S. A. White and J. B. Jackson, *Biochemistry*, 2004, **43**, 10952–10964.
78. G. S. Prasad, V. Sridhar, M. Yamaguchi, Y. Hatefi and C. D. Stout, *Nat. Struct. Biol.*, 1999, **6**, 1126–1131.
79. S. A. White, S. J. Peake, S. McSweeney, G. Leonard, N. P. Cotton and J. B. Jackson, *Structure*, 2000, **8**, 1–12.
80. V. Sundaresan, M. Yamaguchi, J. Chartron and C. D. Stout, *Biochemistry*, 2003, **42**, 12143–12153.
81. M. Jeeves, K. J. Smith, P. G. Quirk, N. P. Cotton and J. B. Jackson, *Biochim. Biophys. Acta*, 2000, **1459**, 248–257.
82. P. G. Quirk, M. Jeeves, N. P. J. Cotton, J. K. Smith and B. J. Jackson, *FEBS Lett.*, 1999, **446**, 127–132.
83. C. Johansson, A. Bergkvist, O. Fjellström, J. Rydström and B. G. Karlsson, *FEBS Lett.*, 1999, **458**, 180–184.
84. A. Bergkvist, C. Johansson, T. Johansson, J. Rydström and B. G. Karlsson, *Biochemistry*, 2000, **39**, 12595–12605.
85. O. Fjellstrom, C. Johansson and J. Rydstrom, *Biochemistry*, 1997, **36**, 11331–11341.
86. S. J. Peake, J. D. Venning, N. P. Cotton and J. B. Jackson, *Biochim. Biophys. Acta*, 1999, **1413**, 81–91.
87. T. Olausson, T. Hultman, E. Holmberg, J. Rydstrom, S. Ahmad, N. A. Glavas and P. D. Bragg, *Biochemistry*, 1993, **32**, 13237–13244.
88. J. Meuller, X. Hu, C. Bunthof, T. Olausson and J. Rydstrom, *Bba-Bioenergetics*, 1996, **1273**, 191–194.
89. X. Hu, J. Zhang, O. Fjellstrom, T. Bizouarn and J. Rydstrom, *Biochemistry*, 1999, **38**, 1652–1658.
90. O. Fjellstrom, M. Axelsson, T. Bizouarn, X. Hu, C. Johansson, J. Meuller and J. Rydstrom, *J. Biol. Chem.*, 1999, **274**, 6350–6359.
91. C. Johansson, A. Pedersen, B. G. Karlsson and J. Rydstrom, *Eur. J. Biochem.*, 2002, **269**, 4505–4515.
92. J. B. Jackson, S. A. White, P. G. Quirk and J. D. Venning, *Biochemistry*, 2002, **41**, 4173–4185.
93. A. Pedersen, T. Johansson, J. Rydstrom and B. Goran Karlsson, *Biochim. Biophys. Acta*, 1707, **2005**, 254–258.
94. N. P. Cotton, S. A. White, S. J. Peake, S. McSweeney and J. B. Jackson, *Structure*, 2001, **9**, 165–176.
95. V. Sundaresan, J. Chartron, M. Yamaguchi and C. D. Stout, *J. Mol. Biol.*, 2005, **346**, 617–629.
96. T. Bhakta, S. J. Whitehead, J. S. Snaith, T. R. Dafforn, J. Wilkie, S. Rajesh, S. A. White and J. B. Jackson, *Biochemistry*, 2007, **46**, 3304–3318.
97. P. S. Padayatti, J. H. Leung, P. Mahinthichaichan, A. Ishchenko, V. Cherezov, S. M. Soltis, J. B. Jackson, E. Tajkhorshid, C. D. Stout, R. B. Gennis and Q. Zhang, *Structure*, 2017, **25**, 1111–1119.e3.
98. P. D. Bragg and C. Hou, *Arch. Biochem. Biophys.*, 2000, **380**, 141–150.

99. E. Holmberg, T. Olausson, T. Hultman, J. Rydstrom, S. Ahmad, N. A. Glavas and P. D. Bragg, *Biochemistry*, 1994, **33**, 7691–7700.
100. R. C. W. Tong, N. A. Glavas and P. D. Bragg, *Biochim. Biophys. Acta*, 1991, **1080**, 19–28.
101. D. C. Phelps and Y. Hatefi, *Biochemistry*, 1985, **24**, 3503–3507.
102. D. C. Phelps and Y. Hatefi, *Biochemistry*, 1984, **23**, 4475–4480.
103. J. D. Venning, R. L. Grimley, T. Bizouarn, N. P. Cotton and J. B. Jackson, *J. Biol. Chem.*, 1997, **272**, 27535–27538.
104. C. P. Lee, N. Simard-Duquesne, L. Ernster and H. D. Hoberman, *Biochim. Biophys. Acta*, 1965, **105**, 397–409.
105. D. E. Griffiths and A. M. Roberton, *Biochim. Biophys. Acta*, 1966, **118**, 453–464.
106. R. L. Hanson, *J. Biol. Chem.*, 1979, **254**, 888–893.
107. T. M. Lever, T. Palmer, I. J. Cunningham, N. P. Cotton and J. B. Jackson, *Eur. J. Biochem.*, 1991, **197**, 247–255.
108. T. Bizouarn, R. L. Grimley, N. P. Cotton, S. N. Stilwell, M. Hutton and J. B. Jackson, *Biochim. Biophys. Acta*, 1995, **1229**, 49–58.
109. J. D. Venning and J. B. Jackson, *Biochem. J.*, 1999, **341**(Pt 2), 329–337.
110. J. D. Venning, T. Bizouarn, N. P. Cotton, P. G. Quirk and J. B. Jackson, *Eur. J. Biochem.*, 1998, **257**, 202–209.
111. T. J. Pinheiro, J. D. Venning and J. B. Jackson, *J. Biol. Chem.*, 2001, **276**, 44757–44761.
112. T. Palmer and J. B. Jackson, *FEBS Lett.*, 1990, **277**, 45–48.
113. D. Stock, C. Gibbons, I. Arechaga, A. G. W. Leslie and J. E. Walker, *Curr. Opin. Struct. Biol.*, 2000, **10**, 672–679.
114. J. B. Jackson, *J. Bioenerg. Biomemembr.*, 1991, **23**, 715–741.
115. P. D. Bragg and C. Hou, *Arch. Biochem. Biophys.*, 1999, **363**, 182–190.
116. T. Bizouarn, M. Althage, A. Pedersen, A. Tigerstrom, J. Karlsson, C. Johansson and J. Rydstrom, *Biochim. Biophys. Acta*, 2002, **1555**, 122–127.
117. K. Enander and J. Rydstrom, *J. Biol. Chem.*, 1982, **257**, 14760–14766.
118. W. Junge and N. Nelson, *Ann. Rev. Biochem.*, 2015, **84**, 631–657.
119. A. Zhou, A. Rohou, D. G. Schep, J. V. Bason, M. G. Montgomery, J. E. Walker, N. Grigorieff and J. L. Rubinstein, *eLife*, 2015, **4**, e10180.
120. J. E. Walker, *Biochem. Soc. Trans.*, 2013, **41**, 1–16.
121. T. Bizouarn, G. I. van Boxel, T. Bhakta and J. B. Jackson, *Biochim. Biophys. Acta*, 1708, **2005**, 404–410.
122. T. Bizouarn, C. Diggle and J. B. Jackson, *Eur. J. Biochem.*, 1996, **239**, 737–741.
123. R. L. Grimley, P. G. Quirk, T. Bizouarn, C. M. Thomas and J. B. Jackson, *Biochemistry*, 1997, **36**, 14762–14770.
124. C. Diggle, M. Hutton, G. R. Jones, C. M. Thomas and J. B. Jackson, *Eur. J. Biochem.*, 1995, **228**, 719–726.
125. M. Yamaguchi and Y. Hatefi, *J. Biol. Chem.*, 1993, **268**, 17871–17877.

CHAPTER 6

The Na^+-Translocating NADH: Ubiquinone Oxidoreductase (Na^+-NQR)

KATHERINE MEZIC,[a] MASATOSHI MURAI,[b] HIDETO MIYOSHI[b] AND BLANCA BARQUERA*[a]

[a] Department of Biological Sciences and Center for Biotechnology and Interdisciplinary Studies, Rensselaer Polytechnic Institute, Rm 2239 Biotech. 110 8th St., Troy, NY 12180, USA; [b] Division of Applied Life Sciences, Graduate School of Agriculture, Kyoto University, Sakyo-ku, Kyoto 606-8502, Japan

*Email: barqub@rpi.edu

6.1 Introduction

Among the known respiratory enzymes, only one conserves energy by pumping Na^+ instead of H^+ across the membrane; the Na^+-translocating NADH: ubiquinone oxidoreductase, Na^+-NQR.[1,2] This enzyme is found in many marine, soil, gut, and pathogenic bacteria, including *Vibrio cholerae*, *Pseudomonas aeruginosa*, *Neisseria gonohorrea*, *Yersinia pestis* and *Bacteroides fragilis*.[3–5] For many of these bacteria, maintaining a low internal Na^+ concentration is critical to their survival in high saline environments. Na^+-NQR actively pumps Na^+ from the cytoplasm, across the membrane into the periplasm, against its concentration gradient. In addition to removing Na^+ from the cell, the key function of Na^+-NQR is to create an electrochemical gradient that the cell can use for a variety of energy requiring functions, such as import of nutrients,

Chemical Biology No. 5
Mechanisms of Primary Energy Transduction in Biology
Edited by Mårten Wikström

Published by the Royal Society of Chemistry, www.rsc.org

extrusion of toxins and antibiotics, as well as motility, allowing them to adapt to a great variety of environments.[4–8] Recently, inactivation of Na^+-NQR has been shown to cause a decrease in overall production of the cholera toxin in *Vibrio cholerae*, underlining the link between toxicity and basic cellular functions, which deserves to be studied in greater detail.[9]

In order to transport Na^+ across the membrane, Na^+-NQR utilizes the energy released by the redox reaction between NADH and ubiquinone-8 (quinone).[1,10–12] This is the first step of the electron transport chain of aerobic respiration, the same redox reaction catalyzed by Complex I or Nuo. Complex I is a proton pump, and besides the shared redox reaction there are no other known similarities in structure or mechanism between these two respiratory enzymes.[13,14]

6.2 The Electron Pathway

In Na^+-NQR, the electrons move from NADH to quinone along a linear pathway consisting of five redox cofactors distributed across at least three of the enzyme's six subunits, NqrA-F, as shown in Figure 6.1.[2,15–18] The electron transport pathway begins in NqrF, a subunit with a large soluble domain on the cytoplasmic side of the membrane anchored by one transmembrane helix (Figure 6.2A). NADH binds at a conserved site in NqrF and electrons

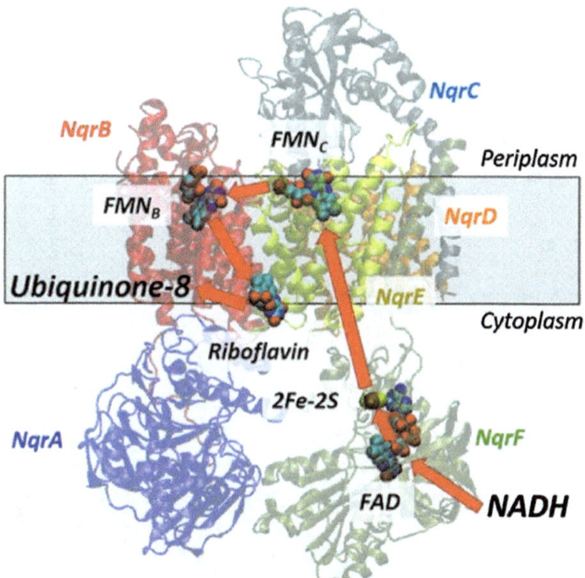

Figure 6.1 Electron transport Pathway of Na^+-NQR. The sequence of the electron transfer reactions is shown superimposed on the crystallographic model of Na^+-NQR. The pathway is depicted by red arrows connecting the substrate NADH to the cofactors in the order or the reaction: NADH$\rightarrow$ FAD$\rightarrow$2Fe–2S center$\rightarrow$FMN$_C$$\rightarrow$FMN$_B$$\rightarrow$riboflavin to the final electron acceptor ubiquinone-8.

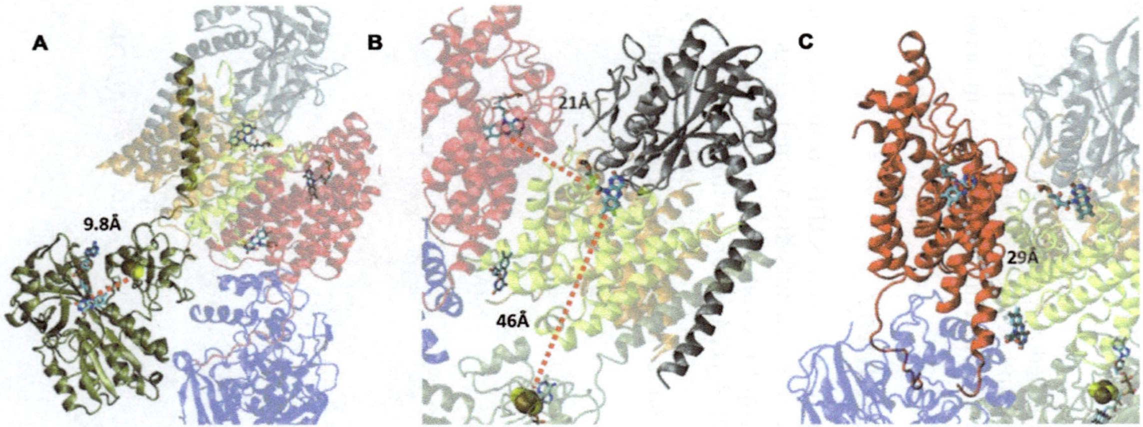

Figure 6.2 Close-up views of subunits NqrF, NqrB and NqrC of Na^+-NQR based on PDB file 4P6V. **(A)** NqrF (dark green); the 2Fe–2S center is shown as brown (Fe) and yellow (S) spheres, while the FAD molecule is shown in cyan (C), red (O), and blue (N). The shortest path between FAD and the 2Fe–2S center is 9.8 Å (red dashed line). **(B)** NqrC (dark grey) showing FMN_C (same color scheme as FAD). The distance between 2Fe–2S center and FMN_C is 46 Å; the distance between FMN_C and FMN_B is 21 Å. **(C)** NqrB (bold red) showing FMN_B (same color scheme as FAD). The shortest path between FMN_B and Riboflavin is 29 Å (grey dashed line).

enter the enzyme through a non-covalently bound FAD cofactor, also in NqrF. This FAD is attached at a well-defined motif in the cytosolic domain, first identified by Rich *et al.*,[19] and subsequently confirmed by point mutation studies.[15] FAD accepts two electrons from NADH and binds two protons, but with sufficiently fast measurement, a one-electron intermediate consisting of a neutral flavosemiquinone, can be resolved.[10]

Also, located in the cytosolic domain of NqrF is the 2Fe–2S center, a one-electron carrier, bound to a well-defined binding motif, also originally identified by Rich *et al.*,[19] and confirmed through mutagenesis studies together with Electron Paramagnetic Resonance spectroscopy.[15,20] The 2Fe–2S center accepts electrons from the FAD cofactor, located approximately 10 angstroms away, according to the crystallographic model, recently published by Steuber *et al.*[21] This distance is well within range to be consistent with the electron transfer, which occurs on the millisecond time scale.[17,22] Neither of the first two electron transfer steps—reduction of FAD and then the 2Fe–2S center, is rate limiting for catalytic turnover, and neither step is $[Na^+]$ dependent.[11,23,24] The first Na^+ dependent step, and likely the rate limiting step for turnover as determined by stopped flow kinetics, is the electron transfer from the 2Fe–2S center to an FMN cofactor located in NqrC[25] (Figure 6.2B). NqrC is anchored to the membrane by a single transmembrane helix, but in contrast to NqrF, its major soluble domain is located in the periplasm (according to the crystallographic model). FMN_C is covalently bound in a region of the soluble domain close to the membrane. FMN_C is bound to a conserved threonine residue by a phosphoester bond—a very unusual linkage. Since the attachment does not involve the isoalloxazine ring, it allows for maximum flexibility.[2,16] This type of covalent FMN binding is only known to occur in three enzymes: in Na^+-NQR, in the NqrC and NqrB subunits, in RNF, the closest related enzyme to Na^+-NQR, where it is found in the corresponding RnfD and RnfG subunits, and in NosR, an essential enzyme for the reduction of nitrous oxide.[2,26,27] According to the crystallographic model, FMN_C is located over 40 angstroms away from the 2Fe–2S center, twice the distance that would be consistent with effective electron transfer.[28] However, since this is both the rate limiting and Na^+ dependent step, this electron transfer is likely to be associated with conformational changes, that not only bring the 2Fe–2S center and FMN_C closer together, but also facilitate Na^+ uptake.

From FMN_C electrons are transferred to a second FMN cofactor, located in NqrB, and thus known as FMN_B (Figure 6.2C). NqrB has a substantial structure within the membrane with at least 10 transmembrane helices, the most of any subunit in Na^+-NQR. FMN_B is located on the periplasmic side of the membrane close to the ends of several of the transmembrane helices. As described above, like FMN_C, FMN_B is covalently bound through a conserved threonine residue. During reduction of the enzyme, both FMN_C and FMN_B are converted from flavoquinones to anionic flavo-semiquinones, which have characteristic visible spectra as well as individually distinct EPR spectra.[20] Most likely FMN_C is then subsequently reduced to the flavo-hydroquinone form. These two FMN's are located just over 20 angstroms apart, which would place them just outside of the range predicted from the electron transfer rate, but, given the resolution of the structure and the flexible nature of their linkage, these cofactors could potentially approach each other closely enough without the need for large conformational changes.

From FMN_B, electrons are transferred to the last, and possibly the most unusual cofactor found in Na^+-NQR, riboflavin.[25] Riboflavin, or vitamin B2, usually functions as the precursor to the FMN and FAD cofactors, as illustrated in Figure 6.3, but this is the only known instance in which riboflavin itself is a *bona fide* redox cofactor. The riboflavin cofactor was first discovered through HPLC analysis of the small molecules that are released when the Na^+-NQR complex becomes unfolded in denaturing agents.[29] Subsequent studies, involving the individual removal of other flavin cofactors, showed that in the oxidized (as prepared) form of Na^+-NQR, the riboflavin cofactor is a stable neutral flavo-semiquinone, which gives rise to a characteristic EPR signal; upon reduction, riboflavin is converted to a flavo-hydroquinone which is EPR-silent.[18]

The recently published crystallographic model localizes the riboflavin cofactor on the interface between NqrB and NqrE.[21] In this structure, only part of the isoalloxazine ring is contained within the putative binding pocket and the tail is completely free. It is possible that the location of this cofactor may be revised as higher resolution and more complete structures become available. The currently proposed location is nearly 30 angstroms from FMN_B, which would put it out of range for electron transfer given

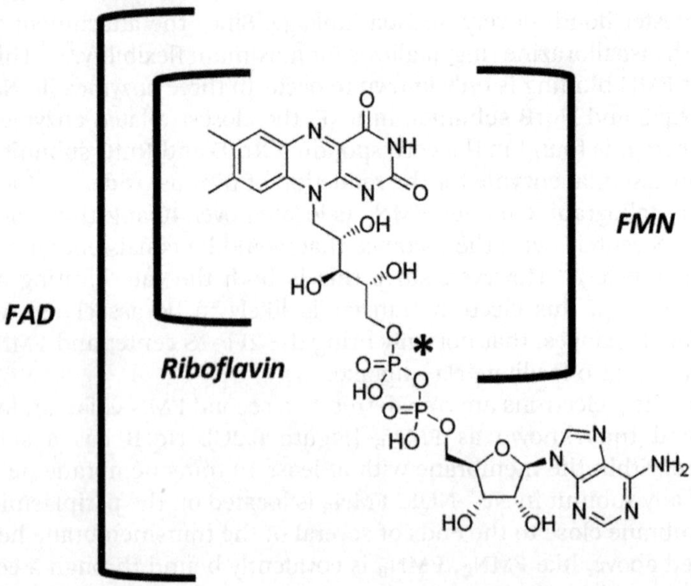

Figure 6.3 Chemical structures of riboflavin, FMN, and FAD. All structures include the isoalloxazine ring. The addition of phosphate groups at the 5′ hydroxyl side chain of riboflavin produces FMN and the addition of an ADP group produces FAD. *Denotes the phosphate at which FMN_B and FMN_C are covalently attached to their respective subunits.

experimentally determined rates. However, the electron transfer from FMN_B to riboflavin is the step that is linked to generation of membrane potential, and is likely associated with transport of Na^+ across the membrane and its release to the periplasm.[24] It is possible that the coupling between this redox step and Na^+ transport involves a conformational change. One result of this motion could be to bring the riboflavin close enough to FMN_B for electron transfer to occur. It is interesting to note that the uptake and release of Na^+ are linked to two separate redox steps that do not share any common cofactor. This suggests to us that the putative conformational changes involved in Na^+ translocation could take place in separate steps, as different cofactors are reduced. This could have important implications for the mechanism of Na^+ translocation, discussed later.

In the final redox step, Riboflavin donates its electrons to the electron accepting substrate, ubiquinone-8. The work of Zhou *et al.*[30] showed that catalytic turnover of Na^+-NQR does not generate a pH gradient; for this to be the case, since the NADH binding site is located in the cytoplasm, the quinone binding site must also be in protonic contact with the cytoplasmic side.[30] Hayashi and Unemoto originally proposed that ubiquinone binds near NqrB-G140, as a spontaneous mutation at this location conferred resistance to koromicin, an inhibitor of Na^+-NQR that is non-competitive with respect to quinone.[31] Further characterization of mutations at NqrB-G140 as well as the adjacent NqrB-G141 showed that these residues are critical for the binding and reaction of ubiquinone.[32,33] Tuz *et al.*[34] also characterized mutations of nearby aromatic residues and showed that NqrB-F211 and NqrB-F213 are also essential for the binding of quinone. However, Casutt *et al.*[35] proposed that the ubiquinone binds to NqrA, with evidence from both photoactivable quinone derivatives and saturation-transfer difference NMR studies.[35] They also found, using surface plasmon resonance, that the steady state binding behavior is best fit by a two-binding-site model, suggesting that there could be quinone binding sites at both locations. Nedielkov *et al.* have also confirmed the binding of two quinone analogs to NqrA, by observing the quenching of tryptophan fluorescence.[36,37]

A recent photoaffinity labeling study has considerably clarified the character of quinone binding to Na^+-NQR. The study confirms the presence of a binding site in NqrA but also shows that there is a binding site in NqrB. The latter site is located close to the point where a long hydrophilic stretch at the N-terminus connects to a transmembrane helix. Aurachin-type inhibitors and presumably korormicin bind at the site, causing inhibition of electron transfer from riboflavin to ubiquinone. Three residues located in transmembrane helix II of NqrB, Gly140, Gly 141 and Glu144, all appear to be involved in this inhibition since mutations at these sites can lead to resistant to aurachin and korormicin.[38] It should be noted that this ubiquinone location creates an unusual pathway for the electrons, in which they have to cross the membrane twice as they travel through the enzyme from NADH to ubiquinone, once, in order to reach FMN_C and FMN_B on the periplasmic side from the cytosolic NADH binding site, and again to return to quinone on the cytosolic side. The first crossing of the membrane would be downhill in terms of the electrical gradient, and the transport of a negative charge could assist with the translocation of the positively charged Na^+ across the membrane. When the electrons cross the membrane they would be moving up-hill, and this would contribute to the gradient and conserve energy. This presence of these two steps could be important for the overall function of the enzyme.

6.3 A Possible Sixth Cofactor

The recent publication of a crystallographic model of Na^+-NQR not only reveals distances between established redox cofactors, but also suggests the presence of a sixth cofactor in Na^+-NQR.[21] The authors propose that there is a single Fe bound by four cysteine ligands from NqrD and NqrE that facilitates electron transfer between the 2Fe–2S center and FMN_C. The location of this Fe is depicted in Figure 6.4. However, the existence of this putative cofactor is still under debate. If this cofactor exists, Na^+-NQR would contain three equivalents of iron. However, the best published measurements on the iron content of the enzyme by mass spectrometry are unable to distinguish between two or three equivalents of Fe.[39] A recent paper by Belevich *et al.*,[40] using an ultra-fast microfluidic stopped-flow instrument to determine rate constants of the electron transfer reactions in Na^+-NQR, was able to visualize the reduction of this Fe as a component of the kinetic phase including both the full reduction of the 2Fe–2S center and the disappearance of the neutral FAD radical. This kinetic phase, containing the Fe, is the first

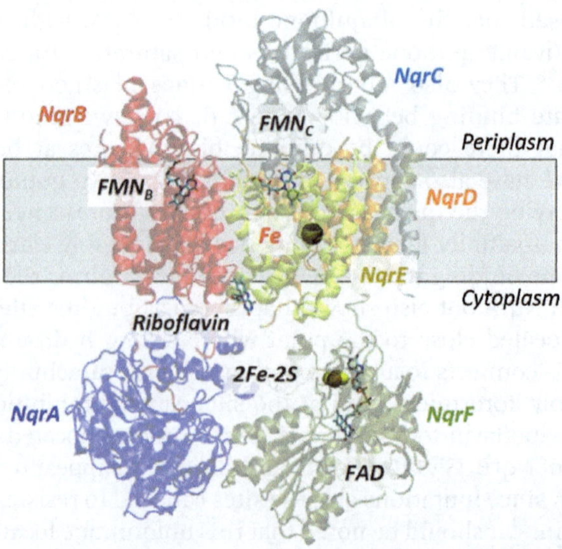

Figure 6.4 Location of the putative single Fe cofactor. The single Fe cofactor is proposed to be located between NqrD and NqrE and coordinated by four cysteines. The black sphere, representing the Fe, is located about 40 Å from the 2Fe–2S center and 8 Å from FMN_C. This view was created using PDB file 4P6V.

Na^+ dependent phase. Given its location in the crystallographic model, Belevich *et al.*[40] suggested that the reduction of this Fe is the step which is coupled to the binding of Na^+ to the enzyme. However, it is possible that Fe may be more of an electron facilitator than a *bona fide* cofactor, as this kinetic phase containing the Fe becomes unresolvable from the subsequent kinetic phase when Na^+ is added. There is work currently being done to resolve this question. Interestingly, even if the Fe cofactor is present, it only shortens the electron transfer distance by 8 angstroms out of the 40+ angstroms between the 2Fe–2S center and FMN_C, which still suggests that conformational changes are needed to bring the cofactors close enough to achieve the observed electron transfer rates.

6.4 Defining Na^+ Translocation Through Acidic Residues

The coupling between the electron transfer mechanism of Na^+-NQR and Na^+ translocation across the membrane is not yet defined. As mentioned in the Electron Pathway section, when stopped-flow kinetic measurements resolved individual steps in the redox reaction, electron transfer from the 2Fe–2S center to FMN_C was determined to be the first Na^+-dependent step.[25] However, still very little is known about the actual mechanism of Na^+ translocation in Na^+-NQR.

The efforts to identify the machinery of Na^+ translocation of Na^+-NQR began while the only structural information available was topological models of the Na^+-NQR subunits. These models were developed using computer algorithms and confirmed by means of reporter-group fusion experiments in which either green fluorescent protein (GFP) or bacterial alkaline phosphatase (PhoA) was genetically fused to the C-terminus of one of the Na^+-NQR subunits, which determined cytoplasmic and periplasmic localization, respectively. The topological models of the six subunits of Na^+-NQR guided the localization of the cofactors in the enzyme with respect to the membrane and identified conserved acidic residues located in the transmembrane helices.[41]

From studies on other transport enzymes it is known that Na^+ binding sites are typically made up of six oxygen ligands, in an octahedral arrangement,[42–44] as shown in Figure 6.5. The acidic residues, Glu and Asp, are the most commonly found ligands for Na^+, since they have two R-group oxygens and a backbone oxygen atom that can all potentially serve as ligands, as well as a negative charge that can stabilize the positively charged Na^+. Thus, Glu and Asp are the most likely candidates to be part of the Na^+ translocation machinery. In order to facilitate Na^+ translocation, residues need to be located inside of the membrane, where they can stabilize the charged Na^+ ion within the hydrophobic milieu. In many secondary Na^+ transporters and channels, and in the Na^+-ATPase, there is a clear path of acidic residues that provide a series of stepping-stones where Na^+ can bind as it passes through the membrane.[45,46] In Na^+-NQR, there are 17 conserved acidic residues in the transmembrane helices of Na^+-NQR but, in contrast to most other Na^+ transporters,

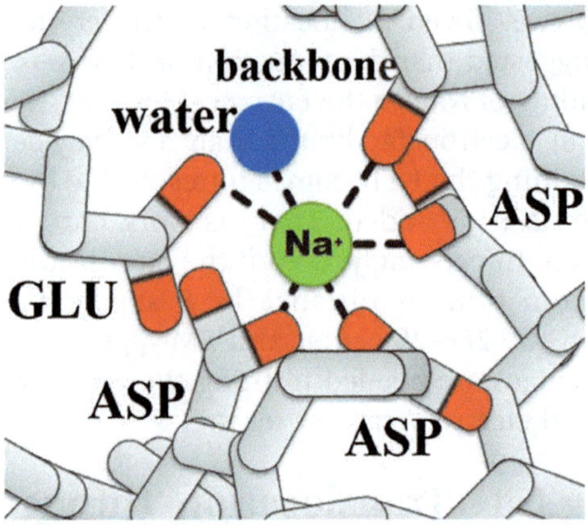

Figure 6.5 Scheme illustrating the optimal binding site for Na^+. The consensus from other transport enzymes is that most Na^+ binding sites have a six-oxygen-ligand coordination. The most common Na^+ coordinating residues are Glu and Asp, since they both have three potential oxygen ligands (shown in red): two R-group oxygens and a backbone oxygen atom. Backbone oxygens from other residues as well as water molecules in the structure (shown in blue) can also coordinate the Na^+ atom.

all of these residues are located at the ends of helices, with no clear path across the membrane (Figure 6.6). This is consistent with other evidence suggesting that Na^+-NQR has a more complex ion translocation mechanism.

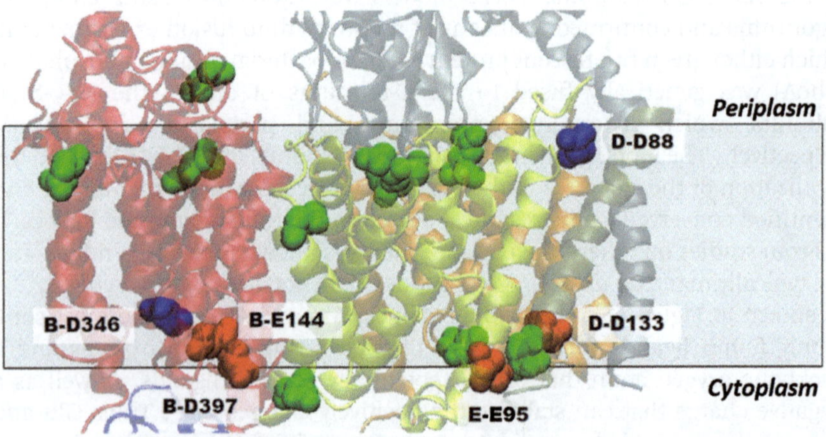

Figure 6.6 Conserved Acidic Residues Located in the transmembrane helices of Na^+-NQR. All of the conserved acidic residues located within the transmembrane helices of the enzyme are depicted by spheres in either green, red or blue (with the exception of NqrE-E28 which is not resolved in the crystal structure.) The residues in green are those which do not have an apparent role in Na^+. The residues in red, NqrB-NqrE144, NqrB-D397, NqrD-D133 and NqrE-E95, are involved in Na^+ uptake, while those in blue, NqrB-E346 and NqrD-D88, are involved in Na^+ translocation or release.

In order to clarify the role of these 17 conserved acidic residues in the mechanism of Na^+-NQR, mutants were constructed in which each was individually replaced with a smaller, nonpolar residue, alanine or leucine. This substitution eliminates both the charge and polarity of the side chain, making it a poor Na^+ ligand. Subsequent activity measurements found that 7 of the 17 residues are critical for Na^+ transport in Na^+-NQR. Four of the acidic residues, NqrB-D397, NqrB-E144, NqrD-D133, and NqrE-E95, were judged to be linked to Na^+ uptake, on the basis of a decrease in Na^+ affinity and a corresponding decrease in ubiquinone reductase activity upon substitution. Three acidic residues, NqrB-E28, NqrB-E346, and NqrD-D88, were linked to Na^+ release or translocation, on the basis of decreased Na^+ sensitivity without a significant change in Na^+ affinity.[47]

6.5 Defining Na^+ Binding Sites

Originally Na^+-NQR was believed to be selective exclusively for Na^+.[48] Since the isolation and purification of the enzyme, it has been shown that Na^+-NQR can translocate Li^+ across the membrane with a similar affinity to Na^+, but Li^+ stimulates the ubiquinone reductase activity to a lesser degree than Na^+.[49,50] This was not completely unexpected as Li^+ with its smaller ionic radius has been shown to be utilized in other Na^+ binding proteins.[51] When ^{23}Na NMR measurements indicated that Na^+-NQR had more than one binding site[50,52] the competition between Na^+ and Li^+ was utilized to define the cooperativity of the binding sites in steady-state kinetics measurements. These studies demonstrated that Na^+-NQR has at least two catalytically relevant Na^+ binding sites that show positive cooperativity, meaning that the binding of the first Na^+ increases the affinity for the second Na^+.[50] Equilibrium ^{22}Na binding experiments have indicated that there are three Na^+ binding sites, but they do not reveal the same cooperativity observed in the kinetic measurements.[50] One possible explanation for this apparent discrepancy is that the cooperativity observed could arise if conformational changes linked to the redox reaction occur when they are triggered by Na^+ binding, leading to a change in Na^+ affinity at the second site.

6.6 Discovery of the First Na^+ Binding Site

When NqrB-D397 was replaced by alanine in the initial acidic residue screening, the resulting mutant was found to be almost completely insensitive to Na^+.[53] In the wild-type enzyme, the rate of ubiquinone reduction increases approximately eight-fold when 100 mM Na^+ is introduced, but in the NqrB-D397A mutant this Na^+ sensitivity disappears almost completely.[53] The $K_{m(app)}$, an inverse measure of Na^+ affinity of the enzyme, is at least two orders of magnitude greater for the mutant than the wild-type $K_{m(app)}$, significantly greater than any other mutant in that study.[53] It appeared that NqrB-D397A was showing unsaturable behavior for Na^+, in contrast to the saturable, Michaelis–Menten behavior found in the wild-type enzyme.

Furthermore, the Na^+-dependence of enzyme activity shows no curvature over the accessible Na^+ concentration range, indicating either that the binding is extremely weak, or that saturation (Michaelis–Menten) kinetics are not applicable. These dramatic changes suggest that the NqrB-D397 residue either plays a role in the direct binding of Na^+ or is a part of the gateway to the binding site.

In order to define which properties of the NqrB-D397 acidic group were most important for function, a subsequent study constructed and investigated the properties of a series of mutants in which NqrB-D397 was replaced by different amino acids in order to vary the size, charge or the number of potential oxygen ligand atoms at that site.[54] For this work, NqrB-D397 was replaced by: glutamic acid, which is larger but still a carboxylic acid; asparagine, where the carboxylic acid is substituted by an amide group of the same general size; lysine, a basic amino acid; serine, a smaller polar residue with one fewer oxygen ligand; and cysteine, which is similar to aspartate in size and polarity, but has no oxygens in its side chain (Figure 6.7). In all of the mutants, the steady-state kinetics of quinone reduction is significantly hindered. It was concluded that both the size and the charge of NqrB-D397 are important for the function of this residue.

Subsequent studies suggest more specifically that NqrB-D397 is involved in Na^+ uptake and possibly binding. Stopped flow kinetics data revealed that mutating NqrB-D397 to an uncharged alanine or positively charged lysine slowed the rate of the first Na^+ dependent electron transfer step and lowered Na^+ affinity.[47,53] When Fourier transform infrared spectroscopy (FTIR) coupled to an electrochemical cell was used to directly study the changes in secondary structure of Na^+-NQR and the NqrB-D397E mutant in the fully oxidized and fully reduced state of the enzyme, clear changes in the Na^+ dependent nature of several spectral features corresponding to the

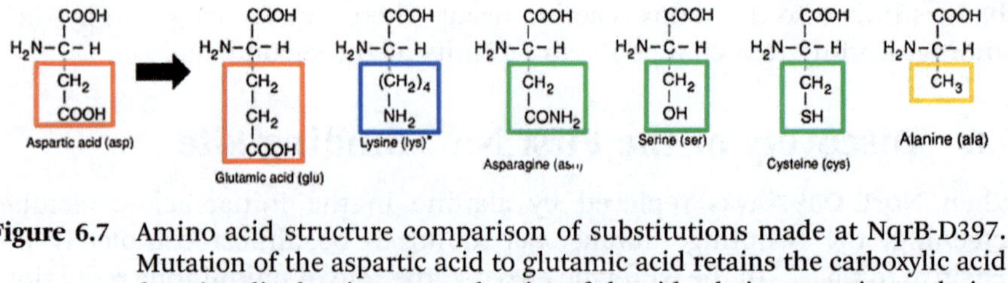

Figure 6.7 Amino acid structure comparison of substitutions made at NqrB-D397. Mutation of the aspartic acid to glutamic acid retains the carboxylic acid functionality but increases the size of the side chain; mutation to lysine changes the negative charge to a positive charge; mutation to asparagine replaces the carboxylic acid group with an acid amide, removing one potential oxygen ligand; mutation to serine removes one potential oxygen ligand; mutation to cysteine retains side and polarity but removes one potential oxygen ligand; mutation to alanine removes two potential oxygen ligands. The colors of the boxes enclosing the side chains indicate their properties: red, acidic; blue, basic; green, polar; yellow, aliphatic, non-polar.

carboxylic acid groups of acidic amino acids were observed.[55] This also suggests that NqrB-D397 is directly involved in Na^+ binding in Na^+-NQR.

Mutating NqrB-D397 to cysteine made the enzyme sensitive to iodoacetamide, a thiol-modifying agent. When iodoacetamide was incubated with this mutant there was a significant decrease in steady-state turnover activity. However, this inactivation by iodoacetamide could be prevented with the addition of Na^+. Since Na^+ protects this cysteine from labeling, it is likely that NqrB-D397 is part of a Na^+ binding site. Moreover, the saturation kinetics of NqrB-D397C differed from the Michaelis–Menten behavior found in the wild-type enzyme.[47] Through careful kinetic modeling, it was determined that this mutant has negative cooperativity between two independent Na^+ binding sites, with separate Na^+ pathways. It was concluded that in the wild-type enzyme, where both binding sites are functional, there is positive cooperativity; hence the binding sites are indistinguishable. Not only does this corroborate the idea that NqrB-D397 is part of a Na^+ binding site, it also suggests that NqrB-D397 only participates in one of two binding sites, which could be connected to separate translocation pathways. A more thorough characterization of NqrB-D397N revealed this mutant could be stimulated by K^+ without being able to pump K^+ across the membrane.[54] This is the only known instance where the redox activity of Na^+-NQR is uncoupled from the pumping activity. Therefore, NqrB-D397 may have a role in both ion selectivity as well as the coupling mechanism of the enzyme.

6.7 Localization of the First Na^+ Binding Site in the Crystal Structure

When the recent crystallographic model of Na^+-NQR was published,[21] the authors proposed a pathway for Na^+ passing the membrane through the middle of NqrB, as shown in Figure 6.8. The authors identify a series of six residues which they propose provide ligands for Na^+ binding through either their R group, NqrB-D346, NqrB-F342 and NqrB-F338, or their backbone carbonyl, NqrB-V161, NqrB-I164 and NqrB-L168. They propose that the two phenylalanines create a constriction point in the pathway, which opens upon the reduction of the riboflavin cofactor as helixes are shifted.[21] This is consistent with previous kinetic data showing that this step in the catalytic cycle is linked to generation of membrane potential, and thus Na^+ translocation across the membrane, and likely its release to the periplasm.[25] However, the residues in this proposed pathway have yet to be investigated by mutation studies.

While there are currently no biochemical or kinetic data with relevance to the pathway proposed by Steuber *et al.*,[21] the residues previously determined to be important in the mutational studies, including NqrB-D397, can be compared to this pathway in the structural model. NqrB-D397 is located about 15 angstroms away from the pathway proposed by Steuber *et al.*,[21] so it is possible that NqrB-D397 could be part of this pathway. However, in the crystallographic model, NqrB-D397 is located only 10 Angstroms away from

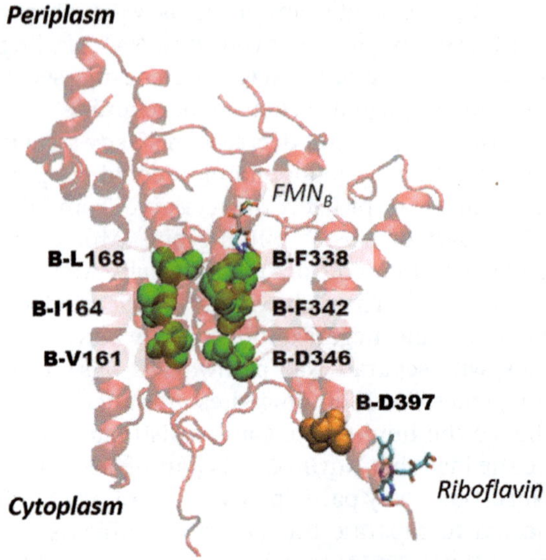

Figure 6.8 Na^+ pathway in NqrB proposed by Steuber *et al.* The six residues that make up the proposed Na^+ translocation pathway, NqrB-L161, NqrB-164, NqrB-168, NqrB-F338, NqrB-F342 and NqrB-D346, are depicted using green spheres. NqrB-D397, which has been shown by kinetic evidence to be part of a Na^+ uptake site,[56] located about 14 Å away from the start of the proposed pathway and is depicted using orange spheres. FMN_B and Riboflavin cofactors are shown using the same color scheme described in Figure 6.2.

the riboflavin cofactor, and it is possible that Na^+ binding to NqrB-D397 could be linked to the reduction of riboflavin. NqrB-D397 is positioned on the lower third of helix X on the cytoplasmic side of the membrane (Figure 6.9A). At the top of this helix, buried in the membrane near the cytoplasmic side, is the cofactor FMN_B, while at the bottom of this helix the riboflavin cofactor is bound, according to the crystallographic model. Upon the reduction of FMN_B, Steuber *et al.*[21] suggest that helix X may be pulled toward FMN_B through dipole interactions. This movement could help bring the riboflavin cofactor within electron transfer range of FMN_B. Moreover, we propose that since NqrB-D397 appears to be at a 'kink,' or bend in helix X, Na^+ binding at this location could potentially straighten the helix, moving riboflavin a few angstroms closer, aiding in the rate of electron transfer (Figure 6.9B). This would also explain how the substitution of NqrB-D397 to asparagine could allow for the weak binding of K^+, creating a situation where K^+ stimulates the electron transfer reactions, without transport across the membrane, as demonstrated by Shea *et al.*[54]

If this is true, there are two possibilities regarding the translocation of Na^+ in this subunit. On one hand, if the binding of Na^+ in the NqrB-D397 binding site coincides with the helical movement proposed by Steuber *et al.*,[21] the Na^+ that binds is not the Na^+ being translocated. Since Na^+-NQR

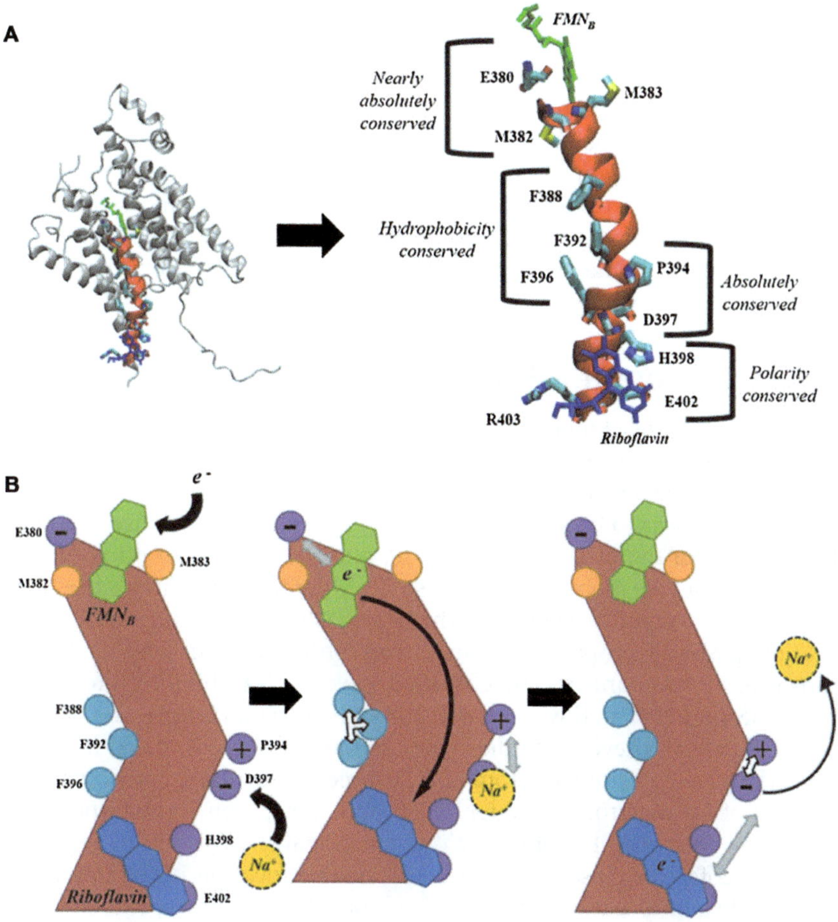

Figure 6.9 Our proposed Na^+ pathway through NqrB. **(A)** Crystallographic model of NqrB. In this isolated structure of NqrB (grey), helix X is highlighted in red, with Riboflavin shown in blue and FMN_B shown in green. Important residues of helix 10 are shown in cyan (C), red (O), and blue (N). Helix X is then enlarged, with the same color scheme. The type of conservation of each of the identified residues is indicated. **(B)** Schematic of helix bending in NqrB. Helix X is shown in red, with the Riboflavin in blue and FMN_B in green; the hydrophobic phenylalanines shown in teal, the methionines shown in yellow and the polar residues shown in purple, with charged residues indicated. As Na^+ binds at NqrB-D397 and an electron reduces the FMN_B, the helix shortens as shown in the second panel. This change is stabilized by the hydrophobic interactions of the phenylalanines, shown as white arrows, as well as repulsions, shown as grey arrows, of NqrB-E380 with the reduced FMN_B and NqrB-P394 with NqrB-D397. Once riboflavin is reduced, Na^+ is released, and the original helical arrangement is stabilized by the interaction of NqrB-P394 with NqrB-D397, as well as the repulsion of the reduced Riboflavin with NqrB-D397 and NqrB-E402.

has three Na^+ binding sites, but only two are catalytically relevant, this is plausible.[50] On the other hand, the reduction of Riboflavin may force the 'kink' back in helix X, releasing the Na^+ from the binding site, from where it can continue along the negatively charged vestibule into the proposed Na^+ pathway, with the constriction site open. In this scenario, there would need to be some mechanism to prevent Na^+ backflow, which could come in the form of helical rearrangements upon the reduction of riboflavin. At this point, these theories are speculation based on the crystallographic model, so more biochemical data are needed to clarify the physical links between Na^+ transport and the catalytic cycle.

6.8 Location of a Second Na^+ Binding Site

As discussed earlier, kinetic data support the idea that there are two Na^+ uptake/binding sites in Na^+-NQR, possibly leading to two separate translocation pathways.[47] Bogachev *et al.* determined the stoichiometry of Na^+-NQR from *Vibrio alginolyticus* to be about one Na^+/e^-, by observing the change in pH in the presence of the protonophore CCCP, and the absence of other ions that could penetrate the membrane besides H^+; under these conditions the Na^+/e^- and H^+/e^- are equal.[57] Therefore, the NqrB-D397 binding site and the translocation pathway proposed in NqrB may only represent half of the Na^+ translocation mechanism. There may be a second Na^+ binding site and possibly a second Na^+ pathway located through NqrD and NqrE, starting with NqrE-E95, which previous studies proposed as a residue involved in Na^+ uptake. The electron transfer from the Fe–S center to FMN_C must traverse a distance of 40 angstroms through NqrD and NqrE, even if the transfer is facilitated by the single Fe cofactor. Conformational changes in this region that bring the cofactors closer together could also be involved in the transport of Na^+. While a clear pathway for Na^+ has not yet been shown, two of the residues identified in the original screening of conserved acidic residues as likely to be involved in Na^+ uptake, NqrE-E95 and NqrD-D133, as well as a residue likely to be involved in Na^+ release, NqrD-D88, are located in this region.[53]

Furthermore, the arrangement of NqrD and NqrE is reminiscent of that in other transport proteins. These two subunits share 32% sequence identity and each contains six transmembrane helices, but they are inverted with respect to one another in their membrane orientation (Figure 6.10). This type of inverted structure is commonly found in the transport domains of major facilitator superfamily (MSF) transporters.[58,59] MSF transporters utilize electrochemical gradients to transport small molecules, such as sugars, and can be uniporters, symporters or antiporters. The transport domains of MSF transporters cycle between inward and outward facing conformations, where the binding sites are alternatively exposed to the two sides of the membrane. Based on the crystallographic model, Belevich *et al.*[40] propose that Na^+-NQR could use the same type of mechanism to transport Na^+. They propose that there is a conformation change that hinges around the recently proposed Fe

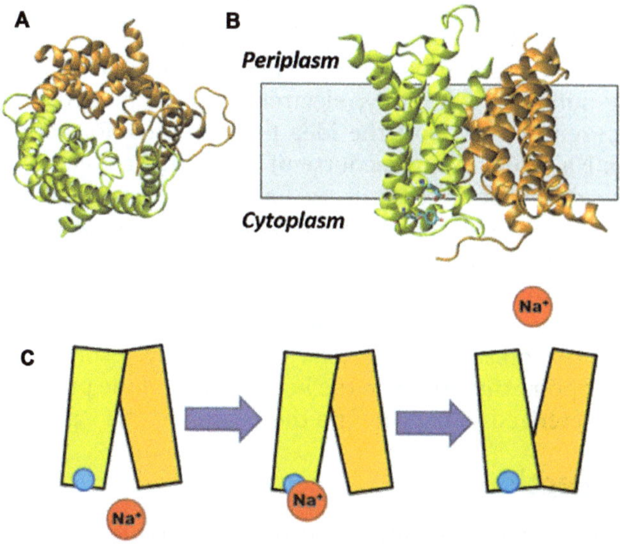

Figure 6.10 Proposed Na^+ pathway through NqrD and NqrE. **(A)** In this view from the cytoplasm, NqrD, in orange, and NqrE, in yellow, are shown without the other subunits. The symmetry between the subunits is clear, as well as an opening between the two. **(B)** In this view from the membrane, the proposed binding site is shown on the cytoplasmic side of NqrE in cyan (C), red (O), and blue (N). **(C)** A schematic of how the subunits may move with respect to each other to possibly allow for Na^+ translocation. A Na^+ (red) binds to the proposed binding site on the cytoplasmic side of NqrE (blue). This binding, along with the reduction of FMN_C and FMN_B may cause a conformational change in NqrD and NqrE that allow for the release of Na^+ into the periplasm.

cofactor, in the middle of NqrD and NqrE. More specifically, based on the location of the Fe in the crystallographic model, they suggest that Na^+ binding in the inward facing conformation is coupled to the reduction of the single Fe, the switch to the outward facing conformation and Na^+ release is thermodynamically linked to the reduction of FMN_C.[40] The coupling between the reduction of the single Fe cofactor and Na^+ uptake does not necessarily conflict with previous kinetic data, but it has been shown that the electron transfer from FMN_B to riboflavin is the first mechanistic step linked to generation of membrane potential, and thus Na^+ translocation, and likely its release.[25] Again, it is important to note that, as with the pathway proposed through NqrB, this pathway through NqrD and NqrE is based on examination of the crystallographic model and is not supported by biochemical or kinetic data. Therefore, much more biochemical data, including point mutations in the four cysteines of NqrD and NqrE, are needed to prove the existence of this pathway.

A comparison of the two proposed pathways shows that they would be linked to two quite different types of mechanisms of Na^+ transport. In the case of the pathway proposed through NqrB, Na^+ translocation would be facilitated by fairly localized movements of two phenylalanines when riboflavin becomes reduced. In the case of the pathway through NqrD/NqrE, Na^+ translocation would depend on much larger conformational changes of the surrounding subunits. Currently, electrochemically-induced FTIR spectroscopic measurements support the idea that large conformational changes take place in Na^+-NQR over the course of the catalytic cycle.[55] Interestingly, one pathway does not necessarily exclude the other, which means both pathways could be working in parallel to pump Na^+ across the membrane. This does not conflict with the determined stoichiometry of one Na^+ per electron, as NADH donates two electrons to the system.[57] Obtaining additional structural data on Na^+-NQR in both oxidized and reduced forms, as well as in the presence of various cations may resolve some of these questions, but point mutations in the key residues of the proposed pathways must be characterized to truly define these possible Na^+ pathways.

6.9 Perspectives

Much more information is needed in order to define the conformational changes that occur during the catalytic cycle of Na^+-NQR and how they are related to the redox reactions and to Na^+ transport. Structural data on the enzyme in different states may be able to provide a more complete framework within which to interpret data obtained from the functioning, moving enzyme by kinetic, electrochemical and spectroscopic experiments. Between the presence of riboflavin as a unique cofactor, the unusual covalent binding of FMN's, and the possibility of two independent Na^+ transport pathways as a part of the catalytic cycle, Na^+-NQR promises to provide greater insights to the diversity of transport and respiratory enzymes. Furthermore, not much is known about the physiological role and regulation of Na^+-NQR. With our current knowledge we now have enough information to begin to understand how and when this enzyme operates in many diverse bacteria.

Acknowledgements

The authors would like to thank Dr Joel Morgan for the critical reading of the manuscript. The authors were supported by the NSF MCB 1052234 and the NIGMS NIH Training Grant J11501.

References

1. M. Hayashi, Y. Nakayama and T. Unemoto, Recent progress in the Na+-translocating NADH-quinone reductase from the marine Vibrio alginolyticus, *Biochim. Biophys. Acta – Bioenerg.*, 2001, **1505**, 37–44.
2. B. Barquera, C. C. Hase and R. B. Gennis, Expression and mutagenesis of the NqrC subunit of the NQR respiratory Na+ pump from Vibrio cholerae with covalently attached FMN, *FEBS Lett.*, 2001, **492**, 45–49.
3. H. Tokuda and T. Unemoto, The Na(+)-motive respiratory chain of marine bacteria, *Microbiol. Sci.*, 1985, 2(65–66), 69–71.

4. S. Kojima, K. Yamamoto, I. Kawagishi and M. Homma, The polar flagellar motor of Vibrio cholerae is driven by an Na+ motive force, *J. Bacteriol.*, 1999, **181**, 1927–1930.
5. C. C. Häse, N. D. Fedorova, M. Y. Galperin and P. A. Dibrov, Sodium ion cycle in bacterial pathogens: evidence from cross-genome comparisons, *Microbiol. Mol. Biol. Rev.*, 2001, **65**, 353–370, table of contents.
6. K. Krishnan and M. J. Duncan, Role of Sodium in the RprY-Dependent Stress Response in Porphyromonas gingivalis, *PLoS One*, 2013, **8**, e63180.
7. S. G. Dashper *et al.*, Sodium ion-driven serine/threonine transport in Porphyromonas gingivalis, *J. Bacteriol.*, 2001, **183**, 4142–4148.
8. H. Fukuoka, T. Wada, S. Kojima, A. Ishijima and M. Homma, Sodium-dependent dynamic assembly of membrane complexes in sodium-driven flagellar motors, *Mol. Microbiol.*, 2009, **71**, 825–835.
9. Y. Minato, S. R. Fassio, R. L. Reddekopp and C. C. Häse, Inhibition of the sodium-translocating NADH-ubiquinone oxidoreductase [Na+-NQR] decreases cholera toxin production in Vibrio cholerae O1 at the late exponential growth phase, *Microb. Pathog.*, 2014, **66**, 36–39.
10. A. V. Bogachev, N. P. Belevich, Y. V. Bertsova and M. I. Verkhovsky, Primary steps of the Na+-translocating NADH:ubiquinone oxidoreductase catalytic cycle resolved by the ultrafast freeze-quench approach, *J. Biol. Chem.*, 2009, **284**, 5533–5538.
11. A. V. Bogachev, Y. V. Bertsova, B. Barquera and M. I. Verkhovsky, Sodium-dependent steps in the redox reactions of the Na+-motive NADH:quinone oxidoreductase from Vibrio harveyi, *Biochemistry*, 2001, **40**, 7318–7323.
12. B. Barquera, The sodium pumping NADH:quinone oxidoreductase (Na+-NQR), a unique redox-driven ion pump, *J. Bioenerg. Biomembr.*, 2014, **46**, 289–298.
13. S. Stolpe and T. Friedrich, The Escherichia coli NADH:ubiquinone oxidoreductase (complex I) is a primary proton pump but may be capable of secondary sodium antiport, *J. Biol. Chem.*, 2004, **279**, 18377–18383.
14. J. Steuber, W. Krebs and P. Dimroth, The Na+-translocating NADH: ubiquinone oxidoreductase from Vibrio alginolyticus–redox states of the FAD prosthetic group and mechanism of Ag+ inhibition, *Eur. J. Biochem.*, 1997, **249**, 770–776.
15. B. Barquera *et al.*, Mutagenesis study of the 2Fe-2S center and the FAD binding site of the Na(+)-translocating NADH:ubiquinone oxidoreductase from Vibrio cholerae, *Biochemistry*, 2004, **43**, 12322–12330.
16. M. Hayashi *et al.*, FMN is covalently attached to a threonine residue in the NqrB and NqrC subunits of Na+-translocating NADH-quinone reductase from Vibrio alginolyticus, *FEBS Lett.*, 2001, **488**, 5–8.
17. O. Juarez, J. E. Morgan and B. Barquera, The Electron Transfer Pathway of the Na+-pumping NADH: Quinone Oxidoreductase from Vibrio cholerae, *J. Biol. Chem.*, 2009, **284**, 8963–8972.
18. O. Juárez, M. J. Nilges, P. Gillespie, J. Cotton and B. Barquera, Riboflavin is an active redox cofactor in the Na+-pumping NADH: quinone

oxidoreductase (Na+-NQR) from Vibrio cholerae, *J. Biol. Chem.*, 2008, **283**, 33162–33167.

19. P. R. Rich, B. Meunier and E. B. Ward, Predicted structure and possible ionmotive mechanism of the sodium-linked NADH-ubiquinone oxidoreductase of Vibrio alginolyticus, *FEBS Lett.*, 1995, **375**, 5–10.
20. B. Barquera *et al.*, X- and W-band EPR and Q-band ENDOR studies of the flavin radical in the Na+-translocating NADH:quinone oxidoreductase from Vibrio cholerae, *J. Am. Chem. Soc.*, 2003, **125**, 265–275.
21. J. Steuber *et al.*, Structure of the V. cholerae Na+-pumping NADH:quinone oxidoreductase, *Nature*, 2014, **516**, 62–67.
22. A. V. Bogachev, Y. V. Bertsova, D. A. Bloch and M. I. Verkhovsky, Thermodynamic properties of the redox centers of Na(+)-translocating NADH:quinone oxidoreductase, *Biochemistry*, 2006, **45**, 3421–3428.
23. A. V. Bogachev, Y. V. Bertsova, E. K. Ruuge, M. Wikström and M. I. Verkhovsky, Kinetics of the spectral changes during reduction of the Na+-motive NADH:Quinone oxidoreductase from Vibrio harveyi, *Biochim. Biophys. Acta – Bioenerg.*, 2002, **1556**, 113–120.
24. O. Juárez, J. E. Morgan and B. Barquera, The Electron Transfer Pathway of the Na+-pumping NADH:Quinone Oxidoreductase from Vibrio cholerae, *J. Biol. Chem.*, 2009, **284**, 8963–8972.
25. O. Juárez, J. E. Morgan, M. J. Nilges and B. Barquera, Energy transducing redox steps of the Na+-pumping NADH:quinone oxidoreductase from Vibrio cholerae, *Proc. Natl. Acad. Sci. U. S. A.*, 2010, **107**, 12505–12510.
26. P. Wunsch and W. G. Zumft, Functional domains of NosR, a novel transmembrane iron-sulfur flavoprotein necessary for nitrous oxide respiration, *J. Bacteriol.*, 2005, **187**, 1992–2001.
27. J. Backiel *et al.*, Covalent binding of flavins to RnfG and RnfD in the Rnf complex from Vibrio cholerae, *Biochemistry*, 2008, **47**, 11273–11284.
28. M. Tao, G. Fritz and J. Steuber, The Na+-translocating NADH:quinone oxidoreductase (Na+-NQR) from Vibrio cholerae enhances insertion of FeS in overproduced NqrF subunit, *J. Inorg. Biochem.*, 2008, **102**, 1366–1372.
29. B. Barquera, W. Zhou, J. E. Morgan and R. B. Gennis, Riboflavin is a component of the Na+-pumping NADH-quinone oxidoreductase from Vibrio cholerae, *Proc. Natl. Acad. Sci. U. S. A.*, 2002, **99**, 10322–10324.
30. W. D. Zhou *et al.*, Sequencing and preliminary characterization of the Na+-translocating NADH:ubiquinone oxidoreductase from Vibrio harveyi, *Biochemistry*, 1999, **38**, 16246–16252.
31. M. Hayashi, N. Shibata, Y. Nakayama, K. Yoshikawa and T. Unemoto, Korormicin insensitivity in Vibrio alginolyticus is correlated with a single point mutation of Gly-140 in the NqrB subunit of the Na+-translocating NADH-quinone reductase, *Arch. Biochem. Biophys.*, 2002, **401**, 173–177.
32. O. Juárez *et al.*, The role of glycine residues 140 and 141 of subunit B in the functional ubiquinone binding site of the Na+-pumping NADH:quinone oxidoreductase from Vibrio cholerae, *J. Biol. Chem.*, 2012, **287**, 25678–25685.

33. M. Strickland *et al.*, The conformational changes induced by ubiquinone binding in the Na+-pumping NADH:Ubiquinone oxidoreductase (Na+-NQR) are kinetically controlled by conserved glycines 140 and 141 of the NqrB subunit, *J. Biol. Chem.*, 2014, **289**, 23723–23733.
34. K. Tuz, C. Li, X. Fang, D. A. Raba, P. Liang, D. D. Minh and O. Juarez, Identification of the catalytic ubiquinone-binding site of Vibrio cholerae sodium-dependent NADH dehydrogenase: a novel ubiquinone-binding motif, *J. Biol. Chem.*, 2017, 20009–20021.
35. M. S. Casutt *et al.*, Localization of Ubiquinone-8 in the Na+-pumping NADH:Quinone Oxidoreductase from Vibrio cholerae, *J. Biol. Chem.*, 2011, **286**, 40075–40082.
36. R. Nedielkov, W. Steffen, J. Steuber and H. M. Möller, NMR reveals double occupancy of quinone-type ligands in the catalytic quinone binding site of the Na+-translocating NADH:Quinone oxidoreductase from Vibrio cholerae, *J. Biol. Chem.*, 2013, **288**, 30597–30606.
37. M. S. Casutt, S. Wendelspiess, J. Steuber and G. Fritz, Crystallization of the Na+-translocating NADH:quinone oxidoreductase from Vibrio cholerae, *Acta Crystallogr. Sect. F Struct. Biol. Cryst. Commun*, 2010, **66**, 1677–1679.
38. T. Ito, *et al.*, Identification of the Binding site for ubiquinone and inhibitors in the Na+ pumping NADH-ubiquinone oxidoreductase from Vibrio cholerae by photoaffinity labeling, *Biochemistry*, 2017, **292**, 7727–7742.
39. M. S. Fadeeva, Y. V. Bertsova, M. I. Verkhovsky and A. V. Bogachev, Site-directed mutagenesis of conserved cysteine residues in NqrD and NqrE subunits of Na+-translocating NADH:quinone oxidoreductase, *Biochemistry*, 2008, **73**, 123–129.
40. N. P. Belevich, Y. V. Bertsova, M. L. Verkhovskaya, A. A. Baykov and A. V. Bogachev, Identification of the coupling step in Na+-translocating NADH:quinone oxidoreductase from real-time kinetics of electron transfer, *Biochim. Biophys. Acta – Bioenerg.*, 2016, **1857**, 141–149.
41. E. B. Duffy and B. Barquera, Membrane topology mapping of the Na+-pumping NADH: quinone oxidoreductase from Vibrio cholerae by PhoA-green fluorescent protein fusion analysis, *J. Bacteriol.*, 2006, **188**, 8343–8351.
42. Y. Kim, T. T. Nguyen and D. G. Churchill, *Met. Ions Life Sci.*, 2016, 1–10.
43. M. M. Harding, Geometry of metal-ligand interactions in proteins, *Acta Crystallogr. Sect. D Biol. Crystallogr.*, 2001, **57**, 401–411.
44. M. M. Harding, Metal-ligand geometry relevant to proteins and in proteins: Sodium and potassium, *Acta Crystallogr. Sect. D Biol. Crystallogr.*, 2002, **58**, 872–874.
45. A. Schneeberger and H. J. Apell, Ion selectivity of the cytoplasmic binding sites of the Na,K-ATPase: II. Competition of various cations, *J. Membr. Biol.*, 2001, **179**, 263–273.
46. D. V. Dibrova, M. Y. Galperin and A. Y. Mulkidjanian, Characterization of the N-ATPase, a distinct, laterally transferred Na+-translocating form of the bacterial F-type membrane ATPase, *Bioinformatics*, 2010, **26**, 1473–1476.

47. O. Juarez, M. E. Shea, G. Makhatadze and B. Barquera, Aspartic acid 397 in subunit B of the Na+-pumping NADH:quinone oxidoreductase from Vibrio cholerae forms part of a sodium-binding site, is involved in cation selectivity, and affects cation-binding site cooperativity, *J. Biol. Chem.*, 2011.
48. T. Unemoto and M. Hayashi, Na(+)-translocating NADH-quinone reductase of marine and halophilic bacteria, *J. Bioenerg. Biomembr.*, 1993, **25**, 385–391.
49. B. Barquera *et al.*, Purification and characterization of the recombinant Na+-translocating NADH: quinone oxidoreductase from vibrio cholerae, *Biochemistry*, 2002, **41**, 3781–3789.
50. O. Juárez, M. E. Shea, G. I. Makhatadze and B. Barquera, The role and specificity of the catalytic and regulatory cation-binding sites of the Na+-pumping NADH:quinone oxidoreductase from Vibrio cholerae, *J. Biol. Chem.*, 2011, **286**, 26383–26390.
51. E. Gouaux and R. Mackinnon, Principles of selective ion transport in channels and pumps, *Science*, 2005, **310**, 1461–1465.
52. A. V. Bogachev, Y. V. Bertsova, O. Aitio, P. Permi and M. I. Verkhovsky, Redox-dependent sodium binding by the Na(+)-translocating NADH:quinone oxidoreductase from Vibrio harveyi, *Biochemistry*, 2007, **46**, 10186–10191.
53. O. Juárez, K. Athearn, P. Gillespie and B. Barquera, Acid residues in the transmembrane helices of the Na+-pumping NADH:quinone oxidoreductase from Vibrio cholerae involved in sodium translocation, *Biochemistry*, 2009, **48**, 9516–9524.
54. M. E. Shea, K. G. Mezic, O. Juárez and B. Barquera, A mutation in Na+-NQR uncouples electron flow from Na+ translocation in the presence of K+, *Biochemistry*, 2015, **54**, 490–496.
55. Y. Neehaul, O. Juárez, B. Barquera and P. Hellwig, Infrared spectroscopic evidence of a redox-dependent conformational change involving ion binding residue NqrB-D397 in the Na(+)-pumping NADH:quinone oxidoreductase from Vibrio cholerae, *Biochemistry*, 2013, **52**, 3085–3093.
56. M. E. Shea, O. Juárez, J. Cho and B. Barquera, Aspartic Acid 397 in Subunit B of the Na+-pumping NADH: Quinone Oxidoreductase from *Vibrio cholerae* Forms Part of a Sodium-binding Site, Is Involved in Cation Selectivity, and Affects Cation-binding Site Cooperativity, *J. Biol. Chem.*, 2013, **288**, 31241–31249.
57. A. V. Bogachev, R. A. Murtazina, V. P. Skulachev and A. V. Bogachev, Murtazina RA, S. V. The Na+/e- stoichiometry of the Na+-motive NADH:quinone oxidoreductase in Vibrio alginolyticus, *FEBS Lett.*, 1997, **409**, 475–477.
58. D. Drew *et al.*, Rapid topology mapping of Escherichia coli inner-membrane proteins by prediction and PhoA/GFP fusion analysis, *Proc. Natl. Acad. Sci. U. S. A.*, 2002, **99**, 2690–2695.
59. S. Radestock and L. R. Forrest, The alternating-access mechanism of MFS transporters arises from inverted-topology repeats, *J. Mol. Biol.*, 2011, **407**, 698–715.

CHAPTER 7

The bc_1 *Complex: A Physicochemical Retrospective and an Atomistic Prospective*

ANTONY R. CROFTS,*[a,b] CHARLES A. WILSON,[b] STUART W. ROSE,[b] SERGEI A. DIKANOV[c] AND RODNEY L. BURTON[a,†]

[a] Department of Biochemistry, University of Illinois at Urbana-Champaign, Urbana, IL 61801, USA; [b] Center for Biophysics and Quantitative Biology, University of Illinois at Urbana-Champaign, Urbana, IL 61801, USA; [c] Department of Veterinary Clinical Medicine, University of Illinois at Urbana-Champaign, Urbana, IL 61801, USA
*Email: crofts@illinois.edu

7.1 Introduction

In a previous book in this series a dozen years ago,[1] one of us asked of the Q-cycle mechanism, what is there left to argue about? Fortunately, the answer has been "plenty". In this review, we look back at the insights that have illuminated our understanding since, and then forward to what new perspectives atomistic approaches can provide. Excitement about the future

[†]Current address: MSU-DOE Plant Research Laboratory, Michigan State University, 612 Wilson Road, East Lansing, MI, USA

Chemical Biology No. 5
Mechanisms of Primary Energy Transduction in Biology
Edited by Mårten Wikström

Published by the Royal Society of Chemistry, www.rsc.org

comes from the foretaste provided by a flurry of atomistic papers over the last few years,[2–10] several of them in a recent volume honoring the memory of Klaus Schulten, and his seminal contributions to computational approaches in this area.

In a 2004 personal perspective,[11] the senior author had reviewed the development of the modified Q-cycle, and we will take that as a starting point. Points of contention since have included: (i) the involvement of the dimeric structure in modifying the mechanism, which was at that time essentially monomeric; (ii) the related question of communication between monomers; (iii) the question of how the mechanism has evolved to minimize generation of reactive oxygen species (ROS); (iv) the related question of the occupancy of the semiquinone intermediate of the bifurcated reaction (SQ_o); and (v) the general question of how these all pan out at the atomistic level in understanding the mechanism in the structural context.

7.2 Current Paradigm: The Monomeric Q-cycle Mechanism of the bc_1 Complex

The bc_1 complex and related proteins are the central enzymes of all major energy transducing pathways of the biosphere.[12–14] In eukaryotes, the bc_1 complex functions as Complex III of the respiratory chain of mitochondria, the "power-houses" of the cell. We have studied the complex in photosynthetic bacteria derived from common ancestral forms, where the advantage of photoactivation allows facile measurement of kinetic parameters. The bacterial and mitochondrial complexes work through the same mechanism, but the bacterial system is smaller, simpler, and easier to manipulate or modify by molecular engineering. The complex in both mitochondria and bacteria is a homodimer, with a catalytic core of three subunits in each monomer, cyt b, cyt c_1, and the Rieske iron-sulfur protein (ISP). In some bacterial complexes, no other subunits are structurally defined, but in *Rb. sphaeroides* when grown photosynthetically, the isolated complex has an additional subunit (SU IV) of uncertain function. Mitochondrial complexes have up to 8, many also of uncertain function.[15–18]

The interest of the Q-cycle (Figure 7.1) is that it doubles the energy saved from each turnover compared to that if the ubiquinol (QH_2), generated either by substrate oxidation in respiratory systems, or in the photochemistry, directly reduced the cytochrome (cyt) c (or c_2) oxidized. Instead, QH_2 is oxidized in a bifurcated reaction at the ubiquinol oxidizing site (Q_o-site)[19–24] that directs one electron *via* a high-potential chain (2Fe2S cluster and heme c_1) to cyt c_2, leaving an intermediate semiquinone (SQ) at the Q_o-site (SQ_o). This is oxidized by a different, low-potential chain (hemes b_L and b_H), which delivers the electron to reduce ubiquinone (Q) at the quinone-reducing site (Q_i-site) on the other side of the membrane. Since this takes two electrons, the Q_o-site reaction must turnover twice, with the electron from the first stored as SQ_i, which is then reduced to QH_2

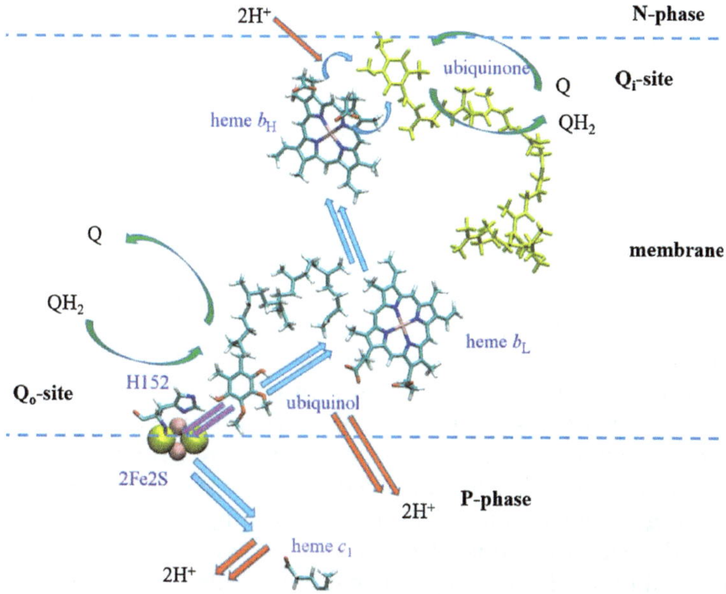

Figure 7.1 Consensus version of the Q-cycle. The catalytic core of a monomer of the *Rb. sphaeroides bc*$_1$ complex showing the main players. The reactions are represented by magenta arrows (H-transfers), green arrows (Q, QH_2 exchange at catalytic sites), cyan arrows (electron transfers) and red arrows (H^+ transfers). The prosthetic groups indicate locations of substrates in *ES*-complexes at the Q_o- and Q_i-sites (from an MD trajectory). See text for details.

on the second turnover. This QH_2 is recycled to replace one of those oxidized at the Q_o-site. Since protons are released on one side and taken up on the other, and electrons cross the insulating phase through the low-potential chain, the Q-cycle pumps $2H^+/QH_2$ oxidized across the membrane. The electrogenic process contributes to the electrical component of the proton gradient used to drive ATP synthesis. Coupling to the ΔpH component of the proton gradient is through release of 2 scalar H^+ from the bifurcated reaction to the P-phase, release of $2H^+$ to the P-phase and uptake of $2H^+$ at the Q_i-site from the N-phase, representing transport. The overall reaction is given below, together with the overall thermodynamic equation.

$$\begin{aligned}&QH_2 + 2\,\text{ferricyt}\,c_2 + 2H_N^+ \rightleftharpoons Q + 2\,\text{ferrocyt}\,c_2 + 2H_P^+ + 2H_{\text{scalar}}^+\\&\Delta G' = \Delta G'_{\text{scalar\_chem}} + \Delta G'_{\text{transport}} = \Sigma\nu_i\mu'_i + \Sigma\nu_i\bar{\mu}_i = -2F\Delta E' + 2\Delta\bar{\mu}_{H^+}\end{aligned} \tag{7.1}$$

At static head, approximated by the equilibrium condition, $\Delta G' \sim 0$, and $F\Delta E' \sim \Delta\bar{\mu}_{H^+}$, with the chemical driving force of the redox reaction balanced by the back-pressure of the proton gradient. Note that this simple solution masks a substantial complexity; the kinetic model,[14,25] which simulates the truncated turnover in the presence of antimycin, parses the $\Delta E'$ term into ~15 partial processes, and the $\Delta\bar{\mu}_{H^+}$ term also parses into separate processes, and to separate energy terms associated with electrical and chemical components of the proton gradient, $\Delta\bar{\mu}_{H^+} = F\Delta\psi - 2.303RT\Delta pH$.

7.3 Control and Gating

The bifurcated reaction at the Q_o-site is the key to understanding control of the Q-cycle. Minimizing the occupancy of SQ_o requires a compromise between two effects. As the first electron transfer is endergonic, the back reaction is more rapid than the forward one. The second electron transfer then must compete with the back reaction for the SQ_o. However, since rate depends on the occupancy, the more endergonic the first step, the higher the rate constant needed in the second. Two features of special interest are the role of ISP in determining rate (discussed in the context of the Marcus–Brønsted treatment below), and the importance of rapid removal of SQ_o in controlling bypass reactions.[13,14,23,26] The turnover of the Q_o-site is most easily assayed by measuring the rate of reduction of heme b_H, the acceptor for the second electron when the Q_i-site is blocked by antimycin. This gives a truncated Q-cycle in which the thermodynamics of the bifurcated reaction can be readily measured by sampling the poise of reactants in the two chains. Since within a chain the electron equilibrates rapidly, thermodynamic parameters for all components can be readily calculated from known $\Delta G^{o\prime}$ values by measurement of one of them.[25] For the first QH_2 oxidized,

$$\Delta G^{o\prime} = -F\sum_{i}^{z} E_i^{o\prime} = -F(E^{o\prime}_{ISP} + E^{o\prime}_{b_H} - 2E^{o\prime}_{QH_2})$$
$$E^{o\prime}_{cytb_H} \sim 40\,\text{mV}$$
$$E^{o\prime}_{cytb_L} \sim -90\,\text{mV} \tag{7.2}$$

which, with $E^{o\prime}_{cytb_H} \sim 40$ mV, gives $K_{eq} \sim 480$ at pH 7. For the second QH_2 oxidized, since heme b_H is now reduced (leaving heme b_L as acceptor), and with $E^{o\prime}_{cytb_L} \sim -90$ mV,

$$\Delta G^{o\prime} = -F(E^{o\prime}_{ISP} + E^{o\prime}_{b_L} - 2E^{o\prime}_{QH_2}) \tag{7.3}$$

which gives $K_{eq} \sim 3$. When the reactions are measured following flash-activation of *Rb. sphaeroides*, the extent of turnover is determined by relative stoichiometry of RC to bc_1 complex monomers, and the above difference in K_{eq}. The outcome is that, after a saturating flash, the work from the photochemistry is sufficient to drive the first QH_2 oxidation to completion and reduce heme b_H, but, since the most favorable acceptors in the two chains are by then consumed, there is not enough work left to significantly drive the second turnover. The second QH_2 oxidation can be driven by re-oxidizing the high potential chain on a second flash, to drive reduction of heme b_L. Understanding this separation of the partial processes by their different K_{eq} values was an important component of our earlier demonstration of the modified Q-cycle.[21,27] What mechanisms limit the occupancy of SQ_o? The primary mechanism in play is to make the first step rate-limiting and slow, and second step fast, so that SQ_o is removed as soon as it is formed. Our understanding of how this is achieved continues to evolve, but the last decade has seen significant progress.

Interest in control mechanisms was rekindled by Osyczka *et al.*,[28] who noticed a curious paradox. In kinetic models using rate constants estimated using the Moser–Dutton equation,[29,30] an important determinant was the distance for electron transfer. In a static model, the substrate for the second electron transfer, SQ_o, was restricted to the position of QH_2 in the enzyme-substrate (*ES*-) complex where it was generated, 11.5 Å from heme b_L, the acceptor for its electron in the forward reaction. However, SQ_o was also the substrate in a bypass reaction that came into play under backed-up conditions, in which SQ_o could act as acceptor of electrons from ferroheme b_L, with the same distances, and similar occupancies and driving forces. Consequently, similar rates would be expected, leading to massive decoupling from the proton-pumping function. Since this was not observed, the Q-cycle had to be "fixed" to explain the discrepancy. Suggested remedies were a reaction in which the two electron transfers were concerted, so that no SQ_o was generated, or, in two-step mechanisms, reactivity of the SQ_o was gated so that the forward reaction could occur only when both ISP and heme b_L were oxidized.

In considering mechanisms involving SQ_o, we could explain the slow first step through a Marcus–Brønsted treatment for the first electron transfer sketched out below, in which proton distribution along the H-bond would lower the rate constant (by up to 10^{-5}) expected from the Moser–Dutton approach.[13,31] However, because the first step is rate limiting, other partial processes are constrained to that rate, so that we can only use indirect evidence to estimate parameters. Using the early structures as a guide to electron transfer distances,[32–34] we had explored physicochemical properties of the bifurcated reaction in the context of the primary control function above,[13,35,36] with the aid of a simple computer simulation that calculated the poise of all reactants, and the driving forces relevant to derivation of rate constants from a Moser–Dutton–Marcus treatment.

In the antimycin-inhibited complex after excitation by a group of six closely spaced flashes (similar results are seen on continuous illumination), the two chains are strongly out of equilibrium, with the *b*-hemes both reduced, and the high-potential chain oxidized, with a difference in redox potential of $\Delta E' \sim 820$ mV between the chains. The occupancy of SQ_o had to be incorporated into the above framework to understand its mechanistic role. Victoria *et al.*[25] discussed this problem in an exploration of parameters for the equilibria of eqn 7.1 in wildtype and strains mutated at Glu-295 of the -PEWY- loop of the Q_o-site (conserved across the bacterial/mitochondrial divide); this included measurement of SQ_o occupancy under conditions in which we could directly measure the equilibria, including the occupancy of the ferriheme b_L state. Including SQ_o is straightforward, because the two SQ couples are in equilibrium with two different chains, so that:

$$E'_{SQ/QH_2} \sim E'_{\text{high potential chain}} \sim 500\ \text{mV}$$

and

$$E'_{Q/SQ} \sim E'_{\text{low potential chain}} \sim -320\ \text{mV}$$

$$E^{o\prime}_{SQ/QH_2} \sim 570\ \text{mV}$$

$$E^{o\prime}_{Q/SQ} \sim -390\ \text{mV}$$

and SQ_o is a reactant in both steps, so that $[SQ_o]$ must have the same value in determining the poise of the half-reactions. Then, measurement of the $[SQ_o]$ (an occupancy of 0.06) provided $E^{o\prime}_{SQ/QH_2} \sim 570$ mV, and $E^{o\prime}_{Q/SQ} \sim -390$ mV (see ref. 25 for details).

Hong *et al.*[35] had concluded that a concerted mechanism was not feasible, as Forquer *et al.*[37] have more recently confirmed. From the thermodynamic and kinetic characteristics then available, all plausible scenarios demanded a low occupancy, and the treatment summarized above put realistic quantities on this, but presented a challenge. The occupancy above was measured under strongly forcing conditions. Although the parameters show a very unstable SQ_o, a much lower occupancy would be expected under the maximal flux of the normal forward reaction. Hong *et al.* noted that with the 11.5 Å distance involved, estimations of rate at the low end of the plausible range would lead to too slow a second step.[35] We had earlier pointed out that if SQ_o could move in the site from the distal position to the volume occupied by myxothiazol or MOA-type inhibitors, more proximal to heme b_L,[33,38] the rate constant would be increased >1000-fold to allow operation at very low occupancies, and we have continued to favor this type of mechanism. Over the last decade, we have developed a kinetic model (currently with 15 partial processes) that accounts well for known features of the forward chemistry, incorporating complexities introduced by different PCET mechanisms for the two reactions, and movement of SQ_o in the site, as detailed in.[14,25] The question of the rate limiting step raises an interesting point. As the first step is highly endergonic, the rate measured actually reflects the removal of SQ_o needed to pull the reaction over. No matter how fast that is, the rate observed can't go faster than the limiting first step. However, it can get slower if the second step is inhibited, as in many mutant strains. Whenever the observed rate is inhibited so that it falls below the limit from the first electron transfer, it must always represent a substantial inhibition of the second step. By including these features, the kinetic model shows how, in normal flux, SQ_o can be rapidly removed to minimize deleterious bypass reactions without resorting to unrealistic assumptions. By slowing the second step, the model can simulate the behavior in mutant strains. For example, in the E295 mutants, the pH dependence seemed to show that the effects attributed to pK_{ox1} of ISP were lost, suggesting that the rate limit had shifted to the second electron transfer.

Although the published model includes a coulombic gating, we are aware that this does not adequately address all the gating needed under the scenario proposed by Osyczka *et al.*[28] A realistic explanation of how this might work has proved elusive. The two PCET processes open many possibilities for gating based on the involvement of domain movement in ISP, two spatially and temporally distinct transfers with very different properties, diffusion of SQ_o, and possible coulombic effects. However, the partial processes involved are not readily accessible to conventional measurement, and this is an incentive to development of novel methods, including atomistic computational approaches.

7.4 The Marcus–Brønsted Mechanism for First Electron Transfer of the Q-cycle

Our understanding of the role of ISP has depended on characterization of parameters in strains modified at key residues, allowing us to map out the partial processes, and define physicochemical parameters for these (*cf.* ref. 16, 23, 24, 26, 39, 40). Two properties of ISP are found to be important in determining the reaction profile. The redox potential, E'^{o}, and pK_{ox1} of H152 (cluster ligand), can be changed in mutant strains, both affecting the rate. The overall rate depends on pH, with an apparent $pK_{app} \sim 6.5$. This value changed in parallel with changes in pK_{ox1} in mutant strains.[40–42] The latter is measured in the isolated protein from the dependence of E'^{o} on pH, with a value of 7.6 in wildtype. The correlation between pK_{app} and pK_{ox1} (Figure 7.2, top) shows that the dissociated form of H152 is involved in the H-bond stabilizing the *ES*-complex, and the difference (~1.1 pH units) shows the strength of the bond.[13,40,42,43] The reaction proceeds from this complex through a proton-first-then-electron mechanism,[41,43] with a rate constant depending on both E'^{o}, and pK_{ox1}. A modified Marcus treatment (the Marcus–Brønsted equation) accounts neatly for the rate:

$$\log_{10}k = 13 - \frac{\beta}{2.303}(R-3.6) - \gamma\frac{\left(\Delta G^{o}_{ET} + \lambda_{ET}\right)^2}{\lambda_{ET}} - (pK_{QH_2} - pK_{app}) \qquad (7.4)$$

Since E'^{o} contributes to ΔG^{o}_{ET}, and pK_{ox1} determines pK_{app}, the equation predicts how k will vary as these parameters are changed in mutant strains, and the outcome is in agreement with experiment.[40] The Moser–Dutton

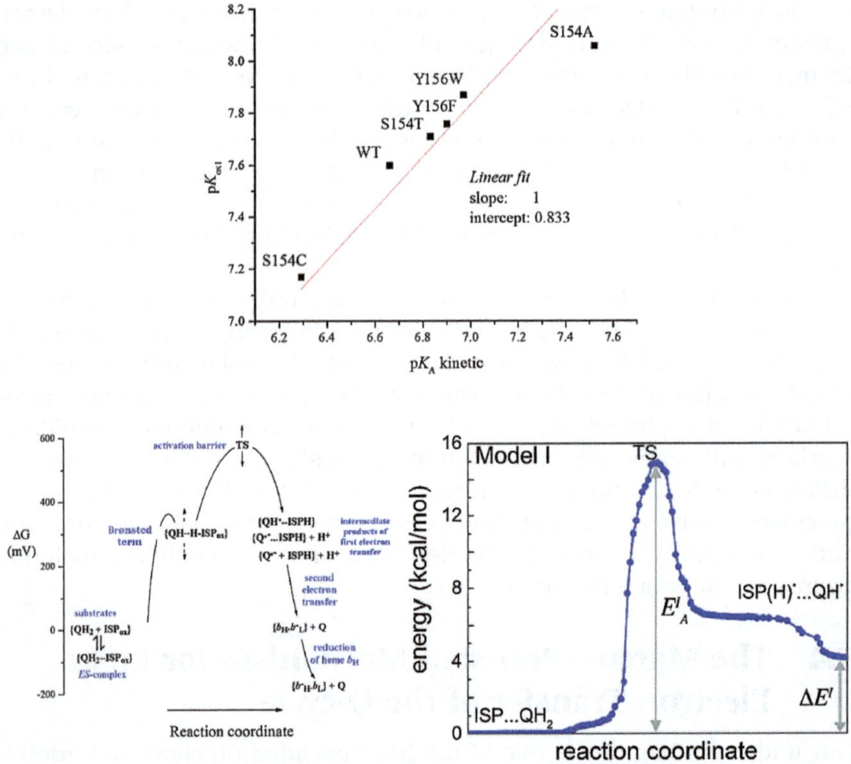

Figure 7.2 Top: correlation between pK_{app} and pK_{ox1}. Data from ref. 40. Reprinted with permission from ref. 42. Copyright 2015 American Chemical Society. Left: energy profile of the Q_o-site reaction. Values for ΔG are calculated from experimental data. The vertical dotted arrows show limits on estimated values. Reprinted with permission from ref. 3. Copyright 2017 American Chemical Society. Right: energy profile from QM calculation. The *ES*-complex (ISP…QH_2) started with QH_2 as H^+ donor to H152 N_ε in the H-bond, so that the reaction follows the path expected from the Marcus-Brønsted approach. Reprinted with permission from ref. 9. Copyright 2016 American Chemical Society.

distance-dependence and a conventional value for λ_{ET} lead to expectation of $k{\sim}10^8\ s^{-1}$, but the ΔpK value is an opposing term representing the weak probability (Brønsted barrier, $\Delta pK{\sim}5$) for positioning the proton close to H152 before the electron can transfer. This lowers the rate constant to the $10^3\ s^{-1}$ range observed.

Both the *ES*-complex and its transition through the first electron transfer have recently been simulated in MD/QC calculations in our collaboration with the Schulten and Solov'yov labs.[9,42] The outcome of these calculations show a gratifying agreement with the mechanism suggested above, but with a new wealth of atomistic detail. Free-energy profiles of the reaction coordinate from both approaches are similar (Figure 7.2), including energy levels, and endergonic nature. In the physicochemical profile (bottom left), the contribution of the Brønsted barrier is shown as a distinct state, but in the equivalent linear form, the product of the last two terms of eqn 7.4 would represent activation probability, and, with the rapid rate constants involved, would be a composite term. The approach through a QM superposition of states is essentially probabilistic, and simulates not only the profile, but also this composite feature, in the landscape from QC calculations (right).

7.5 The Dimer Interface

Complexities from the dimer are associated with coulombic effects, or traffic across the interface between the two monomers. The distance between the two b_L hemes is ~11 Å (between CBC carbons of the heme vinyl groups, which contribute to the conjugate structure), or 21.6 Å between heme irons, varying by ~1 Å during MD simulation. The dimer interface shows several interesting features.

7.5.1 The ISP Subunit and Its Clamp

The structures show that the ISP subunit has its hydrophobic helical anchor in one monomer, but the 2Fe2S cluster of the extrinsic domain, acting as acceptor of electrons from QH_2 in cyt *b*, and then as donor to heme c_1 in cyt c_1, in the other monomer.[44] The extrinsic head is connected by a tethering span to the anchor; in structures with different positions for the head, the anchor has the same structure (as does most of the head), and the action is mainly in the tether. Where the helical anchor transitions to tether, the helix is held in a clamp contributed by both cyt *b* and cyt c_1 subunits from both monomers, through forces including van der Waals contacts and H-bonds. The two clamps are separated by the width of the dimer. In our MD simulation, the protein is quite dynamic, and this distance varies between ~34.5 and 39.5 Å. This interface has received new attention from its participation in the spring-loaded scenario,[45–47] as discussed below.

7.5.2 Electron Transfer Between the b_L Hemes

The closest path for electron transfer across the interface between conjugate structures of the hemes would allow rapid electron transfer if distance were the main determinant.[23,24,48,49] Alternatively, a through-bond path would be between the iron atoms, involving His-198 (liganding the Fe) and Tyr-199 from both monomers, then a through-space jump between the two tyrosines. Whether or not this electronic traffic between monomers contributes a significant fraction of normal flux has been much contested. A significant literature has been devoted to a scenario in which electron transfer across the dimer inter face can carry a substantial flux in normal forward chemistry, discussed at some length in.[23] To accommodate this, a major revision of the Q-cycle mechanism would be required. The evidence for such a flux was strongly supported in three papers claiming to demonstrate high rates of electron transfer across a "bus bar" between the b_L hemes across the dimer interface. These all used protocols in which selective mutations enforced such a path.[50–52] Three different experimental approaches were reported, each depending on expression of two different copies of the cyt *b* subunit in the same cell, mutated so as to force this pathway in heterodimers incorporating the two different subunits. Different mutations in the monomers led to blocking the Q_o-site in one, or blocking the exit through heme b_H in the other, so that neither monomer could function separately, and activity in different combinations (six in ref. 52) was tested and compared to wildtype. The three groups agreed on the general conclusion that a substantial flux across the dimer interface was detected experimentally, but these results were obviously in conflict with our own earlier work demonstrating that no such rapid flux could be observed.[23] In order to help us understand the system, we constructed strains in *Rb. sphaeroides* equivalent to those used by Świerczek *et al.*[52] Our preliminary results seemed to confirm the earlier reports, but more detailed examination of the underlying molecular biology showed that these could all be explained by cross-over recombination, in which the bacteria reconstruct the native sequence from the 'good' spans of DNA in the heterodimeric construct.[53] We suggested that such processes could account for the results from other labs, although no such problems were noted in the earlier reports. Subsequent papers from two of the three labs later detailed similar problems encountered in that work, but they also claimed that this did not invalidate their earlier claims. For example, one paper reporting data from the Osyczka group using a well-characterized set of isolated proteins,[54] claimed that "...The cross-inactivated form corresponding to cytochrome bc_1 with disabled complementary parts of each monomer retains the enzymatic activity at the level that...demonstrates that inter-monomer electron transfer through the bridge effectively sustains the enzymatic turnover...". However, the figure presenting the most significant data showed turnover rates as a function of [cyt *c*], used as acceptor, in which, compared to the wildtype (WT) strain (or the constructed strain with WT sequence), strains with one or other disabling mutation in one monomer (but with one WT monomer) showed approximately half the rate, and the strain with both disabling mutations in the same monomer (but with one WT monomer) showed $\sim\frac{1}{3}$ the rate, reasonably interpreted as showing that a functional monomer carried half the flux of the dimer. However, data showing the activity of the strain enforcing inter-monomer electron transfer (~14% of WT rate) also showed that the activity was independent of [cyt *c*], and must therefore have been almost entirely non-enzymatic. Although the data were claimed to support the above model, any enzymologist would recognize that it showed the opposite.

7.5.3 Coulombic Interactions Across the Interface

From the distance between the b_L hemes noted above, coulombic forces would be expected if the charge on the hemes changed on reduction. Crofts *et al.*[23] (see also Holland[55]) noted that redox titrations, usually interpreted as showing a single $n=1$ component for both heme b_H ($E_{m,7}=\sim 40$ mV) and heme b_L ($E_{m,7}=-90$ mV), could in both hemes be better fit by two components, and suggested that this might reflect a coulombic interaction. Since, from $\left[\Delta E_m=\dfrac{z_A q\cdot z_B q}{4\pi\varepsilon_0\varepsilon r_{AB}}\right]$ the force would depend on the dielectric of the phase separating the charges, the effect expected (ΔE_m) could be calculated only if the dielectric constant was known. For the interaction between heme b_H and heme b_L, we could estimate a value of $\varepsilon\sim 7.8$ from the ΔE_m of heme b_L on reduction of heme b_H ($\Delta E_m\sim -80$ mV, from the difference of E_m in strains with heme b_H present, or lost on mutation of one of the histidine ligands[56]), using $r_{AB}=20.5$ Å. If a similar value for ε were appropriate for the dielectric constant between the b_L hemes, a similar ΔE_m might be expected. Bhaduri *et al.*[57] have recently rediscovered this treatment, and discussed such interactions in the context of the sequence of reduction of the *b*-hemes in *Rb. capsulatus bc*$_1$ complex, and in the b_6f complex of oxygenic photosynthesis, measured both by absorbance spectroscopy and circular dichroism (CD) in the heme Soret band. The absorbance changes are proportional to the reduced hemes, and CD changes are largely contributed by excitonic interaction between the reduced b-hemes in a monomer, so appear only when heme b_H is reduced. The reduction of the bc_1 complex hemes on addition of dithionite followed the sequence expected from equilibrium thermodynamics (the sequence seen in redox titration). The time-course seen in the absorbance changes was also that anticipated (fast heme b_H reduction (>80% complete in 12 s), slow b_L reduction (50% complete in 100 s)). The major CD changes came in only as heme b_H became reduced. Somewhat surprisingly, in the b_6f complex, the absorbance and CD changes more closely coincided, and kinetic fits suggested that both hemes of one monomer were reduced before those in the other monomer. This is in contrast with results previously reported for the b_6f complex measures *in situ* in *Chlorella* and chloroplasts, in which, when care was taken to minimize photo-induced oxidation of heme b_L, the standard equilibrium sequence was observed.[58,59] Since the measurements in ref. 57 were made over a 10 min timescale after addition of dithionite, it would be surprising under any scenario if the hemes of one monomer equilibrated at a different rate from those in the other.

7.5.4 The 'void' Between the Monomers Seen in Crystallographic Structures

All crystallographic structures of vertebrate mitochondrial or bacterial bc_1 complexes show a void, located above the apposed interface between monomers at the level of the b_L hemes. The void is supported by a scaffolding of helices, including transverse helix 2 (residues 24–34), and transmembrane helix 10 (residues 187–220) from cyt *b* of each monomer. How has the void been treated in setting up MD simulations? In the yeast bc_1

complex, the void is occupied by a cardiolipin molecule, and in simulation of the *Rhodobacter capsulatus* bc_1 complex, the Róg group have used a cardiolipin to fill the volume.[5,60] This then provided a stable structure, which they used to simulate catalytic behaviour at the Q_o- and Q_i-sites. In other MD models of bacterial complexes, it was assumed that "the physics" of MD simulation would solve the problem. In one case, this led to protein displacements to collapse the structure, leading to substantial unravelling at one Q_i-site. In a different simulation, the void was filled by an influx of waters. Neither of these "solutions" is realistic, but both involve structure at some distance from the Q_o-site, the main feature studied, which was assumed to behave normally. In our simulation, we filled the void with two phosphatidylglycerol (PG) molecules (because *Rb. sphaeroides* is fully active in the absence of cardiolipin[61]). This generated a stable structure, which has allowed us to model *ES*-complexes at both Q-sites (see below).

7.6 Recent Developments

7.6.1 A New Gating Mechanism in the First Electron Transfer

Recent studies of *C. elegans* in collaboration with the Kaeberlein group[62] have highlighted an important role of the Rieske ISP in aging. A point mutation close to the 2Fe2S cluster confers a 3-fold increase in lifetime. Likely, the primary mutation slows the flux of reducing equivalents into the Q_o-site reaction to limit generation of ROS. Longevity is reversed by suppressor mutations; remarkably, all these were also in the ISP subunit, and located in a span distant from the primary site, involved in a dynamic tethering of the mobile cluster domain to its helical anchor. The site of both actions in the same subunit focuses attention on the role of ISP.[39–41,63–66] We suggested a mechanism to explain the effects; the primary mutation slows the flux into the Q_o-site reaction by steric obstruction of the *ES*-complex, thereby limiting generation of ROS, and mutations in the tether reverse this by modulating a spring-loaded mechanism. The binding forces associated with formation of the *ES*-complex stretch out the tether, to provide a counteracting force towards re-contraction.[47,62] Of course, the "spring" here is chemical, not physical; and the forces come largely from zipping or unzipping H-bonds. Analysis of earlier mutational studies in *Rhodobacter* systems[47,67–69] allowed us to estimate values for the counteracting forces. A major interest of the *C. elegans* work was that many pleotropic effects typically associated with longevity were also suppressed.[45] This shows that the inhibition of flux into the Q_o-site (the primary role in energy metabolism), translates to profound secondary phenotypes. These include participation in arbitration of the life or death of the cell through apoptosis, autophagy, *etc*. Many of these pleotropic effects can be linked (though indirectly) to the primary lesion.[45] Identification of the primary cause therefore provides a secure base for further exploration of these medically notable features.

7.6.2 Dissecting the Second Electron Transfer

In exploring the rapidity of SQ_o removal in the second electron transfer, focus shifted to Glu-295 in the conserved -PEWY- loop of the Q_o-site. The strong inhibition of rate, and a changed pH dependence of partial processes in mutants at E295,[25,26] have highlighted an essential role in catalysis of proton exit.[25,33] When referred to the much more rapid intrinsic rate of the second step, inhibition is always substantial, and is sufficient to shift the rate limiting step from the first to the second electron transfer.[25] Early models based on inhibitor binding assumed that Glu-295 would help to stabilize the *ES*-complex through a H-bond to QH_2.[33,70] However, our MD/QC simulations[42] showed that Tyr-147 rather than E295 is more likely to H-bond, suggesting a different *ES*-complex stabilized by a H-bond from the tyrosine -OH to QH_2.[42] In later QC simulations, Barragan *et al.*[9] used an extensive *ES*-complex in which a second H-bond from Y147 served as a relay for proton transfer to Glu-295 in its critical role in proton exit. However, in our improved MD model, although both H-bonds were observed, they did not both form at the same time, suggesting that the *ES*-complex is stabilized mainly through the H-bond to H152 of $ISP_{ox,}$ and that Y147 and E295 participate through a more stochastic mechanism.[3]

7.6.3 The Semiquinone Intermediate of the Bifurcated Reaction

A SQ_o species was first detected when its oxidation was blocked (by antimycin inhibition at the Q_i-site, for example), and a driving force to pull the first electron transfer reaction over was available in the high potential chain.[25,71,72] High resolution EPR showed that this SQ_o was trapped in the Q_o-site but not involved in H-bonding.[71,73] In the E295W mutant, a similar "free" SQ_o still accumulated, but its oxidation was so strongly inhibited that heme b_L remained oxidized. From its accumulation in E295W, we could quantify concentrations for SQ_o and ferriheme b_L, and measure rates, to allow calculation, from $k = v/[SQ_o.\text{ferriheme } b_L]$, of a rate constant for oxidation of SQ_o trapped in the distal domain. The value was too slow by at least 3 orders of magnitude to account for the normal forward flux. If the same rate constant applied in wildtype, the rapid forward rate could only be explained by movement of SQ in the Q_o-site closer to heme b_L,[25] as predicted from our earlier work.[13,32,33,35] Our kinetic model for the Q_o-site reaction was based on parameters gleaned from this work.[25]

7.6.4 A New Intermediate SQ_o Complex

In more recent work, we have characterized a new intermediate state, a SQ_o.ISPH complex, which is formed when heme b_L is reduced.[3] The same complex is formed in the antimycin inhibited complex, when the kinetics show a lag determined by the number of turnovers needed to reduce the *b* hemes: no lag if the hemes are pre-reduced; >1 ms if both hemes *b* are initially oxidized; ~200 μs if heme b_H is initially reduced. In a Glu-295 mutant (E295Q), the same complex accumulated without a lag, even when

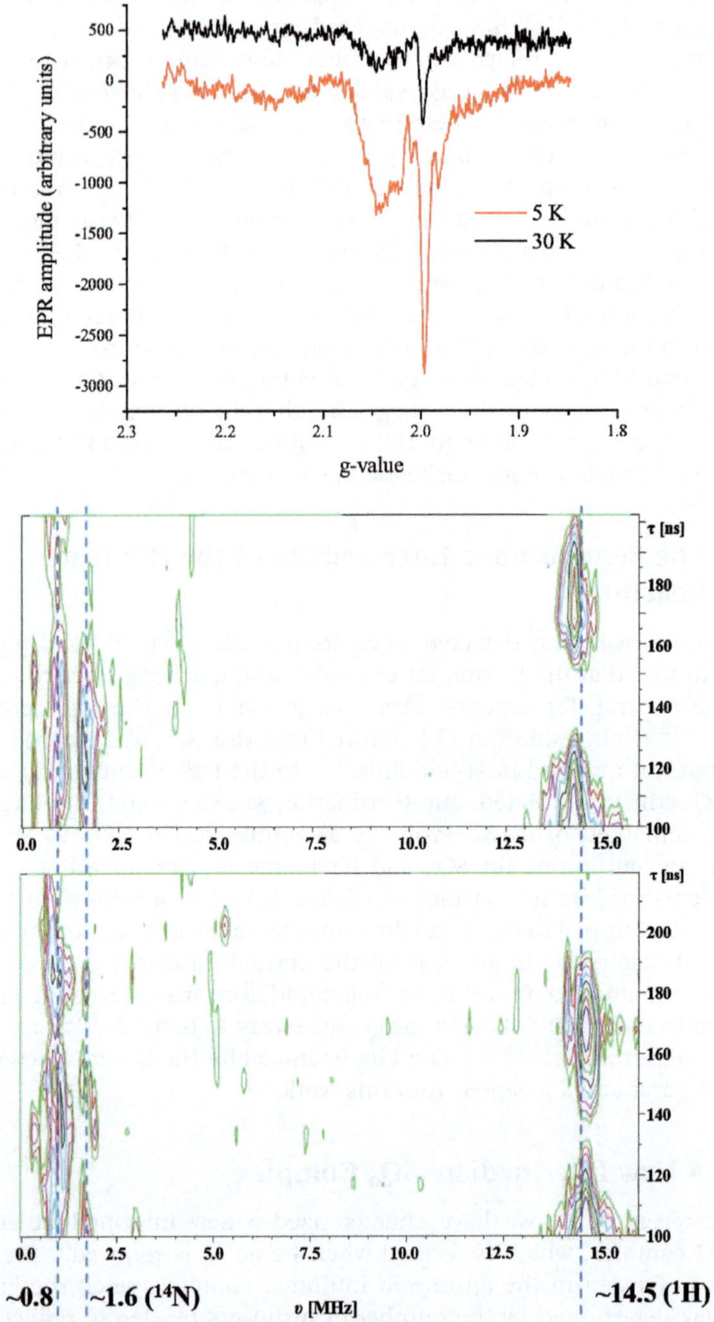

heme b_L is oxidized, reflecting the inhibition of SQ_o oxidation previously seen in electron transfer kinetics.[25] Although the kinetics of SQ_o formation closely match those expected from the model, other properties of the SQ_o species observed in these experiments are surprising, because they differ in several important respects from those previously reported (Figure 7.3). The EPR spectrum shows several lines not seen in previous work. A different spin-coupled state, although initially described as a triplet state, has been reported on prolonged incubation under strongly reducing conditions,[2,74] but the long timescale, and the fact that heme b_L was oxidized in this state, means that it cannot be an intermediate. In our $SQ_o^{\bullet}$.ISPH$^{\bullet}$ state, the properties indicate spin interaction between the SQ_o and a neighbouring paramagnetic species. Power saturation and temperature dependence also differ from previous species. The signal saturates only at much higher power, indicating faster relaxation through spin coupling. Similar saturation effects are revealed in field-swept ESE spectra taken at different temperatures and times (τ) along the decay curve. A prominent peak around $g=2.06$

Figure 7.3 Semiquinone intermediate formed on turnover of the Q_o-site with heme b_H initially reduced. Top:. CW X-band EPR spectra of a sample trapped early in the accumulation of SQ (at 280 μs after mixing), as heme b_L, initially oxidized, became reduced by turnover when heme b_H was initially reduced. The shape of the signal shows that the normal symmetrical narrow line-shape of the SQ signal is distorted by electron spin–spin interaction with a neighboring paramagnetic metal center. The temperature dependence determined at 5 K and 30 K (the 30 K trace at top is offset by 500 units for clarity) shows that this substantially enhances the spin relaxation. The interaction is also seen in power saturation studies, and in the temperature and time dependence of field-sweep spectra (not shown). Lower frames: The ESEEM spectra of the bc_1 complex (sampled 4 ms after mixing), collected at $g=2.06$ (top) and $g=2.005$ (bottom). The spectra show contour plots of frequencies, υ (from the modulus of Fourier transform as time T increases), at different times, τ, using a 3-pulse sequence $\pi/2\ldots\tau\ldots\pi/2\ldots T\ldots\ \pi/2\ldots\ \tau\ldots$echo. The initial time, τ, was 100 ns and was increased by 16 ns in successive traces. The lines at ~1.6 and 0.8 MHz in spectra are from ^{14}N, and likely produced by interaction with a histidine nitrogen. The ^{1}H line in these spectra (at ~14.5 MHz) is produced by protons. The periodical variations of line intensity are due to dependence on time τ at stable frequencies. For both samples, the microwave frequency was 9.633 GHz, and temperature was 15 K. The magnetic field was 334.1 mT (top) and 343.2 mT (bottom). The spectra show small differences in temperature sensitivity and different kinetics of decay, but nuclear spin interactions for ^{14}N, likely through H-bonding, were at the same frequencies (~1.6 and 0.8 MHz), so were likely with the same N-atom. These values are consistent with the pure nuclear quadrupole frequencies of the protonated nitrogen in an imidazole residue, and suggest that the electron spins contributing to the ESE signals interacts with N_ε of His-152 of ISPH$^{\bullet}$, the same N-atom as involved in the H-bond stabilizing the *ES*-complex. This set of characteristics is very different from those of the dissociated SQ determined in earlier work.[2–4] becomes obvious as the temperature of measurement is lowered from 70 K to 5 K, and swamps the ESE SQ signal at $g=2.005$ at 5 K. This same component is obvious in the CW spectrum at lower temperatures, but is barely detectable at 30 K (Figure 7.3).

Additional information about the local environment comes from 3-pulse ESEEM data. Figure 7.3 shows data obtained from a sample, quenched 2 ms after mixing, which had bc_1 complex at higher concentration to obtain greater signal strength. The data were measured at the $g=2.005$ peak of the CW EPR spectrum dominated by SQ_o. Two ^{14}N ESEEM frequencies ~1.6 and 0.8 MHz are observed, consistent with pure nuclear quadrupole frequencies of protonated N_ε of histidine (*cf.* ref. 75), most probably that of His-152 of ISPH•, as expected from its role in stabilizing the *ES*-complex. Spectra taken at $g=2.06$ (dominated by ISPH) show the same frequencies, but different spin relaxation in field-sweep spectra (not shown). The ESEEM spectra also showed a line at ~15 MHz from ^1H,[3] resulting from hyperfine interactions with protons in the SQ_o.ISPH electron spin environment.

7.6.5 How Is the Electron Pair in the SQ_o•.ISPH• Complex Entangled?

A feature of the SQ_o•.ISPH• state is that signal amplitudes in the $g=2.005$ line are considerably higher than expected from an $S=\frac{1}{2}$ free radical. The most straightforward interpretation is spin sharing with a neighboring metal center. A likely candidate, consistent with the configuration of the *ES*-complex, and the properties summarized above, is the $S=2$ spin of the nearest Fe-atom of the cluster. In the reduced cluster with $S=\frac{1}{2}$, (Fe(III) has $S=\frac{5}{2}$, and Fe(II) has $S=2$, the iron spins are antiferromagnetically coupled, the spin of Fe(II) is polarized negatively, and the His ligands are coordinated with Fe(II)). In,[74] the spectrum of the "triplet" state clearly showed all components of the EPR spectrum from the Rieske cluster, but in our case the EPR spectrum shows only a broad line at $g=2.06$, and "wings" around the SQ line at $g=2.005$, indicating significant disturbance of Fe(III) and Fe(II) electronic states. One can suggest that sufficiently strong interaction of the radical spin with $S=2$ disturbs antiferromagnetic coupling of two iron spins, and that part of the polarization transfers onto the radical; it might be that all three spins should be considered in analysis of the SQ_o•.ISPH• complex. In the *ES*-complex at the Q_o-site, the QH_2 and ISP_{ox} both have electrons paired in separate molecular orbitals; electron transfer from QH_2 to ISP_{ox} then generates two paramagnetic species in the spin-coupled state, SQ_o•.ISPH•. The pair of electrons must initially have had opposite spins, but all further evolution is local and in a complicated condensed phase, so the question of interest is whether spin coherence continues as the state evolves over its >4 ms lifetime. If the two electrons occupied the same orbital in the superposition, the exclusion principle would require that coherence be maintained, and the paramagnetic property would be lost. On the other hand, since the state is EPR visible, the paramagnetic population must have electron spins in a Boltzmann distribution, and the many properties accessible to direct measurement through spectroscopy (Figure 7.3, and see ref. 76) must be accounted for. In forward chemistry, the separate spins are revealed as the $ISPH^{\bullet}$ and $SQ_o^{\bullet}$ separate to chemical species with well-characterized discrete paramagnetic properties. In the quantum chemical calculations by Barragan *et al.*,[9] the evolution has been modelled, but the intermediate states of the calculation have not yet been examined from this perspective.

As discussed further below, the state examined in the quantum chemical treatment was set up to allow proton release from the neutral state, $SQ_o^{\bullet}.ISPH^{\bullet}$, expected as an early intermediate. The additional residues modelled to facilitate proton release may have been somewhat unnatural, but alternative models can be readily explored. From the rapid kinetics of formation, the long lifetime of the $SQ_o^{\bullet}.ISPH^{\bullet}$ state at room temperature, and the accessibility to measurement, the state might be of interest to the entanglement community, but until we have a deeper understanding of the quantum chemistry of the state, and a more complete quantification of the spin energies and amplitudes (currently not possible), further speculation is unjustified.

7.7 Proton Release Associated with Q_o-site Turnover

Oxidation of SQ_o in the site must involve proton exit, rotational displacements of key residues (Y147, E295) to open the volume, diffusion if SQ_o into this open space, and electron transfer. Most of these processes have not previously been visible, but we have recently refined a protocol using neutral red for measuring the release of H^+ into the P-phase (inside of thylakoids and chromatophores).[77–80] By use of specific inhibitors, we can isolate particular partial processes, as demonstrated in Figure 7.4 (left two panels). In analyzing the data, we take advantage of characteristic properties of the Q-cycle operating in the chromatophore system. As a consequence of the relative stoichiometry of reaction centers and the bc_1 complex (2:1) noted above, 2 RC per bc_1 complex monomer are excited by each flash, generating two oxidizing equivalents for each heme c_1. In the absence of inhibitors, each Q_o-site oxidizes $2QH_2$, as seen in the yield of H^+, or in the electrogenic flux (right two panels). In the presence of antimycin, only $1QH_2$ is oxidized on each flash; turnover is constrained on reduction of heme b_H ($E_{m,7} \sim 40$ mV) by the thermodynamics of the Q_o-site. On a second flash, heme b_L ($E_{m,7} \sim -90$ mV) is reduced, driven by reoxidation of the high potential chain to pull the reaction over.[21,81,82] In general, the pattern of release of protons on oxidation of QH_2 is then consistent with the behavior known from electron transfer studies, and can be represent more formally through reaction equations.

Flash excitation (↯) generates two oxidizing equivalents for each bc_1 complex monomer, initially oxidizing cyt c_2, and QH_2

$$2P.Q_AQ_B + 2\text{ferrocyt } c_2 \,↯\!\Rightarrow 2P^+.Q_A^-\ Q_B^- + 2\text{ferrocyt } c_2 + 2H_N^+$$
$$\rightleftharpoons 2P.Q_A + QH_2 + 2\text{ferricyt } c_2$$

In the presence of myxothiazol (myxo) (or of asoxystrobin), one proton is released on oxidation of ISPH ($pK_{red} \sim 12.5$ in the initially reduced high potential chain), by ferriheme c_1 generated after the flash:

$$E.(\text{myxo}).\text{ISPH.ferro } c_1^- + 2\text{ferricyt } c_2 \rightleftharpoons E.(\text{myxo}).\text{ISP}_{ox}.\text{ferri } c_1 + 2\text{ferrocyt } c_2 + H_P^+ \quad (7.5)$$

With stigmatellin (not shown), no H^+ release is seen, consistent with binding of ISPH in a tight complex with the inhibitor.

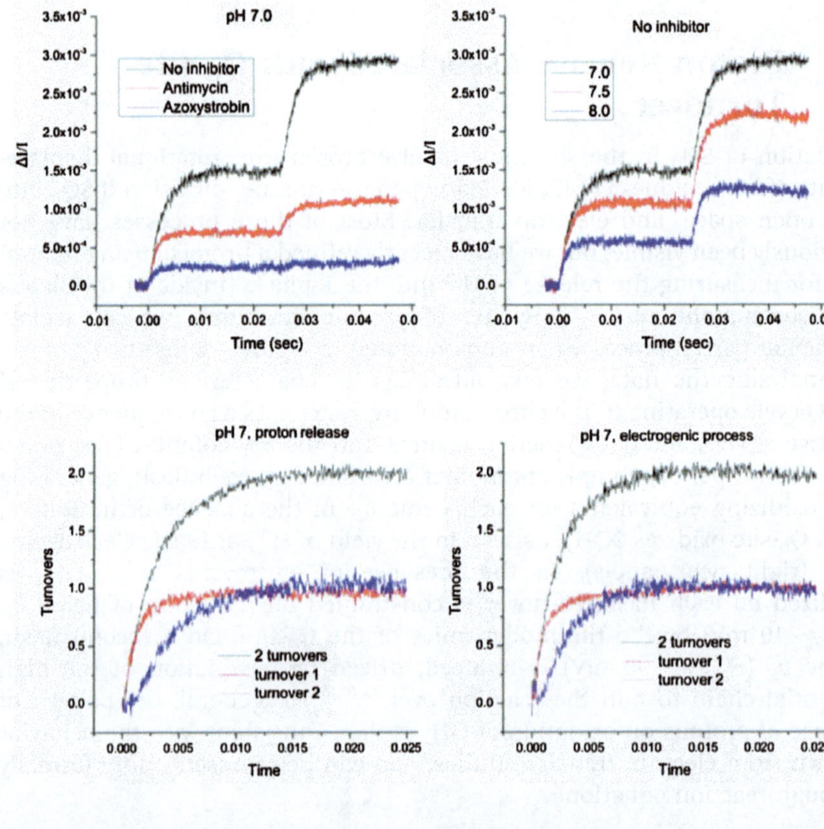

With the quinone pool half-reduced, most bc_1 complexes have the $g_x = 1.80$ complex (Q.ISPH, an enzyme product (*EP*-) complex) at the Q_o-site. This is replaced by the *ES*-complex when ISP_{ox} is generated as oxidizing equivalents reach the complex.[22,83] In the presence of antimycin, with the two *b*-hemes initially oxidized, 2 H^+ are released on oxidation QH_2. The first is that from oxidation of ISPH (as above), but ISPH is regenerated in the first electron transfer on oxidation of QH_2 by ISP_{ox}:

$$\begin{array}{ll} g_x = 1.80 \text{ (EP-)complex} & \text{ES-complex} \\ E\text{.Q.ISPH.ferro } c_1^- + 2\text{ferricyt } c_2 + QH_2 \rightleftharpoons & E.QH_2.ISP_{ox}.\text{ferri } c_1 \\ & + Q + 2\text{ferrocyt } c_2 + H^+_{scalar} \end{array} \quad (7.6)$$

$$\begin{array}{c} \text{intermediate complex} \\ E.QH_2.ISP_{ox}.\text{ferri } c_1 \ldots b_L.b_H \text{ (ant)} \rightleftharpoons E.QH^{\bullet}.\text{ISPH.ferri } c_1 \ldots b_L.b_H \text{ (ant)} \end{array} \quad (7.7)$$

Figure 7.4 Proton release to chromatophore interior. Proton release assayed using the pH-indicator dye neutral red[77–80] with valinomycin present to dissipate $\Delta\psi$. Traces were taken at 542 nm, with the proton release determined by subtracting traces in the absence from those in the presence of dye. Changes in pH of the external aqueous phase are highly buffered, and not seen in the neutral red changes. Left panels. Left, basic approach. In the absence of inhibitor (black trace), each flash causes the bc_1 complex to turn over twice, releasing $4H^+$ to the P-phase. Partial processes were selected by specific inhibition. The Q_i-site inhibitor antimycin prevents exit of electrons from the low-potential chain (red); the change on the first flash is limited to one turnover, associated with reduction of heme b_H, releasing 2 H^+ to the P-phase; the change on the second flash is associated with reduction of heme b_L. The Q_o-site inhibitor azoxystrobin prevents any Q_o-site turnover, but allows release of H^+ on oxidation of ISPH by heme c_1 on the first flash (blue). In the presence of stigmatellin, no proton release occurred. Right, pH dependence of turnover in the absence of inhibitor. The amplitude dependence reflects mainly the pK of neutral red.[77,78] Normalization to the change on the first flash would then show that the uninhibited proton release is not pH dependent in this range. Actinic flashes were given at $t = 0$ and 25 ms. Right panels. Left. The protons released on a single flash in the absence of inhibitor (black) ($4H^+$), or in the presence of antimycin (red) ($2H^+$). Initial rates show the same kinetic parameters, but the half amplitude shows the restriction to one turnover in the presence of antimycin. The difference shows a lag before release of $2H^+$ on a second turnover in the absence of inhibitor. Right. Electron transfer across the membrane generates an electrical potential, $\Delta\psi$, measured in the absence of valinomycin through the electrochromic carotenoid shift[22] (black). The kinetics of electron transfer are shown by the reduction of heme b_H in the presence of antimycin (red), reflecting a single turnover. The difference (blue) shows the electrogenic process on the second turnover in the absence of inhibitor, showing the same general behavior, and the same lag.

The second H^+ comes from oxidation of $QH^\bullet$, the neutral SQ intermediate generated in the first electron transfer, by heme b_L, which is then rapidly oxidized by heme b_H.

$$E.QH^\bullet.ISPH.c_1...b_L.b_H\ (\text{ant}) \rightleftharpoons E.Q.ISPH.c_1...b_L^-.b_H\ (\text{ant}) \rightleftharpoons E.Q.ISPH.c_1...b_L.\ b_H^-\ (\text{ant}) + H_P^+ \quad (7.8)$$

Further turnover is then arrested by the equilibria discussed above. The oxidizing equivalent in the high potential chain is shared between ISP and heme c_1: $ISP_{ox}.\text{ferro}\ c_1^- \rightleftharpoons ISPH.\text{ferri}\ c_1$, $K_{eq} \sim 3$.

In the absence of inhibitor, ferroheme b_H^- is oxidized at the Q_i-site, and, after exchange of product Q, by QH_2 a second QH_2 is oxidized at the Q_o-site, consuming the second oxidizing equivalent generated in the photochemistry, releasing 2 more H^+ to the P-phase, and passing its electron to the Q_i-site to complete reduction of the quinone.

$$QH_2 + E.Q.ISPH.c_1...b_L.\ b_H^-.Q_i \rightleftharpoons E.QH_2.ISP_{ox}.c_1^-...b_L.\ b_H.Q_i^{\bullet-}(H^+) + Q \quad (7.9)$$

$$E.QH_2.ISP_{ox}.c_1^-...b_L.\ b_H.Q_i^{\bullet-}H^+) \rightleftharpoons E.Q.ISPH.\text{ferro}\ c_1^-...b_L.\ b_H + QH_2 + 2H_P^+ \quad (7.10)$$

Except for the H_P^+ generated on oxidation of ISPH, which is released with $t_{\frac{1}{2}} \sim 150\ \mu s$, all rates of H^+ release measured in wildtype reflect the rate-limiting first electron transfer (with the constraints implicit in the Marcus–Brønsted equation), although convoluted with turnover as the second QH_2 is oxidized. Since electrogenic processes and release of the other proton follow the first electron transfer, the intrinsic rates of partial processes must all be fast compared to the limiting rate observed. The second turnover follows the first without any hiatus, showing that exit of product Q, and replacement by QH_2 must also be rapid compared to the limiting step. Dissection of the partial processes leading to release of the four protons is illustrated in Figure 7.5 (see legend for explanation), and this provides a template for investigation of proton release in mutant strains, currently underway.

7.8 A Model of the *Rb. sphaeroides* bc_1 Complex for MD Simulation in a Native Membrane

Molecular dynamics (MD) simulations most commonly start from the crystallographic structures. The atoms are frozen in a lattice constrained by contacts with neighboring components of the unit cell, and information on dynamic features is limited. For membrane proteins, the prison is even more unnatural; there is no membrane, but instead ancillary lipids, detergent

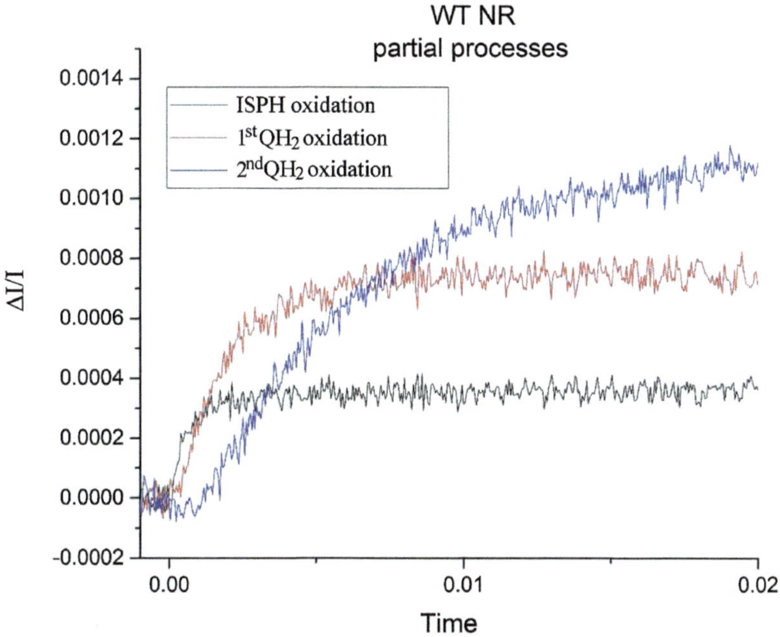

Figure 7.5 Dissection of partial processes in proton release to the P-phase. The H^+ release on oxidation of ISPH (black trace, a scalar proton, $\tau \sim 150$ μs) in the presence of myxothiazol, *eq.* 3a, precedes (note lag) turnover of the Q_o-site (formation of the *ES*-complex, *eq.* 3b, and oxidation of the 1st QH_2, *eq.* 3c, d (red trace, proton release in the presence of antimycin minus that in the presence of myxothiazol). In the second turnover of the Q_o-site, these two process are convoluted together, and with the exchange of the Q product of the 1st turnover for QH_2, *eq.* 3e (the curve without inhibitors minus the curve with antimycin, blue trace). The three traces, when summed, would show the protons released on this timescale in the two turnovers of the Q_o-site (black traces in Figure 7.4).

molecules, waters, *etc.*, fill the interstices. The MD simulation is set up to liberate the native structure from this prison by releasing it into a native reaction milieu, and this release process is an essential preliminary to mechanistic exploration. We have previously explored atomistic features through collaborations with the Schulten group, applying MD and QC approaches to several different mechanistic problems using MD models based on mitochondrial or *Rhodobacter* bc_1 complexes.[14,42,84] Building on this experience, we have developed a new MD model using the *Rb. sphaeroides* bc_1 complex,[3] in which several artificial features of previous models have been avoided, and the protein environment has been modeled in a native membrane (including ubiquinone) (Figure 7.6). Our concern was that in previous simulations, distortions of the structure compared to the crystallographic model might have led to artificialities in the dynamics modelled. The main improvement comes from insertion of 2 phospholipids in the void volume in

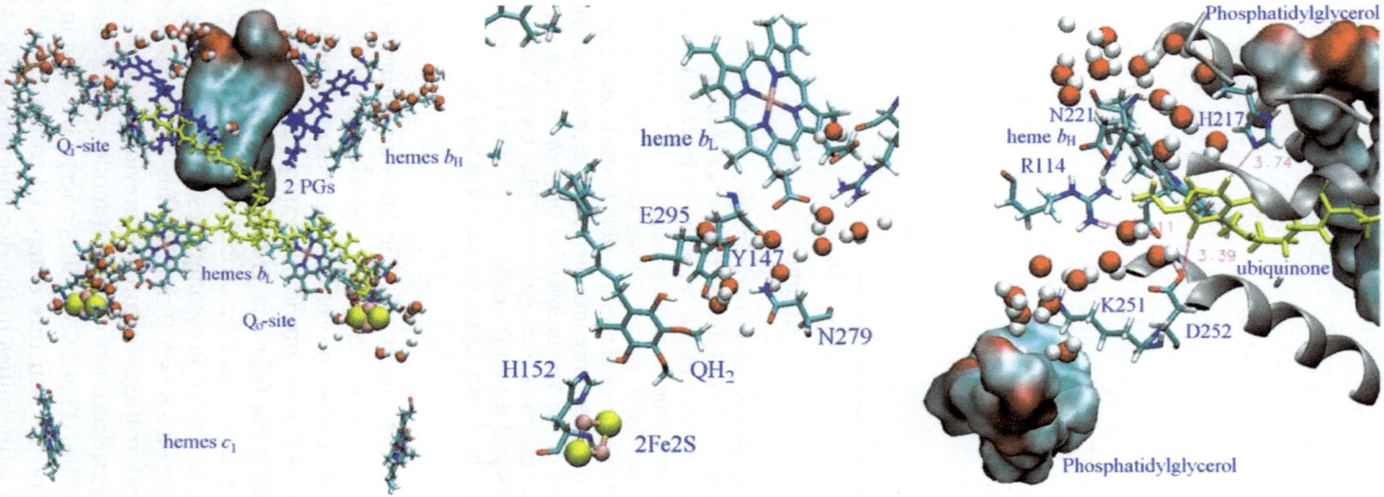

Figure 7.6 Left. Cross-section through the MD model of the *Rb. sphaeroides* bc_1 complex showing the dancers in the Q-cycle ballet. The hemes (colored CPK), QH_2 (at the Q_o-site, yellow), antimycin (at the Q_i-site, blue) are stick models, two PGs filling the void are shown by surfaces, a third PG at the Q_i-site on the left (stick model) is involved in reduction of Q when antimycin is absent. The 2Fe2S clusters, and waters facilitating H^+ transfers into and out of the Q-sites, are shown as VDW spheres. Center. The *ES*-complex of ISP_{ox} with QH_2 at the Q_o-site. Right. The *ES*-complex with Q at the Q_i-site (in the absence of antimycin). At both sites, water chains stabilized by conserved residues allow H^+ equilibration with aqueous phases.

the crystallographic models, noted above, but which in yeast bc_1 complex, is filled by a cardiolipin. In the absence of lipids, MD excursions eliminated the void, either by flooding it with waters, or by unnatural protein displacements that disrupt the neighboring Q_i-site. To avoid such artifacts, we used two PG molecules to fill the void, chosen because we needed to take account of the full activity observed in the absence of cardiolipin.[3,61] Our initial report and further details of the model were recently published,[3] and show that *ES*-complexes are formed at both Q_o- and Q_i-sites, in excellent conformity with expectations from experimental evidence. However, both *ES*-complexes differ in important details from those previously reported. Figure 7.6 shows frames from trajectories with the *ES*-complex for QH_2 oxidation at the Q_o-site highlighted (center), and the *ES*-complex for Q reduction at the Q_i-site highlighted (right). More recently, we have modeled the antimycin inhibited complex (Figure 7.6, left).

7.8.1 The Q_o-site

Molecular dynamics explores stochastic processes governed by classical Newtonian physics and coulombic interactions. A state in equilibrium will remain in equilibrium. To explore process, a state must be set up out of equilibrium. In the simulations shown, the *ES*-complex at the Q_o-site is metastable, with QH_2 and ISP_{ox} as reactants, poised for electron transfer with chains displaced from equilibrium. However, MD cannot explore the electron transfer process, which must be treated quantum chemically. A number of previous reports have used this approach, but with very different result. This comes from different choices for the configuration for the *ES*-complex.[9,85,86] As noted above, Barragan *et al.*[9,42] have simulated the *ES*-complex expected from the large body of experimental evidence supporting the Marcus–Brønsted treatment, and they have also explored the quantum chemical evolution of the first electron transfer starting from that *ES*-complex. The MD and QC simulations, in line with that treatment, match the experimental data. None of the other treatments, all of which used different configurations for the *ES*-complex, can make such a claim.

The papers by Barragan *et al.*[7,42] started from the *ES*-complex modeled with $QH_2.ISP_{ox}$, but they included in their QC calculations an association of other residues (the 2Fe2S ligands, and neighboring polar sidechains, including Y147 and E295). This larger complex may be subject to revision (see below), but it remains a good starting point for exploring the role of these residues in release of the H^+, because inclusion of E295 in the QC model allows calculation of a reaction coordinate including H^+ release. The initial intermediate product of the first electron transfer must involve a neutral semiquinone that remains bound at the site, likely $QH^{\bullet}.ISPH^{\bullet}$. However, in normal forward chemistry, that state is expected to rapidly dissociate to $ISPH^{\bullet}$ and either $Q^{\bullet -}$ and H^+, or $QH^{\bullet}$. Once separate, these product species would be free to move apart to reach their respective acceptors. In the studies of Barragan *et al.*,[7,9] dissociation of the H^+ to give $Q^{\bullet -}.ISPH^{\bullet}$ leads to a transitional state in which $Q^{\bullet -}$ remains in association with ISPH. However, in the time range explored in these calculations the direct experimental evidence provides no useful basis for choice between these different models, since all observed rates are determined by the limiting step (~1 ms). What different scenarios need to be considered? Three possibilities are summarized in the following schemes for the Q_o-site reaction:[3]

heme b_L...QH_2.ISP_{ox}.E295 $\rightleftharpoons$ heme b_L...$QH^{\bullet}$.$ISPH^{\bullet}$.E295
heme b_L...$QH^{\bullet}$.$ISPH^{\bullet}$.E295 $\rightleftharpoons$ heme b_L...$Q^{\bullet -}$.$ISPH^{\bullet}$. E295(H^+)
heme b_L...$Q^{\bullet -}$.$ISPH^{\bullet}$.E295(H^+) $\rightleftharpoons$ heme b_L...$Q^{\bullet -}$...$ISPH^{\bullet}$ $\rightarrow$ heme b_L^-(H^+)...$ISPH^{\bullet}$ + Q+ E295

Scheme 7.1 (Barragan, Solov'yov, Schulten mechanism).

heme b_L...QH_2.ISP_{ox} ....E295 $\rightleftharpoons$ heme b_L...$QH^{\bullet}$.$ISPH^{\bullet}$ E295
heme b_L...$QH^{\bullet}$.$ISPH^{\bullet}$ ....E295 $\rightleftharpoons$ heme b_L...$Q^{\bullet -}$.$ISPH^{\bullet}$ ....E295(H^+)
heme b_L...$Q^{\bullet -}$...$ISPH^{\bullet}$ + E295 (H^+) $\rightleftharpoons$ heme b_L^-(H^+)...$ISPH^{\bullet}$ + Q + E295

Scheme 7.2 (From ref. 3).

heme b_L...QH_2.ISP_{ox} $\rightleftharpoons$ heme b_L...$QH^{\bullet}$.$ISPH^{\bullet}$
heme b_L...$QH^{\bullet}$.$ISPH^{\bullet}$ $\rightleftharpoons$ heme b_L...$Q_oH^{\bullet}$...$ISPH^{\bullet}$
heme b_L...$Q_oH^{\bullet}$...$ISPH^{\bullet}$ $\rightarrow$ (E295-mediated) $\rightarrow$ heme b_L^-(H^+)...$ISPH^{\bullet}$ + Q_o

Scheme 7.3

Barragan *et al.*[9] observed no dissociation of the $Q^{\bullet -}$.$ISPH^{\bullet}$.E295(H^+) state in the 350 ns of their trajectory, so the partial processes in the third line of reaction Scheme 7.1 are conjectural. The authors emphasize that, although dissociation of $ISPH^{\bullet}$ had not occurred on that timescale, this might be unremarkable in view of the ms range of experimental turnover. The experimentally determined lag in the kinetics of heme b_H reduction limits SQ_o occupancy to <0.02 in forward chemistry, suggesting that dissociation and movement of SQ_o must occur faster than 10 μs. In our kinetic model, occupancies in the forward reaction are much lower, and diffusion of SQ_o to heme b_L requires $\tau \sim 100$ ns. In our MD simulations, we have observed several instances in which quinone species diffuse substantial distances along entry or exit channels, and from those displacements can calculate diffusion times. Excursions over a few Å can occur at rates ~40 Å/ns; in one somewhat artificial case, the mean rate for a 15 Å excursion was 0.5 Å/ns. From these values, $\tau \sim 100$ ns for a 5.5 Å excursion seems quite feasible, and faster rates would be expected if coulombic forces steered the reaction forward.[14]

In Scheme 7.2, the Q_o-site *ES*-complex is stabilized predominantly through the H-bond from QH_2 to His-152 of ISP, and the H-bond exchanges between partners connecting this complex to E295 are stochastic, and relatively weak. This pattern is strongly suggested from the evolution of distances between potential partners. These are essentially the same during trajectories from simulations with either Q or antimycin at the Q_i-site, as reported in.[3] Y147 spends more time interacting with E295 and N279 than with QH_2, all three are in communication through a small pool of H_2O molecules with a heme b_L propionate, and its partner, R94, and thereby linked *via* a stochastic chain of waters to the P-phase. It is tempting to speculate that the dissociation state of the propionate is dependent on the redox state of the heme (the E_m value shows an appropriate pH dependence), and that interactions might be exploited in gating.

More extensive revision from the Barragan *et al.* model could be justified (Scheme 7.3); it is possible that the earlier modeling[32,33,38,83] of the *ES*-complex as involving E295 as a direct ligand to QH_2 has biased our thinking towards an emphasis on a direct role in H^+ release. If so, an alternative scenario in which the product state $QH^{\bullet}$.$ISPH^{\bullet}$ dissociates to release the neutral SQ to diffuse in the site needs to be considered. E295 would then be involved only in transfer of the H^+ to the heme propionate after the $QH^{\bullet}$ is close enough to transfer the electron, thus facilitating an electrostatically linked PCET. The stochastic proton exchanges among this group of residues and associated waters would enable transfers fast enough to make these last two models indistinguishable experimentally from the earlier model. One consideration in deciding between these scenarios is the need to constrain the SQ_o to the Q_o-site volume. Preference might depend on a simple physical principle; the low probability (high energy cost) of solvating a charged species favors mechanisms involving $Q^{\bullet-}$ as the liberated form, rather than $QH^{\bullet}$, since the former would have a much lower probability of escape into the lipid phase *via* the hydrophobic entrance channel with decoupling from proton pumping when the SQs are disproportionate. Another consideration is the possibility of coulombic bias that would favor diffusion of the charged species. The reconfiguration of the Q_o-site, seen when comparing myxothiazol-containing structures to those with stigmatellin,[43] which we called the -PEWY- see-saw. Similar conformational changes would likely accompany diffusion of SQ_o from distal to proximal locations. These displacements lead to closure of the access port through which ISP contacts the occupant, and to enlargement of the volume proximal to the heme. Others have noted additional displacement[18] contributing to the reconfiguration of the site. Does occupancy distort the structure, or are the structural changes driven by longer range forces (coulombic, for example) that favor the occupancy? Perhaps extrinsic forces could lead to volume changes sucking the head group into the proximal domain!

The *ES*-complex of Schemes 7.2 and 7.3 would not be suitable for QC calculation of the reaction profile because the linkage to proton release is stochastic. If proton release is exoergic, the neutral form would be less stable, and the overall endoergicity of the first step would provide even tighter constraints on occupancy of the intermediate state, making it even less accessible to population on simulation. This would necessitate a more complicated set of protocols for exploring the reaction coordinate than the grand superposition of the Barragan *et al.*[9,42] model, which at present appears to be the most practicable approach.

One additional point of note is that the spring-loaded effect (above) would favor dissociation[45,46] of the intermediate state. The associated work-term, and its mechanistic expression in configurational changes in the tether (~9 H-bonds broken on extension[45,46]), could be explored by the physics of the MD simulation if properly implemented, an obvious target for future studies. If the SQ_o.ISPH• complex is an early intermediate after electron transfer from the *ES*-complex of Figure 7.3, its lifetime in forward chemistry would have to be much shorter (<100 ns) than under the conditions in which we detect it, where it lasts at least 4 ms. This state is formed only when the electron from SQ has nowhere to go (because the acceptor, heme b_L, is reduced). The proton may also have nowhere to go, for example, if E295(H^+) had been unable to pass on its H^+ during the previous turnover. The failure to dissociate could reflect either loss of acceptor of the electron, or the proton, or both, and any combination would prevent SQ_o oxidation. It is also easy to imagine constraints on sidechain movement in mutant strains, or from local fields, that would limit mobility in the Q_o-site. These are all features that can be explored in MD simulation.

7.8.2 The Q_i-site

The Róg group used a similar MD model, extended to include QC calculations, in exploring the Q_i-site reactions.[5,6,60] They have a nice discussion of mechanistic features, in which one focus was on a specific role for cardiolipin in H^+-uptake. Since electron transfer through the bc_1 complex in *Rb. sphaeroides* can occur normally in the absence of cardiolipin,[61] we used two PG molecules in our MD model (with an equivalent molecular volume) to fill the void, and also found that PG diffused from the membrane to spontaneously occupy the binding sites interacting with K251 at the Q_i-site, to fulfilled the functions said to require cardiolipin (Figure 7.6, right).[3]

In the uninhibited complex, the forward reaction on the first turnover of the Q_o-site leads to electron transfer through the *b*-hemes to reduce Q to SQ_i at the Q_i-site, and, on the second turnover, to the reduction of SQ_i to QH_2. Kolling *et al.*[87] had pointed out that the different occupants in the normal cycle, Q, SQ, and QH_2 might be expected to require different H-bonding patterns: Q with H-bond donors at each $\geqslant$C=O, QH_2 with H-bond acceptors at each >C–OH, and SQ_i with a mix, depending on protonation state. Redox titration of SQ_i shows a maximum for the bell-shaped curve at $E_{h,7} \sim 150$ mV, significantly displaced from the $E_{m,7} \sim 90$ mV of the pool, indicating that the E_m of the bound Q/QH_2 couple in the disproportionation used to formally describe SQ formation, had a much tighter binding of QH_2 than Q. We therefore simulated formation of the *ES*-complex at the site with Asp-252 protonated, and His-217 dissociated, a less favorable configuration (heme b_H was oxidized). These two residues are seen in structures to form either direct or H_2O-linked H-bonds to Q, and high-resolution pulsed EPR studies show spin couplings to N in a direct H-bond to H217. Additional residues thought to participate are Lys-251 (next in sequence to Asp-252), Arg-114 (which H-bonds to the propionate of heme b_H which is "bent" across the face of the heme), and Asn-221. In the MD simulation, the Q engages in a dynamic exploration of these partners. H217 at one end, and D252 at the other, favor H_2O-bridged H-bonds with the two $\geqslant$C=O groups; N221 participates through stabilization of the water cluster around H217; and K251 and R114 are involved in a H-bonded water chain leading from the "bent" propionate and D252 to a PG electrostatically engaged with the K251–NH_3^+. As noted above, the Róg group have previously explored possible mechanistic features in a combined MD/QC study, with a similar configuration for the complex with Q, but with cardiolipin instead of PG interacting with K251.

In the antimycin inhibited simulation, the behavior at the Q_o-site is essentially the same as in the uninhibited complex, as expected from the kinetics of turnover. However, configurations of the antimycin inhibited site differ from those previously reported.[88] In particular, Asn-221 is rotated so as to communicate with the second heme propionate, rather than interacting with His-217 (and associated waters) in liganding Q. We have previously reported effects of mutation at this site,[23,89,90] and noted that N221 mutants show an inhibited rate of electron transfer from heme b_H to the Q_i-site acceptor. This behavior would be compatible with a role of N221 in H^+ uptake at the site. A remarkable additional feature, a dramatic shift of redox potential to lower values seen only in the presence of antimycin, marked in some mutant strains, suggests that the changed configuration connecting to the heme propionate might show a role in equilibration of H^+ with the heme, unavailable in the mutants, to explain the effects on mutation at this residue.[23]

Acknowledgements

We are grateful for support from the Department of Biochemistry for research supplies and teaching stipends for C.A.W. and S.W.R. Pulsed EPR studies were supported in part by Grant DE-FG02-08ER15960 from Chemical Sciences, Geosciences and Biosciences Division, Office of Basic Energy Sciences, Office of Sciences, US DOE (S.A.D. and A.R.C.), and NCRR/NIH Grants S10-RR15878 and S10-RR025438 for pulsed EPR instrumentation. Computational resources were provided through an XSEDE start-up grant MCB150083 and XSEDE Research Request MCB160130 to A.R.C.

References

1. A. R. Crofts, in *Biophysical and Structural Aspects of Bioenergetics*, ed. M. Wikström, Royal Society of Chemistry Publishing, Cambridge, 2005, pp. 123–155.
2. M. Sarewicz, Ł. Bujnowicz, S. Bhaduri, S. K. Singh, W. A. Cramer and A. Osyczka, *Proc. Natl. Acad. Sci. U. S. A.*, 2017, **114**, 1323–1328.

3. A. R. Crofts, S. W. Rose, R. L. Burton, A. V. Desai, P. J. A. Kenis and S. A. Dikanov, *J. Phys. Chem. B*, 2017, **121**, 3701–3717.
4. A. Singharoy, A. M. Barragan, S. Thangapandian, E. Tajkhorshid and K. Schulten, *J. Am. Chem. Soc.*, 2016, **138**, 12077–12089.
5. P. A. Postila, K. Kaszuba, P. Kuleta, I. Vattulainen, M. Sarewicz, A. Osyczka and T. Róg, *Sci. Rep.*, 2016, **6**, 33607.
6. P. Kuleta, M. Sarewicz, P. Postila, T. Róg and A. Osyczka, *Biochim. Biophys. Acta*, 2016, **1857**, 1661–1668.
7. P. Husen and I. A. Solov'yov, *J. Am. Chem. Soc.*, 2016, **138**, 12150–12158.
8. P. Husen and I. A. Solov'yov, *J. Phys. Chem. B*, 2017, **121**, 3308–3317.
9. A. M. Barragan, K. Schulten and I. A. Solov'yov, *J. Phys. Chem. B*, 2016, **120**, 11369–11380.
10. M. A. Hagras, T. Hayashi and A. A. Stuchebrukhov, *J. Phys. Chem. B*, 2015, **119**, 14637–14651.
11. A. R. Crofts, *Photosynth. Res.*, 2004, **80**, 223–243.
12. W. A. Cramer, H. Zhang, J. Yan, G. Kurisu and J. L. Smith, *Biochemistry*, 2004, **43**, 5921–5929.
13. A. R. Crofts, *Biochim. Biophys. Acta*, 2004, **1655**, 77–92.
14. A. R. Crofts, S. Hong, C. Wilson, R. Burton, D. Victoria, C. Harrison and K. Schulten, *Biochim. Biophys. Acta*, 2013, **1827**, 1362–1377.
15. E. A. Berry, H. De Bari and L.-S. Huang, *Biochim. Biophys. Acta*, 2013, **1827**, 1258–1277.
16. E. A. Berry, D.-W. Lee, L.-S. Huang and F. Daldal, in *The Purple Phototrophic Bacteria*, ed. C. N. Hunter, F. Daldal, M. C. Thurnauer and J. T. Beatty, Springer, Dordrecht, The Netherlands, 2009, ch. 22.
17. C. Hunte, S. Solmaz, H. Palsdóttir and T. Wenz, *Bioenergetics*, 2008, **45**, 253–278.
18. D. Xia, L. Esser, W.-K. Tang, F. Zhou, Y. Zhou, L. Yu, C.-A. Yu and B. Meunier, *Biochim. Biophys. Acta*, 2013, **1827**, 1278–1294.
19. M. K. F. Wikström and J. A. Berden, *Biochim. Biophys. Acta*, 1972, **283**, 403–420.
20. P. Mitchell, *J. Theor. Biol.*, 1976, **62**, 327–367.
21. A. R. Crofts, S. W. Meinhardt, K. R. Jones and M. Snozzi, *Biochim. Biophys. Acta*, 1983, **723**, 202–218.
22. A. R. Crofts, V. P. Shinkarev, D. R. J. Kolling and S. Hong, *J. Biol. Chem.*, 2003, **278**, 36191–36201.
23. A. R. Crofts, J. T. Holland, D. Victoria, D. R. Kolling, S. A. Dikanov, R. Gilbreth, S. Lhee, R. Kuras and M. G. Kuras, *Biochim. Biophys. Acta.*, 2008, **1777**, 1001–1019.
24. A. Osyczka, C. C. Moser, F. Daldal and P. L. Dutton, *Nature*, 2004, **427**, 607–612.
25. D. Victoria, R. Burton and A. R. Crofts, *Biochim. Biophys. Acta*, 2012, **1827**, 365–386.
26. A. R. Crofts, S. Lhee, S.B. Crofts, J. Cheng and S. Rose, *Biochim. Biophys. Acta*, 2006, **1757**, 1019–1034.

27. A. R. Crofts and S. W. Meinhardt, *Biochem. Soc. Trans.*, 1982, **10**, 201–203.
28. A. Osyczka, C. C. Moser and P. L. Dutton, *Trends Biochem. Sci.*, 2005, **30**, 176–182.
29. C. C. Moser, C. C. Page, R. Farid and P. L. Dutton, *J. Bioenerg. Biomembranes*, 1995, **27**, 263–274.
30. C. C. Moser, T. A. Farid, S. E. Chobot and P. L. Dutton, *Biochim. Biophys. Acta*, 2006, **1757**, 1096–1109.
31. A. R. Crofts, M. Guergova-Kuras, N. Ugulava, R. Kuras and S. Hong, *Proc. XIIth Congress of Photosynthesis Research*, Brisbane, Australia, 2002, p. 6.
32. A. R. Crofts, M. Guergova-Kuras, L.-S. Huang, R. Kuras, Z. Zhang and E. A. Berry, *Biochemistry*, 1999, **38**, 15791–15806.
33. A. R. Crofts, S. J. Hong, N. Ugulava, B. Barquera, R. Gennis, M. Guergova-Kuras and E. A. Berry, *Proc. Natl. Acad. Sci. U. S. A.*, 1999, **96**, 10021–10026.
34. A. R. Crofts, B. Barquera, R. B. Gennis, R. Kuras, M. Guergova-Kuras and E. A. Berry, in *The Phototrophic Prokaryotes*, ed. G. A. Peschek, W. Loeffelhardt and G. Schmetterer, Plenum Publishing Corporation, New York; London; Washington, DC; Boston, 1999, pp. 229–239.
35. S. J. Hong, N. Ugulava, M. Guergova-Kuras and A. R. Crofts, *J. Biol. Chem.*, 1999, **274**, 33931–33944.
36. A. R. Crofts, M. Guergova-Kuras, N. Ugulava, R. Kuras and S. Hong, *Proc. XIIth Congress of Photosynthesis Research: Brisbane, Australia*, 2002, p. 6.
37. I. Forquer, R. Covian, M. K. Bowman, B. L. Trumpower and D. M. Kramer, *J. Biol. Chem.*, 2006, **281**, 38459–38465.
38. A. R. Crofts, B. Barquera, R. B. Gennis, R. Kuras, M. Guergova-Kuras and E. A. Berry, *Biochemistry*, 1999, **38**, 15807–15826.
39. D. J. Kolling, J. S. Brunzelle, S. Lhee, A. R. Crofts and S. K. Nair, *Structure*, 2007, **15**, 29–38.
40. S. Lhee, D. R. Kolling, S. K. Nair, S. A. Dikanov and A. R. Crofts, *J. Biol. Chem.*, 2010, **285**, 9233–9248.
41. Y. Zu, M. M.-J. Couture, D. R. J. Kolling, A. R. Crofts, L. D. Eltis, J. A. Fee and J. Hirst, *Biochemistry*, 2003, **42**, 12400–12408.
42. A. M. Barragan, A. R. Crofts, K. Schulten and I. A. Solov'yov, *J. Phys. Chem. B*, 2015, **119**, 433–447.
43. A. R. Crofts, M. Guergova-Kuras, R. Kuras, N. Ugulava, J. Li and S. Hong, *Biochim. Biophys. Acta*, 2000, **1459**, 456–466.
44. Z. Zhang, L.-S. Huang, V. M. Shulmeister, Y.-I. Chi, K.-K. Kim, L.-W. Hung, A. R. Crofts, E. A. Berry and S.-H. Kim, *Nature*, 1998, **392**, 677–684.
45. G. Jafari, B. M. Wasko, M. Kaeberlein and A. R. Crofts, *Worm*, 2016, DOI: 10.1080/21624054.2016.1174803.
46. G. Jafari, B. M. Wasko, A. Tong, N. Schurman, C. Dong, Z. Li, R. Peters, E.-B. Kayser, J. N. Pitt, P. G. Morgan, M. M. Sedensky, A. R. Crofts and M. Kaeberlein, *Proc. Natl. Acad. Sci. U. S. A.*, 2015, **112**, E6148–E6157.

47. A. R. Crofts, V. P. Shinkarev, S. A. Dikanov, R. I. Samoilova and D. Kolling, *Biochim. Biophys. Acta*, 2002, **1555**, 48–53.
48. A. R. Crofts and E. A. Berry, *Curr. Opin. Struc. Biol.*, 1998, **8**, 501–509.
49. B. L. Trumpower, *Biochim. Biophys. Acta*, 2002, **1555**, 166–173.
50. P. Lanciano, D.-W. Lee, H. Yang, E. Darrouzet and F. Daldal, *Biochemistry*, 2011, **50**, 1651–1663.
51. M. Castellani, R. Covian, T. Kleinschroth, O. Anderka, B. Ludwig and B. L. Trumpower, *J. Biol. Chem.*, 2010, **285**, 502–510.
52. M. Świerczek, E. Cieluch, M. Sarewicz, A. Borek, C. C. Moser, P. L. Dutton and A. Osyczka, *Science*, 2010, **329**, 451–454.
53. S. Hong, D. Victoria and A. R. Crofts, *Biochim. Biophys. Acta*, 2012, **1817**, 1053–1062.
54. M. Czapla, A. Borek, M. Sarewicz and A. Osyczka, *Biochemistry*, 2012, **51**, 829–835.
55. J. T. Holland, Ph.D., University of Illinois at Urbana-Champaign, 2007.
56. C.-H. Yun, A. R. Crofts and R. B. Gennis, *Biochemistry*, 1991, **30**, 6747–6754.
57. S. Bhaduri, V. Stadnytskyi, S. D. Zakharov, S. Saif Hasan, Ł. Bujnowicz, M. Sarewicz, S. Savikhin, A. Osyczka and W. A. Cramer, *J. Phys. Chem. B*, 2017, **121**, 975–983.
58. D. M. Kramer and A. R. Crofts, *Biochim. Biophys. Acta*, 1993, **1184**, 193–201.
59. P. Joliot and A. Joliot, *Biochim. Biophys. Acta Bioenerg.*, 1988, **933**, 319–333.
60. S. Pöyry, O. Cramariuc, P. A. Postila, K. Kaszuba, M. Sarewicz, A. Osyczka, I. Vattulainen and T. Róg, *Biochim. Biophys. Acta*, 2013, **1827**, 769–778.
61. X. Zhang, C. Hiser, B. Tamot, C. Benning, G. E. Reid and S. M. Ferguson-Miller, *Biochemistry*, 2011, **50**, 3891–3902.
62. G. Jafari, B. M. Wasko, A. Tonge, N. Schurman, C. Dong, Z. Li, R. Peters, E.-B. Kayser, J. N. Pitt, P. G. Morgan, M. M. Sedensky, A. R. Crofts and M. Kaeberlein, *Proc. Natl. Acad. Sci. U. S. A.*, 2015, **112**, E6148–E6157.
63. M. Guergova-Kuras, R. Kuras, N. Ugulava, I. Hadad and A. R. Crofts, *Biochemistry*, 2000, **39**, 7436–7444.
64. D. R. Kolling, R. I. Samoilova, A. A. Shubin, A. R. Crofts and S. A. Dikanov, *J. Phys. Chem. A*, 2009, **113**, 653–667.
65. R. I. Samoilova, D. Kolling, T. Uzawa, T. Iwasaki, A. R. Crofts and S. A. Dikanov, *J. Biol. Chem.*, 2002, **277**, 4605–4608.
66. V. P. Shinkarev, D. R. J. Kolling, T. J. Miller and A. R. Crofts, *Biochemistry*, 2002, **41**, 14372–14382.
67. E. Darrouzet, M. Valkova-Valchanova and F. Daldal, *J. Biol. Chem.*, 2002, **277**, 3464–3470.
68. E. Darrouzet and F. Daldal, *Biochemistry*, 2003, **42**, 1499–1507.
69. E. Darrouzet, M. Valkova-Valchanova, C. C. Moser, P. L. Dutton and F. Daldal, *Proc. Natl. Acad. Sci. U. S. A.*, 2000, **97**, 4567–4572.
70. A. R. Crofts, *Annu. Rev. Physiol.*, 2004, **66**, 689–733.

71. J. L. Cape, M. K. Bowman and D. M. Kramer, *Proc. Natl. Acad. Sci. U. S. A.*, 2007, **104**, 7887–7892.
72. H. Zhang, A. Osyczka, P. L. Dutton and C. C. Moser, *Biochim. Biophys. Acta*, 2007, **1767**, 883–887.
73. P. R. Vennam, N. Fisher, M. D. Krzyaniak, D. M. Kramer and M. K. Bowman, *Chem. Biochem.*, 2013, **14**, 1745–1753.
74. M. Sarewicz, M. Dutka, S. Pintscher and A. Osyczka, *Biochemistry*, 2013, **52**, 6388–6395.
75. S. A. Dikanov, J. T. Holland, B. Endeward, D. R. Kolling, R. I. Samoilova, T. F. Prisner and A. R. Crofts, *J. Biol. Chem.*, 2007, **282**, 25831–25841.
76. R. L. Burton, Ph. D., University of Illinois at Urbana-Champaign, 2015.
77. A. Y. Mulkidjanian and W. Junge, *FEBS Lett.*, 1994, **353**, 189–193.
78. W. Ausländer and W. Junge, *FEBS Lett.*, 1975, **59**, 310–315.
79. S. Saphon and A. R. Crofts, *Z. Naturforsch.*, 1977, **32c**, 617–626.
80. W. Junge, W. Ausländer, A. McGeer and T. Runge, *Biochim. Biophys. Acta*, 1979, **546**, 121–141.
81. S. W. Meinhardt and A. R. Crofts, *Biochim. Biophys. Acta*, 1983, **723**, 219–230.
82. A. R. Crofts, in *The Enzymes of Biological Membranes*, ed. A. N. Martonosi, Plenum Publ. Corp., New York., 1985, vol. 4, pp. 347–382.
83. A. R. Crofts, S. Hong, Z. Zhang and E. A. Berry, *Biochemistry*, 1999, **38**, 15827–15839.
84. S. Izrailev, A. R. Crofts, E. A. Berry and K. Schulten, *Biophys. J.*, 1999, **77**, 1753–1768.
85. D. R. Martin, D. N. LeBard and D. V. Matyushov, *J. Phys. Chem. Lett.*, 2013, **4**, 3602–3606.
86. P. A. Postila, K. Kaszuba, M. Sarewicz, A. Osyczka, I. Vattulainen and T. Róg, *Biochim. Biophys. Acta*, 2013, **1827**, 761–768.
87. D. R. J. Kolling, R. I. Samoilova, J. T. Holland, E. A. Berry, S. A. Dikanov and A. R. Crofts, *J. Biol. Chem.*, 2003, **278**, 39747–39754.
88. O. Kokhan and V. P. Shinkarev, *Biophys. J.*, 2011, **100**, 720–728.
89. S. A. Dikanov, D. R. J. Kolling, B. Endeward, R. I. Samoilova, T. F. Prisner, S. K. Nair and A. R. Crofts, *J. Biol. Chem.*, 2006, **281**, 27416–27425.
90. S. Hong, W. B. de Almeida, A. T. Taguchi, R. I. Samoilova, R. B. Gennis, P. J. O'Malley, S. A. Dikanov and A. R. Crofts, *Biochemistry*, 2014, **53**, 6022–6031.

CHAPTER 8

Advances in Understanding Mechanism and Physiology of Cytochromes bc

ARKADIUSZ BOREK, ROBERT EKIERT AND ARTUR OSYCZKA*

Jagiellonian University, Department of Molecular Biophysics,
Faculty of Biochemistry, Biophysics and Biotechnology, Gronostajowa 7,
30-387 Kraków, Poland
*Email: artur.osyczka@uj.edu.pl

8.1 Introduction

8.1.1 Overview of Cytochrome *bc* Complexes

Cytochrome *bc*-type complexes (*bc* complexes) are the key proteins of respiratory (cytochrome bc_1) and photosynthetic (cytochromes bc_1 and b_6f) electron transport chains involved in conservation of energy, powering cell metabolism.[1]

The *bc* complexes transfer electrons between membranous electron/proton carriers (ubiquinone, menaquinone, plastoquinone) and water-soluble electron carriers (cytochrome *c*, plastocyanin, high potential iron–sulphur protein).[2] These two types of mobile carriers form two redox pools (described further as the Q pool for quinone and the c pool for cytochrome *c*) securing communication between membranous complexes of the electron transport chain. The flow of electrons between the Q pool and the c pool through the cofactor chains of *bc* complexes powers the catalytic cycle, known as

Chemical Biology No. 5
Mechanisms of Primary Energy Transduction in Biology
Edited by Mårten Wikström

Published by the Royal Society of Chemistry, www.rsc.org

the Q-cycle[3,4] which results in net translocation of protons across the energy conserving membranes (bioenergetic membranes). This way, the *bc* complexes contribute directly to the generation of the proton-motive force (PMF) utilized by ATP synthase to produce ATP. At the same time, these complexes, by virtue of connecting two redox pools of electron transport chains and restoring the appropriate redox state of substrates for other complexes, maintain the smooth operation of the entire electron transport chains. This can be regarded as their indirect contribution to generation of PMF. Increasing concentration of ATP results in disequilibrium between ATP, ADP, and P_i, which is at the basis of the energy conversion in cellular metabolism.

The indirect contribution of *bc* complexes to generation of PMF is particularly important for the chains in which these enzymes are the only points of electronic connection of the Q- and c-pools. This is a case for the mitochondrial respiratory chain where cytochrome bc_1 (also named mitochondrial complex III, or ubiquinol-cytochrome *c* oxidoreductase) is the only enzyme linking the quinone-reducing complexes (mitochondrial complexes I and II) with cytochrome *c*-oxidizing mitochondrial complex IV (cytochrome *c* oxidase) (Figure 8.1A). Under cellular ATP demand, the net operation of complex III results in oxidation of quinol and reduction of cytochrome *c*.

8.1.2 Structure of Cytochrome bc_1

Cytochrome bc_1 is a homodimeric membranous protein complex. The number of subunits of cytochrome bc_1 varies among the species. Mitochondrial complex III contains up to 11 subunits per monomer:[5–7] three subunits forming a so-called catalytic core, two core transmembrane proteins and six low-molecular weight proteins, while some bacterial cytochrome bc_1 complexes may contain as few as three catalytic subunits (without the supernumerary subunits).[8,9] The catalytic core consists of cytochrome *b*, cytochrome c_1 and iron–sulfur protein (ISP) (Figure 8.1B) which share structural similarity among the species.

Within the catalytic core there are three catalytic sites (Figure 8.1C): (1) quinol oxidation site (Q_o site) localized at the cytochrome *b*-ISP interface near the positive side of the membrane (inter-membrane space); (2) cytochrome *c* binding site localized at the extramembranous surface of cytochrome c_1 subunit at the positive side of the membrane; and (3) quinone reduction site (Q_i site) confined to cytochrome *b* and localized near the negative side of the membrane (mitochondrial matrix or bacterial cytoplasm).

The redox cofactors of the catalytic core assemble into two chains: high-potential c-chain and low-potential b-chain (Figure 8.1C) which connect the catalytic sites of cytochrome bc_1. In the c-chain, the Q_o site is connected with the cytochrome *c* binding site by two cofactors: Rieske cluster ([2Fe–2S] cluster) embedded in ISP and heme c_1 of cytochrome c_1. Within this chain, transfer of electrons is made possible by movement of the head domain of iron–sulfur protein (ISP-HD) between a position close to the Q_o site

(Q_o position) and a position on the cytochrome c_1 interface (c_1 position).[7,10–13] This movement does not limit the net operation of cytochrome bc_1.[13] The interaction of cytochrome c_1 with diffusible cytochrome c is electrostatic in nature.[14–17] Under physiological conditions, the complexes between these two proteins are short-lived and the frequency of collision between protein surfaces is high.[18–21] The inter-protein electron transfer occurs coincidentally with one of these collisions.

In the b-chain the Q_o site is connected with the Q_i site by heme b_L and heme b_H embedded in cytochrome b allowing the transfer of electrons across the membrane. In a dimer, the cofactors of two monomers are separated

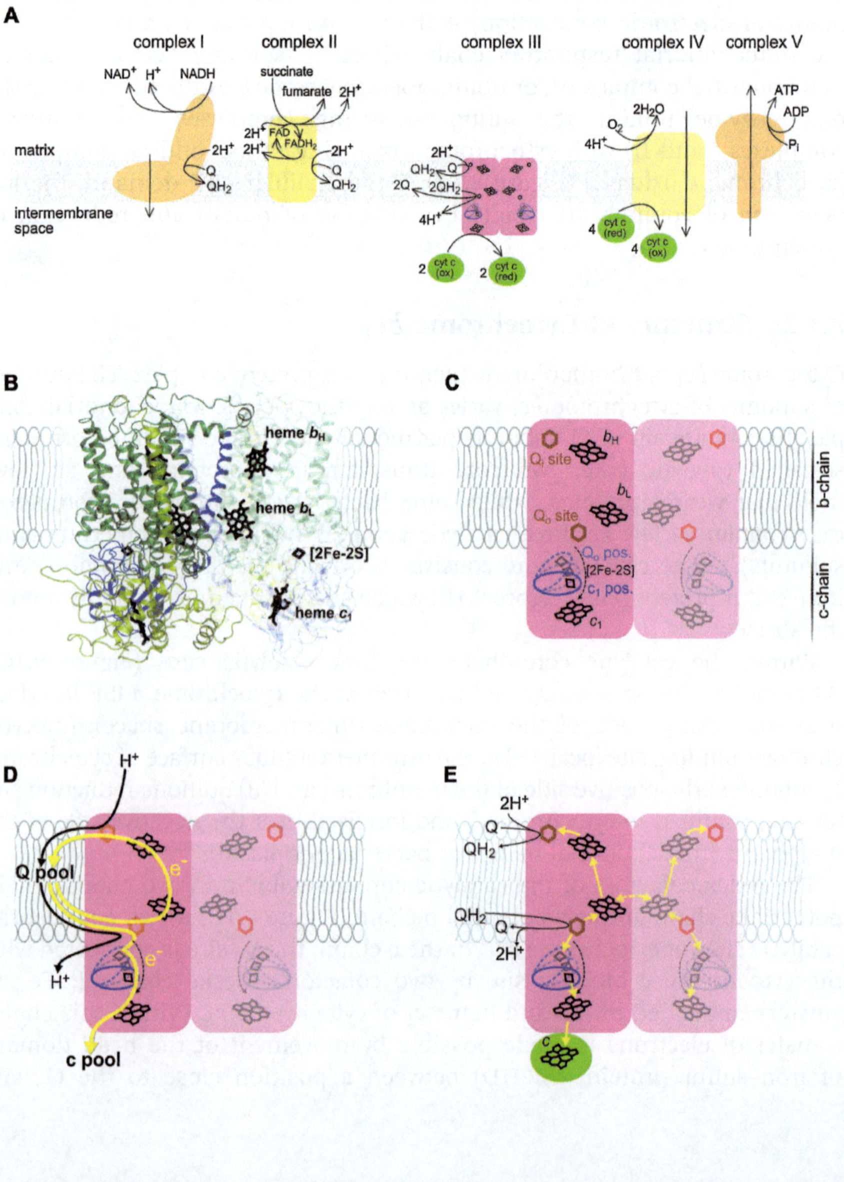

by large distances except for the two hemes b_L which, separated by approximately 14 Å, form a bridge providing additional connection between the catalytic sites (Figure 8.1E).

8.1.3 Catalytic Cycle

The catalytic Q cycle links the oxidation of quinol at the Q_o site associated with release of protons to the intermembrane space with the reduction of quinone at the Q_i site associated with uptake of protons from mitochondrial matrix (Figure 8.1D, E).

The two-electron/two-proton oxidation of quinol at the Q_o site involves a rather unusual in biology bifurcation reaction separating the routes for electrons: one electron reduces the [2Fe–2S] cluster and is transferred *via* c-chain to cytochrome *c*, while the other electron reduces heme b_L and travels along the b-chain to the Q_i site. Since the reduction of quinone at this site is also a two-electron/two-proton reaction, one electron delivered from the Q_o site can reduce quinone to semiquinone or semiquinone to quinol. Thus, a complete reduction of quinone to quinol at the Q_i site requires two turnovers of the Q_o site.

In addition to energy-conserving catalytic reactions of the Q cycle, under some conditions, energy-wasting side reactions disclose themselves. These involve short-circuits or leaks of electrons to molecular oxygen with formation of superoxide radical.[22–24]

Figure 8.1 Structure, cofactors and catalytic cycle of cytochrome bc_1 complex. (A) General scheme of reactions carried out by protein complexes of the respiratory chain. Vertical arrows represent proton pumping across the membrane through proton channels in complexes I, IV and V. Complex III translocates protons across the membrane in oxidation/reduction reactions of quinone molecules. (B) Structure of the subunits of the catalytic core of cytochrome bc_1. Cytochrome *b* containing heme b_L and heme b_H, cytochrome c_1 containing heme c_1 and ISP containing a Rieske cluster are marked in green, yellow and blue, respectively. The cofactors are marked in black. To simplify the scheme one of the monomers is transparent. (C) The heme cofactor chains and catalytic sites of bc_1 complex. The c-chain connects the quinol oxidation site (Q_o site) with cytochrome *c* reduction site by two cofactors: a Rieske cluster ([2Fe–2S] cluster) and heme c_1. The b-chain connects the Q_o site with a quinone reduction site (Q_i site) by heme b_L and heme b_H. ISP-HD, a domain containing a Rieske cluster (black diamond) moves (black dotted arrow) between Q_o position (Q_o pos.) and c_1 position (c_1 pos.). To simplify the scheme: the catalytic core including cytochrome *b*, cytochrome c_1 and ISP is shown as pink rectangle (the same applies for D and E). (D) General electron (yellow arrows) and proton (black arrows) transfers through the cytochrome bc_1 during the Q-cycle at the level of one monomer. (E) Individual reactions of the Q-cycle. Yellow arrows show the path of the electron transfer within the protein. The yellow dotted arrow indicates electron transfer between two hemes b_L in the dimeric cytochrome bc_1 complex. The scheme does not consider stoichiometry of the Q cycle. To simplify the scheme, the Q cycle reactions are highlighted only for one monomer.

8.1.4 Purpose of This Review

Extensive studies that were and are still conducted constantly broaden our understanding of catalytic mechanisms of *bc* complexes and their contribution to physiological processes. At the same time they bring new ideas and identify issues that remain to be elucidated. Here are some examples:

(a) What are the properties of intermediate radical states of catalysis and what is their role?
(b) How are protons transferred to/from the catalytic sites?
(c) What is the molecular mechanism of superoxide production at the Q_o site and how can the levels of reactive oxygen species (ROS) be modulated?
(d) What is the reason behind the existence of the electronic connection between the monomers of the dimeric enzyme?

In this review we will reflect on the recent findings that relate to these questions. We will discuss new structural and kinetic elements of action of catalytic sites inferred from experimental studies and MD simulations. In particular, we will focus on the mechanism of superoxide generation in the context of recently discovered intermediate states of catalysis.

8.2 New Insights into Operation of the Q_o Catalytic Site

The Q_o site can be considered a heart of cytochrome bc_1 and of the entire mitochondrial respiratory chain. This is because the Q_o site harvests quinol molecules derived from complexes I, II and the Q_i site of complex III and connects these complexes with complex IV *via* cytochrome *c*. Once the quinol molecule enters the Q_o site, it is flanked by the two redox active cofactors, which typically engage in the catalytic electron bifurcation. However, this unique environment also creates a risk of side reactions, some of which may lead to ROS formation.

8.2.1 Electron and Proton Routes at the Q_o Site

Under most physiological conditions (ATP generation conditions) cytochrome bc_1 operates in a forward direction (forward reaction) with the bifurcation of two electrons at the Q_o site and overall electron flow from quinol to cytochrome *c* pool. This results in net translocation of protons from matrix to the intermembrane space (Figure 8.1C). The bifurcation reaction can, in principle, be realized in two ways: as a sequential[22,23,25–27] or concerted[22,28–31] reaction (Figure 8.2A).

The sequential model of a forward reaction considers two time-separated one-electron reactions occurring sequentially. According to this model, as a result of the first one-electron reaction, quinol molecule undergoes oxidation to semiquinone intermediate (SQ_o) with an electron transfer to Rieske

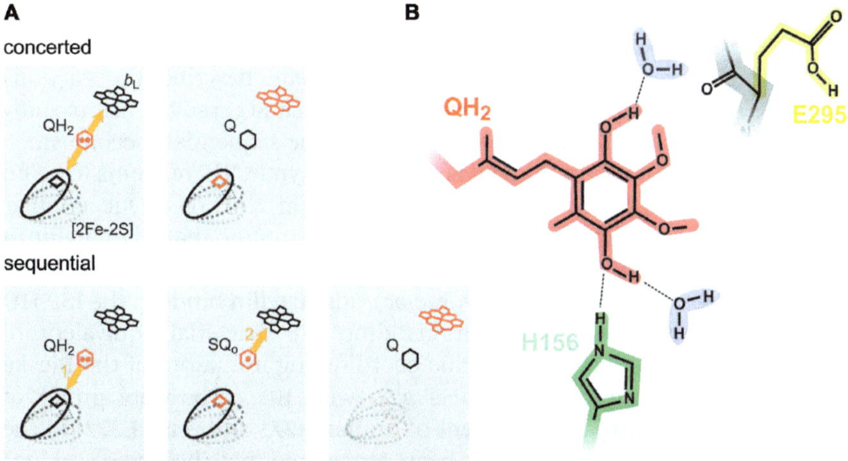

Figure 8.2 Reactions at the Q_o catalytic site. (A) The comparison of two possible mechanisms of electron bifurcation in a forward reaction of the Q_o site. In the concerted mechanism, the electrons derived from quinol (QH_2) are transferred simultaneously to heme b_L and [2Fe–2S] cluster with formation of quinone (Q). In sequential mechanism, quinol molecule undergoes oxidation in the two sequential steps marked as 1 and 2, involving a semiquinone intermediate (SQ_o). The cofactors marked in red represent their reduced state, cofactors marked in black their oxidized state. Red dots represent electrons on quinone molecules. (B) Protonable residues and water molecules in the QH_2 binding pocket. Quinol molecule and water molecules are marked in red and light blue, respectively. Two amino acids E295 of cytochrome *b* and H156 of ISP are marked in yellow and green, respectively. A fragment of the polypeptide backbone formed by E295 is marked in grey.

cluster of ISP. The second reaction is oxidation of SQ_o to quinone with transfer of a second electron to heme b_L. The concerted model of a forward reaction assumes that electrons derived from quinol at the Q_o site are transferred to Rieske cluster and heme b_L almost simultaneously (*i.e.* the time separation between the two electron transfer steps is shorter than the time of atomic rearrangements[22]) with no formation of SQ_o.

Under high membrane potential (at high ATP/ADP ratio), cytochrome bc_1 can transfer electrons in reverse direction (from cytochrome *c* to quinone).[32,33] This results in reverse translocation of protons. Interestingly, such reverse electron and proton flow occurs in lithoautotrophic bacteria, *Acidithiobacillus ferrooxidans*, and is necessary to sustain bacterial growth on Fe^{2+}.[34,35]

Considering the mechanism of reverse electron flow at the level of the Q_o site the same two reaction schemes (*i.e.* sequential and concerted) are possible. For the sequential reverse mechanism, the first reaction is reduction of quinone to SQ_o by heme b_L, followed by reduction of SQ_o to quinol by Rieske cluster. In the concerted reverse mechanism both reactions occur almost simultaneously.

Given all available information on catalytic and side reactions (in particular in the context of ROS generation; see discussion in Section 8.2.2), the question as to whether the sequential or concerted scheme better describes the reactions at this site still remains open. However, the fact that SQ_o radical has recently been detected (Section 8.2.2) argues in favour of the sequential mechanism.

The transfer of protons at this site is even more cryptic. There seems to be no direct experimental evidence for any of the proton reactions. One popular model for proton reactions is inferred from the assumption that stigmatellin, a potent inhibitor of the Q_o site, mimicks interaction of the substrate with the site.[36] In crystal structures of cytochrome bc_1, stigmatellin binds to the ISP-HD by forming a hydrogen bond with histidine 156 (bacterial equivalent of human mitochondrial H161), a residue coordinating iron atom of the Rieske cluster.[11,36–40] This inhibitor interacts also with the carboxylate group of glutamic acid 295 (bacterial equivalent of human E271 and yeast E272) in the PEWY motif of cytochrome *b*. It has been considered that these two residues act as primary acceptors of protons from quinol undergoing oxidation at the site. However, the proposed key role of Glu295 was not supported experimentally as replacing this amino acid with a non-protonable one did not block the operation of the Q_o site.[41–44] Furthermore, the PEWY motif is not absolutely conserved. It appears to be an evolutionary adaptation to the oxygenic atmosphere as it is present in bacteria that use high-potential Rieske cluster and ubiquinone or plastoquinone species. Bacteria using menaquinone and low-potential Rieske contain PDWY motif, while some bacteria use PVWY, APWY and GPWY or even PPWF in the case of Archaea.[45] Thus the postulated mechanism with protonation of Glu from PEWY motif would not hold for all species. In particular, it was noted that with the presence of proline at the second position of this motif, the quinol cannot be hydrogen bonded by the amino acid side chain, but it could be bound through water-mediated hydrogen bonds as seen in cytochrome bc_1 structure with the bound inhibitor HHDBT.[46]

The role of water molecules acting as direct acceptors of both protons from quinol, even when PEWY motif is present, (Figure 8.2B) is implicated from recent MD simulations showing that binding at the Q_o site involves a coordinated water molecules positioned close to the hydroxyls of the quinol ring.[47] In this model cytochrome bc_1 does not need to maintain His156 deprotonated prior to the oxidation of quinol. The putative involvement of water molecule in the proton transfer at the Q_o site was later confirmed by a combination of MD and quantum chemical calculations.[48] However, these authors suggest a key role of Tyr147 that acts as an intermediary in proton transfer to Glu295. Side chain of this amino acid perform a turn during the simulations, that could not be observed in the crystal structures of the bc_1 complexes.

8.2.2 Semiquinone at the Q_o Site and Its Involvement in Superoxide Generation

In the sequential scheme, the reaction at the Q_o site proceeds through the obligatory semiquinone intermediate state. Given the specific nature of electronic bifurcation in the environment of the Q_o site, the models commonly assume that SQ_o is highly unstable.[49–54] It is also commonly assumed that this highly unstable SQ_o can react with oxygen, generating superoxide. Indeed, the long-standing difficulties in detection of SQ_o under equilibrium conditions suggested the lack of semiquinone or its extremely short lifetime.[22,25,27,31,49,55–58] However, more recent work indicated that SQ_o can be detected under non-equilibrium conditions.[59–63] The semiquinones identified so far at the Q_o site appear to have different spectroscopic properties, which leaves space for various interpretations and mechanistic considerations (recently reviewed in Ref. 64).

In the reversibly operating Q_o site, SQ_o can be generated according to two different mechanisms: "semiforward" and "semireverse" (Figure 8.3).[2,65] In the "semiforward" reaction, which can be regarded as an initial part of oxidation of quinol at the Q_o site (forward reaction), SQ_o is formed when ISP withdraws an electron from quinol bound to this site.[23,66] In "semireverse" reaction, which can be regarded as an initial part of quinone reduction at the

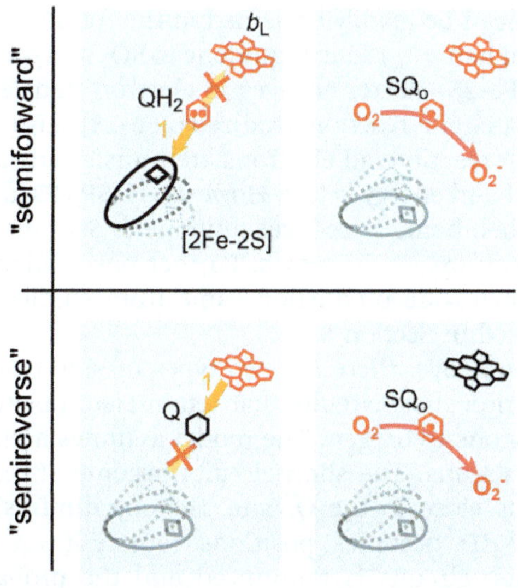

Figure 8.3 Two possible mechanisms of generation of semiquinone at the Q_o site in relation to ROS generation. In a "semiforward" mechanism, which is a part of a forward reaction, semiquinone (SQ_o) is generated when one electron is transferred from quinol molecule (QH_2) to [2Fe–2S] cluster (rhombus) (left panel top). In "semireverse" mechanism, which is a part of reverse reaction, generation of SQ_o occurs when heme b_L donates an electron to quinone (Q) (left panel bottom). While SQ_o generated in both mechanisms (right panel top and bottom) can in principle react with oxygen with formation of superoxide radical (red arrows), this reaction has higher probability in "semireverse" mechanism (right panel bottom). The cofactors marked in red represent their reduced state, cofactors marked in black their oxidized state, red dots represent electrons in quinone molecules.

Q_o site (reverse reaction), SQ_o is formed when reduced heme b_L donates an electron to quinone bound to this site.[65,67,68] Both mechanisms can, in principle, explain the production of the superoxide radical by the cytochrome bc_1 when the flow of electrons through the b-chain is impeded by blocking oxidation of heme b_H at the level of the Q_i site (*e.g.* in the presence of the Q_i site inhibitor, antimycin). This is because in both mechanisms the reduced state of heme b_L promotes the formation of SQ_o (Figure 8.3). However, only the "semireverse" mechanism explains the observation that the maximum rate of superoxide generation in antimycin-inhibited cytochrome bc_1 is detected when the Q pool is partly oxidized.[67,69,70] This is because in "semireverse" mechanism, unlike in "semiforward" mechanism, quinone acts as a substrate for semiquinone generated at the Q_o site.

It appears that the probability of superoxide generation at the Q_o site depends on the position of ISP-HD at the time of SQ_o formation. This was first indicated by the observation that mutants, which arrest the ISP-HD at the Q_o site,[13] generally suppress production of superoxide and this effect is larger for the mutant that arrests the ISP-HD for seconds, comparing to the mutant that arrests it for milliseconds (2Ala and 1Ala, respectively, see Figure 8.4A).[21,68] Such suppression can be readily explained in the frame of the "semireverse" mechanism. When heme b_L reduces quinone to SQ_o and the ISP-HD is present at the site, the [2Fe–2S] cluster engages in electron transfer with SQ_o, which may lead to a short-circuit reaction (oxidized [2Fe–2S] cluster takes an electron from SQ_o). In such reaction, all electrons are transferred *via* a high potential chain to the *c* pool and energy is lost. However, if ISP-HD is not present at the site at the time when heme b_L reduces quinone to SQ_o, the probability of the reaction of SQ_o with oxygen increases. Further observations supporting this mechanism of superoxide production came from studies of mitochondrial mutations discussed in Section 8.2.3.

Considering the above, there are two types of side reactions dissipating energy at the Q_o site: short-circuits that retain electrons within the enzyme and leaks of electrons to oxygen. The model assumes a kinetic competition between these reactions. The short-circuits become effective when ISP-HD occupies positions close to the Q_o site, thereby diminishing the electron leaks. When ISP-HD occupies positions remote from the Q_o site, the probability of short-circuits is minimized and the probability of electron leakage on oxygen is enhanced. This competition may be of physiological relevance: despite the fact that the short-circuits dissipate energy, under certain conditions, they may be beneficial in helping to diminish the ROS production.[21,24,68]

Recent MD simulations indicate that oxygen molecules can diffuse spontaneously into the Q_o site of cytochrome bc_1 and may react with semiquinone bound at this site.[71] This means that SQ_o does not have to leave the catalytic site to react with oxygen. Moreover, oxygen molecules may also occasionally reach the vicinity of hemes *b* (within 5 Å of the central iron atom of the hemes b_L and b_H) and it was suggested that these hemes may also be directly involved in reaction with oxygen.[71]

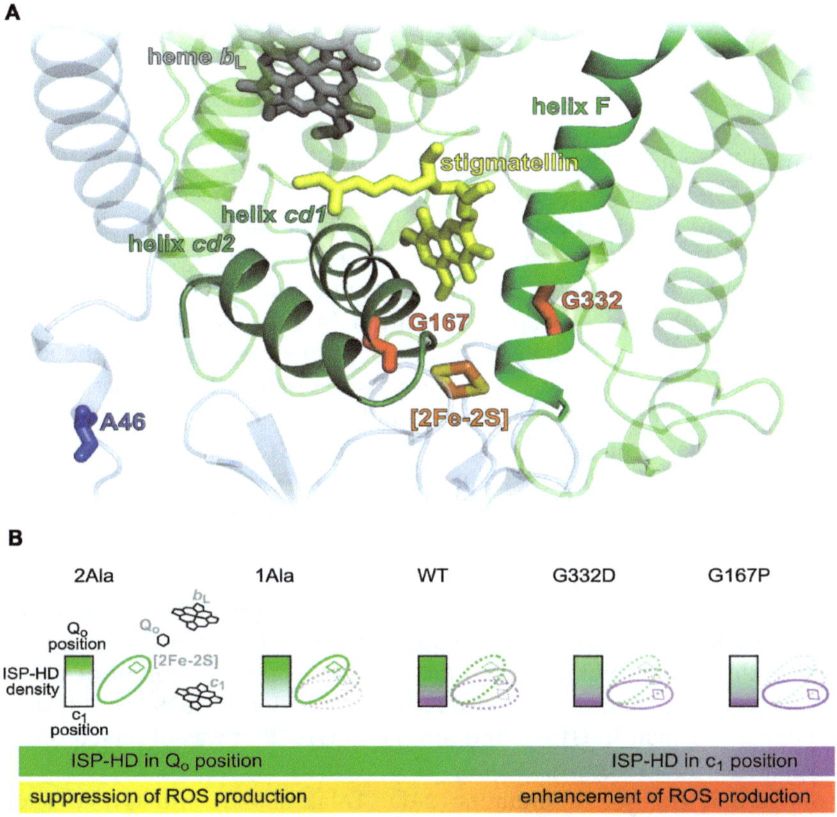

Figure 8.4 Mutations in cytochrome *b* and ISP affecting the motion of ISP-HD and generation of ROS. (A) Crystal structure of *Rb. capsulatus* cytochrome bc_1 (PDB ID: 1ZRT) showing the vicinity of the Q_o site and positions of various mutations. Yellow sticks represent the Q_o site inhibitor, stigmatellin. G167 (red sticks) is localized at the end of *cd1* helix (dark green) of cytochrome *b* (light green) while G332 (red sticks) is localized on helix F (green) of cytochrome *b*. Both mutations are located at the entry to the Q_o site, at the cytochrome *b*-iron–sulfur protein (ISP) interface. A46 (blue sticks) located in the neck region of ISP (light blue) indicates the position where one (1Ala) or two (2Ala) alanine residues are inserted. Dark gray and light-orange sticks indicate heme b_L and Rieske cluster, respectively. (B) Scheme illustrating the difference in equilibrium distribution of ISP-HD position in various mutants in comparison to WT. In 2Ala, ISP-HD stays captured at the Q_o site for seconds, and its average position is shifted toward the Q_o site in relation to WT. The presence of ISP-HD at the Q_o site suppresses the production of ROS. In the native cytochrome bc_1, ISP-HD moves freely between Q_o and c_1 positions. G167P has an effect opposite to 2Ala: the average position of ISP-HD is more remote from the Q_o site than in WT. This shift enhances the production of ROS. The molecular effects of 1Ala and G332D mutants are moderate (1Ala falls between 2Ala and WT; G332D falls between G167P and WT). Horizontal rectangles depict the density of ISP-HD at different positions: green and purple denote Q_o and c_1 positions of ISP-HD, respectively.

8.2.3 Testing the "Semireverse" Model of ROS Production Using Mitochondrial Mutations

As described in the previous paragraph, the kinetic competition between the short-circuits and leaks of electrons at the Q_o site was deduced from the remarkable suppression in ROS generation observed in the mutants that arrest the ISP-HD at the Q_o site (1Ala, 2Ala).[65,68,69] Macroscopically, the effect on motion of ISP-HD in these mutants, can be viewed as a shift in the average ISP-HD position towards positions close to the Q_o site.[68,72] In this context, the "semireverse" model of superoxide production predicts that shifting the average ISP-HD position in the opposite direction (*i.e.* towards the positions remote from the Q_o site) should enhance the superoxide production. Recently, it became possible to test this prediction using the bacterial mimics of certain mitochondrial mutations.

The equivalents of mitochondrial mutations S151P and G290D in cytochrome *b* of *Rb. capsulatus* cytochrome bc_1, G167P and G332D, respectively (reported for patients suffering from exercise intolerance[73,74] and localized at the entry of the Q_o catalytic site (Figure 8.4A)) were found to change the average position of ISP-HD in such a way that it stays more remote from the Q_o site than in the native enzyme.[69,70] Indeed, as predicted by the model, both mutations led to an increased propensity of cytochrome bc_1 to generate superoxide. Furthermore, as the effect of G332D on the motion of ISP-HD was more subtle than that of G167P, the production of ROS in G332D was less enhanced than in G167P and, unlike in G167P, it was observed only in the presence of antimycin.[69,70]

Altogether, the set of mutants: 2Ala, 1Ala, G332D, and G167P shows remarkable correlation between the average ISP-HD position and the level of superoxide production, in agreement with the "semireverse" mechanism of ROS generation (Figure 8.4B).

8.2.4 Physiological Aspects of Superoxide Generation at the Q_o Site

As a result of the side reactions, part of the energy released from oxidation of quinol is dissipated, which reduces the efficiency of separation of electrons at the Q_o site into two cofactor chains, reduces the efficiency of proton translocation and therefore reduces efficiency of oxidative phosphorylation. The contribution of side reactions of the Q cycle is considered negligible under high respiration rate.[75,76] Nevertheless, factors limiting the rate of respiration (high membrane potential, nearly inactive complex V, high NADH : NAD^+ ratio or hypoxia) appear to favour superoxide production which originates mainly from complexes I and III.[75–79] Production of superoxide by complex III under these conditions can be explained in the frame of "semireverse" mechanism of superoxide generation, given that under low respiration rate, the reduction levels of cofactors increase, which in turn increases the probability of the "semireverse" reaction occurring at the Q_o site. The same logic can be used to explain the influence of antimycin[23,80–82] and several mutations that impede the electron flow through the cofactor chains of cytochrome bc_1 (including mitochondrial mutations and cofactor knock-outs)[65,69,70] on the superoxide production by this enzyme.

Generally, ROS are considered deleterious as they could cause damage to certain macromolecules including lipids, proteins and DNA. However, it depends on the level of ROS production. It is postulated that below a certain threshold free radicals may be beneficial. According to the theory of mitohormesis,[83] a low, continuous production of ROS by mitochondria may cause an adaptive response by activating the long-term defence against higher levels of ROS produced during oxidative stress. This mechanism could relate to the extended lifespan of organisms.[84] Limited ROS may also act as signalling molecules.[85–89] Considering the respiratory chain, ROS may be released to two sides of the membrane: matrix and intermembrane space.[90] It seems that ROS released to the latter side may play a more important role in signalling.[91] Cytochrome bc_1, as one of the major mitochondrial ROS producers[92] and the only respiratory complex releasing ROS to the intermembrane space,[90] is most likely involved in these processes.

8.2.5 Possible Role of Metastable State of Semiquinone: Rieske Cluster in Protection Against ROS

Traditionally, the semiquinone formed at the Q_o site has been expected to exhibit an electron paramagnetic resonance (EPR) transition of $g = 2.0$, typical for free radicals. Studies reporting SQ_o were indeed consistent with this notion.[59–61,63,93] However, Sarewicz *et al.* discovered that, in addition to that form of SQ_o, cytochrome bc_1 generates another form in which semiquinone is coupled to the reduced Rieske cluster *via* spin–spin exchange interaction.[61,62] These two forms of SQ_o will be referred to thereafter as a "free SQ_o" and a "SQ-FeS", respectively. The SQ-FeS was recognized by its distinct EPR spectrum with a transition of $g = 1.94$. Both forms of SQ_o were observed in antimycin-inhibited enzyme, when the Q_o site exposed to excess of substrates was turning over albeit at lower rates (comparing to non-inhibited enzymes). The signals $g = 2.0$ and $g = 1.94$ were transient and present only under non-equilibrium conditions. SQ-FeS emerged as a new state that should be taken into consideration as one of the intermediates of reactions taking place at the Q_o site.

Interesting insights into the possible role of the SQ-FeS state were provided by the recent discovery that it can also be generated in cytochrome b_6f,[62] a key component of oxygenic photosynthesis analogous to cytochrome bc_1 function. SQ-FeS was also transient and required an excess of substrates. However, in cytochrome b_6f, unlike in cytochrome bc_1, the SQ-FeS state was detected in a non-inhibited enzyme, which indicated that the probability of the enzyme residing in the SQ-FeS state is higher in cytochrome b_6f. Remarkably, it was observed that the yield of generated SQ-FeS in both enzymes was not significantly sensitive to the presence or absence of oxygen.[62]

To explain these two observations (the larger tendency of cytochrome b_6f to reside in the SQ-FeS state and the general lack of the sensitivity of the signals to the presence of oxygen) a model was proposed,[62] a simplified version of which is presented in Figure 8.5. A central feature of this model is an inclusion of a metastable state SQ-FeS (state D), in which the energy level lies below the energy levels of the state of the Q_o site with "free" SQ_o and a reduced [2Fe–2S] cluster (state B) or the state of the Q_o site with quinone (Q) immediately after the complete oxidation of QH_2 (both [2Fe–2S] cluster and heme b_L are reduced) (state C). The energetic level of state D also lies below the level of the O_2^- generating state F. Thus, in state D, unlike in state B,

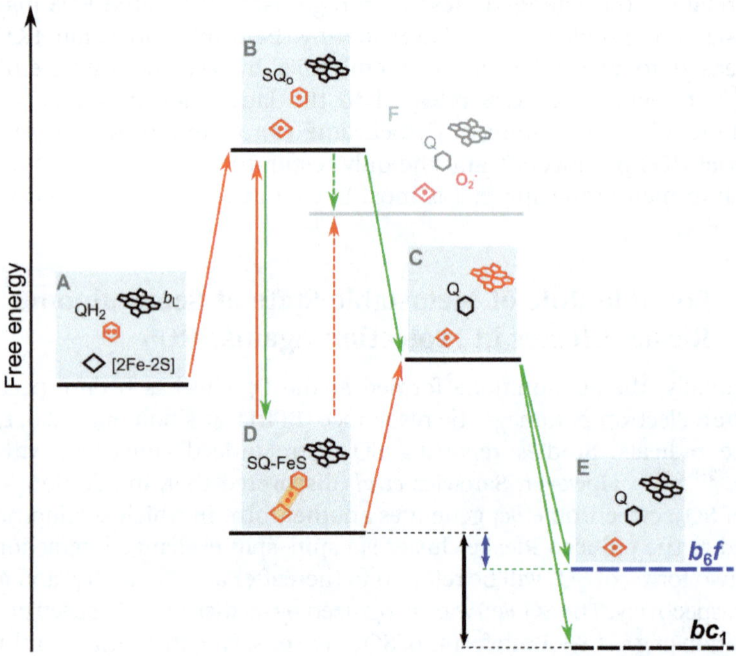

Figure 8.5 Diagram of relative energy levels at different stages of the Q_o site reactions. (A) Initial state of QH_2 oxidation with QH_2 bound and both [2Fe–2S] cluster and heme b_L oxidized. (B) Intermediate state with highly unstable SQ_o and reduced [2Fe–2S] cluster. (C) State with Q bound and both [2Fe–2S] cluster and heme b_L reduced. (D) metastable state containing semiquinone coupled to [2Fe–2S] cluster *via* spin–spin exchange interaction (SQ-FeS). (E) State with oxidized heme b_L after the electron transfer to heme b_H, for simplicity E represents all steps of enzymatic turnover subsequent to C. (F) State in which SQ_o generates O_2^-. Red dot represent electron on the [2Fe–2S] cluster (rhombus) or quinone molecule (hexagon). Cofactors ([2Fe–2S] cluster and heme b_L) and quinone molecule when being in a reduced state are marked in red. Double arrows (blue and black) depict energetic gap between states D and E, which differs within the members of *bc*-family. Red and green arrows represent an energetically uphill and downhill reactions, respectively. All reactions should be considered reversible.

electrons are safely held at a local energetic minimum and do not react with oxygen. In this way, state D serves the role of a ''buffer'' for electrons that are unable to be relegated from the Q_o site by catalytic reactions, protecting the enzyme against ROS formation. At the same time, the energy level of state D lies above the energy levels of the subsequent states of enzymatic turnover (represented here as state E in which heme b_L is oxidized by heme b_H). However, the energetic gap between states D and E differs within the *bc*-family, which may effectively influence the probability of the enzyme being in state D. In cytochrome b_6f, the gap is smaller, compared to cytochrome bc_1, thus the probability of the enzyme being in the ROS protective state D is higher. This could be beneficial for cytochrome b_6f, given that it operates under higher oxygen tension compared to cytochrome bc_1. Therefore, based on the scheme in Figure 8.5 one can suggest a mechanism, in which a modulation of the energy level of state D on the energy landscape in photosynthetic versus respiratory enzymes allows efficient catalysis under different oxygen tensions.

It is worth mentioning that in the case of cytochrome bc_1, the SQ-FeS state was detected in purple bacterial enzyme, a natural component of anoxygenic photosynthesis. This suggests that stabilization of SQ by spin-coupling to FeS is an inherent feature of all *bc* complexes (operating in either oxygenic or anoxygenic environment). However, one may envisage that this feature became beneficial when organisms faced the challenge of the appearance of oxygen in the atmosphere.

As far as the evolutionary adaptation of *bc* complexes to the presence of oxygen is concerned, an interesting mechanism was proposed recently by Bergdoll *et al.*, who compared the redox midpoint potential of hemes *b* and Rieske clusters in organisms that use different quinones and live in aerobic or anaerobic conditions.[94] The authors suggested that organisms responded to the presence of oxygen by an up-shift of the redox potential of the pool of quinone by 150 mV (menaquinone versus ubi/plastoquinone), together with a commensurable increase in the entire set of redox potentials of the cofactors of *bc* complexes. Such a shift is proposed to account for thermodynamic constraints that acted as an evolutionary selection pressure to avoid deleterious ROS generation. Intriguingly, despite this shift all *bc* complexes appear to preserve a large difference in redox midpoint potentials between heme b_L and heme b_H. The reason for this preservation remains unclear in light of recent mutational studies documenting that the difference in redox midpoint potentials between heme b_L and heme b_H can be markedly increased or decreased (to the point that the hemes are almost isopotential) with little impact on functionality of enzymes *in vitro* and *in vivo*.[95]

8.3 New Insights into Proton Transfer at the Q_i Site

When cytochrome bc_1 works in a forward direction, the Q_i site catalyzes two sequential, one-electron reductions of quinone *via* heme b_H where both electrons originate from the Q_o site. The first electron is utilized to reduce quinone to semiquinone intermediate (SQ_i) and, after transferring the second electron to the Q_i site, SQ_i is reduced to quinol (Figure 8.6A). As it is clearly a sequential reaction, SQ_i has to be stabilized until it receives the second electron. Indeed, the stable SQ_i is readily detected by EPR spectroscopy.[96–99] The observation that this semiquinone is in anionic form[98,99] suggests a scheme, in which the attachment of two protons to the benzoquinone ring requires transfer of the second electron to the Q_i site. It is of note, that the so far detected SQ_i comes from partial reversal of the reaction at the Q_i site (oxidation of QH_2 to SQ_i by heme b_H), as it is observed together with reduced heme b_H.

Detailed information concerning amino acids engaged in the substrate binding at the Q_i site comes from the crystal structure of cytochrome bc_1

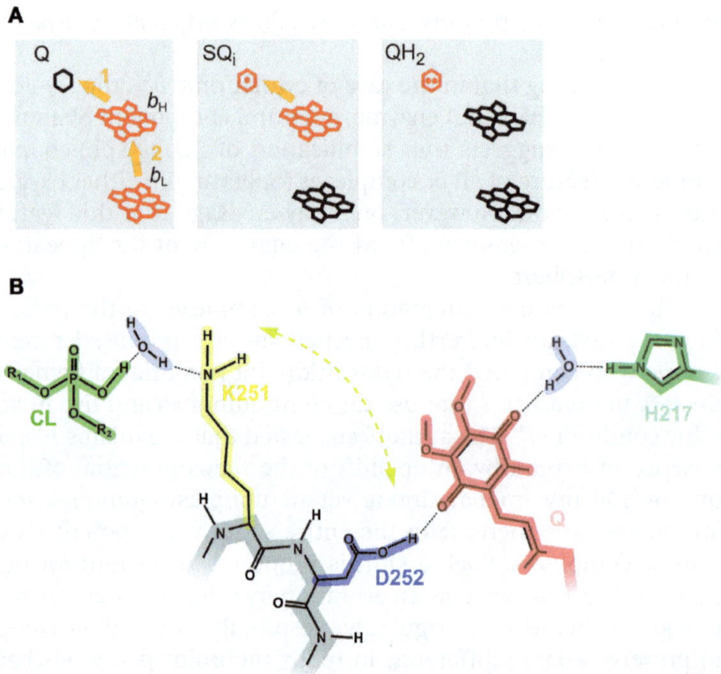

Figure 8.6 Reactions at the Q_i catalytic site. (A) The sequence of reduction of quinone (Q) at the Q_i site. Q undergoes reduction to quinol (QH_2) *via* stable semiquinone (SQ_i). The cofactors marked in red represent their reduced state, cofactors marked in black their oxidized state. Red dots represent electrons in quinone molecules. (B) Amino acid residues of cytochrome *b* and water molecules proposed to be involved in the transfer of protons to the Q_i site. Quinone molecule (red) can receive protons *via* E/R pathway (including H217 (dark green)) and CL/K-pathway (cardiolipin CL (light green), K251 (yellow), and D252 (dark blue)). Postulated rotation of the K251 side-chain is indicated by yellow arrow. A fragment of the polypeptide backbone chain is marked in grey. Water molecules are shown in light blue.

with substrate (quinone) bound to this site.[36] According to the crystal structure, two amino acid residues, Asp229 and His202 (yeast equivalents of *Rb. capsulatus* D252 and H217), were proposed to stabilize the protein-substrate complex.[36,97,100,101] It was postulated that one of the carbonyl oxygen can form a hydrogen bond directly with Asp229 (bacterial D252) or with participation of a nearby water molecule. The second carbonyl oxygen forms a hydrogen bond with the water molecule stabilized by His202 (bacterial H217).[28] Besides these two residues, several key protonable residues also associated with the operation of the Q_i site, such as Asn208 and Lys228 (N223 and K251 in *Rb. capsulatus*, respectively) were identified.[28,102,103]

With these details of substrate binding, two possible pathways for uptake of protons were considered, the E/R-pathway and CL/K-pathway[28] (Figure 8.6B).

The E/R-pathway is located on the side of His202. It was suggested that, besides this residue, Arg218 is particularly important. This highly conserved residue mediates the interaction between an array of the water molecules found in the vicinity of the Q_i site, with the water molecules at the matrix side. It was proposed that upon reduction of quinone at the Q_i site, a proton is abstracted from Arg218 and transferred *via* the water molecules to the quinone binding pocket.

The CL/K-pathway is oriented on the side of Asp229. In this path, highly conserved Lys228 (yeast equivalent of *Rb. capsulatus* K251), from one side provides a connection with Asp229 *via* water molecules and, from the other side, provides a connection with cardiolipin (CL) molecule *via* only one water molecule, which is tightly bound to the protein surface. The connection provided by Lys228 allows communication to the Q_i site with aqueous solvent. It was proposed that upon the reduction of quinone at the Q_i site, a proton is abstracted from Lys228. After that, the proton can be supplied to Lys228 from the matrix side by reversible deprotonation of CL or proton exchange occurring *via* water molecule located between CL and membranous phospholipids. This allows the coupling of the Q_i site with the proton source at the membrane surface from the matrix side. This key role of CL in proton uptake is in agreement with the MD simulations, in which it was proposed that cardiolipin molecules spontaneously diffuse to the cytochrome bc_1 interface in close proximity to the Q_i site.[104,105]

Recent MD simulations suggest that the role of K251 is more direct than it was speculated in the CL/K-pathway hypothesis.[106] After acquiring a proton from the CL, the positively charged K251 can rotate into the Q_i site (Figure 8.6B) to form a salt bridge with negatively charged D252, which is in a deprotonated state after donating a proton to quinone.

Recent experimental studies have indicated that the proton path in K251/D252 pair is disabled only when both protonable groups are removed.[107] Removing only one protonable group from this dyad is insufficient to completely block the entrance of protons to the Q_i catalytic site. This suggests that protons can travel through parallel routes, possibly involving water molecules.

8.4 Cytochrome bc_1 as a Functional Dimer

An interesting feature of all enzymes belonging to the cytochrome *bc* family is that they always form homodimers. As each of the monomers is equipped with all cofactors necessary to conduct the enzymatic catalysis on its own, the functional role of the dimer was debated.[108–116] The distance between two hemes b_L, each present on separate monomers, in bc_1 and b_6f enzymes are within the limit allowing for the transfer of electrons between them on a catalytic timescale.[12,22] In recent years the experimental evidence for such inter-monomer electronic connection was provided.[117–122] From the mechanistic point of view, this inter-monomer electron transfer converts independent cofactor chains of monomers into an H-shaped electron transfer system which connects all the four catalytic sites of the dimer (Figure 8.1E). The physiological meaning of this design remains an open question related to several issues including regulatory function of cytochrome bc_1, protection from ROS formation, and protection from mutational damage. These issues are discussed in recent reviews.[2,116]

The H-shaped electron transfer system can also be discussed in the context of the supercomplexes, that are postulated to be formed between various complexes of the mitochondrial respiratory chain.[123–125] Extended connection between the catalytic sites of cytochrome bc_1 may offer a communicational advantage in larger protein assemblies, within which, the individual catalytic sites may not be equally accessible to the pools of substrates. However, we note that while the possible physiological meaning of supercomplexes is still debated[126–129] some of the recently proposed mechanistic models of their operation are difficult to accept.[130–132] The most preposterous example includes the proposal, unsubstantiated by any kinetic and functional data, of the functional electron flow from the Q_i site to cytochrome *c* involving a direct electron transfer between heme b_L and Rieske cluster.[130] In our view, such "revolutionary" concepts should be formed with more caution.

8.5 Concluding Remarks

In a book "Biophysical and Structural Aspects of Bioenergetics" published by the Royal Society of Chemistry more than a decade ago, the chapter on cytochrome bc_1 had a provocative title: "The bc_1 complex: what is there left to argue about?".[133] We believe that in recent years remarkable progress has been made in this area. The issues discussed here leave no doubt that this progress exposed new important mechanistic and physiologic aspects related to the operation of cytochrome bc_1 that remain to be characterized and understood. We anticipate that the most exciting discoveries will concern the regulatory function of these enzymes (for various physiological processes) in which the newly recognized states of the enzymes may play important roles.

Acknowledgements

A.O. acknowledges The Wellcome Trust for International Senior Research Fellowship and National Science Centre for Maestro grant.

References

1. *Cytochrome Complexes: Evolution, Structures, Energy Transduction, and Signaling*, ed. W. A. Cramer and T. Kallas, Springer Science, Dordrecht, 2016.
2. M. Sarewicz and A. Osyczka, *Physiol. Rev.*, 2015, **95**, 219–243.
3. P. Mitchell, *FEBS Lett.*, 1975, **59**, 137–139.
4. A. R. Crofts, S. W. Meinhardt, K. R. Jones and M. Snozzi, *Biochim. Biophys. Acta*, 1983, **723**, 202–218.
5. H. Schägger, T. A. Link, W. D. Engel and G. von Jagow, *Methods Enzymol.*, 1986, **126**, 224–237.
6. D. Xia, L. Esser, W.-K. Tang, F. Zhou, Y. Zhou, L. Yu and C.-A. Yu, *Biochim. Biophys. Acta*, 2013, **1827**, 1278–1294.
7. S. Iwata, J. W. Lee, K. Okada, J. K. Lee, M. Iwata, B. Rasmussen, T. A. Link, S. Ramaswamy and B. K. Jap, *Science*, 1998, **281**, 64–71.
8. E. A. Berry, L.-S. Huang, L. K. Saechao, N. G. Pon, M. Valkova-Valchanova and F. Daldal, *Photosynth. Res.*, 2004, **81**, 251–275.
9. T. Kleinschroth, M. Castellani, C. H. Trinh, N. Morgner, B. Brutschy, B. Ludwig and C. Hunte, *Biochim. Biophys. Acta*, 2011, **1807**, 1606–1615.
10. H. Kim, D. Xia, C. A. Yu, J. Z. Xia, A. M. Kachurin, L. Zhang, L. Yu and J. Deisenhofer, *Proc. Natl. Acad. Sci. U. S. A.*, 1998, **95**, 8026–8033.
11. Z. Zhang, L. Huang, V. M. Shulmeister, Y.-I. Chi, K. K. Kim, L.-W. Hung, A. R. Crofts, E. A. Berry and S.-H. Kim, *Nature*, 1998, **392**, 677–684.
12. D. Xia, C. A. Yu, H. Kim, J. Z. Xia, A. M. Kachurin, L. Zhang, L. Yu and J. Deisenhofer, *Science*, 1997, **277**, 60–66.
13. E. Darrouzet, M. Valkova-Valchanova, C. C. Moser, P. L. Dutton and F. Daldal, *Proc. Natl. Acad. Sci. U. S. A.*, 2000, **97**, 4567–4572.
14. G. Engstrom, R. Rajagukguk, A. J. Saunders, C. N. Patel, S. Rajagukguk, T. Merbitz-Zahradnik, K. Xiao, G. J. Pielak, B. Trumpower, C.-A. Yu, L. Yu, B. Durham and F. Millett, *Biochemistry*, 2003, **42**, 2816–2824.
15. J. Hall, A. Kriaucionas, D. Knaff and F. Millett, *J. Biol. Chem.*, 1987, **262**, 14005–14009.
16. S. Güner, A. Willie, F. Millett, M. S. Caffrey, M. A. Cusanovich, D. E. Robertson and D. B. Knaff, *Biochemistry*, 1993, **32**, 4793–4800.
17. J. Janzon, A. C. Eichhorn, B. Ludwig and F. Malatesta, *Biochim. Biophys. Acta*, 2008, **1777**, 250–259.
18. S. Gupte, E. S. Wu, L. Hoechli, M. Hoechli, K. Jacobson, A. E. Sowers and C. R. Hackenbrock, *Proc. Natl. Acad. Sci. U. S. A.*, 1984, **81**, 2606–2610.
19. C. R. Hackenbrock, B. Chazotte and S. S. Gupte, *J. Bioenerg. Biomembr.*, 1986, **18**, 331–368.

20. R. Pietras, M. Sarewicz and A. Osyczka, *J. Phys. Chem. B*, 2014, **118**, 6634–6643.
21. M. Sarewicz, A. Borek, F. Daldal, W. Froncisz and A. Osyczka, *J. Biol. Chem.*, 2008, **283**, 24826–24836.
22. A. Osyczka, C. C. Moser, F. Daldal and P. L. Dutton, *Nature*, 2004, **427**, 607–612.
23. F. Muller, A. R. Crofts and D. M. Kramer, *Biochemistry*, 2002, **41**, 7866–7874.
24. A. W. Rutherford, A. Osyczka and F. Rappaport, *FEBS Lett.*, 2012, **586**, 603–616.
25. U. Brandt, *Biochim. Biophys. Acta*, 1998, **1365**, 261–268.
26. A. R. Crofts, S. Hong, N. Ugulava, B. Barquera, R. Gennis, M. Guergova-Kuras and E. A. Berry, *Proc. Natl. Acad. Sci. U. S. A.*, 1999, **96**, 10021–10026.
27. T. A. Link, *FEBS Lett.*, 1997, **412**, 257–264.
28. C. Hunte, H. Palsdottir and B. L. Trumpower, *FEBS Lett.*, 2003, **545**, 39–46.
29. C. H. Snyder, E. B. Gutierrez-Cirlos and B. L. Trumpower, *J. Biol. Chem.*, 2000, **275**, 13535–13541.
30. B. L. Trumpower, *Biochim. Biophys. Acta*, 2002, **1555**, 166–173.
31. J. Zhu, T. Egawa, S.-R. Yeh, L. Yu and C.-A. Yu, *Proc. Natl. Acad. Sci. U. S. A.*, 2007, **104**, 4864–4869.
32. B. Chance and G. Hollunger, *J. Biol. Chem.*, 1961, **236**, 1562–1568.
33. T. Miki, M. Miki and Y. Orii, *J. Biol. Chem.*, 1994, **269**, 1827–1833.
34. A. Elbehti, G. Brasseur and D. Lemesle-Meunier, *J. Bacteriol.*, 2000, **182**, 3602–3606.
35. G. Brasseur, G. Levican, V. Bonnefoy, D. Holmes, E. Jedlicki and D. Lemesle-Meunier, *Biochim. Biophys. Acta*, 2004, **1656**, 114–126.
36. C. Hunte, J. Koepke, C. Lange, T. Roßmanith and H. Michel, *Structure*, 2000, **8**, 669–684.
37. D. Stroebel, Y. Choquet, J.-L. Popot and D. Picot, *Nature*, 2003, **426**, 413–418.
38. E. A. Berry and L. Huang, *FEBS Lett.*, 2003, **555**, 13–20.
39. L. Esser, B. Quinn, Y.-F. Li, M. Zhang, M. Elberry, L. Yu, C.-A. Yu and D. Xia, *J. Mol. Biol.*, 2004, **341**, 281–302.
40. L. Esser, M. Elberry, F. Zhou, C.-A. Yu, L. Yu and D. Xia, *J. Biol. Chem.*, 2008, **283**, 2846–2857.
41. A. Osyczka, H. Zhang, C. Mathé, P. R. Rich, C. C. Moser and P. L. Dutton, *Biochemistry*, 2006, **45**, 10492–10503.
42. T. Wenz, P. Hellwig, F. MacMillan, B. Meunier and C. Hunte, *Biochemistry*, 2006, **45**, 9042–9052.
43. N. Seddiki, B. Meunier, D. Lemesle-Meunier and G. Brasseur, *Biochemistry*, 2008, **47**, 2357–2368.
44. D.-W. Lee, Y. El Khoury, F. Francia, B. Zambelli, S. Ciurli, G. Venturoli, P. Hellwig and F. Daldal, *Biochemistry*, 2011, **50**, 4263–4272.
45. W.-C. Kao and C. Hunte, *Genome Biol. Evol.*, 2014, **6**, 1894–1910.

46. H. Palsdottir, C. G. Lojero, B. L. Trumpower and C. Hunte, *J. Biol. Chem.*, 2003, **278**, 31303–31311.
47. P. A. Postila, K. Kaszuba, M. Sarewicz, A. Osyczka, I. Vattulainen and T. Róg, *Biochim. Biophys. Acta*, 2013, **1827**, 761–768.
48. A. M. Barragan, K. Schulten and I. A. Solov'yov, *J. Phys. Chem. B*, 2016, **120**, 11369–11380.
49. A. Osyczka, C. C. Moser and P. L. Dutton, *Trends Biochem. Sci.*, 2005, **30**, 176–182.
50. K.-I. Takamiya and P. L. Dutton, *Biochim. Biophys. Acta*, 1979, **546**, 1–16.
51. S. Jünemann, P. Heathcote and P. R. Rich, *J. Biol. Chem.*, 1998, **273**, 21603–21607.
52. H. Ding, C. C. Moser, D. E. Robertson, M. K. Tokito, F. Daldal and P. L. Dutton, *Biochemistry*, 1995, **34**, 15979–15996.
53. H. Zhang, S. E. Chobot, A. Osyczka, C. A. Wraight, P. L. Dutton and C. C. Moser, *J. Bioenerg. Biomembr.*, 2008, **40**, 493–499.
54. A. R. Crofts and Z. Wang, *Photosynth. Res.*, 1989, **22**, 69–87.
55. U. Brandt and B. Trumpower, *Crit. Rev. Biochem. Mol. Biol.*, 1994, **29**, 165–197.
56. U. Brandt, *J. Bioenerg. Biomembr.*, 1999, **31**, 243–250.
57. A. Y. Mulkidjanian, *Biochim. Biophys. Acta – Bioenerg.*, 2005, **1709**, 5–34.
58. J. L. Cape, M. K. Bowman and D. M. Kramer, *Trends Plant Sci.*, 2006, **11**, 46–55.
59. H. Zhang, A. Osyczka, P. L. Dutton and C. C. Moser, *Biochim. Biophys. Acta*, 2007, **1767**, 883–887.
60. P. R. Vennam, N. Fisher, M. D. Krzyaniak, D. M. Kramer and M. K. Bowman, *ChemBioChem*, 2013, **14**, 1745–1753.
61. M. Sarewicz, M. Dutka, S. Pintscher and A. Osyczka, *Biochemistry*, 2013, **52**, 6388–6395.
62. M. Sarewicz, Ł. Bujnowicz, S. Bhaduri, S. K. Singh, W. A. Cramer and A. Osyczka, *Proc. Natl. Acad. Sci. U. S. A.*, 2017, **114**, 1323–1328.
63. J. L. Cape, M. K. Bowman and D. M. Kramer, *Proc. Natl. Acad. Sci. U. S. A.*, 2007, **104**, 7887–7892.
64. R. Pietras, M. Sarewicz and A. Osyczka, *J. R. Soc. Interface*, 2016, **13**, 1–11.
65. A. Borek, M. Sarewicz and A. Osyczka, *Biochemistry*, 2008, **47**, 12365–12370.
66. D. M. Kramer, A. G. Roberts, F. Muller, J. Cape and M. K. Bowman, *Methods Enzymol.*, 2004, **382**, 21–45.
67. S. Dröse and U. Brandt, *J. Biol. Chem.*, 2008, **283**, 21649–21654.
68. M. Sarewicz, A. Borek, E. Cieluch, M. Świerczek and A. Osyczka, *Biochim. Biophys. Acta*, 2010, **1797**, 1820–1827.
69. A. Borek, P. Kuleta, R. Ekiert, R. Pietras, M. Sarewicz and A. Osyczka, *J. Biol. Chem.*, 2015, **290**, 23781–23792.
70. R. Ekiert, A. Borek, P. Kuleta, J. Czernek and A. Osyczka, *Biochim. Biophys. Acta*, 2016, **1857**, 1102–1110.
71. P. Husen and I. A. Solov'yov, *J. Am. Chem. Soc.*, 2016, **138**, 12150–12158.

72. M. Sarewicz, M. Dutka, W. Froncisz and A. Osyczka, *Biochemistry*, 2009, **48**, 5708–5720.
73. F. Legros, E. Chatzoglou, P. Frachon, H. Ogier De Baulny, P. Laforêt, C. Jardel, C. Godinot and A. Lombès, *Eur. J. Hum. Genet.*, 2001, **9**, 510–518.
74. M. F. Bouzidi, H. Schägger, J. M. Collombet, H. Carrier, F. Flocard, S. Quard, B. Mousson and C. Godinot, *Neuromuscul. Disord.*, 1993, **3**, 599–604.
75. V. Adam-Vizi and C. Chinopoulos, *Trends Pharmacol. Sci.*, 2006, **27**, 639–645.
76. M. P. Murphy, *Biochem. J.*, 2009, **417**, 1–13.
77. R. D. Guzy and P. T. Schumacker, *Exp. Physiol.*, 2006, **91**, 807–819.
78. G. Solaini, A. Baracca, G. Lenaz and G. Sgarbi, *Biochim. Biophys. Acta*, 2010, **1797**, 1171–1177.
79. S. S. Korshunov, V. P. Skulachev and A. A. Starkov, *FEBS Lett.*, 1997, **416**, 15–18.
80. C. L. Quinlan, A. A. Gerencser, J. R. Treberg and M. D. Brand, *J. Biol. Chem.*, 2011, **286**, 31361–31372.
81. A. Boveris and E. Cadenas, *FEBS Lett.*, 1975, **54**, 311–314.
82. M. Ksenzenko, A. A. Konstantinov, G. B. Khomutov, A. N. Tikhonov and E. K. Ruuge, *FEBS Lett.*, 1983, **155**, 19–24.
83. M. Ristow and K. Zarse, *Exp. Gerontol.*, 2010, **45**, 410–418.
84. M. Ristow and S. Schmeisser, *Free Radic. Biol. Med.*, 2011, **51**, 327–336.
85. T. Finkel, *J. Cell. Biol.*, 2011, **194**, 7–15.
86. T. Finkel, *J. Biol. Chem.*, 2012, **287**, 4434–4440.
87. R. B. Hamanaka and N. S. Chandel, *Trends Biochem. Sci.*, 2010, **35**, 505–513.
88. A. A. Starkov, *Ann. N. Y. Acad. Sci.*, 2008, **1147**, 37–52.
89. M. P. Murphy, A. Holmgren, N.-G. Larsson, B. Halliwell, C. J. Chang, B. Kalyanaraman, S. G. Rhee, P. J. Thornalley, L. Partridge, D. Gems, T. Nyström, V. Belousov, P. T. Schumacker and C. C. Winterbourn, *Cell Metab.*, 2011, **13**, 361–366.
90. J. St-Pierre, J. A. Buckingham, S. J. Roebuck and M. D. Brand, *J. Biol. Chem.*, 2002, **277**, 44784–44790.
91. L. Bleier and S. Dröse, *Biochim. Biophys. Acta – Bioenerg.*, 2013, **1827**, 1320–1331.
92. Q. Chen, E. J. Vazquez, S. Moghaddas, C. L. Hoppel and E. J. Lesnefsky, *J. Biol. Chem.*, 2003, **278**, 36027–36031.
93. S. de Vries, S. P. Albracht, J. A. Berden and E. C. Slater, *J. Biol. Chem.*, 1981, **256**, 11996–11998.
94. L. Bergdoll, F. ten Brink, W. Nitschke, D. Picot and F. Baymann, *Biochim. Biophys. Acta – Bioenerg.*, 2016, **1857**, 1569–1579.
95. S. Pintscher, P. Kuleta, E. Cieluch, A. Borek, M. Sarewicz and A. Osyczka, *J. Biol. Chem.*, 2016, **291**, 6872–6881.
96. D. E. Robertson, R. C. Prince, J. R. Bowyer, K. Matsuura, P. L. Dutton and T. Ohnishi, *J. Biol. Chem.*, 1984, **259**, 1758–1763.

97. K. A. Gray, P. L. Dutton and F. Daldal, *Biochemistry*, 1994, **33**, 723–733.
98. D. R. J. Kolling, R. I. Samoilova, J. T. Holland, E. A. Berry, S. A. Dikanov and A. R. Crofts, *J. Biol. Chem.*, 2003, **278**, 39747–39754.
99. S. A. Dikanov, J. T. Holland, B. Endeward, D. R. J. Kolling, R. I. Samoilova, T. F. Prisner and A. R. Crofts, *J. Biol. Chem.*, 2007, **282**, 25831–25841.
100. E. A. Berry, Z. Zhang, L.-S. Huang and S.-H. Kim, *Biochem. Soc. Trans.*, 1999, **27**, 565–572.
101. A. R. Klingen, H. Palsdottir, C. Hunte and G. M. Ullmann, *Biochim. Biophys. Acta*, 2007, **1767**, 204–221.
102. A. Crofts, B. Hacker, B. Barquera, C.-H. Yun and R. Gennis, *Biochim. Biophys. Acta*, 1992, **1101**, 162–165.
103. B. Hacker, B. Barquera, A. R. Crofts and R. B. Gennis, *Biochemistry*, 1993, **32**, 4403–4410.
104. S. Pöyry, O. Cramariuc, P. A. Postila, K. Kaszuba, M. Sarewicz, A. Osyczka, I. Vattulainen and T. Róg, *Biochim. Biophys. Acta*, 2013, **1827**, 769–778.
105. C. Arnarez, J.-P. Mazat, J. Elezgaray, S.-J. Marrink and X. Periole, *J. Am. Chem. Soc.*, 2013, **135**, 3112–3120.
106. P. A. Postila, K. Kaszuba, P. Kuleta, I. Vattulainen, M. Sarewicz, A. Osyczka and T. Róg, *Sci. Rep.*, 2016, **6**, 33607.
107. P. Kuleta, M. Sarewicz, P. Postila, T. Róg and A. Osyczka, *Biochim. Biophys. Acta - Bioenerg.*, 2016, **1857**, 1661–1668.
108. V. P. Shinkarev and C. A. Wraight, *FEBS Lett.*, 2007, **581**, 1535–1541.
109. A. R. Crofts, J. T. Holland, D. Victoria, D. R. J. Kolling, S. A. Dikanov, R. Gilbreth, S. Lhee, R. Kuras and M. G. Kuras, *Biochim. Biophys. Acta*, 2008, **1777**, 1001–1019.
110. A. R. Crofts, S. Hong, C. Wilson, R. Burton, D. Victoria, C. Harrison and K. Schulten, *Biochim. Biophys. Acta*, 2013, **1827**, 1362–1377.
111. A. Y. Mulkidjanian, *Biochim. Biophys. Acta*, 2010, **1797**, 1858–1868.
112. W. A. Cramer, S. S. Hasan and E. Yamashita, *Biochim. Biophys. Acta*, 2011, **1807**, 788–802.
113. B. Khalfaoui-Hassani, P. Lanciano, D.-W. Lee, E. Darrouzet and F. Daldal, *FEBS Lett.*, 2012, **586**, 617–621.
114. J. N. Bazil, K. C. Vinnakota, F. Wu and D. A. Beard, *Biophys. J.*, 2013, **105**, 343–355.
115. S. S. Hasan, S. D. Zakharov, A. Chauvet, V. Stadnytskyi, S. Savikhin and W. A. Cramer, *J. Phys. Chem. B*, 2014, **118**, 6614–6625.
116. M. Sarewicz, R. Ekiert and A. Osyczka, in *Cytochrome Complexes: Evolution, Structures, Energy Transduction, and Signaling*, ed. W. A. Cramer and T. Kallas, Dordrecht, 2016, pp. 281–294.
117. M. Świerczek, E. Cieluch, M. Sarewicz, A. Borek, C. C. Moser, P. L. Dutton and A. Osyczka, *Science*, 2010, **329**, 451–454.
118. P. Lanciano, D.-W. Lee, H. Yang, E. Darrouzet and F. Daldal, *Biochemistry*, 2011, **50**, 1651–1663.

119. P. Lanciano, B. Khalfaoui-Hassani, N. Selamoglu and F. Daldal, *Biochemistry*, 2013, **52**, 7196–7206.
120. M. Czapla, A. Borek, M. Sarewicz and A. Osyczka, *Biochemistry*, 2012, **51**, 829–835.
121. M. Czapla, E. Cieluch, A. Borek, M. Sarewicz and A. Osyczka, *Biochim. Biophys. Acta*, 2013, **1827**, 751–760.
122. R. Ekiert, M. Czapla, M. Sarewicz and A. Osyczka, *Biochem. Biophys. Res. Commun.*, 2014, **451**, 270–275.
123. H. Schägger and K. Pfeiffer, *EMBO J.*, 2000, **19**, 1777–1783.
124. R. Acín-Pérez, P. Fernández-Silva, M. L. Peleato, A. Pérez-Martos and J. A. Enriquez, *Mol. Cell*, 2008, **32**, 529–539.
125. S. Cogliati, C. Frezza, M. E. Soriano, T. Varanita, R. Quintana-Cabrera, M. Corrado, S. Cipolat, V. Costa, A. Casarin, L. C. Gomes, E. Perales-Clemente, L. Salviati, P. Fernandez-Silva, J. A. Enriquez and L. Scorrano, *Cell*, 2013, **155**, 160–171.
126. J. A. Enríquez, *Annu. Rev. Physiol.*, 2016, **78**, 533–561.
127. M. Trouillard, B. Meunier and F. Rappaport, *Proc. Natl. Acad. Sci. U. S. A.*, 2011, **108**, E1027–E1034.
128. G. Lenaz and M. L. Genova, *Am. J. Physiol. Cell Physiol.*, 2007, **292**, 1221–1239.
129. D. Milenkovic, J. N. Blaza, N. G. Larsson and J. Hirst, *Cell Metab.*, 2017, **25**, 765–776.
130. M. Wu, J. Gu, R. Guo, Y. Huang and M. Yang, *Cell*, 2016, **167**, 1598–1609.
131. J. S. Sousa, D. J. Mills, J. Vonck and W. Kühlbrandt, *Elife*, 2016, **5**, 1–17.
132. J. A. Letts, K. Fiedorczuk and L. A. Sazanov, *Nature*, 2016, **537**, 644–648.
133. A. R. Crofts, in *Biophysical and Structural Aspects of Bioenergetics*, ed. M. Wikström, Royal Society of Chemistry, Cambridge, 2005, pp. 123–156.

CHAPTER 9

Life and Death of Cytochrome c Oxidase: Influence of Subunit III on the D pathway, Proton Backflow and Suicide Inactivation

SHELAGH FERGUSON-MILLER[a] AND JONATHAN HOSLER*[b]

[a] Department of Biochemistry, Michigan State University, East Lansing, MI 48824, USA; [b] Department of Cell and Molecular Biology, University of Mississippi Medical Center, 2500 N. State St., Jackson, MS 39216, USA
*Email: jhosler@umc.edu

9.1 Introduction

For over a hundred years, researchers have sought to understand the complex function of cytochrome *c* oxidase (C*c*O).[1] As a critical player in supporting the survival of most living organisms, it is of great interest to comprehend how the powerful exergonic process of oxygen reduction to water can be carried out and controlled. Much progress has been made in defining the chemistry of oxygen reduction[2] and the fascinating question of how oxygen reduction is harnessed to proton pumping continues to garner much interest.[3–6] Nonetheless, it remains to be solved how the whole process is regulated to control the rate and efficiency of steady-state C*c*O activity. It is clear that there exist a plethora of regulatory mechanisms that need to be

Chemical Biology No. 5
Mechanisms of Primary Energy Transduction in Biology
Edited by Mårten Wikström

Published by the Royal Society of Chemistry, www.rsc.org

more completely understood, including regulation by the nuclear-encoded subunits,[7] the roles of tissue-specific and metabolism-specific subunit isoforms,[7–9] the consequences of phosphorylation and acetylation of C*c*O[7,10–12] and the binding of regulatory ligands.[13–15] Understanding the regulation of C*c*O is not a trivial exercise, especially since it is now understood that the mitochondrion is an important control center of the cell, far more influential than formerly thought. Although the mitochondrion continues to be the obedient servant supplying energy on cue, mitochondria also send and receive signals related to damage, disease and metabolic state, and they participate in cell death and clean up processes, such as mitophagy.[16] It is recognized that many, if not most, disease states, including aging, involve mitochondrial malfunction. Mitochondria are a major producer of reactive oxygen species (ROS), which cause cellular damage and function in signal transduction.[17] Although C*c*O itself is not a direct source of ROS, any compound or condition that leads to inhibition of its activity can in turn lead to accumulation of reducing equivalents upstream in the respiratory chain, promoting the production of ROS.[18] Conversely, increased synthesis or activation of C*c*O may protect cells from ROS, as well as drive respiration by favoring the exit of electrons.

This chapter focuses on one of the three largest subunits in C*c*O, subunit III, which lacks metal centers and therefore is not a player in the oxygen chemistry. It can even be removed from the C*c*O complex with the retention of O_2 reduction and proton pumping, at least for a while. Yet subunit III, along with the central subunit I, is highly conserved from bacteria to mammals. Also, subunit III first appears in the A-type heme-Cu oxidases, coincident with the appearance of the D pathway for proton transfer in subunit I. With subunit III present, the D pathway is altered to provide C*c*O with higher activity at physiologic pH (~8). In addition, subunit III confers far less susceptibility to turnover-based suicide inactivation and higher efficiency of proton pumping. Thus, subunit III may be seen as the first of many later additions of non-catalytic subunits and isoforms whose presence tunes and adapts the activity of the aa_3-type C*c*Os to meet physiological demands.[7,8]

In the sections below, we will elaborate on how subunit III alters the activity of the D pathway and how it protects C*c*O from suicide inactivation, along with explaining what suicide inactivation is and proposing a chemical mechanism for the inactivation event. We will discuss how the functions of subunit III may relate to its unique structure, part protein and part integral lipid. Finally, we will discuss two other proposed functions for subunit III: a role as a proton antenna for the D pathway and a facilitator for the process of proton backflow from the outer surface of C*c*O when necessary as a relief valve for electron build up in the electron transfer chain.

9.2 A Short Primer on Cytochrome *c* Oxidase

Readers unfamiliar with C*c*O are directed to reviews,[2,3,7,19–22] websites or texts that explain C*c*O function. Briefly, the catalytic core of the heme

aa_3-type C*c*O consists of three subunits: I, II and III. These subunits are encoded in mitochondrial DNA in eukaryotes, while ten additional subunits are encoded in the nucleus and imported into mitochondria. Bacterial C*c*Os consist of the three core subunits plus a small subunit IV with a single transmembrane helix. Cytochrome *c* binds on the external domain of subunit II on the positive (P) side of the membrane to deliver its electron to the di-copper Cu_A center, located in subunit II near the interface with subunit I. From Cu_A, the electron is transferred 19 Å into the transmembrane region of subunit I to buried six-coordinate (bis–histidine) heme *a*. Heme *a* then transfers the electron into the nearby heme a_3–Cu_B center, also buried in subunit I. Heme a_3 is coordinated by a single histidine; the sixth coordination site binds O_2. Cu_B is bound by three histidine ligands. One of these is cross-linked to a tyrosine residue that forms the third redox center in the active site. Two pathways transfer protons into subunit I from the negative (N) side of C*c*O. The K pathway begins at a glutamate residue on the surface of subunit II and terminates at the Cu_B center. The D pathway leads from an aspartate residue accessible from the surface of subunit I and terminates at a glutamate residue located between heme *a* and heme a_3–Cu_B. From this glutamate, D pathway protons flow to either the heme a_3–Cu_B center or to the proton loading site of the proton pump, located above the active site near the D-propionate of heme a_3. The exit pathway to the P side of C*c*O for pump protons remains to be defined.

Oxygen enters C*c*O through the transmembrane region of subunit III[3] and binds to heme a_3, once the active site is fully reduced (R→A) (Figure 9.1). A concerted four electron reduction of O_2 by the heme a_3, Cu_B and the histidine-tyrosine redox centers of the active site ($A \rightarrow P_M$) splits the O=O bond and leaves each O with eight electrons in its outer shell, *i.e.* as reduced as O in H_2O.[2] The $P_M \rightarrow F$, $F \rightarrow O_H$, $O_H \rightarrow E_H$ and $E_H \rightarrow R$ reactions are the one electron reductions of C*c*O that re-generate the reduced, deoxygenated state, R, release H_2O and pump protons. Each of these reactions is initiated by electron input (green arrows in Figure 9.1) into the active site from cytochrome *c*, *via* Cu_A and heme *a*. The second event is the transfer of a proton to the pump proton loading site *via* the D pathway (first light blue arrows in Figure 9.1), followed by the transfer of a substrate proton to the heme a_3–Cu_B center by either the K pathway ($H^+{}_K$) or the D pathway ($H^+{}_D$). The arrival of the substrate proton in the active site drives the transfer of the pump proton to the P surface of C*c*O *via* the exit pathway (second light blue arrows).

In situ, the $P_M \rightarrow F$, $F \rightarrow O_H$, $O_H \rightarrow E_H$ and $E_H \rightarrow R$ reactions generate a membrane potential by proton pumping, as explained, plus an equal amount of potential energy by vectorial charge transfer. During each reaction a negative charge (an electron) and a positive charge (a proton) are each transferred part way across the membrane. In the active site, the negative and positive charges of the electron and proton annihilate as they are incorporated into H_2O. The net result is one complete charge transfer across the membrane per reaction.

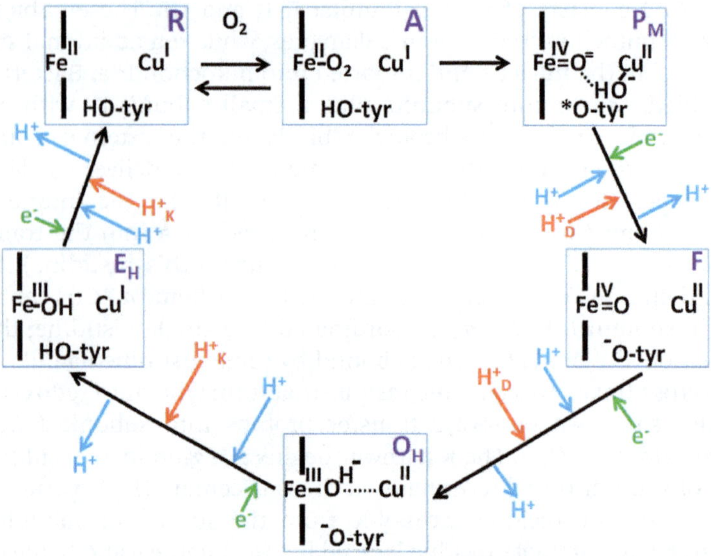

Figure 9.1 Catalytic cycle of C*c*O during steady-state activity, adapted from ref. 3. The rectangles represent the heme a_3–Cu$_B$ active site in subunit I. The histidine–tyrosine ligand of Cu$_B$ is abbreviated tyr. Other ligands of heme a_3 and Cu$_B$ are not shown. The black arrows represent the reaction leading to each intermediate. Red arrows represent protons supporting O_2 chemistry, with the subscripts showing their uptake by either the K or the D pathway. The blue arrows represent pump protons, first being transferred to the PLS by the D pathway and then being transferred to the P-surface of C*c*O by the exit pathway.

9.3 Subunit III Structure

Subunit III is part of the three-subunit catalytic core of C*c*O. Like subunit I, subunit III is largely buried in the membrane. Subunit III and subunit I are both highly conserved (45% *vs.* 47% identity between the α-proteobacter *Rhodobacter sphaeroides* and human subunits III and I, respectively) and considerably more conserved than subunit II (29% identity between subunit II of *R. sphaeroides* and humans). Lacking metal centers and therefore having no direct role in electron transfer, subunit III is nevertheless well resolved in crystal structures of mitochondrial C*c*O, indicating structural stability. The structure mainly used in this chapter is PDB 2DYR, of oxidized bovine C*c*O at a resolution of 1.8 Å.[23] However, subunit III has only been retained in three published structures of the aa_3-type bacterial oxidases; one from *Paracoccus denitrificans* at 3.0 Å resolution (PDB 1QLE)[24] and two from *Rhodobacter sphaeroides:* the oxidized enzyme at 2.3 Å resolution (PDB 1M56)[25] and the reduced enzyme at 3.4 Å (PDB 5WEH).[26] The many structures of C*c*O, with and without subunit III, provide a wealth of information regarding the structural basis of C*c*O function and support the use of powerful computational approaches that provide new insight.

9.3.1 Subunit III Binds Lipid in an Internal Cleft

The crystal structures show that subunit III contains seven transmembrane helices that are split into two domains; a bundle of five helices (3–7) is separated from a helix 1–2 pair by a deep V-shaped cleft or groove (Figure 9.2). The front of the cleft is open to the lipid bilayer, while the back wall is formed by helices 3, 4, and 5 of subunit I. Phospholipids are bound within the cleft of subunit III, by conserved residues that bind the lipids in specific orientations (Figure 9.3).[23–25,27] Two cleft lipids, here termed PL1 and PL2 (Table 9.1), are roughly in register with the lipids of the N-side leaflet. In *R. sphaeroides* C*c*O structure 1M56 these lipids are identified as phosphatidylethanolamine (PE),[25] while in bovine C*c*O structure 2DYR they are known to be phophatidylglycerol (PG).[23] Only PL2 resolves in a crystal structure of the C*c*O of *Paracoccus denitrificans*, where it is assigned as phosphatidylcholine (PC).[24] A conserved arginine, Arg-137$_{\mathrm{I}}$, extends from subunit I to form an ion pair with phosphate group of PL1 (all residue numbers used in the text are those of *R. sphaeroides* C*c*O. Table 9.1 provides the numbers of the cognate residues in bovine C*c*O). Conserved Arg-226$_{\mathrm{III}}$ of subunit III forms a similar ion pair with PL2. Two conserved tryptophan

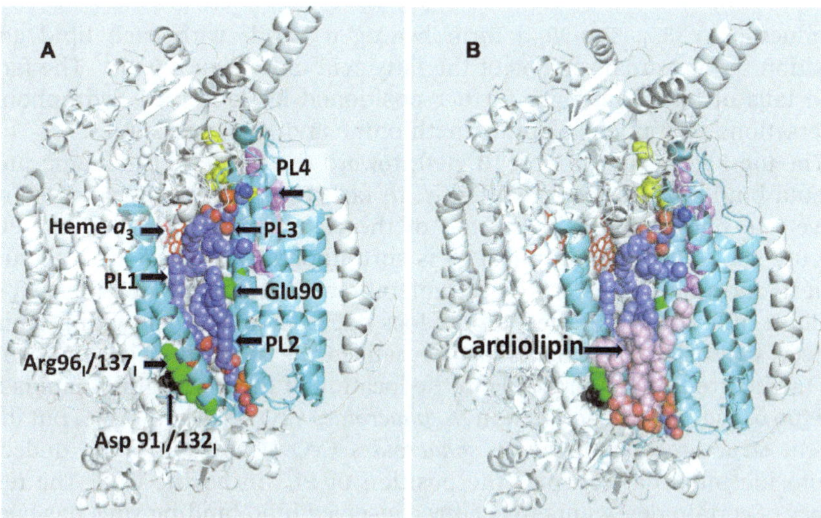

Figure 9.2 A ribbon diagram of a monomer of bovine C*c*O with subunit III shown in cyan, all other subunits in gray. PL1, PL2 and PL3 (purple) are shown in the cleft of subunit III, while PL4 (in light magenta) is shown on the outer surface. Heme a_3 is visible, marking the active site that is accessible to oxygen through the lipid plug. (A) Important ionic/hydrogen bonding interactions between subunits I and III involve PL1 and Arg-137$_{\mathrm{I}}$ (green, shielding D132 (black)), and PL4 with residues of subunits I and II (cluster at top in yellow). (B) Same components as A with the addition of the cardiolipin molecule (light pink) that covers the subunit III cleft in bovine C*c*O (from PDB 2EIL).

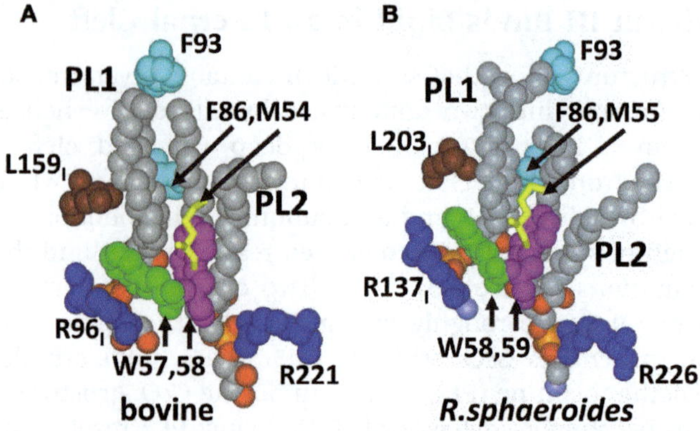

Figure 9.3 Highly conserved lipid binding sites in the cleft of subunit III of (A) Bovine CcO (PDB 2DYR)[23] and (B) *R. sphaeroides* CcO (PDB 1M56).[25] PL1 is the lipid closest to subunit I in the cleft of subunit III. Some of the conserved residues of subunits I and III that form hydrophilic and hydrophobic interactions to bind and position PL1 and PL2 (described in Section 9.3.2) are shown. Bovine and *R. sphaeroides* residue numbers are used in the corresponding panels.

residues (Trp-58_{III}; Trp-59_{III}) form hydrogen bonds with each lipid and position the proximal regions of the fatty acid tails (Figure 9.3).[27] The fatty acid tails of both lipids are further positioned by conserved hydrophobic interactions with phenylalanine, methionine and leucine residues.

The top half of the subunit III cleft, toward the P-side of the membrane, should bind an additional two lipids. In fact, bovine C*c*O structures do resolve one phospholipid in the top of the cleft, a PE,[23] here termed PL3 (Figure 9.2). The headgroup of PL3 is surrounded by residues of both subunit III and subunit VIa, whose extramembrane domain lies atop the cleft of subunit III, but the lipid headgroup does not form an ion pair with protein. A decyl maltoside molecule is also present in the top of the cleft in bovine C*c*O structure 2DYR,[23] probably in the location of the fourth lipid. Lipids in the top of the cleft are not seen in *R. sphaeroides* C*c*O structure 1M56, but in a recent structure of reduced *R. sphaeroides* C*c*O (PDB 5WEH)[26] a dodecyl maltoside molecule occupies the position of PL3 in bovine C*c*O. The tendency of certain detergents to occupy conserved lipid binding sites has been well documented.[28]

The cleft of subunit III of the bacterial C*c*Os opens into the lipid bilayer. In contrast, in bovine C*c*O the cleft opening is plugged by partial insertion of a cardiolipin (CL), one of the two that resolve in each monomer of mammalian C*c*O.[23,29] The CL is in register with the lipid of the N-leaflet and is bound to bovine subunit III by an ion pair to each of the lipid phosphate groups, plus hydrophobic interactions with the fatty acid tails. This cardiolipin has been implicated in forming an interface with Complex III in a mitochondrial supercomplex[30] and in playing a role in facilitating oxygen access to the active site of C*c*O[3] (Figure 9.2B) (see Section 9.11).

Table 9.1 Lipid assignments and correlation of *Cc*O residue numbers used in the text.

Structural feature	Lipid/amino acid nomenclature	
	Rhodobacter sphaeroides PDB 1M56	**Bovine** PDB 2DYR
PL1	PEH2009	PGV266
PL2	PEH2008	PGV267
PL3		PEK264
PL4	PEH2010	PGV268
PL5	PEH2011	Occupied by cholate CHD525
Cardiolipin		CDL270
Subunit I–III/PL1-PL2 interactions in the cleft	Arg-137_I Arg-226_{III} Trp-58_{III} Trp-59_{III} Glu-65_{III} Phe-86_{III} Met-55_{III} Leu-203_I	Arg-96_I Arg-221_{III} Trp-57_{III} Trp-58_{III} Glu-64_{III} Phe-86_{III} Met54_{III} Leu-159_I
Subunit I–II-III/PL4-PL5 interactions	Lys-103_{III} Asp-271_I Arg-234_{II}	His-103_{III} Asp-227_I Arg-178_{II}
Connection between helix 3 of subunit III and the Cu_B center	Trp-99_{III} Trp-331_I His333_I His334_I His-284_I Glu-90_{III}	Trp-99_{III} Trp-288_I His-290_I His-291_I His-240_I Glu-90_{III}
D path entry	Asp-132_I	Asp-91_I
Subunit III residues surrounding the entry to D pathway		Met-10_{III} Val11_{III} Asn-12_{III} Pro-13_{III}
	Ile-11_{III} Leu-12_{III} Pro-13_{III} Pro-14_{III}	
Histidine "proton antenna" residues of D pathway	His-549_I His-10_{III} His-7_{III} His-3_{III}	His-503_I His-9_{III} His-6_{III} His-3_{III}

Analyses of the crystal structures show that the lipids of the subunit III cleft form about twice as many hydrogen bonds and hydrophobic interactions as other lipids bound specifically to *Cc*O, indicating that they are more tightly bound.[23] The fatty acid tails of PL1 and PL2, at least, are more organized by hydrophobic interactions than other associated lipids. Moreover, many of the residues that form the hydrophilic and hydrophobic interactions with PL1 and PL2 are conserved, *i.e.* the lipid binding sites are conserved from *R. sphaeroides* to mammals.[27] Hence, PL1 and PL2, and possibly the lipids in upper region of the subunit III cleft, appear as conserved structural components of *Cc*O, and therefore they are likely required for *Cc*O function.

9.3.2 Lipids of Subunit III Play a Key Role in the Subunit I–III Interaction

Subunits I and III lie adjacent, with hydrophobic interactions involving helices 1 and 3 of subunit III, plus one of the fatty acids of PL1. However, ionic and hydrogen bond interactions between these two subunits are limited. The two prominent hydrophilic contacts each involve protein–protein plus protein–lipid–protein interactions, and all of the interactions are conserved from *R. sphaeroides* *Cc*O to bovine heart *Cc*O. First, Arg-137_{I} extends from helix 3 of subunit I to participate in a constellation of five ion pair and hydrogen bond interactions between the arginine side chain, the phosphate group of PL1 and the carboxyl group of Glu-65_{III} (Figure 9.3). The protein–protein contact occurs in two hydrogen bonds between the side chain of Arg-137_{I} and the carboxyl group of Glu-65_{III}. The protein–lipid–protein interaction begins with an apparent ion pair between Arg-137_{I} and the phosphate group of PL1. Then PL1 is connected to subunit III by three hydrogen bonds between its headgroup and residues of subunit III. The importance of this interaction has been tested *via* mutagenesis.[27] When Arg-137_{I} of *R. sphaeroides* *Cc*O is altered to alanine, and *Cc*O is isolated using an affinity tag on subunit I, 75–95% of subunit III is missing, indicating that a maximum of 25% of synthesized subunit III can successfully assemble with subunit I without the interactions made by Arg-137_{I}.

The second prominent interaction between subunits I and III, also present in both *R. sphaeroides* and bovine *Cc*O, involves two lipids bound on the outer surface of the complex (here termed PL4 and PL5), their headgroups in register with the P surface of the membrane.[23,25] The phosphate groups of PL4 and PL5 interact, apparently *via* hydrogen bonds. In the protein–lipid–protein connection at this site, the pair of interacting phosphate groups form two hydrogen bond/ion pair interactions with residues of subunit III and two with residues of subunit I. One of these interactions shows a fascinating level of complication, bringing together the three core subunits and lipid: Lys-103_{III}, Asp 271_{I}, Arg 234_{II}. The Lys-103_{III} is located near the top of helix 3 of subunit III, the helix that lies closest to subunit I, where it forms an ion pair/hydrogen bond with Asp-271_{I}. Joining this interaction is a conserved arginine of subunit II, Arg-234_{II}, which simultaneously forms an ion pair/hydrogen bond with Asp-271_{I}. The phosphate group of one of the paired lipids is a fourth member of this cluster of bonds, in that the phosphate forms an ion pair/hydrogen bond with both Lys-103_{III} and Arg-234_{II}. This four-way interaction is the only point in C*c*O where subunits I, II and III converge. Modification of either Asp-271_{I} or Arg-234_{II} significantly affects the stability of *Rs*C*c*O.[31]

Alteration of subunit III residues that bind and position PL1 and PL2 within the cleft of subunit III (Figure 9.3) also leads to decreased assembly of subunit III onto subunit I.[27] This suggests that appropriate lipid binding in subunit III is important to maintain subunit III in a conformation that can dock with subunit I, *i.e.* lipid binding in the cleft of subunit III precedes the subunit I–III assembly event.

9.3.3 Connecting Subunit III to the Active Site Region of Subunit I

Immediately below the Lys/His103_{III} residue described above, lies Trp-99_{III}, which docks in a pocket on the surface of subunit I to contact Trp-331_{I}. The tryptophan of subunit I, in turn, contacts the three histidine ligands of Cu_B: His-333_{I} and His334_{I}, and less directly His-284_{I} (Figure 9.4). Hence, the crystal structures show a reasonably direct connection *via* aromatic residues between one point of subunit III and the active site in subunit I. Is there evidence that this connection has functional significance? Perhaps. Glu-90_{III} lies ~12 Å below Trp-99_{III}, also in helix 3 but more distant from the active site in subunit I. Glu-90_{III} gained notoriety in early biochemical studies of C*c*O when it was discovered that modification of this residue with DCCD

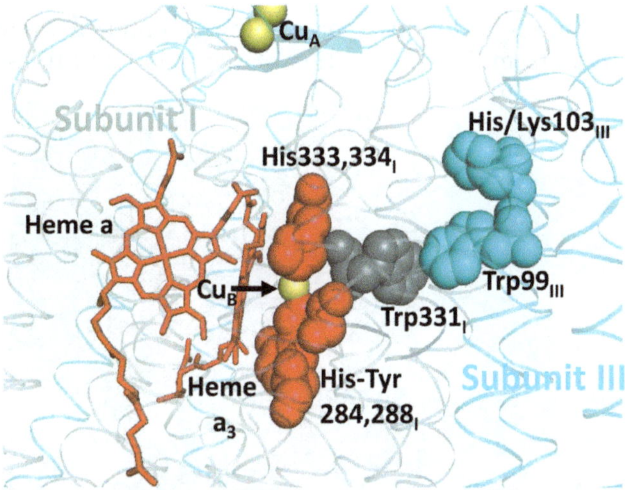

Figure 9.4 Connection between helix 3 of subunit III and the Cu_B ligands in subunit I. Trp-99_{III} (cyan) of helix 3 of subunit III contacts Trp-331_{I} (gray), which in turn contacts the Cu_B ligands His-333_{I} and His-334_{I}, and less directly, His-284_{I} (all red). See Section 9.3.3 and Section 9.8. This connection may mediate one aspect of subunit III's capability to prevent suicide inactivation of C*c*O.

compromised both proton pumping and O_2 reduction by CcO,[32,33] *i.e.* the modification of Glu-90_{III} affects the active site and nearby regions. The bulky modification of Glu-90_{III} by DCCD likely affects the position of helix 3 of subunit III. This effect may be amplified by the disruption of lipid binding in the cleft of subunit III, since fatty acid tails of two of the lipids in the cleft are in van der Waals contact with the carboxyl oxygens of Glu-90_{III} in the normal enzyme. Since helix 3 makes no connection, beyond non-directional van der Waals interactions, with subunit I near Glu-90_{III} or near the active site, except at Trp-99_{III}, it makes sense that structural information is communicated with subunit I *via* this route.

9.3.4 Relationship Between Subunit III and the D Pathway

Residues near the N-terminus of subunit III form part of a circle of residues that surrounds and partially buries Asp-132_I, the initial proton acceptor of the D pathway (Figure 9.5). The significance of this interaction is discussed in Section 9.4.3.

Conserved histidines near the N-terminus of subunit III have been proposed to be local proton antenna for the D pathway.[34] This will be discussed in Section 9.9. Finally, the CcO structures show that one of the fatty acids of PL1 essentially traces the water file of the D pathway, with one protein-helix width of separation. The significance of this, if any, is yet to be experimentally tested.

9.3.5 Removal of Subunit III

Differential scanning calorimetry reveals that subunit III forms a protein domain distinct from the combination of subunits I and II, as subunit III

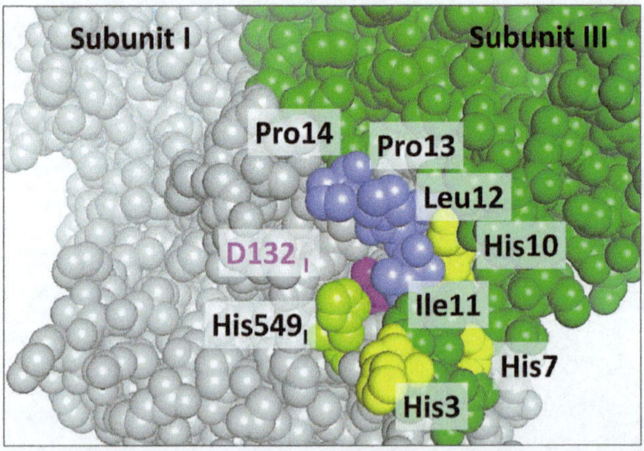

Figure 9.5 The Asp-132_I pocket in *R. sphaeroides* CcO (PDB 1M56). Subunit III is green, subunit I, gray. Subunit III residues forming half of the pocket enclosing Asp-132_I (Section 9.4.3) are Ile-11_{III}, Leu-12_{III}, Pro-13_{III} and Pro-14_{III} (light purple). Conserved histidine residues proposed as proton antenna for Asp-132_I (Section 9.9) are His-549_I, His-7_{III}, His-10_{III} and His-3_{III} (yellow).

unfolds at considerably lower temperature, 46.7 °C for subunit III of *P. denitrificans* C*c*O *vs.* 67 °C for subunits I and II.[35] From the crystal structures, this likely reflects release of lipid from subunit III, causing collapse of its structure. Subunits I and II of the aa_3-type bacterial C*c*Os can be expressed in the absence of subunit III.[36,37] The resulting subunit I–II oxidases are active and all of the metal centers are properly assembled.[35,37]

Detergents can be used to remove subunit III from purified C*c*O, likely because the detergents displace lipids involved in subunit I–III interactions. Once removed, the subunit has never been successfully reconstituted. Triton X-100 is effective in removing subunit III from the bacterial C*c*Os,[35,38] while dodecyl maltoside or lauryldimethylamine N-oxide is capable of removing subunit III from mammalian C*c*O.[39–41] Subunit III has also been removed from bovine C*c*O using limited proteolytic digestion.[42,43] Early characterization of the aa_3-type C*c*O of *P. denitrificans* reported it as a two-subunit enzyme due to the fact that Triton X-100 removed the subunit during isolation.[44] Even now, with solubilization with dodecyl maltoside and purification *via* an affinity tag on subunit I, preparations of *R. sphaeroides* C*c*O are generally mildly deficient in subunit III. C*c*O with accurate subunit stoichiometry can be obtained by using an affinity tag on subunit III,[27] since excess subunit III does not accumulate in the bacterial membrane.[45] For *R. sphaeroides* C*c*O, the C-terminal affinity tag on subunit I and the N-terminal tag on subunit III are in roughly the same place, since the C-terminus of subunit I is adjacent to the N terminus of subunit III. Subunits associated with subunit III are also removed. For the bacterial C*c*Os, the single helix subunit IV that binds atop lipids at the junction of subunits I and III is removed along with subunit III. In mitochondrial C*c*O, additional subunits are removed with subunit III depending upon the method of removal.

With care, subunit III-depleted C*c*O retains up to 80–90% of the initial activity of the starting material.[39,46,47] This ability to remove subunit III from C*c*O with retention of C*c*O activity is a tremendous advantage in studies that interrogate the effects of subunit III on C*c*O activity. In bacteria, subunit III may also be genetically deleted for the expression and purification of active subunit I–II C*c*O forms.[36,37] In this case, the isolated C*c*O generally shows less apparent activity when activity is expressed as the turnover number, and the amount of enzyme is calculated from the heme A spectrum. This occurs because in the absence of subunit III, inactivated C*c*O accumulates in the bacterial membrane, along with the active form of the enzyme.

9.4 Subunit III and the D Pathway

Subunit III appears in heme-Cu oxidases only when the D pathway is also present.[3] Subunit III lies close to the D pathway in the C*c*O complex, although direct interactions only take place in the N-terminal region of subunit III. Nonetheless, the function of the D pathway is considerably influenced by subunit III.

9.4.1 Normal D Pathway Structure and Function

The D pathway transfers protons ~26 Å from Asp-132_I on the interior N surface of subunit I, to Glu-286, whose carboxyl group is buried close to the active site, ~11 Å from the iron of heme a_3 and ~10 Å from Cu_B.[19,20,48] Asp-132_I lies in a pocket, with one carboxyl O facing out and the other facing in. The dimensions of the pocket are such that the outward facing O of Asp-132_I communicates with only one solvent water molecule at a time.[49] Water wires extend into bulk solvent, wires that may function to transfer protons to Asp-132_I.[49,50] The carboxyl group is free to rotate, in order to move a protonated O from the outside to the inside. Once inside, the protonated carboxyl O can form a hydrogen bond with the first internal H_2O of the pathway in order to transfer a proton. This first internal water lies midway between Asp-132_I and Asn-139_I. Above Asn-139_I, 11–12 waters lie between it and Glu-286_I,[51] of which a subset have been argued to form the most likely pathway.[52] A mystery of the D pathway is how protons get past Asn-139_I. While the amide side chain is free to rotate between waters above and below, the chemistry of the amide is not likely to allow rapid protonation and deprotonation of either the carbonyl or the amine constituents. Molecular dynamics simulations suggest an alternative, that the side chain of Asn-139_I is mobile to the extent that one of the waters seen above Asn-139_I in the crystal structures can move down in order to create a continuous water file between Asp-132_I and Glu-286_I.[51] From Glu-286_I, protons are assumed to alternately be directed to the proton loading site of the proton pump (in the vicinity of the D propionate of heme a_3) or to the O_2 reduction site through series of waters that are not seen in the crystal structures.[20,53–55]

9.4.2 Rapid Kinetic Analysis Reveals D Path Function

Much of our knowledge about D pathway function comes from rapid kinetic spectroscopy of the reaction of the fully reduced enzyme with O_2. The 'flow-flash' experimental system was first developed by Gibson and Greenwood[56] and then refined and used extensively by the Brzezinski group,[20,57] by Wikström and Verkhovsky,[58] and others. In this chapter, the reduction of a single O_2 by reduced *Cc*O will be referred to as the 'R→O sequence', in order to distinguish this experimental system from the continuous reduction of O_2. In the R→O sequence experiment, fully reduced *Cc*O (intermediate R-CO; Figure 9.6), with CO bound to heme a_3, is mixed with O_2. A laser flash is used to dissociate CO and O_2 binds, forming intermediate A[59,60] (Figure 9.6). A rapid, concerted transfer of four electrons from heme *a* ($1e^-$), heme a_3 ($2e^-$) and Cu_B ($1e^-$) to O_2 splits the double bond of O_2 and generates intermediate P_M.[60–62] P_M is converted to intermediate F by the transfer of a proton from Glu-286 to the hydroxyl group on Cu_B.[60,63–66] In the final reaction of this experiment, the fourth electron (originally from Cu_A) and another proton from Glu-286 reduce and protonate the oxoferryl of heme a_3 to produce intermediate O.[60,64,66]

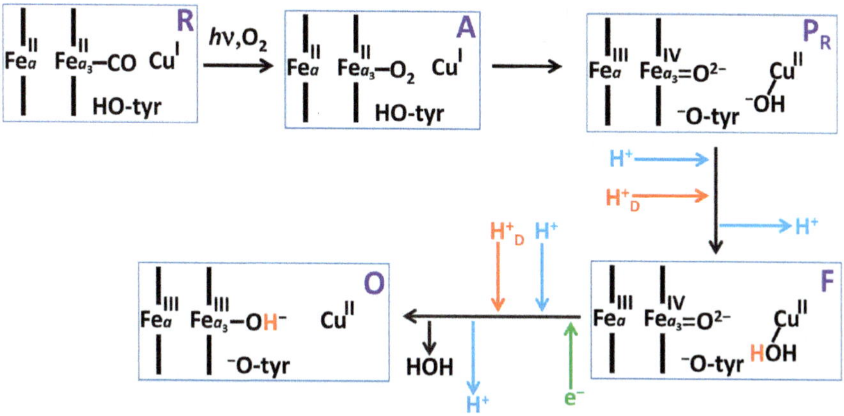

Figure 9.6 Reactions of the R→O sequence. The green arrow represents electron input from CuA, while red arrows represent proton input into the active site *via* the D pathway. Blue arrow designations are as described for Figure 9.1. The reaction series is described in Section 9.4.2.

The reactions of the R→O sequence differ from those of steady-state O_2 reduction (Figure 9.1) in two significant ways. First, intermediate P_M of steady-state O_2 reduction contains a tyrosine radical, while intermediate P_M of the R→O sequence contains a deprotonated tyrosine.[60,67] During the R→O sequence an electron is already present on heme *a*, allowing for rapid electron transfer from heme a^{2+} during the A→P_M reaction.[66,68] During steady-state electron transfer, an electron may not be available on heme *a* for the rapid electron transfers of A→P_M. Therefore, the histidine-tyrosine group of the active site transfers both an electron and a proton to O_2, generating a tyrosine radical, as shown for intermediate P in Figure 9.1.[2,60,62] Second, the P_M→F transition of the R→O sequence involves only a proton transfer from Glu-286 to the active site since the electron is already present.[60,63]

The R→O sequence is particularly useful for studying the D pathway since four protons are taken up during the short series of reactions, all by the D pathway. In both the P_M→F and the F→O reactions, one 'pump' proton is transferred through Glu-286 to the proton loading site (PLS) of the proton pump and one 'substrate' or 'chemical' proton re-protonates Glu-286, which has just transferred a proton to the active site for consumption.[3,20] The pK_a values associated with the P_M→F and F→O reactions have been used to argue the identity of the groups controlling the proton uptake events. Brzezinski and colleagues have assigned a pK_a of 9.4 measured for the P_M→F reaction as the pK_a of Glu-286.[69] This group models the P_M→F reaction as an internal proton transfer from Glu-286 to intermediate P_M, with the simultaneous reprotonation of Glu-286 by the D pathway.[69] In addition, a second proton is transferred from bulk solvent, through Glu-286, to the PLS of the pump.[20] Since the entire P_M→F reaction takes place in 100 to 150 μsec at pH 8 using detergent-solubilized C*c*O of *R. sphaeroides*,[69] the model of the reaction requires that each proton is taken up at a rate of at least 13 000 s^{-1}. Hence, the normal D pathway, in the presence of subunit III, is capable of extremely rapid rates of proton uptake. Wikström and Verkovsky[48] have proposed an alternative explanation for the pK_a of 9.4 measured for the $P_M \rightarrow F$ reaction: they suggest that it represents the aggregate pK_a of proton antenna groups on the inner surface of *Cc*O that supply protons to Asp-132, the initial proton acceptor of the D pathway. Proton antenna residues assigned to the D pathway are discussed further in Section 9.8.

The F → O reaction, ~ten-fold slower than the $P_M \rightarrow F$ reaction, is modeled as the simultaneous delivery of an electron and a proton to intermediate F. Earlier work from the Brzezinski group showed a single pK_a for the F → O reaction of ~8.6.[70] Extension and refinement of the pH dependence measurements of the F → O reaction reveals two pK_a values. One, of 8.9, is assigned to Glu-286, while a pK_a ~ 6.3 is suggested to arise from the cluster of groups that comprise the PLS of the proton pump.[64,71]

9.4.3 The pH Dependence of the D Pathway is Strongly Shifted in the Absence of Subunit III. Why?

Upon the removal of subunit III, the most rapid rate of proton uptake measured in the R → O sequence drops from >10 000 s^{-1} to ~350 s^{-1} (at pH 8.0).[72] Importantly, rapid proton uptake is restored to subunit III-depleted *Cc*O at low pH (<6), *i.e.* the rate of proton uptake at low pH is rapid for both *Cc*O (III+) and *Cc*O (III−).[72] Therefore, the absence of subunit III does not alter the D pathway per se, rather its pH profile is shifted more acid. The apparent pK_a of proton uptake by the D pathway is measured in the R → O sequence by plotting the pH dependence of the rate of the F → O reaction, which requires the transfer of a proton from bulk water to the active site, through Glu-286_I. The pK_a of D pathway proton uptake drops from 8.5 to ~7 with the removal of subunit III.[72] Thus, at a typical pH used for experiments, *e.g.* pH 7.4, the D pathway of *Cc*O (III−) is considerably slower than the D pathway of *Cc*O (III+).

Two possibilities, which are not mutually exclusive, have been put forth to explain the drop of ~1.5 pH units in the pK_a of proton uptake into the D pathway in the absence of subunit III. One argument is that the removal of subunit III removes residues that function as proton antenna for the D pathway, and the loss of these residues can only be overcome at low pH.[72] A corollary of this argument is that the pK_a observed for D pathway proton uptake by *Cc*O (III+) in the R → O sequence may be that of the antenna residues, since the aggregate antenna now becomes the protonatable 'group' that controls the rate of proton uptake. Indeed, it has been explicitly proposed that the pK_a of 9.4 measured in the R → O sequence is not that of Glu-286_I, but rather the aggregate pK_a of proton antenna groups near the entrance of the D pathway.[48] Attempts to experimentally examine if specific histidines near Asp-132_I function as proton antenna will be described in Section 9.9.

Another explanation for the altered pK_a of proton uptake is that the presence of subunit III has a direct effect on the thermodynamic pK_a of Asp-132_I. Asp-132_I is located at a junction of subunits I and III, and four residues of subunit III comprise approximately half of the residues surrounding Asp-132_I, as it lies in its well (Figure 9.5).[23,25,26] In the absence of subunit III, Asp-132_I is more exposed to bulk water; the greater exposure to bulk water should act to stabilize the anionic form of the carboxyl, *i.e.* the pK_a of Asp-132_I should decline. There exists experimental and structural evidence for a greater exposure of the entry of the D pathway to water in the absence of subunit III. When Asp-132_I is removed from C*c*O (III−), by altering Asp-132_I to alanine, D pathway proton transfer from bulk solvent to Glu-286I becomes facilitated entirely by water.[73] With water substituting for a carboxyl group at the entrance of the D pathway of D132A (III−), the pK_a of proton uptake in the R→O sequence shifts again, from ∼7 to greater than 10. Proton uptake by the D pathway of D132A (III−) may be controlled by Glu-286_I, the sink of the D pathway, since measurements indicate that its pK_a increases to >10 in the absence of subunit III.[72] Although we are currently focused on the entry region of the D pathway, the increased pK_a of Glu-286_I in C*c*O (III−) suggests that subunit III exerts a structural influence on regions in subunit I near the active site.

If Asp-132_I is responsible for the pK_a of proton uptake by the D pathway of C*c*O (III−), replacing Asp-132_I with a different protonatable group may further alter the pK_a of proton uptake. When exogenous arachidonic acid (Aa) is added to detergent-solubilized D132A (III−), some of the fatty acid binds in such a way that its carboxyl group becomes the initial proton acceptor of the D pathway. While proton uptake by D132A (III−) has a pK_a of >10 and a rate of ∼400 $H^+ s^{-1}$ at pH 7.4, the pK_a of D132A (III−) + Aa is 7.6 with a rate of ∼900 $H^+ s^{-1}$.[73] The pK_a of 7.6 differs from the pK_a of 7 obtained for C*c*O with Asp-132_I,[72] presumably due to slightly different environments for the carboxyl of the amino acid and the carboxyl of the fatty acid. However, the pK_a of proton uptake is apparently determined by the initial proton acceptor in both cases.

9.4.4 Steady-state Proton Uptake in the Absence of Subunit III Can Report the pK_a and the Rate of D Pathway Proton Uptake

The steady-state activity of C*c*O (III+) shows a clear pH dependence. C*c*O (III+) of *R. sphaeroides* shows high activity at pH 6.5, which then declines with increasing pH over a broad range.[38,74] The apparent pK_a of steady-state activity by C*c*O (III+) is 8.4 for the detergent-solubilized form of the enzyme,[74] as well as when the complex is incorporated into a lipid bilayer of soybean phospholipid and assayed under the condition of rapid proton equilibration across the membrane (Hosler, unpublished). Surprisingly, perhaps, there still exists no clear definition of what protonatable group or groups give rise to the pK_a of 8.4. The pH dependence of steady-state *Cc*O (III+) activity may arise from the K pathway, the D pathway or the exit pathway for pumped protons, or combinations of these.

The R→O sequence experiments[72,73] indicate that proton uptake into the D pathway is slower for WT *Cc*O (III−) than for WT *Cc*O (III+) at pH values >7. If the rate of proton uptake by the D pathway is rate limiting for *Cc*O activity (keeping in mind that with multiple substrates, *i.e.* protons, electrons and O_2, there can simultaneously be more than one rate limiting reaction) then the rate of steady-state O_2 reduction activity can report the rate of proton uptake by the D pathway and estimate the apparent pK_a of the initial proton acceptor. Experimental evidence indicates that this is, indeed, the case.[38,47,72–74] The entire pH profile of steady-state O_2 reduction by WT (III−) closely matches the pH profile for D pathway proton uptake during the R→O sequence.[38,47,72,73] The pK_a of O_2 reduction activity by WT (III−) is 7.0–7.3,[74] similar to the pK_a ∼7 measured for proton uptake during the R→O sequence.[72] Moreover, the rate of steady-state O_2 reduction closely matches the rate of D pathway proton uptake measured during the F→O reaction of the R→O sequence.[72,74] This similarity also holds true when the carboxyl of arachidonic acid replaces Asp-132I as the initial proton acceptor. The pH dependence of steady-state activity of D132A (III−) + Aa yields a pK_a of 7.9–8.0,[38] similar to the pK_a of D pathway proton uptake measured for this form.[73] Again the rate of proton uptake measured during the F→O reaction of the R→O sequence closely matches the rate of *Cc*O activity at the various pH values.[38,73] Thus, in the absence of subunit III, a pH dependent rate-limiting process for steady-state *Cc*O activity can be identified: proton uptake into the D pathway. The comparison of the pK_a values of steady-state *Cc*O activity in the presence (8.4) and absence (7.0–7.3) of subunit III allows a second conclusion, that the presence of subunit III extends the pH range of *Cc*O activity well into the physiologic range.

Correspondence of the pK_a and rate of steady-state activity with the pK_a and rate of proton uptake by the D pathway does not occur in some *Cc*O (III−) forms. For D132A (III−), in which the carboxylic proton acceptor is replaced by water, the D pathway has a pK_a of >10.[73] Hence, the rate of ∼400 s^{-1} for D pathway proton uptake is observed at both pH 6 and pH 10. However, the pK_a values of the K pathway and/or the exit pathway are clearly lower than 10, since steady-state activity by D132A (III−) reaches near zero at pH 9.5.[38] Interestingly, the high pK_a of the D pathway does confer a broad pH range of activity to D132A (III−), similar to *Cc*O (III+).[38] This suggests that the pK_a of Asp-132_I is higher in *Cc*O (III+) than it is in *Cc*O (III−), consistent with arguments presented above.

Manipulation of proton entry into the D pathway can be used to study the K pathway and, presumably, the exit pathway for pump protons. For example, placing two carboxyls into the D pathway of *Cc*O (III−), by placing aspartic acid at position 139 and retaining Asp-132_I, creates a *Cc*O form with a D pathway that delivers protons to the active site at a rate at least 20% faster than normal *Cc*O (III+).[47] Since the more rapid steady-state activity of

N139D (III−) requires an equal increase in activity by the unaltered K pathway, the result indicates that the K pathway of similarly isolated CcO cannot be rate-limiting for the maximal activity of CcO at low pH. No conclusions can be reached about the exit pathway for pump protons, since N139D is reported to lack proton pumping capability;[75,76] but see ref. 77.

9.5 Subunit III & Proton Pumping

Subunit III has long been implicated in proton pumping by CcO. Several independent reports indicate that proton pumping in the absence of subunit III declines by ~50% (at pH 7.4), for both the bacterial and mitochondrial aa_3-type CcOs.[38,42,78] The reason for this loss of pumping stoichiometry has yet to determined. One relevant fact is that in the absence of subunit III, Glu-286$_I$ has a pK_a of >10, as revealed varying the pH in R→O sequence experiments[72] since the P→F reaction *is* the deprotonation of Glu-286$_I$. A simplistic explanation for the lower stoichiometry of proton pumping by CcO (III−) is that the higher pK_a of Glu-286 decreases the probability of proton transfer to the PLS of the pump, but not to the high pK_a proton acceptors of the active site. A similar explanation has been proposed for the lower stoichiometry of proton pumping by some CcO forms containing alterations in the D pathway.[79]

9.6 Subunit III & Suicide Inactivation

Suicide inactivation (SI) of enzymes is classically induced by the addition of a suicide substrate, often a compound that is oxidized during catalysis to a radical species, which then modifies or cross-links amino acid residues in the active site. However, enzymes that activate O_2, and particularly those that utilize iron and copper to bind O_2, must utilize strategies to avoid inactivating themselves.[80] CcO has an innate tendency to undergo turnover-based inactivation, or suicide inactivation, *i.e.* the spontaneous loss of activity during normal turnover. Studies have shown that one of the important roles of subunit III is to lower the probability of SI.

Suicide inactivation was first reported as the exponential loss of activity during continuous O_2 consumption in mitochondrial (rat liver)[81] and bacterial (*P. denitrificans*)[35] CcOs, both lacking subunit III. These results were confirmed using the aa_3-type CcO of *R. sphaeroides* and a system for describing the kinetics of inactivation was derived.[82] It was also found that subunit III-depleted bovine CcO exhibits the same kinetics and pH dependence of SI as the *R. sphaeroides* enzyme (Hosler & Hill, unpublished).

The single exponential kinetics of SI[82] indicate that a CcO molecule is either fully active or fully inactive, without any partially active intermediates. The kinetics also indicate that CcO may undergo permanent inactivation during any given catalytic cycle. The probability of SI depends upon the elapsed number of catalytic cycles, and not upon the time between catalytic cycles. Hence, a useful measurement of SI cannot be time-based, like a $t_{1/2}$.

The chosen measurement is the CC_{50}, also termed 'catalytic lifespan' or 'catalytic half-life', which is defined as the number of catalytic cycles (one $CC = O_2 \rightarrow 2H_2O$) required for the activity of a population of *Cc*O to drop to half. A greater CC_{50} indicates a longer catalytic half-life as well as a lower probability of SI during any catalytic cycle.

Catalytic half-life varies strongly with pH, becoming progressively shorter as pH increases.[82,83] The pH dependence of catalytic half-life closely matches the pH dependence of proton uptake into the D pathway of *Cc*O (III−), as measured in the R→O sequence.[72,83] Both bacterial and mitochondrial *Cc*Os are most active in the presence of dodecyl maltoside plus detergent-solubilized phospholipids. Under these conditions, and at pH 7.4, *R. sphaeroides* *Cc*O (III+) shows a CC_{50} of $>3.5*10^6$, while *Cc*O (III−) exhibits a CC_{50} of approximately 10 000.[74] Therefore, at pH 7.5, the catalytic half-life of subunit III-depleted *Cc*O is less than 0.3% that of *Cc*O that contains subunit III. During catalytic turnover in the presence of a membrane potential, the CC_{50} of *Cc*O (III−) declines further, to ~1000.[84]

The increased probability of SI in the absence of subunit III suggests an explanation for the strong conservation of subunit III, even though the subunit contains no redox centers and it is not absolutely required for the chemistry of O_2 reduction or proton pumping. As discussed in Section 9.3, the primary sequence of subunit III is highly conserved from bacteria to humans, significantly more so than subunit II and similar to subunit I. A power plant is successful only if its energetic output is considerably greater than the cost of construction. This logic should hold true for a cell as much as it does for human society. Just the synthesis of the peptide chains of mitochondrial *Cc*O requires at least 7200 ATP, leaving aside the energetic requirements of the rest of the elaborate assembly process for complex IV. If fully assembled Complex IV produces the theoretical maximum of 2 ATP per catalytic cycle, 3600 catalytic cycles are required simply to pay for the cost of peptide synthesis. Inactivation after 10 000 catalytic cycles, as measured for *R. sphaeroides* *Cc*O (III−) in detergent solution, would be a poor return on investment. Inactivation after 1000 catalytic cycles, as measured for *Cc*O (III−) in the presence of a membrane potential,[84] would fail to recoup the energetic investment, an untenable situation for a cell. Therefore, evolving, and then conserving, a subunit that lowers the probability of SI should be a powerful selective force.

9.7 Mechanism of Suicide Inactivation

As a first step toward understanding the mechanism of SI, structural characteristics of inactivated *Cc*O were compared to active *Cc*O.[82] Spectroscopic analysis of inactivated *Cc*O showed structural alterations at the heme a_3–Cu_B active site. In particular, Cu_B appears to be absent, while heme a_3 is present. The first indication that Cu_B was absent from inactivated *Cc*O was the finding that CN^- bound to oxidized, inactive enzyme but not to the reduced form. In active *Cc*O, CN^- binding to reduced heme a_3 requires Cu_B, while the binding of cyanide to oxidized heme a_3 does not involve Cu_B.[85] The absence of Cu_B in inactivated *Cc*O was confirmed by metal analyses and EPR spectroscopy (Hosler, unpublished). Metal analysis of *Cc*O showed the absence of one of the three coppers of *Cc*O, while EPR indicated that the two coppers of Cu_A were present. Hence, the missing copper is Cu_B.

Consistent with this finding, suicide inactivated *Cc*O no longer produces the P and F intermediates upon the addition of CO/O_2 or H_2O_2.[82] The production of these intermediates by ligand addition requires the presence of Cu_B. Even though heme a_3 is retained in inactivated *Cc*O, its redox potential declines dramatically and resonance Raman analysis indicates that heme a_3 of inactivated *Cc*O can adopt a variety of slightly different orientations within its binding pocket. Both heme *a* and Cu_A are present, spectroscopically normal and can be reduced by cytochrome *c*, indicating that the damage leading to inactivation is limited to the heme a_3–Cu_B active site. Reconstitution of *Cc*O (III−) into lipid vesicles has no protective effect and the elimination of exogenous reactive oxygen species (by adding catalase and superoxide dismutase) also fails to prevent inactivation.

The probability of SI by *Cc*O (III−) is strongly influenced by the bulk pH, in that its catalytic half-life (CC_{50}) declines with increasing pH.[82] At pH 8.5 the CC_{50} is ~400, compared to ~$2.5*10^5$ at pH 6.2,[74] indicating that during any given catalytic cycle suicide inactivation is 625 times more probable at the higher pH.

The pH profile of catalytic half-life[74,83] is similar to that of the pH profile of proton uptake into the D pathway of *Cc*O (III−),[72] which declines with increasing pH with a pK_a of ~7. This suggests that proton uptake into the D pathway is linked to SI. This relationship is confirmed by demonstrations that site-directed mutations that slow proton transfer through the D pathway specifically increase the probability of SI.[84] Moreover, mutations that speed proton transfer through the D pathway, to rates greater than normal *Cc*O, strongly decrease the probability of SI.[74] Thus, the catalytic half-life of *Cc*O increases as the rate of proton uptake by the D pathway increases.

The D pathway provides both substrate protons, those used in O_2 reduction, and pump protons. Of these, it is the rate of substrate proton delivery that determines the probability of SI, for at least two reasons. One, SI affects Cu_B of the active site, making it most likely that a proton destined for the active site will have an effect. Second, SI takes place in *Cc*O forms that do not pump protons, such as D132A (III−).[38,73] Our conclusion of the preceding paragraph can now be revised: the catalytic half-life of *Cc*O increases as the rate of *proton delivery to the active site*, *via* the D pathway, increases. The relationship is exponential, *i.e.* catalytic half-life increases exponentially with the rate of proton delivery into the active site.[74]

An important question regarding the mechanism of SI is whether other processes that slow the steady-state activity of *Cc*O also promote SI. For example, catalytic turnover in the presence of a membrane potential further increases the probability of SI, and the membrane potential slows proton uptake into both the D and the K pathways. But does inhibition of proton uptake specifically into the K pathway lower catalytic half-life as it lowers steady-state O_2 reduction activity? Site-directed mutants provide the test. For example, the alteration of Thr-359 within the K pathway to alanine strongly inhibits proton uptake into the K pathway,[19,86] but not proton pumping efficiency. Using T359A (III−), the K pathway is more limiting for steady-state turnover than is the D pathway. However, the probability of SI is not increased in comparison to WT (III−),[84] indicating the two transitions that depend upon substrate proton uptake by the K pathway (O→E, E→R)[87] are not involved in SI chemistry. Therefore, the increase in inactivation seen in the presence of ΔΨ is due to inhibition of the D pathway.

Another way of limiting C*c*O activity is by modifying the rate of electron delivery *via* cytochrome *c*. The four transitions of the catalytic cycle (O→E, E→R, P→F and F→O) are initiated by the transfer of an electron into the active site (Figure 9.1).[3] However, the rate of electron delivery from cytochrome *c* does not alter the subsequent rapid rates of internal electron transfer through C*c*O to the active site, rather it dictates the time between transitions. When the rate of electron delivery is varied by adjusting the concentration of cytochrome *c*, both low and high rates of electron input yield the same probability of SI.[84] Since slowing the time between transitions does not alter the probability of SI, it suggests that the chemical events of SI take place during one transition. Alternatively, the result may indicate that a deprotonated intermediate that is not part of the normal catalytic cycle is nonetheless stable enough to persist until the arrival of the next electron.

The results described thus far suggest the following model. During one of the transitions of the catalytic cycle, electron entry into the active site generates an active site intermediate capable of initiating the chemistry of SI, *i.e.* the presence of this intermediate raises the probability of SI chemistry. Subsequent proton delivery *via* the D pathway generates the next intermediate, which is benign, *i.e.* the protonated intermediate vastly lowers the probability of SI chemistry.

But during which transition does SI take place? The O→E and E→R transitions are unlikely to be involved since their substrate protons are taken up by the K pathway. The D pathway takes up two substrate protons, those consumed during the P→F and F→O transitions. It is reasonable to suggest that SI might be caused by oxidative damage to the active site by the nucleophilic oxoferryl of heme a_3 and the tyrosine radical (Figure 9.1), particularly when slow proton uptake allows these species to persist for a longer period of time. In this regard, it seems significant that two other oxygen-activating enzymes that also produce heme oxoferryl and protein radical intermediates during their catalytic cycles, prostaglandin H synthase and linoleate diol synthase, exhibit rapid SI during normal turnover.[88,89] From Figure 9.1, it can be seen that slow proton uptake during the P→F transition should extend the lifetime of the oxoferryl intermediate of heme a_3. However, there exist at least two arguments against proposing the oxoferryl or the tyrosine radical as the source of inactivation chemistry. First, intermediate P, containing the oxoferryl of heme a_3 and probably the tyrosine radical, can be formed from oxidized C*c*O by reducing the active site with CO in the presence of O_2.[90] Incubation of C*c*O (III+) or C*c*O (III−) in the P state for several minutes, in the absence of catalytic turnover, does not inactivate the enzyme (Hosler, unpublished). Hence, the existence of these active site oxidants for several milliseconds, as a result of slow proton uptake, is unlikely to cause inactivation. Second, the heme a_3 oxoferryl and the tyrosine radical are first formed as intermediate P_M (Figure 9.1). Delaying *electron* entry into P_M will extend the lifetimes of these species. Thus, if the oxoferryl and the tyrosine radical were the initiators of SI chemistry, then, SI would be more probable with slow *electron* delivery into C*c*O. The fact that SI is not favored by slow electron delivery[84] further argues against the heme a_3 oxoferryl and the tyrosine radical as reactive molecules that cause SI.

Nonetheless, slow proton input into the P→F→O reactions could lead to chemistry that causes the key lesion of SI, namely, the functional, and then physical, loss of Cu_B. An example is proposed in Figure 9.7. P_M-reduced ($P_{M,R}$) is proposed as the intermediate that forms after the electron of the P_M→F reaction has entered the active site but the substrate proton has not. The tyrosine radical has been reduced, while the hydroxyl anion remains bound to Cu_B. Overall, $P_{M,R}$ is electrically neutral and it is proposed to be stable. If the substrate proton arrives on schedule, $P_{M,R}$ is converted to F, as shown in Figures 9.1 and 9.7. However, if the next electron arrives in the active site *before* the substrate proton arrives, *i.e.* with slow proton delivery, intermediate F-reduced (F_R) is produced. The O^{2-} coordinating the iron of heme a_3 of F_R contains three lone pairs of electrons, *i.e.* the O^{2-} should be a potent Lewis base. The nucleophilic nature of the heme a_3 O^{2-} can drive the irreversible conversion of F_R to inactivated C*c*O (C*c*O_{SI}), which results from transfer of the heme O^{2-} to Cu_B along with the incorporation of a nearby water. The active site has, in fact, undergone oxidative damage in that Cu_B now contains two hydroxyl ligands. In the hydrophobic environment of Cu_B

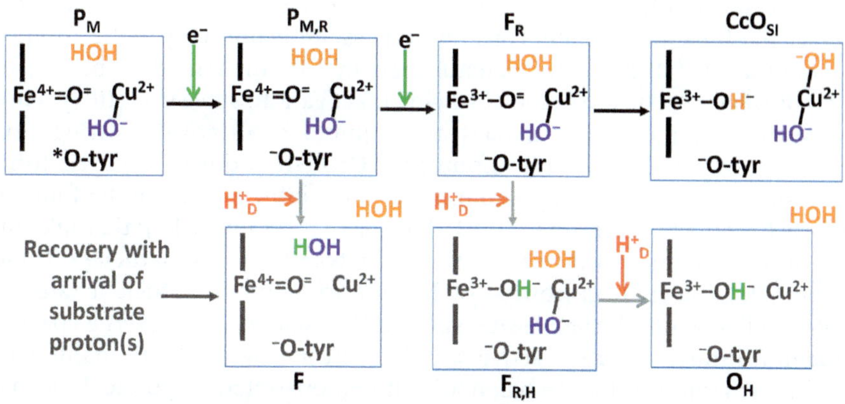

Figure 9.7 A putative scheme for suicide inactivation. The scheme is described in Section 9.7. Arrow designations are as described for Figure 9.1. Free HOH indicates water near heme a_3–Cu_B.

in *Cc*O, di-hydroxide ligation should prevail, leading to dissociation of the histidine ligands and, ultimately, to the loss of Cu_B from the active site. In this putative mechanism, the stable binding of a second hydroxyl ligand to Cu_B^{2+} is proposed as the inactivating event. Note that suicide inactivation can be avoided even from the F_R intermediate, if substrate protons arrive before the conversion to CcO_{SI} occurs. The first proton creates a stable protonated F_R ($F_{R,H}$), while a second substrate proton produces the normal O_H intermediate (Figure 9.7).

Another mechanism proposed to result in SI is the "uncontrolled" supply of O_2 to the active site, with the production of reactive O_2 species (ROS) that damage the active site.[23] Yoshikawa and colleagues have suggested that the absence of subunit III leads to this possibility since the proposed O_2 delivery pathway of *Cc*O leads through the cleft lipids of subunit III.[23,91] This seems an unlikely mechanism, for several reasons. First, SI can take place in the presence of subunit III, with normal O_2 delivery, if the rate of proton delivery to the active site is slow.[84] Second, hydrogen peroxide is not released during the process of SI,[82] and the direct application of millimolar quantities of H_2O_2 to *Cc*O (III−) for several minutes does not recapitulate SI (Hosler, unpublished). Both results indicate that the production of superoxide or hydrogen peroxide at the active site is unlikely to cause SI. Third, there is no change in the CC_{50} of SI by *Cc*O (III−) in buffers that are saturated with O_2 rather than saturated with air (Hosler, unpublished). Hence, neither the ambient O_2 concentration nor the rate of O_2 delivery to the active site alter the probability of SI.

9.8 Structural Influence of Subunit III Helps Prevent SI, and Involves its Bound Lipids

Subunit III clearly influences the structure of subunit I, as expected for two proteins that bind, but the influence on structure is not large.[35,92] For example, all of the metal centers of subunit I assemble normally in the absence of subunit III.[37] Progressive denaturation of subunit I in the absence of subunit III cannot be the cause of suicide inactivation, since *Cc*O (III−) is not thermally unstable;[35] several days are required to inactivate resting *Cc*O (III−) as compared to several minutes for *Cc*O (III−) during catalytic turnover. Moreover, *Cc*O lacking subunit III can exhibit a catalytic half-life as high as *Cc*O (III+), in those *Cc*O (III−) forms (*e.g.* N139D (III−)) that take up D pathway protons at a faster rate than wild-type *Cc*O.[74] Nonetheless, while subunit III protects *Cc*O from SI by helping to maintain sufficient rates of proton delivery into the active site, subunit III also appears to exert a specific structural influence on the active site that further lowers the probability of SI. This structural influence has not been demonstrated by physical means, but it can be inferred from enzymatic analysis of various *Cc*O forms. An example is D132A, the mutant *Cc*O in which Asp-132_I, the initial proton acceptor of the D pathway, is replaced by alanine. With subunit III present,

the surface of the protein by a process of release and recapture, effectively increasing the proton concentration near the initial acceptor of a pathway.[34,94]

Conserved histidine residues near Asp-132$_I$ have been proposed to function as a local proton antenna for the D pathway.[34] Three conserved histidines are present in the N-terminal region of subunit III (Figure 9.5). Both imidazole nitrogens of His-7$_I$ and His-10$_I$ are ~10 Å from the carboxyl of Asp-132$_I$. His-3$_I$ is more distant, at 18 Å. The roles of these three conserved histidines were tested in a recent study in which all three histidines were altered to glutamine.[95] H3,7,10Q (III+) was examined for cytochrome *c* peroxidase activity. Cytochrome *c* peroxidase activity becomes evident at high pH (10), conditions where proton antenna should be most needed, and is only dependent upon proton uptake *via* the D pathway. The peroxidase activity of H3,7,10Q (III+) was significantly lower than that of normal C*c*O (III+) and similar to that of C*c*O (III−). These data are consistent with an antenna-like function for these histidine residues, or a subset of them.

An imidazole nitrogen of His-549$_I$ is located 6.5 Å from the exposed carboxyl O of Asp-132$_I$ in a recent structure of *R. sphaeroides* C*c*O.[26] The same is true for its cognate in bovine C*c*O, His-503$_I$, plus a water is coordinated between His-503$_I$ and Asp-91$_I$ (the cognate of Asp-132$_I$).[49] Thus, the structures place His-549$_I$/His-503$_I$ in an ideal position to assist in proton uptake into the D pathway. Nonetheless, the alteration His-549$_I$ to alanine has no effect on the function of *R. sphaeroides* C*c*O from pH 6.5 to 9 (Hosler & Ferguson-Miller, unpublished results). In order to examine this further, we have examined a double mutant of *R. sphaeroides* C*c*O in which both H549$_I$ and H7$_{III}$ were altered to alanine (Hosler, unpublished results). When this mutant is incorporated into soybean phospholipid vesicles, and assayed with uncoupler present, such that the pH is equilibrated to 7.4 in both the lumen and the exterior of the vesicle, catalytic half-life measurements show that the D pathway of the histidine mutant has the ability to take up substrate protons at the same rate as normal C*c*O (III+), also in the lipid bilayer. This indicates that His-549$_I$ and His-7 III$_I$ do not enhance D pathway proton uptake at pH 7.4. This agrees with calculations in that simple diffusion should be sufficient to maintain high D pathway proton uptake at pH 7.4.[19] With catalytic turnover of C*c*O in a sealed vesicle, the pH at the entry to the D pathway increases by as much as three pH units. Under these conditions, proton uptake into the normal D pathway of (C*c*O III+) is slowed. However, proton uptake into the D pathway of H549A$_I$-H7A$_{III}$ III (+) slows even more, since the catalytic half-life of the mutant declines to ~10% of the normal enzyme under the same conditions (Hosler, unpublished results). Hence, at high pH or high $\Delta\Psi$, His-549$_I$ and/or His-7$_{III}$ appear to help to maintain rapid proton uptake into the D pathway.

Thus far, the limited data available suggests that conserved histidines near Asp-132$_I$ do function as proton antenna at high pH (and, therefore, at high $\Delta\Psi$ in a sealed vesicle). However, detergent solubilized H549A$_I$-H7A$_{III}$ (III+) exhibits the same pH profile of activity as WT (III+) between pH 6.5 and pH 9, suggesting that these residues do not strongly affect D pathway proton uptake at normal physiologic pH (Hosler, unpublished). The latter finding is not consistent with the hypothesis that the removal of subunit III alters the pH profile of C*c*O activity by eliminating local proton antenna residues.[72]

9.10 Proton Backflow and SI

9.10.1 Identification of Proton Backflow

Once cytochrome oxidase has been reconstituted into 25 micron phospholipid vesicles (COVs), a few catalytic turnovers are sufficient to generate a sizeable voltage gradient ($\Delta\Psi$; positive outside) across the membrane. In addition, the pH on the inside of the vesicle increases rapidly as protons are taken up by the D and the K pathways for consumption at the active site and for pumping across the membrane. Both the $\Delta\Psi$ and the high internal pH inhibit proton uptake from the inner surface of the enzyme. Moreover, proton exit from the outer surface of the enzyme is also inhibited, and can only be observed following the collapse of $\Delta\Psi$ by valinomycin. Despite these inhibitions of normal proton uptake and release, a slow rate of O_2 reduction continues (termed controlled turnover) indicating that protons are still flowing to the active site. For the wild-type *aa*$_3$-type C*c*O of *R. sphaeroides*, the rate of proton delivery into the active site during controlled turnover is typically ~100 H^+ s^{-1}.[38,96] At first, it was generally assumed that the protons necessary for controlled turnover were transferred into the vesicle lumen by electrophoretic, non-specific proton leak across the membrane, followed by uptake into the D and K pathways. This assumption was challenged by the finding that micromolar concentrations of zinc or cadmium, but not nickel, inhibited controlled turnover in vesicles.[96] Zinc and cadmium are well known to inhibit proton uptake into proton transfer pathways, particularly those that begin with aspartate, glutamate and histidine.[19] Hence, the inhibition of controlled turnover in COVs by externally-added zinc (reversible with EDTA) argued for the existence of a proton transfer pathway leading from the outer surface of C*c*O to the O_2 reduction site.[96] Since both the D pathway and the K pathway are slowed by the high $\Delta\Psi$ and the high internal pH, it seems likely that proton backflow compensates for both pathways in terms of proton delivery. In fact, when both the K pathway and the D pathway of *P. denitrificans* C*c*O are inhibited by mutagenic alterations, proton backflow to the active site is observed.[97] Thus far, proton backflow has only been measured in bacterial *aa*$_3$-type C*c*Os. However, the rate of O_2 reduction at high $\Delta\Psi$ and high internal pH using lipid vesicles containing mammalian C*c*Os is great enough to suggest that catalytic turnover is supported by proton backflow.

9.10.2 The Mutant D132A as a System to Study Proton Backflow

In the mutant *Cc*O D132A (and also D132N), the carboxylate proton acceptor of the D pathway on the inner surface of the enzyme is removed. As a result, proton uptake into the D pathway is all but eliminated.[73,98] The position of D132 makes it unlikely that its mutation will alter either the structure of the other proton pathways of *Cc*O or the metal centers of subunit I. Indeed, spectral analysis and crystal structures (PDB 3OMI) of the D132A mutant show that the metal centers and the K-path are unaltered.[99] While the rate of proton uptake by the D pathway of D132A is extremely slow ($\sim$5 s^{-1}),[73] the steady-state activity of the detergent solubilized *Cc*O is higher, $\sim$40 H^+ s^{-1}.[38] Thus, an additional path for protons to reach the active site of D132A must exist. Several lines of evidence indicate that proton backflow from the outer surface of D132A provides this route, as opposed to proton uptake by the K pathway. First, when D132A *Cc*O is incorporated into lipid vesicles, the pre-steady-state rate of soluble cytochrome *c* oxidation matches the rate of alkanization on the outside of the COVs, indicating that both protons and electrons are being taken from the outside.[96] Second, the rate of steady-state O_2 reduction by D132A in lipid vesicles is significantly enhanced in the presence of a membrane potential, as expected if proton uptake is from the outer surface and driven by the negative-inside membrane potential.[100] The source of the membrane potential must be the transfer of protons from the inner lumen of the vesicles to the active site *via* the K pathway, since the D pathway is largely inactive and D132A does not pump protons. Third, the alteration of some residues near the outer surface of *Cc*O, and distant from the K pathway, inhibit proton backflow. Arg-481, for example, lies above and between the hemes of subunit I. The R481K *Cc*O exhibits a slow rate of turnover in the presence of a membrane potential, indicative of inhibition of proton backflow.[101] Finally, combining the R481K mutation (inhibition of proton backflow) with the D132A mutation (inhibition of D pathway proton uptake) should create a *Cc*O form with a rate of proton delivery to the active site that is slower than either alteration separately. This was shown to be true, and also confirmed by using catalytic half-life as a measure of proton delivery to the active site.[84]

9.10.3 Proton Backflow *versus* Proton Exit

The relationship between the normal exit pathway for pumped protons and the proton backflow pathway, if any, is unknown. Both the proton exit pathway and the proton backflow pathway are accessible on the outer surface of *Cc*O, but the former leads *from* the pump proton loading site (near heme a_3 in the bacterial aa_3-type *Cc*Os) while the latter leads *to* the site of O_2 reduction. The pathways could share common elements and then diverge, or they may be completely separate. One line of evidence suggests that they may be separate. The addition of zinc to the external medium in reconstituted *Cc*O strongly inhibits *Cc*O activity in the presence of a membrane potential, *i.e.* controlled turnover that requires proton backflow.[96] However, under these conditions but during uncontrolled turnover, *i.e.* conditions when proton exit should be operative, external zinc does not inhibit nearly as strongly.[96,102] On the other hand, there is also evidence (under different conditions) for zinc inhibition in the uncoupled state[103] and for inhibition of proton exit in the P$\rightarrow$F reaction step,[104] as well as zinc inhibition of a double mutant of the D and K paths whose activity appears to require reversal of the exit pathway.[97] These observations favor the possibility that the proton exit and proton backflow pathways share some common elements.

9.10.4 Functional Significance of Proton Backflow *In vivo*

Proton backflow becomes active when proton uptake from the inner surface of *Cc*O is slowed by a high membrane potential. In this condition and in the presence of reduced substrates, the respiratory electron transfer chain becomes more reduced, driving electron leak to O_2 to produce superoxide and downstream reactive oxygen species, resulting in cellular oxidative damage.[105] Proton backflow may partially mitigate this problem by acting as a release valve, maintaining a flow of electrons through *Cc*O to O_2 uncoupled from proton pumping.

In Section 9.7, we have discussed how the rate of proton delivery to the active site is a strong determinant for the onset of suicide inactivation chemistry. Thus, the ability of proton backflow to maintain a basal level of proton delivery to the active site during periods of high membrane potential will help to lower the extent of suicide inactivation of *Cc*O.

9.10.5 Subunit III Appears to be Required for a Functional Proton Backflow Pathway

The removal of subunit III also slows controlled electron transfer, to a similar extent as (and additive to) the addition of zinc.[38] WT (III+) exhibits a rate of controlled activity of ~100 s^{-1},[38,96] while WT (III−), under the same conditions, shows a rate of ~10 s^{-1}.[38] From this, it has been proposed that subunit III plays a role in maintaining an active proton backflow, either directly or by structural influence on a backflow pathway in subunit I.[38,83] A simpler alternative has also been proposed,[3] that in the presence of subunit III, the extended pH range of proton uptake by the D pathway maintains a higher rate of controlled activity.[3] However, a problem with this simpler explanation is that in the presence of external zinc, the controlled activity of WT (III+) drops from ~100 s^{-1} to ~15 s^{-1}.[96] This indicates that 85% of the protons used during controlled turnover, in the absence of zinc, arrive *via* proton backflow, and only 15% from the D pathway of WT (III+). Hence, the decline in controlled activity from 100 s^{-1} to 10 s^{-1} upon the removal of subunit III must arise primarily from a slower rate of proton backflow in the absence of subunit III.

Further evidence for the role of subunit III in proton backflow comes from consideration of the mutant *Cc*O D132A, which lacks the carboxyl proton acceptor at the entrance of the D pathway. As described in Section 9.4.4, D132A (III+) has little D pathway activity (~5 s^{-1} at pH 7.4), but this increases to ~400 s^{-1} when the removal of subunit III allows water to reconstitute the entrance to the D pathway.[73] Moreover, the D pathway of D132A (III−) retains its activity at high pH.[73] During controlled turnover, protons are delivered to the active site of D132A (III+) by proton backflow plus the residual activity of the K pathway at high internal pH; the slow K pathway activity also generates $\Delta\Psi$ and ΔpH. Upon the removal of subunit III from D132A, K pathway activity is unaffected and D pathway activity *increases* considerably. Nevertheless, when subunit III is removed, controlled activity by D132A decreases by 75%.[38] This can only be explained as inhibition of proton backflow in the absence of subunit III. Determining *how* subunit III influences proton backflow remains a challenge.

9.11 Subunit III and O_2 Delivery

The most likely pathway for the delivery of O_2 to the active site of *Cc*O proceeds through the lipids of the subunit III cleft, as detailed in a recent review.[3] The protein portion of subunit III has been proposed to fend off other integral membrane proteins, so that O_2 entry will not be occluded (Figure 9.2).[3] A recent study of fish exposed to different degrees of hypoxia shows a correlation between increased tolerance to hypoxia and stronger binding of cardiolipin (Figure 9.2B) at the entrance of the cleft of subunit III.[106] This adaptation may help to keep the cleft lipids in place in order to ensure rapid O_2 delivery.

9.12 Summary

This chapter details the experimental evidence for the functions of subunit III, in the context of its unique structure that includes integral lipids. The appearance of subunit III in the heme-Cu oxidases coincides with that of the D pathway for proton transfer. Although the D pathway is located in subunit I, subunit III makes direct contact with the pathway at its entrance. By shielding the initial proton acceptor from bulk water, subunit III raises the pK_a of D pathway proton uptake into the physiologic pH range (~pH 8). By maintaining sufficient rates of proton uptake at physiologic pH, subunit III partly fulfills what may be its most important function: preventing turnover-based SI of the enzyme. Slow proton delivery into the active site after O_2 binds increases the probability of side reactions that irreversibly inactivate the active site by damaging the Cu_B center. Specific chemistry for the inactivation event is proposed here. Two other ways by which subunit III appears to prevent SI are discussed: by exerting structural influence on the Cu_B center and by enhancing proton delivery to the active site from the P-side of the membrane at high PMF. Finally, we evaluate experimental evidence testing the hypotheses that subunit III provides proton antenna residues important for proton uptake into the D pathway and is an important player in controlling oxygen access to the active site. The set of functions for which evidence is described above are arguably critical to preventing premature death of CcO, an otherwise untenable event for a cell, if it cannot recoup the energetic cost of CcO synthesis and assembly before the enzyme ceases to function.

References

1. R. Snyder, *Toxicol. Sci.*, 2000, **58**, 3–4.
2. V. R. Kaila, M. I. Verkhovsky and M. Wikström, *Chem. Rev.*, 2010, **110**, 7062–7081.
3. M. Wikström, V. Sharma, V. R. Kaila, J. P. Hosler and G. Hummer, *Chem. Rev.*, 2015, **115**, 2196–2221.
4. S. Yoshikawa, K. Muramoto and K. Shinzawa-Itoh, *Biochim. Biophys. Acta*, 2011, **1807**, 1279–1286.
5. E. Svahn, K. Faxen, R. B. Gennis and P. Brzezinski, *J. Inorg. Biochem.*, 2014, **140**, 6–11.
6. S. Ferguson-Miller, C. Hiser and J. Liu, *Biochim. Biophys. Acta*, 2012, **1817**, 489–494.
7. B. Kadenbach and M. Huttemann, *Mitochondrion*, 2015, **24**, 64–76.
8. C. A. Sinkler, H. Kalpage, J. Shay, I. Lee, M. H. Malek, L. I. Grossman and M. Huttemann, *Oxid Med Cell Longev*, 2017, **2017**, 1534056.
9. D. Pierron, D. E. Wildman, M. Huttemann, G. C. Markondapatnaikuni, S. Aras and L. I. Grossman, *Biochim. Biophys. Acta*, 2012, **1817**, 590–597.
10. S. Helling, M. Huttemann, R. Ramzan, S. H. Kim, I. Lee, T. Muller, E. Langenfeld, H. E. Meyer, B. Kadenbach, S. Vogt and K. Marcus, *Proteomics*, 2012, **12**, 950–959.
11. M. Huttemann, I. Lee, L. I. Grossman, J. W. Doan and T. H. Sanderson, *Adv. Exp. Med. Biol.*, 2012, **748**, 237–264.
12. I. Liko, M. T. Degiacomi, S. Mohammed, S. Yoshikawa, C. Schmidt and C. V. Robinson, *Proc. Natl. Acad. Sci. U. S. A.*, 2016, **113**, 8230–8235.
13. V. Frank and B. Kadenbach, *FEBS Lett*, 1996, **382**, 121–124.
14. L. Qin, D. A. Mills, L. Buhrow, C. Hiser and S. Ferguson-Miller, *Biochemistry*, 2008, **47**, 9931–9933.
15. C. Hiser, L. Buhrow, J. Liu, L. Kuhn and S. Ferguson-Miller, *Biochemistry*, 2013, **52**, 1385–1396.
16. W. X. Ding and X. M. Yin, *Biol. Chem.*, 2012, **393**, 547–564.
17. M. P. Murphy, *Biochem. J.*, 2009, **417**, 1–13.
18. M. Huttemann, S. Helling, T. H. Sanderson, C. Sinkler, L. Samavati, G. Mahapatra, A. Varughese, G. Lu, J. Liu, R. Ramzan, S. Vogt, L. I. Grossman, J. W. Doan, K. Marcus and I. Lee, *Biochim. Biophys. Acta*, 2012, **1817**, 598–609.

71. L. Salomonsson, G. Branden and P. Brzezinski, *Biochim. Biophys. Acta*, 2008, **1777**, 343–350.
72. G. Gilderson, L. Salomonsson, A. Aagaard, J. Gray, P. Brzezinski and J. Hosler, *Biochemistry*, 2003, **42**, 7400–7409.
73. P. Adelroth and J. Hosler, *Biochemistry*, 2006, **45**, 8308–8318.
74. L. Varanasi and J. P. Hosler, *Biochim. Biophys. Acta*, 2012, **1817**, 545–551.
75. U. Pfitzner, K. Hoffmeier, A. Harrenga, A. Kannt, H. Michel, E. Bamberg, O. M. Richter and B. Ludwig, *Biochemistry*, 2000, **39**, 6756–6762.
76. A. S. Pawate, J. Morgan, A. Namslauer, D. Mills, P. Brzezinski, S. Ferguson-Miller and R. B. Gennis, *Biochemistry*, 2002, **41**, 13417–13423.
77. V. Rauhamaki and M. Wikström, *Biochim. Biophys. Acta*, 2014, **1837**, 999–1003.
78. P. Sarti, M. G. Jones, G. Antonini, F. Malatesta, A. Colosimo, M. T. Wilson and M. Brunori, *Proc. Natl. Acad. Sci. U. S. A.*, 1985, **82**, 4876–4880.
79. H. J. Lee, E. Svahn, J. M. Swanson, H. Lepp, G. A. Voth, P. Brzezinski and R. B. Gennis, *J. Am. Chem. Soc.*, 2010, **132**, 16225–16239.
80. J. P. Klinman, *Acc. Chem. Res.*, 2007, **40**, 325–333.
81. D. A. Thompson, Ph.D. Thesis, Michigan State University, 1984.
82. M. R. Bratton, M. A. Pressler and J. P. Hosler, *Biochemistry*, 1999, **38**, 16236–16245.
83. J. P. Hosler, *Biochim. Biophys. Acta*, 2004, **1655**, 332–339.
84. D. A. Mills and J. P. Hosler, *Biochemistry*, 2005, **44**, 4656–4666.
85. R. Mitchell, A. J. Moody and P. R. Rich, *Biochemistry*, 1995, **34**, 7576–7585.
86. J. P. Hosler, J. P. Shapleigh, D. M. Mitchell, Y. Kim, M. A. Pressler, C. Georgiou, G. T. Babcock, J. O. Alben, S. Ferguson-Miller and R. B. Gennis, *Biochemistry*, 1996, **35**, 10776–10783.
87. V. Sharma and M. Wikström, *Biochim. Biophys. Acta*, 2016, **1857**, 1111–1115.
88. I. Song, T. M. Ball and W. L. Smith, *Biochem. Biophys. Res. Commun.*, 2001, **289**, 869–875.
89. C. Su, M. Sahlin and E. H. Oliw, *J. Biol. Chem.*, 1998, **273**, 20744–20751.
90. M. Iwaki, A. Puustinen, M. Wikström and P. R. Rich, *Biochemistry*, 2006, **45**, 10873–10885.
91. S. Riistama, A. Puustinen, A. Garcia-Horsman, S. Iwata, H. Michel and M. Wikström, *Biochim. Biophys. Acta*, 1996, **1275**, 1–4.
92. I. Echabe, T. Haltia, E. Freire, F. M. Goni and J. L. Arrondo, *Biochemistry*, 1995, **34**, 13565–13569.
93. V. Sharma, P. Ala-Vannesluoma, I. Vattulainen, M. Wikström and T. Rog, *Biochim. Biophys. Acta*, 2015, **1847**, 690–697.
94. M. Gutman, E. Nachliel and R. Friedman, *Biochim. Biophys. Acta*, 2006, **1757**, 931–941.

95. K. S. Alnajjar, J. Hosler and L. Prochaska, *Biochemistry*, 2014, **53**, 496–504.
96. D. A. Mills, B. Schmidt, C. Hiser, E. Westley and S. Ferguson-Miller, *J. Biol. Chem.*, 2002, **277**, 14894–14901.
97. T. Meyer, F. Melin, O. M. Richter, B. Ludwig, A. Kannt, H. Muller, H. Michel and P. Hellwig, *FEBS Lett*, 2015, **589**, 565–568.
98. I. A. Smirnova, P. Adelroth, R. B. Gennis and P. Brzezinski, *Biochemistry*, 1999, **38**, 6826–6833.
99. J. Liu, L. Qin and S. Ferguson-Miller, *Proc. Natl. Acad. Sci. U. S. A.*, 2011, **108**, 1284–1289.
100. J. Fetter, M. Sharpe, J. Qian, D. Mills, S. Ferguson-Miller and P. Nicholls, *FEBS Lett.*, 1996, **393**, 155–160.
101. D. A. Mills, L. Geren, C. Hiser, B. Schmidt, B. Durham, F. Millett and S. Ferguson-Miller, *Biochemistry*, 2005, **44**, 10457–10465.
102. L. Qin, D. A. Mills, C. Hiser, A. Murphree, R. M. Garavito, S. Ferguson-Miller and J. Hosler, *Biochemistry*, 2007, **46**, 6239–6248.
103. T. V. Vygodina, W. Zakirzianova and A. A. Konstantinov, *FEBS Lett.*, 2008, **582**, 4158–4162.
104. K. Faxen, L. Salomonsson, P. Adelroth and P. Brzezinski, *Biochim. Biophys. Acta*, 2006, **1757**, 388–394.
105. L. D. Gauthier, J. L. Greenstein, S. Cortassa, B. O'Rourke and R. L. Winslow, *Biophys. J.*, 2013, **105**, 1045–1056.
106. G. Y. Lau, M. Mandic and J. G. Richards, *Mol. Biol. Evol.*, 2017, **34**, 2153–2162.

CHAPTER 10

Computational Means of Assessing Proton Pumping in Cytochrome c *Oxidase (Complex IV)*

JESSICA M. J. SWANSON

University of Chicago, Department of Chemistry, 5735 S. Ellis Ave., Chicago, IL 60637, USA
Email: jmswanson@uchicago.edu

10.1 Introduction

Enzymes that couple chemical reactions to long-range charge transport play a primary role in bioenergetic transformations, especially processes such as photosynthesis and respiration. Cytochrome *c* oxidase (C*c*O, Figure 10.1) is one such enzyme and a paradigmatic proton pump. As the terminal enzyme in the respiratory chain of mitochondria and many bacteria, C*c*O couples the exergonic reduction of molecular oxygen to the endergonic movement of protons across a membrane.[1–4] The resulting transmembrane electrochemical gradient is used for multiple purposes, including the synthesis of the majority of the cell's ATP from ADP and inorganic phosphate (Pi) by ATP-synthase.[2,5]

C*c*O is spectacular for at least two reasons. First, because it couples long-range charge transport to a redox reaction that involves reactive oxygen

Chemical Biology No. 5
Mechanisms of Primary Energy Transduction in Biology
Edited by Mårten Wikström

Published by the Royal Society of Chemistry, www.rsc.org

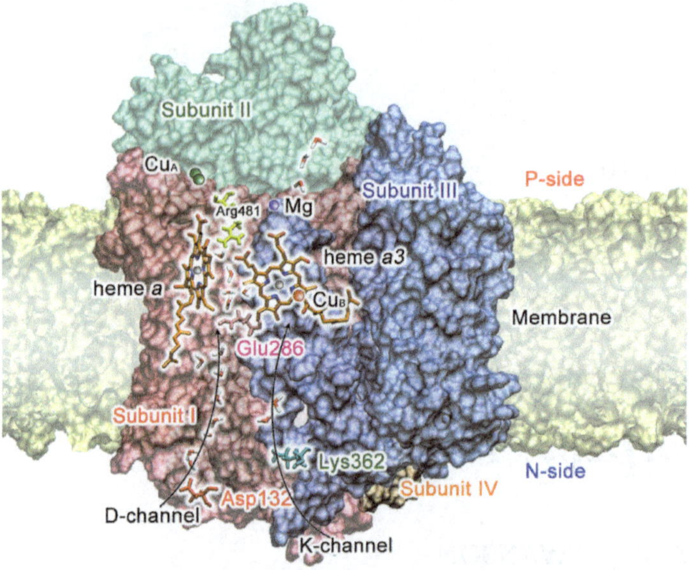

Figure 10.1 Illustration of CcO from *Rhodobactor sphaeroides* highlighting the four subunits, metal centers, the D- and K-channels including key residues and internal water molecules, and Arg481, which forms a salt bridge with PRDa_3.

intermediates without releasing measurable amounts of reactive oxygen species. The overall reaction catalyzed by CcO is:

$$8H^+(\text{N-side}) + 4e^-(\text{P-side}) + O_2 \rightarrow 2H_2O + 4H^+(\text{P-side}) \quad (10.1)$$

where 'N-side' denotes the negatively charged inside of the membrane, and 'P-side' denotes the positively charged outside. This reaction can be broken down into four one-electron steps (see Figure 10.2) that involve the transient formation of several reactive oxygen species, including superoxide ($O_2{}^{*/-}$), hydrogen peroxide (H_2O_2), and the hydroxyl radical (OH*). Despite these toxic intermediates being relatively long-lived, very few are released to the surrounding media, unlike those made by complexes I and III in the respiratory electron transport chain.[3,5]

Second, and likely more widely recognized, CcO is spectacular because in addition to coupling long range charge transport to a redox reaction, it simultaneously pumps protons across a membrane against an electrochemical proton gradient. Many of the proteins that couple oxidoreductase reactions to the creation of a transmembrane proton gradient (*e.g.*, photosystems I and II and respiratory complex III) simply use a vectorial arrangement of redox chemistry (*i.e.*, the consumption and production of a proton on opposing sides of the membrane), in line with Mitchell's chemiosmostic theory.[6]

In contrast, respiratory complexes I and IV (CcO) act as *true proton pumps* by translocating protons across the membrane against a proton gradient.[5]

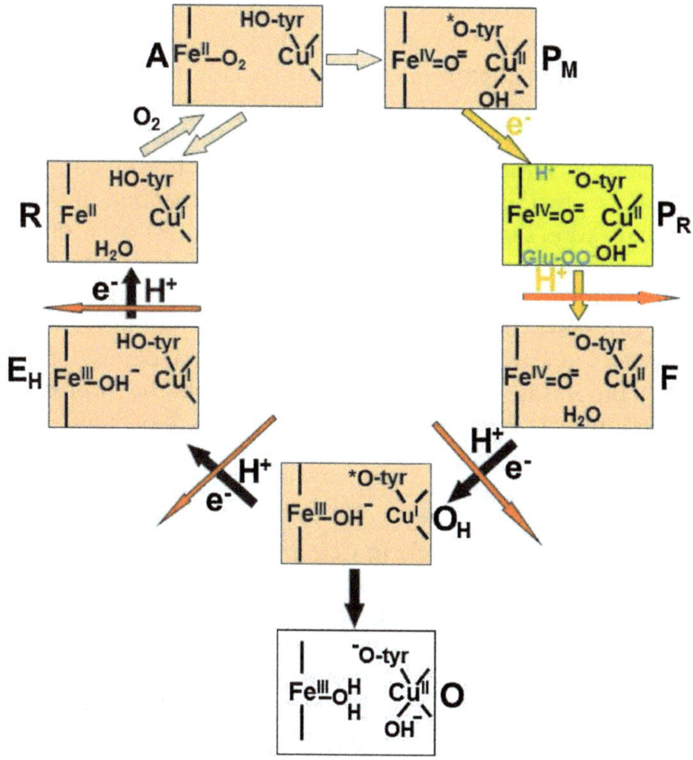

Figure 10.2 CcO catalytic cycle. Each box contains the heme *a*3-copper binuclear center (BNC) for one of seven transition states, including the tyrosine and histidine ligands of Cu_B (shown as tyr and lines, respectively). Oxygen binds the reduced enzyme (R state) producing the oxygenated A state. During normal turnover four electrons are taken from the BNC to form the P_M state in *ca.* 300 μs. In the fully reduced enzyme, the fourth electron (shown in yellow) is instead taken from a reduced heme *a* to convert A directly to the P_R state in *ca.* 30 μs (yellow box). Protons (H^+) are taken up from the N-side of the membrane with red arrows indicating pumping to the P-side of the membrane. States O, O_H, and E_H are tentative.
Adapted with permission from ref. 5. Copyright (2017) American Chemical Society.

As indicated in eqn (10.1), during each reaction cycle four electrons are taken from the mobile electron carrier cytochrome *c* (or ubiquinone for some bacteria) on the positively-charged (P-side) outside of the membrane, and passed to the catalytic binuclear center (BNC), consisting of heme a_3 and the Cu_B complex. These electrons are joined by four protons (herein called 'chemical protons'), which are taken up from the negatively-charged (N-side) inside of the membrane and consumed in the chemical reduction of O_2 to H_2O. This process already creates a transmembrane electrochemical proton gradient. In addition, for each chemical proton that is passed to the BNC, another proton (herein called a 'pumped proton') is translocated from the N-side to the P-side of the membrane, more than doubling the resulting transmembrane electrochemical gradient. Perhaps more intriguing is the finding that the pumped protons use the same proton uptake pathway (the D-channel; Figure 10.1) as at least two of the four chemical protons (the remaining chemical protons are taken up through the K-channel).[7,8] How the enzyme moves two protons for each electron consumed in the chemical reaction, is a question that has captivated researchers for over four decades.[1] This chapter focuses on our current understanding of the *Cc*O proton pumping mechanism, particularly the proton transport (PT) processes therein, from the lens of computational modeling and simulations. More extensive reviews on the chemistry, known intermediates, and electron transfer (ET) processes that go hand-in-hand with the proton pumping during a reaction cycle can be found elsewhere.[3–5,9–11]

10.2 Catalytic Cycle

Many aspects of the catalytic cycle in *Cc*O are fairly well agreed upon at this point thanks to thorough experimental analysis, including exquisite time-resolved spectroscopic studies, with additional insight from quantum calculations (see ref. 5, 11–14 for reviews). As depicted in Figure 10.2, the oxidative phase begins with the binding of O_2 to the doubly-reduced BNC in the R state to form either a ferrous iron Fe[II]–O_2 or a ferric iron-superoxide Fe[III]–$O_2^{\bullet-}$ (not shown) called compound A. This is followed by a four-electron reduction of O_2, splitting the O–O bond to form the P_M state in ~300 μs during normal turnover. In this case, the active site tyrosine is thought to be a neutral radical and all four electrons needed to split the O–O bond come from the BNC. However, in flow-flash experiments[3,4] two additional electrons have been pre-loaded (one on Cu_A and one on heme *a*) to create the fully reduced enzyme. Thus, heme *a* contributes the requisite fourth electron, converting state A to the P_R state in only ~30 μs. There is no proton pumping associated with the A to P_M transition, however the initial step of pumping, involving PT to the pump loading site (PLS; see below for discussion), has been associated with the A to P_R transition.[15] In either case, the following four transitions proceed through the F, O, and E states before regenerating the R state. Each of these transitions is associated with the transfer of one electron and one proton to the BNC, along with the pumping of one proton to the P-side of the membrane.

Despite this impressive level of understanding of the chemical reaction, lingering questions remain regarding the chemical identity of some intermediates, particularly the high-energy O_H and E_H states. These states are thought to explain the enigma of how the early reducing steps (O→E→R) are linked to proton pumping at high proton gradients. Experimentally the reduction potential of these steps, involving Cu(II) and Fe(III), was measured as only slightly exergonic in the absence of a proton gradient.[3,16–18] Introducing a proton gradient would thus make them endergonic, resulting in a reaction rate of hours to days, which is inconsistent with the measured proton pumping rates under high proton gradients.[19,20] The accepted solution to this issue is the existence of activated O_H and E_H states that are higher in energy and thus have larger reducing potentials than the O and E states.[9,21] Since such activated states have been difficult to capture in experimental measurements,[18] quantum calculations have been helpful in suggesting their potential identities and roles in the chemical reaction.[11,13,14,22]

Sharma and colleagues used density functional theory (DFT) calculations of clusters of the active site combined with experimental data to suggest that the active site tyrosine has a partial neutral radical character that explains the E_H and O_H large reduction potentials.[11] Similarly, Blomberg and colleagues have combined hybrid DFT calculations with the experimentally measured exergonicity of O_2 reduction to present the thermodynamics of the full catalytic cycle.[14] Their presented relative energies of the activated states again explain how proton pumping occurs during the $O \rightarrow E \rightarrow R$ transitions in the absence and presence of a large electrochemical gradient. This explanation hinges on the reduction potential for Cu(II) being larger than that deduced from equilibrium measurements, although it is argued that the required increase is still within the experimentally reported uncertainty.[14]

A critical element in the high-energy E_H and O_H states is the active site tyrosine residue, not only because its identity is still debated, but also because it seems to influence whether the substrate proton is taken up through the D- or K-channel. Recent work has shown how PT through the K-channel is effectively gated by the protonation of the active site tyrosine.[13] During the oxidative phase ($P_M \rightarrow F \rightarrow O_H$) there is a thermodynamic and kinetic cost of protonating this tyrosine, kinetically blocking PT through the K-channel. Only during the reductive phase ($O \rightarrow E \rightarrow R$), when Cu_B is cuprous, does the pKa of the active site tyrosine rise above 7, enabling PT through the K-channel. Sharma et al. suggested that the active site tyrosine is deprotonated in the O state,[11] in agreement with FTIR data,[23,24] and that substrate proton uptake occurs through the K-channel in the $O/O_H \rightarrow E/E_H$ step,[7,9,23] and also possibly in the $E/E_H \rightarrow R$ transition.[25] In contrast, Blomberg concluded that the active site tyrosine remains unprotonated with radical character until the very last reduction step ($E \rightarrow R$).[14,22] This supports the notion of a high energy one electron reduced state, E_H with an unprotonated tyrosine radical, and proton uptake through the D-channel being kinetically favored during the $O \rightarrow E$ transition. In this view, which is not in agreement with experiment (as noted above), proton uptake through the K-channel only occurs during the final reduction step ($E \rightarrow R$).

Collectively this work provides a helpful perspective on control of the K-channel by the active site tyrosine residue, and the role of the K-channel in the proton pumping process. It is particularly interesting to place these conclusions in the context of the B- and C-family oxidases, which lack a D-channel and have decreased proton pumping compared to the A-family oxidases.[26–28] This is potentially explained by their unique proton uptake (K-) channel ending in the active site tyrosine residue.

10.3 Proton Pumping Mechanism

Although the proton pumping process entails many coupled kinetic transitions, each involving a distribution of states evolving over transport barriers, it is often described as a series of charge translocation events. Electrons are transferred from cytochrome *c* through Cu_A and heme *a* to the BNC as described above. Pumped protons, and at least two of the chemical protons, are taken up from the N-side of the membrane into the D-channel, so named because amino acid residue D132 (*Rhodobacter sphaeroides* numbering) lies at the channel entrance.[7,8] They are then transferred *via* Grotthuss shuttling through water molecules to residue E286, one third of the way into the membrane and just below the BNC. Depending on their fate, protons are then shuttled through water molecules in the hydrophobic cavity (HC) either to react with oxygen in the BNC, or to the PLS prior to their release to the P-side of the membrane (see Figure 10.1). The K-channel, named for the conserved K362 residue, is responsible for the uptake of the remaining 1–2 chemical protons during the reductive phase, as described above.

Many aspects of the C*c*O proton pumping mechanism are now understood thanks to four decades of extensive experimental and computational research (see ref. 1, 3–5, 29, 30 and citations therein). For example, high resolution crystal structures of C*c*O in different redox states are available.[31–35] Structures together with site-directed mutagenesis studies have identified proton transfer pathways (the D- and K-channels) from the N-side of the membrane to the BNC.[7,8] Intermediates of the catalytic cycle are mostly known, as described above, revealing reactants and products for specific PT steps. Likely protonation states of residues have been predicted for various redox states of the system through pKa calculations.[36–39] The thermodynamics and kinetics pertaining to some of the elementary electron/proton transfer steps are at least partially understood,[5] and have been used to construct thermodynamic models for the entire cycle.[14,22] Minimal requirements for a functional pump have been outlined with kinetic master equation models, revealing that one redox site and two proton loading sites (one pump loading site and one active site) are essential for minimal pumping, but an additional redox site and kinetic gates are needed for efficient pumping against a proton gradient.[40–42]

Despite this impressive level of understanding, several aspects of the C*c*O proton pumping mechanism are still debated at the mechanistic and atomistic levels. For example, there has been significant controversy over the location of the PLS,[23,39,43–46] the nature of the coupling between the internal PT and ET events,[3,5,10,15,23,39–42,47–49] and the molecular mechanism that drives forward pumping against an electrochemical gradient while limiting proton back flow.[3,5,10,15,47,50,51] As described below, recent computer simulations have been helpful in resolving these issues by characterizing the energetics, dynamics, and resulting kinetics of elementary PT processes. This information in turn suggests how proton pumping is coupled to ET and enabled by kinetic gating.

10.4 Simulating PT in Biomolecular Systems

The mechanistic proposals of the proton pumping mechanism in CcO based on computer simulations, whether they have focused on electrostatics and solvation dynamics,[15,23,52–58] conformational changes of key protonatable residues,[59–61] or explicit simulations of specific PT steps,[43,48,49,62–72] have varied widely. This is largely a consequence of how difficult it is to describe PT *in silico*. Though ubiquitous throughout biology and seemingly simple, hydrated protons are surprisingly complex. They are most aptly described as dynamic electronic charge defects that span multiple molecules and dance through frustrated hydrogen-bonded networks *via* the rearrangement of covalent bonds.[73] To capture this process requires a method that can delocalize the excess charge defect across many molecules and dynamically rearrange bonding topologies. To produce reliable results, the underlying energetic representation must be accurate and the simulation efficiency must be sufficient to sample many microseconds. This kind of approach, in combination with enhanced sampling methods, can be used to calculate the free energy profiles for PT processes, therein revealing rates and mechanisms. As described below, accounting for multiple dimensions, such as solvation and PT, is essential in some cases, making the computational cost larger and the need for an efficient approach more important.

Although classical approaches such as MD simulations cannot be used to describe PT explicitly, they have been used extensively to characterize the hydration of CcO under different redox states. Since water plays a central role in PT, these results have been the basis of many mechanistic proposals, as further discussed below. Approaches that have attempted to explicitly describe PT in CcO include quantum mechanical/molecular mechanical (QM/MM) approaches, using density functional theory (DFT) for the QM region,[43] or the more approximate tight-binding form of DFT called SCC-DFTB.[49] It goes without saying that each approach has its advantages and inherent challenges. For DFT based QM/MM the advantages are ever increasing accuracy with continued development of DFT functionals and the avoidance on system-specific parameterization. The primary challenge is its computational expense, which generally limits the sampling required for converged free energy profiles. The SCC-DFTB based QM/MM is more efficient and is also under development to improve its reliability and accuracy for PT processes in biological systems. Both QM/MM approaches have to additionally deal with boundary issues when molecules, especially water molecules, need to cross the QM/MM boundary. Despite these challenges, both approaches have contributed to our understanding of CcO,[43,49] and are expected to be increasingly useful with their continued development in the characterization of various biomolecular processes.

An alternative approach is based on the empirical valence bond (EVB) method in which PT, or any chemical reaction, is described by a linear combination of diabatic states.[74] Each state represents a different bonding topology (*e.g.*, reactant and product in a simple transfer reaction).

The potential energy for each state is defined by empirical function on the diagonal of a Hamiltonian matrix, while the coupling between states is an empirical function in off-diagonal terms. Diagonalization of this matrix yields a ground-state potential energy function along which nuclei can evolve according to classical, deterministic MD, making and breaking covalent bonds. This insightful interpolation scheme is based on the concept of a superposition state in quantum mechanics. The limitation of the original EVB implementation is that two states alone does not describe the delocalized nature of the excess proton in water. In fact, it can require up to 30 states to capture PT in bulk water.[73] For this reason, the multi-state empirical valence bond (MS-EVB) method was created and extensively developed over 15 years to faithfully represent PT in bulk water and biomolecular systems.[73,75–80] Recently MS-EVB was replaced by the multistate reactive molecular dynamics (MS-RMD) method, which directly bridges quantum mechanical data from QM/MM calculations into a reactive MD algorithm.[81–84] Like MS-EVB, this approach describes the dynamically changing charge defect of the excess proton, including Grotthuss shuttling through amino acid residues and water molecules, in a deterministic MD framework. Thus, it is efficient enough to sample PT in addition to other degrees of freedom, such as hydration (see below). The MS-RMD has recently revealed fascinating PT mechanisms in channels, enzymes, and antiporters.[68,85–88] The most recent work on C*c*O focused on the $P_M' \rightarrow P_R \rightarrow F$ transition in the fully reduced enzyme, which has been extensively studied experimentally and has separate measured rates for electron and proton transfer to the BNC.[68] Although this transition requires flash-photolysis of the fully reduced enzyme, it is thought to be representative of the proton pumping process in the native catalytic cycle. The P_M' state represents the system after flash-photolysis of CO, O_2 binding, and the O–O bond scission in the BNC. Thus, heme *a* and Cu_A are reduced in the P_M' state. Electron transfer from heme *a* to the BNC then generates the P_R state, followed by proton uptake to the BNC to form the F state. Concurrently, there is an internal PT from E286 to the PLS with the same time dependence as ET, and a proton pumped to the P-side by the time the F state has formed (see ref. 3 and 4 for reviews). The following sections focus on the conclusions from MS-RMD simulations of the full four-subunit C*c*O (*Rhodobacter sphaeroides*; see Figure 10.1) embedded in a solvated membrane.[68]

10.5 Hydration and Proton Transport

Water plays an essential, and often hard to pin down, role in PT processes. It is perhaps not surprising then, that the involvement of water during PT through the HC and D-channel has received great attention and debate.[52–55,57,58,61,89–93] Water shuttles the excess proton between ionizable residues—this is not debated. However, since crystal structures are often lacking resolved water molecules along known PT pathways, the balance between water–residue versus residue–residue PT is often questioned.

The lack of resolution of water molecules in protein cavities that transport protons is generally due to their dynamic and/or transient nature. This, however, was not the case for the D-channel of CcO, where crystal structures have revealed 10–11 water molecules.[94,95] Molecular dynamics simulations, often with enhanced sampling of solvation degrees of freedom, have been helpful in characterizing the hydration ensembles for given protonation states. For example, the free energy simulations of Henry *et al.* confirmed that 11–12 water molecules are stable in the D-channel prior to PT from D132 to E286.[54,55] This work further revealed a gating motion of N139 that enables waters to enter the asparagine gate region (N139–N121), and which is altered in the N139A and N139V mutants, setting the stage nicely for further simulations as discussed below (see Section 10.8).

However, the hydration of the HC has been more challenging to pin down due to its largely hydrophobic nature and sensitive dependence on the redox and protonation state of the system. Extensive MD simulations have been used to probe the orientation and connectivity of water chains in the HC in different redox states, leading to the proposed 'water gate' mechanism.[5,15,52,53] In this mechanism the equilibrium distribution of water molecules (their preferred locations and orientations) is dependent on the redox state of the system. Prior to ET, when heme *a* is reduced (*i.e.*, the P_M' state), a preorganized chain of water molecules conducts protons from E286 to the PLS (thought to be PRDa3). Following ET (*i.e.*, in the P_R state), waters redistribute to enable conduction to the BNC. In this view, the short-circuiting (*i.e.*, premature PT to BNC before PT to PLS) is avoided by the absence of a water chain leading from protonated E286 to BNC in the P_M' state. An extension of this mechanism is the 'Glu286-valve' proposal, in which the rapid reorientation of deprotonated E286 to the down (D-channel facing) conformation following PT to the PLS helps prevent proton back flow,[96] although the relative stability of E286 conformations has been debated.[59,60,97] In a competing mechanistic proposal, classical MD and SCC-DFTB simulations have been used to propose a stepwise pumping mechanism in which an excess proton is first transported from E286 to the D-propionate on heme a_3 (PRDa_3) through a poorly hydrated HC, followed by an increase in the HC hydration.[57]

A critical point that is often missed in the community is that the process of excess proton migration *itself* can strongly and dynamically influence the surrounding hydration environment. A poignant example of this is the inverse Grotthuss shuttling of water molecules through an excess proton into a dry environment as recently demonstrated for a nanotube confined environment.[98] While this mechanism is likely limited to particular geometric situations, the strong and dynamic coupling between excess protons and changing hydration is unquestionably important in biological systems. To capture this effect one must explicitly sample the changing hydration in addition to the progress of the excess proton. The MS-RMD method was thus developed to be efficient enough to characterize PT in condensed phase environments, including the coupled changes in hydration. The results described below demonstrate its use, presenting two-dimensional potentials of mean force (2D PMFs; *e.g.*, see Figure 10.3) wherein enhanced sampling has been applied to both the position of the excess proton (tracked *via* the center of excess charge, CEC, horizontal axis) and the number of water molecules along the PT pathway (vertical axis). Tracing the minimum free energy pathways through these 2D PMFs (black line in Figure 10.3), reveals

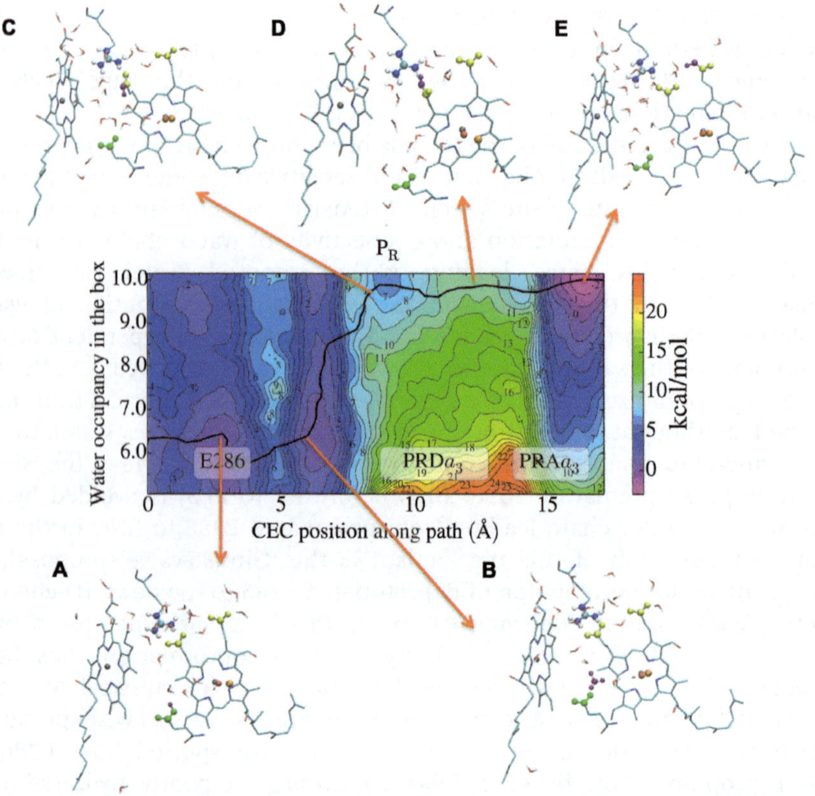

Figure 10.3 Two-dimensional free energy profile (2D-PMF) for PT from E286 to the PLS in the P_R state as a function of the excess proton center of excess charge (CEC) coordinate through the hydrophobic cavity (HC) and the water hydration in the HC. The 1D minimum free energy pathway (black line) is diagonal in nature, indicating the two processes are coupled. The statistical errors of the 2D-PMFs in the range ~0.1–3 kcal mol^{-1}. (A–E) Molecular structures depicting the PT process along the minimum free energy path. E286 is shown in green, the propionate groups of the PLS in yellow, the coordinating arginine in blue, the excess proton in purple, the iron atom of the heme groups in blue, and the copper atom in the BNC in orange. The heme groups are shown as sticks. The transition state structure is shown in (D).
Reprinted with permission from R. Liang, J. M. J. Swanson, Y. Peng, M. Wikström and G. A. Voth, *Proc. Natl. Acad. Sci. U.S.A*, 2016, **113**, 7420.[68]

the coupling between PT and hydration. The 1D PMFs extracted from such 2D PMFs are helpful to compare free energy profiles for different states of the system.

10.6 Transport of the Pumped Proton

Starting in the P_M' state, heme *a* is reduced, the PLS is empty, and E286 is protonated. The O–O bond scission has just occurred in the BNC,[24,99] and the system is primed for ET from heme *a* to the BNC (to create the P_R state) along with PT from E286 to either the BNC or the PLS. Focusing first on PT of the pumped proton from E286 to the PLS, MS-RMD was used to calculate 2D PMFs in the P_M' state (before ET; not shown) and P_R state (after ET; Figure 10.3). The minimum free energy pathway through this 2D PMF (black line in Figure 10.3), shows significant coupling between hydration and PT, with the number of water molecules in the HC increasing while the excess proton travels from E286 to PRDa_3. Prior to PT, when E286 is still protonated, the HC favors ~six water molecules (Figure 10.3A). Note that the tested solvation box includes two waters outside of the inter-heme region. However, as the proton moves to the first water above E286 (Figure 10.3B) it attracts the PRDa_3 side chain, which breaks its salt bridge to R481 and hydrogen bond with W172. This, in turn, allows more waters from above the propionate groups to enter into the HC (Figure 10.3C). The highest barrier in this PT process is reached as the PRDa_3 sidechain rotates to an upward orientation (Figure 10.3D). The excess proton is then transferred to the A-propionate on heme a_3 (PRAa_3), which constitutes the PLS in the P_R state according to these simulations (Figure 10.3E).

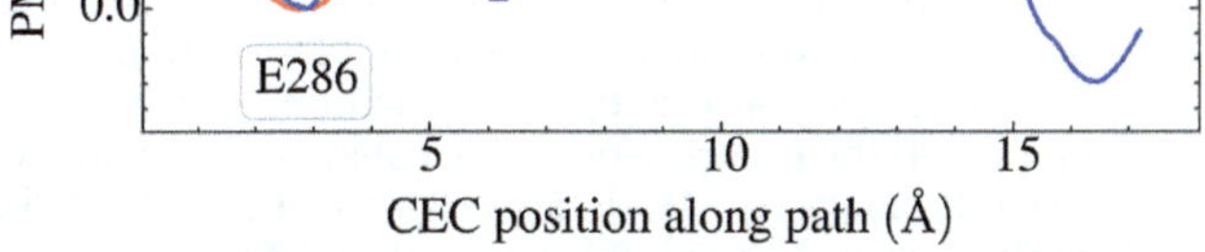

Figure 10.4 1D free energy profiles for PT from E286 to the PLS for the P_M' (red) and P_R (blue) states traced out along the minimum free energy pathways of the full 2D-PMFs. The statistical errors of the 1D free energy profiles are in the range of ~0.1–1 kcal mol^{-1}. The positions of E286, PRDa_3, and PRAa_3 are labeled with text boxes.
Reprinted with permission from R. Liang, J. M. J. Swanson, Y. Peng, M. Wikström and G. A. Voth, *Proc. Natl. Acad. Sci. U.S.A*, 2016, **113**, 7420.[68]

The corresponding 1D free energy curves (Figure 10.4) compare the results for the $P_M{}'$ and P_R states, and reveal that ET significantly changes the free energy profile for the transport of the pumped proton. Prior to ET ($P_M{}'$ state), transport of the pumped proton from E286 to the PLS is unfavorable (*ca.* 8 kcal mol^{-1}), but after ET (P_R state) it is favorable by −3 kcal mol^{-1}. Thus, PT to the PLS is thermodynamically driven by ET to the BNC. Combining this conclusion with experimental results showing that ET is not complete without PT to the PLS in the E286Q mutant,[3] suggests that the two processes are coupled. These 1D free energy curves also suggest that although PRDa_3 plays an important role in transferring the pumped proton, it is PRAa_3 that acts as the dominant PLS. Although this is consistent with a few previous suggestions,[100,101] it is in contrast to many proposals suggesting PRDa_3, a histidine ligand of Cu_B, or even water molecules are the true PLS.[3,5,43,56,102] This is expected to be a topic of continued interest and debate.

The corresponding rates for PT are calculated with transition state theory (Table 10.1; reported as time constants, *i.e.*, the inverse of rate constants). In the $P_M{}'$ state (before ET), forward PT (E286 to PLS) is kinetically facile. It is faster than the experimentally measured 25∼50 µs A→P_R transition,[50,103] and even slightly faster than forward PT in the P_R state (after ET). This is consistent with the water gate mechanism which proposed transport of the pumped proton would be easier when heme *a* is reduced.[52,53] However, the reverse PT (from the PLS back to E286) is even faster, outcompeting forward transport and the reprotonation of E286 through D-channel (Table 10.1 and 10.2; discussed below). Hence, transfer of the pumped proton to the PLS is minimal before the ET. In contrast, after ET (in the P_R state), not only is forward PT (E286 to PLS) still kinetically facile, but now proton back flow (PLS to E286) is significantly slowed due to stabilization at the PLS (*i.e.*, transport of the pumped proton is thermodynamically favored in the P_R state, creating a large barrier for proton backflow). In combination with rapid reprotonation of E286 through the D-channel (discussed below), this

Table 10.1 Calculated time constants (inverse of rate constants) for transport of the pumped proton (E286→PLS), chemical proton (E286→BNC), and back leaked proton (PLS→E286) in the $P_M{}'$ and P_R states, compared with experimental time constants for A→P_R and P_R→F transitions.[50,103] The PT from E286 to BNC in P_R state with PLS protonated is the most physically relevant for the P_R→F transition, and is in quantitative agreement with the experimental time constant. Reprinted with permission from R. Liang, J. M. J. Swanson, Y. Peng, M. Wikström and G. A. Voth, *Proc. Natl. Acad. Sci. U.S.A*, 2016, **113**, 7420.[68]

		E286→BNC (µs)		
State	E286→PLS (µs)	Deprotonated PLS	Protonated PLS	PLS→E286 (µs)
$P_M{}'$	$(2.6\pm0.2)\times10^{-1}$	$(4\pm3)\times10^{6}$	$(7.7\pm0.2)\times10^{8}$	$(1.5\pm0.4)\times10^{-5}$
P_R	2.1 ± 0.3	$(3\pm2)\times10^{2}$	$(1.7\pm0.9)\times10^{2}$	$(1.2\pm0.6)\times10^{3}$
Exp (A→P_R)		25–50		
Exp (P_R→F)		200		

Table 10.2 Calculated time constants (inverse of the rate constants) for PT in the D-channel from protonated D132 to deprotonated E286 in the P_R and F states, compared with the experimental time constant for the $P_R \rightarrow F$ transition.[50] The PLS is protonated in the simulation. Reprinted with permission from R. Liang, J. M. J. Swanson, Y. Peng, M. Wikström and G. A. Voth, *Proc. Natl. Acad. Sci. U.S.A*, 2016, **113**, 7420.[68]

State	D132 → E286 (μs)
P_R	$(5 \pm 2) \times 10^{-1}$
F	$(2 \pm 1) \times 10^{-1}$
Exp ($P_R \rightarrow F$)	200

blocks backflow of the proton from the PLS to E286, and subsequent transport to the BNC, which would short circuit pumping. These results suggest that full loading of the PLS requires ET. If the reverse is also true based on the experiments in ref. 3, then PT to PLS and ET to BNC are coupled, each thermodynamically driving the other to completion. Thus, MS-RMD simulations have both confirmed and expanded upon the water gate and Glu286-valve mechanisms.[52,96] They confirm that PT to the PLS precedes PT to the BNC, and that a kinetic gate is needed to stabilize the proton at the PLS. They further explain why PT to the PLS is preferred after ET (P_R state), even though forward PT from E286 to the PLS is slightly faster before ET ($P_M{}'$ state). This is because ET is required to stabilize the excess proton at the PLS. Thus, the coupling of PT to the PLS and ET provides a kinetic gate (*i.e.*, large barrier) to proton backleak from the PLS to E286. They additionally provide a new perspective on the nature of the PT itself, which will be discussed below.

10.7 Transport of the Chemical Proton

The reprotonation of E286 following the transfer of the pumped proton to the PLS, which is clearly the subsequent step, will be discussed below. Here we focus on the transfer of the chemical proton from E286 to the Cu_B-bound hydroxide in the BNC to form the F state. MS-RMD simulations used to produce 2D PMFs show that the hydration in the HC is stable during the transport of the chemical proton.[68] These free energy profiles were computed for the $P_M{}'$ and P_R states both with and without the PLS (PRAa_3) protonated. Comparing the 1D free energy traces from these 2D PMFs (Figure 10.5) reveals that the protonation of the PLS has little influence on the transport of the chemical proton, but that, just as for the pumped proton, ET has a large influence. Regardless of the protonation state of the PLS, PT from E286 to the BNC is thermodynamically unfavorable (>10 kcal mol^{-1}) in the $P_M{}'$ state, but favorable by *ca.* −5 kcal mol^{-1} in the P_R state. Once again, ET from heme *a* to the BNC is essential to drive the PT, in this case of the chemical proton. Despite this, the molecular mechanism of PT from E286 to the BNC is similar in the $P_M{}'$ and P_R states (P_R shown in Figure 10.5(A–D)).

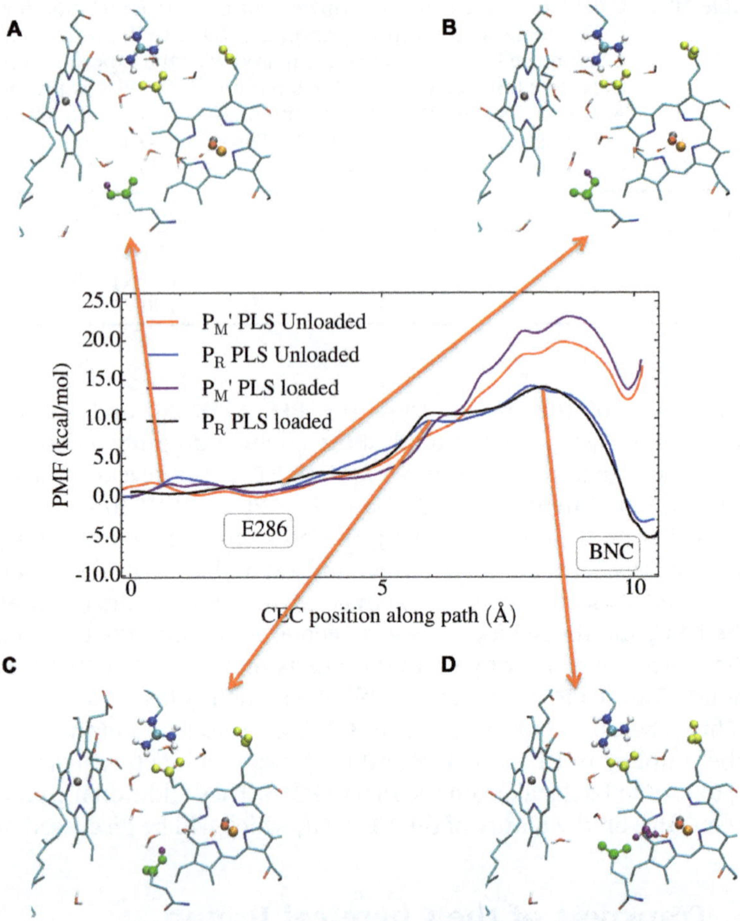

Figure 10.5 1D free energy profiles for PT from E286 to the BNC in the $P_M{}'$ state with and without a proton preloaded in the PLS (purple and red, respectively), and the P_R state (black and blue, respectively) along the minimum free energy pathways of the 2D-PMFs (not shown). PT to the BNC with the PLS preloaded (black curve) is the most physically relevant for the $P_R \rightarrow F$ transition. The statistical errors of the 1D free energy profiles are in the range of $\sim$0.1–1 kcal mol^{-1}. The positions of E286 and the BNC are labeled with text boxes. (A–D) Molecular structures depicting PT from E286 to the BNC in the P_R state with the PLS protonated following the 1D minimum free energy pathway (black line). E286 is shown in green, the propionate groups in yellow, the excess proton in purple, the iron atom of the heme groups in gray, and the copper atom in the BNC in orange. The heme groups are shown as sticks.
Reprinted with permission from R. Liang, J. M. J. Swanson, Y. Peng, M. Wikström and G. A. Voth, *Proc. Natl. Acad. Sci. U.S.A*, 2016, **113**, 7420.[68]

E286 first rotates toward the BNC (Figure 10.5A, B, C), contributing a small amount to the free energy barrier due to intramolecular strain and the disruption of optimal hydrogen bonding. Deprotonation of E286 ($x \simeq 6$ Å) then transfers the excess proton to the single water molecule between E286 and the BNC (Figure 10.5). This water molecule turned hydronium is hydrogen bonded to the carbonyl of G286 in addition to E286 and the Cu_B-bound hydroxide. *This constitutes the transition state of the PT process* due to narrowness of this region (confining the charge distribution to a limited hydrogen bond network) and due to the electrostatic repulsion from the BNC, which is larger for the P_M' than for the P_R state. Finally, Cu_B–OH is protonated to form a water molecule, completing the chemical reaction and transitioning the system to the F state. Reduction of the BNC drives this PT process by reducing the electrostatic repulsion of the proton in the P_R state compared to the P_M' state. Thus, this work explains why transport of the chemical proton to the BNC is favored following ET (in the P_R state), a prediction made in the water gate mechanism. It further shows that although ET alters the water distribution in the HC by slightly increasing the stability of the waters between E286 and the BNC, this has a small influence on the PT free energies. Note that configurations 10.5A and 10.5B were selected without these waters, but they are sampled during the simulation. The dominant influence differentiating the P_M' from P_R free energy profiles is the electrostatic repulsion from the BNC, which is decreased following ET.

The rates deduced from these free energy profiles (Table 10.1) show that it is kinetically prohibitive to transfer the chemical proton from E286 to the BNC before ET (P_M' state). This prevents premature PT to the BNC (*i.e.*, prior to PT to the PLS), which would short-circuit pumping.[53] After ET (P_R state) the calculated rate for transfer of the chemical proton (E286 to the BNC) is in quantitative agreement with the experimentally measured $P_R \rightarrow F$ transition rate.[15,50] Since it is also slower than the transport of the pumped proton (E286 to PLS) and the reprotonation of E286 through the D-channel (discussed below), these results support the widely-accepted notion that transfer of the chemical proton to the BNC is the rate-limiting step for the $P_R \rightarrow F$ transition.

Finally, a comment on the nature of the transport of the pumped and chemical protons. As opposed to rapid PT following the formation of an ideal water network, as proposed in the water gate mechanism, the transport of the pumped proton is shown to be a slow, diffusive process with the excess proton spending time on the water between E286 and PRDa_3. Although this charge translocation is coupled to hydration changes, it is electrostatics (the altered charge distribution following ET) that stabilizes the proton at the PLS and blocks back transfer to E286. The transport of the chemical proton does not have a meta-stable hydronium state, but it does show that the hydration in this region is far more facile than the charge translocation itself. The rate-limiting effect is transitioning the positive charge defect through constricted region—a single water molecule with limited hydrogen bonding partners—to the BNC, while the shifted charge distribution following ET (*i.e.*, electrostatics) is essential for stabilization of the proton in the BNC.

10.8 Proton Transport through the D-Channel

Two protons are taken up from the N-side solution and pass through the D-channel during the $P_M' \rightarrow P_R \rightarrow F$ transition. The first is the chemical proton, which reprotonates E286 after the pumped proton is transported to the PLS. The second reprotonates E286 following transport of the chemical proton to the BNC. The D-channel is often described as starting with D132, but this residue is actually buffered from solution by subunit III. Subunit III in addition to bound lipid molecules have been shown to be essential for WT-like pumping rates at higher pH values and for enzyme stability.[104–106] MD simulations combined with pKa calculations have shown that several histidine residues and specific lipid head groups influence hydration to and increase the pKa of D132, suggesting that they are involved in maintaining rapid proton uptake into the D-channel.[107]

Once protons make it past the subunit III histidines and bound lipids, they do enter the D-channel *via* the explicit protonation of D132. Above D132 the excess proton must pass through a narrow and dehydrated region known as the asparagine gate due to three highly conserved asparagine residues N139, N121, and N207.[108] Two of these residues, N139 and N121, hydrogen bond to each other, breaking the network of water molecules that would otherwise connect D132 and E286 (Figure 10.6). As mentioned above, free energy MD simulations have been helpful in determining the stability of water molecules in the D-channel.[54] These simulations were also used to reveal a gating motion of N139 in the WT system that enables water molecules to enter this region, as well as the perturbation of this gate in two mutants (N139A and N139V). From these results, one might infer a sequential PT mechanism in which the D-channel is first hydrated (the asparagine gate opens and a water wire is formed), and then the excess proton travels from D132 to E286. Such a process of course neglects the cooperative nature of water dynamics and PT, which is clearly demonstrated in MS-RMD simulations of PT through the D-channel.[68] Once again 2D PMFs (shown for the P_R state with the PLS protonated in Figure 10.6) reveal PT through the D-channel and hydration across the asparagine bridge are strongly coupled. The opening of the asparagine gate is unfavorable while D132 is protonated ($x \simeq 35$ Å) (Figure 10.6A). Once the excess proton has dissociated from D132, the N139 side chain rotates down to interact with the hydrated excess charge (Figure 10.6B). This opens the gate and facilitates the formation of a hydrogen-bonded water network past the asparagine residues. Nevertheless, squeezing the excess positive charge through the gate region still constitutes the barrier for PT through the D-channel (Figure 10.6C). Once the excess proton is above the gate, waters leave the asparagine region and the gate closes (Figure 10.6D). The 2D PMF for PT through the D-channel in the F state is very similar (not shown), and the corresponding 1D traces can be compared (Figure 10.7). Once above the asparagine gate, favorable interactions with the pore-lining serine residues ($x \simeq 14$ Å) pull the excess proton towards E286, and most importantly create a

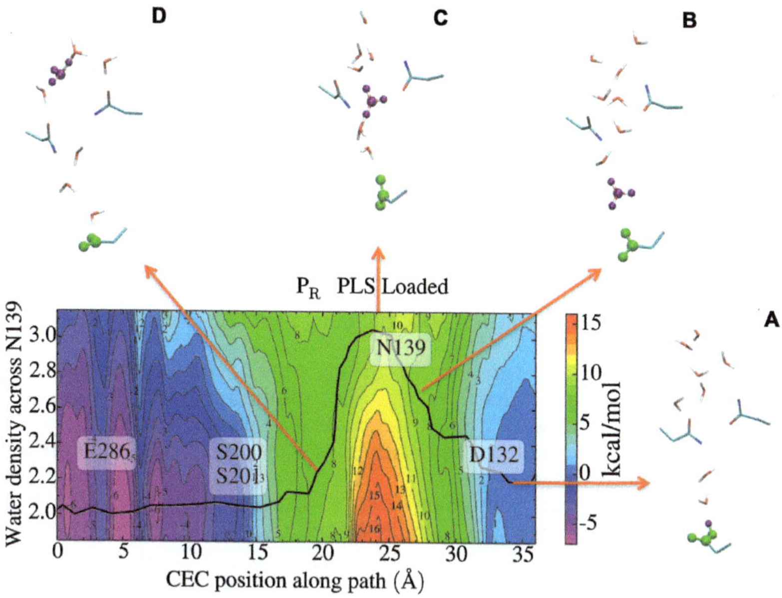

Figure 10.6 Full 2D-PMF for PT from D132 to E286 through the D-channel in the P_R state, with a proton preloaded at the PLS. The 2D-PMF is a function of the CEC coordinate through the D-channel and the water hydration in the asparagine gate region. The statistical errors of the 2D-PMF in D are in the range ~0.1–3 kcal mol^{-1}. The 1D minimum free energy path (black line) highlights the strongly coupled behavior of PT and hydration in the asparagine gate region. (A–D) Molecular structures along the minimum free energy pathway. D132 is shown in green and the excess proton in purple. The N139 and N121 residues are shown as sticks. The transition state structure is shown in (C). The positions of E286, D132, N139, S200 and S201 are labeled with text boxes in the 2D PMF. Reprinted with permission from R. Liang, J. M. J. Swanson, Y. Peng, M. Wikström and G. A. Voth, *Proc. Natl. Acad. Sci. U.S.A*, 2016, **113**, 7420.[68]

large back-flow barrier (further discussed below). E286 is then protonated and rotates from a downward, D-channel-facing conformation ($x \simeq 5$ Å Figure 10.7), to an upward, HC-facing conformation for either pumping or the chemical reaction ($x \simeq 1$ Å Figure 10.7). Calculating the rates from these free energy profiles shows that forward PT (D132 to E286) is much faster than the $P_R \rightarrow F$ transition rate (Table 10.2), verifying that it is not rate limiting. The rate in the P_R state is also faster than back flow of the pumped proton from the PLS to deprotonated E286, which, as discussed above, is essential for pumping.

It is interesting to consider these results in the context of the well-studied D-channel mutants, some of which impair proton pumping and catalysis while others decouple pumping from the chemical reaction altogether.[51,108–118] Some of the decoupling mutants even retain the chemical reaction rate of the WT enzyme. Some mutants increase the pKa of E286,

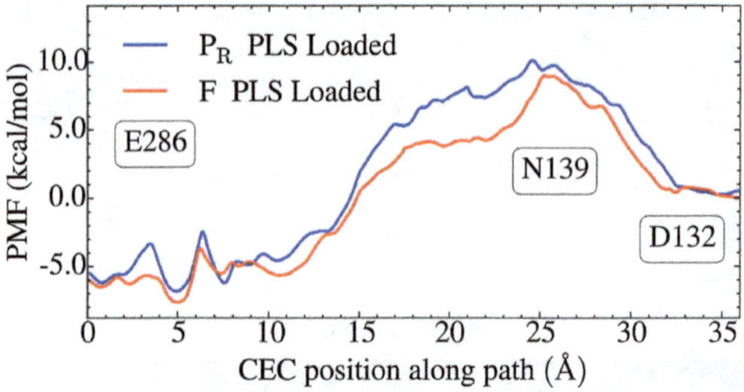

Figure 10.7 1D free energy trace for PT through the D-channel for the P_R (blue) and F (red) states, along the minimum free energy pathway shown in Figure 10.6. The statistical errors are in the range of ~0.1–1 kcal mol^{-1}. The positions of E286, D132, and N139 are labeled with text boxes.
Reprinted with permission from R. Liang, J. M. J. Swanson, Y. Peng, M. Wikström and G. A. Voth, *Proc. Natl. Acad. Sci. U.S.A*, 2016, **113**, 7420.[68]

while others decrease it. The varied observables in these mutants suggests they have quite different mechanistic explanations. The most common site for mutations that result in proton pump impairment and decoupling is N139.[51,109–111,114–117] This may seem surprising given that this residue is more than 25 Å below the HC, where pumping and the chemical reaction occur. However, it is a natural consequence of the primary role that N139 plays in the barrier for both forward and backward PT through the D-channel. It has been suggested that the replacement of N139 with a negatively charged residue (*e.g.*, N139D) decouples the pump by shifting the pK_a of the E286.[118,119] However, decoupling induced by replacement with charge neutral residues (*e.g.*, N139T, N139C and N139L), demands a different decoupling mechanism. Given the results for the WT system described above, it is clear that both the forward and backward barriers through the D-channel are important. Mutations of N139 that weaken the hydrogen bonding with N121 could decrease the barrier for PT in both the forward *and backward* directions. In this case, after the chemical proton has been delivered to the BNC, it would be easier for the pump proton to flow from the PLS back to E286 and through the D-channel to the N-side of the membrane, than for it to flow forward from the PLS to the P-side of the membrane. This trend would of course increase as the transmembrane proton gradient increases and the barrier for release to the P-side of the membrane grows. In contrast, mutants that increase the barriers for forward and backward proton flow through the D-channel could slow down the reprotonation of E286 to the extent that backflow of the pumped proton from the PLS to E286 was favored. This could decouple pumping *and* impair the chemical reaction. Such a mechanism (slowing down PT through the D-channel) is widely anticipated.[3,55,120] Finally, mutations that alter the solvation above the asparagine bridge could shift the pKa of E286. This in turn would change the barriers for proton flow from E286 to the PLS and BNC, potentially slowing down both proton pumping and the chemical reaction. Future MS-RMD simulations will explore the mechanisms of the D-channel mutants.

10.9 Conclusions

Considering the above results collectively leads to the following summary of the proton pumping mechanism in CcO, as exemplified for the $P_M' \rightarrow P_R \rightarrow F$ transition. First, following some chemistry in the BNC during the $A \rightarrow P_R$ transition, there is coupled ET (from heme *a* to the BNC) and PT (from E286 to the PLS). The identity of the PLS has yet to be confirmed, but is potentially PRAa_3 (as described above), PRDa_3, or one of the Cu_B-histidine ligands. As the proton is transported from E286 to the PLS, it disrupts the salt bridge between PRDa_3 and R481, opening a pathway for more waters to interact with the excess positive charge defect and increasing the number of water molecules in the HC. Second, a proton from the N-side of the membrane is taken up through the D-channel to reprotonated E286 at a rate that outcompetes backflow of the pumped proton from the PLS to E286. At this point, it is likely that the hydration in the HC returns to its equilibrium state. Third, the proton on E286 is transported to the reduced BNC where it protonates the Cu_B-bound hydroxide to form water. This constitutes the rate-limiting step, in agreement with previous isotope effect studies.[72,121] Finally, several processes happen at a rate that is faster than the measured rate-limiting step. E286 is once again reprotonated through the D-channel, an electron on Cu_A is partially transferred to heme *a*, and the proton at the PLS is released to the P-side of the membrane.

According to this mechanism kinetic gating enables efficient proton pumping in several ways. First, transport of the pumped proton from E286 to the PLS is faster (*i.e.*, has a smaller forward barrier) than PT to the BNC in both the P_M' and P_R states. At a molecular level, PT to the BNC is slower because it is difficult for the excess proton to pass through the narrow region between E286 and the Cu_B-hydroxide, where a single water molecule has limited hydrogen bonding freedom for charge delocalization. Second, backflow of the proton from the PLS to E286 is slower than reprotonation of E286 from the N-side of the membrane. This is true both before and after PT from E286 to the BNC (*i.e.*, in the P_R and F states, respectively). Molecularly, it is the stabilization of the excess proton on the PLS following ET (in the PR state) that creates a large backflow barrier. Third, backflow of the proton from the PLS to E286 and through the D-channel to the N-side of the membrane is slower than forward transfer from the PLS to the P-side of the membrane. This last point is a consequence of the large backflow barrier in the D-channel, pointing to an additional important role of the asparagine gate. Future work will undoubtedly reveal additional insight into the chemical intermediates, the location of the PLS, the proton release pathway and process, the role of subunit III in proton uptake, and the mechanisms of mutants in addition to the fascinating B- and C-family oxidases, some of which rely on a single proton uptake pathway.

References

1. M. K. F. Wikström, *Nature*, 1977, **266**, 271.
2. S. Ferguson Miller and G. T. Babcock, *Chem. Rev.*, 1996, **96**, 2889.
3. V. R. Kaila, M. I. Verkhovsky and M. Wikström, *Chem. Rev.*, 2010, **110**, 7062.
4. P. Brzezinski and R. B. Gennis, *J. Bioenerg. Biomembr.*, 2008, **40**, 521.
5. M. Wikström, V. Sharma, V. R. I. Kaila, J. P. Hosler and G. Hummer, *Chem. Rev.*, 2015, **115**, 2196.
6. P. Mitchell, *Nature*, 1961, **191**, 144.
7. A. A. Konstantinov, S. Siletsky, D. Mitchell, A. Kaulen and R. B. Gennis, *Proc. Natl. Acad. Sci. U. S. A.*, 1997, **94**, 9085.
8. M. Wikström, A. Jasaitis, C. Backgren, A. Puustinen and M. I. Verkhovsky, *Biochim. Biophys. Acta, Bioenerg.*, 2000, **1459**, 514.
9. D. Bloch, I. Belevich, A. Jasaitis, C. Ribacka, A. Puustinen, M. I. Verkhovsky and M. Wikström, *Proc. Natl. Acad. Sci. U. S. A.*, 2004, **101**, 529.
10. S. Hammes-Schiffer and A. A. Stuchebrukhov, *Chem. Rev.*, 2010, **110**, 6939.
11. V. Sharma, K. D. Karlin and M. Wikström, *Proc. Natl. Acad. Sci. U. S. A.*, 2013, **110**, 16844.
12. M. Wikström, *Biochim. Biophys. Acta*, 2012, **1817**, 468.
13. V. Sharma and M. Wikström, *Biochim. Biophys. Acta, Bioenerg.*, 2016, **1857**, 1111.
14. M. R. Blomberg and P. E. Siegbahn, *Biochim. Biophys. Acta*, 2015, **1847**, 364.
15. I. Belevich, M. I. Verkhovsky and M. Wikström, *Nature*, 2006, **440**, 829.
16. G. T. Babcock and M. Wikström, *Nature*, 1992, **356**, 301.
17. P. Brzezinski, *Trends Biochem. Sci.*, 2004, **29**, 380.
18. D. Jancura, V. Berka, M. Antalik, J. Bagelova, R. B. Gennis, G. Palmer and M. Fabian, *J. Biol. Chem.*, 2006, **281**, 30319.
19. S. J. Ferguson, *Proc. Natl. Acad. Sci. U. S. A.*, 2010, **107**, 16755.
20. M. Wikström and G. Hummer, *Proc. Natl. Acad. Sci. U. S. A.*, 2012, **109**, 4431.
21. M. I. Verkhovsky, A. Jasaitis, M. L. Verkhovskaya, J. E. Morgan and M. Wikström, *Nature*, 1999, **400**, 480.
22. M. R. Blomberg, *Biochemistry*, 2016, **55**, 489.
23. I. Belevich, D. A. Bloch, N. Belevich, M. Wikström and M. I. Verkhovsky, *Proc. Natl. Acad. Sci. U. S. A.*, 2007, **104**, 2685.
24. E. A. Gorbikova, I. Belevich, M. Wikström and M. I. Verkhovsky, *Proc. Natl. Acad. Sci. U. S. A.*, 2008, **105**, 10733.
25. I. Belevich and M. I. Verkhovsky, *Antioxid. Redox Signaling*, 2008, **10**, 1.

26. V. Rauhamaki and M. Wikström, *Biochim. Biophys. Acta*, 2014, **1837**, 999.
27. H. Y. Chang, S. K. Choi, A. S. Vakkasoglu, Y. Chen, J. Hemp, J. A. Fee and R. B. Gennis, *Proc. Natl. Acad. Sci. U. S. A.*, 2012, **109**, 5259.
28. C. von Ballmoos, P. Adelroth, R. B. Gennis and P. Brzezinski, *Biochim. Biophys. Acta*, 2012, **1817**, 650.
29. S. Yoshikawa and A. Shimada, *Chem. Rev.*, 2015, **115**, 1936.
30. P. R. Rich and A. Marechal, *J. R. Soc., Interface*, 2013, **10**, 20130183.
31. T. Tsukihara, K. Shimokata, Y. Katayama, H. Shimada, K. Muramoto, H. Aoyama, M. Mochizuki, K. Shinzawa-Itoh, E. Yamashita, M. Yao, Y. Ishimura and S. Yoshikawa, *Proc. Natl. Acad. Sci. U. S. A.*, 2003, **100**, 15304.
32. T. Tsukihara, H. Aoyama, E. Yamashita, T. Tomizaki, H. Yamaguchi, K. Shinzawa-Itoh, R. Nakashima, R. Yaono and S. Yoshikawa, *Science*, 1996, **272**, 1136.
33. S. Iwata, C. Ostermeier, B. Ludwig and H. Michel, *Nature*, 1995, **376**, 660.
34. J. Abramson, S. Riistama, G. Larsson, A. Jasaitis, M. Svensson-Ek, L. Laakkonen, A. Puustinen, S. Iwata and M. Wikström, *Nat. Struct. Biol.*, 2000, **7**, 910.
35. T. Tiefenbrunn, W. Liu, Y. Chen, V. Katritch, C. D. Stout, J. A. Fee and V. Cherezov, *PLoS One*, 2011, **6**, e22348.
36. Y. Song, E. Michonova-Alexova and M. R. Gunner, *Biochemistry*, 2006, **45**, 7959.
37. J. Quenneville, D. M. Popovic and A. A. Stuchebrukhov, *Biochim. Biophys. Acta*, 2006, **1757**, 1035.
38. A. L. Woelke, G. Galstyan and E. W. Knapp, *Biochim. Biophys. Acta*, 2014, **1837**, 1998.
39. J. Lu and M. R. Gunner, *Proc. Natl. Acad. Sci. U. S. A.*, 2014, **111**, 12414.
40. Y. C. Kim and G. Hummer, *Biochim. Biophys. Acta, Bioenerg.*, 2012, **1817**, 526.
41. Y. C. Kim, M. Wikström and G. Hummer, *Proc. Natl. Acad. Sci. U. S. A.*, 2007, **104**, 2169.
42. Y. C. Kim, M. Wikström and G. Hummer, *Proc. Natl. Acad. Sci. U. S. A.*, 2009, **106**, 13707.
43. S. Supekar, A. P. Gamiz-Hernandez and V. R. Kaila, *Angew. Chem. Int. Ed.*, 2016, **55**, 11940.
44. E. Fadda, C. H. Yu and R. Pomes, *Biochim. Biophys. Acta*, 2008, **1777**, 277.
45. D. M. Popovic and A. A. Stuchebrukhov, *J. Am. Chem. Soc.*, 2004, **126**, 1858.
46. V. R. Kaila, V. Sharma and M. Wikström, *Biochim. Biophys. Acta*, 2011, **1807**, 80.
47. P. E. M. Siegbahn and M. R. A. Blomberg, *Chem. Rev.*, 2010, **110**, 7040.
48. T. Yamashita and G. A. Voth, *J. Am. Chem. Soc.*, 2012, **134**, 1147.
49. P. Goyal, S. Yang and Q. Cui, *Chem. Sci.*, 2015, **6**, 826.

50. K. Faxen, G. Gilderson, P. Adelroth and P. Brzezinski, *Nature*, 2005, **437**, 286.
51. H. Lepp, L. Salomonsson, J. P. Zhu, R. B. Gennis and P. Brzezinski, *Biochim. Biophys. Acta*, 2008, **1777**, 897.
52. M. Wikström, M. I. Verkhovsky and G. Hummer, *Biochim. Biophys. Acta, Bioenerg.*, 2003, **1604**, 61.
53. V. Sharma, G. Enkavi, I. Vattulainen, T. Róg and M. Wikström, *Proc. Natl. Acad. Sci. U. S. A.*, 2015, **112**, 2040.
54. R. M. Henry, C. H. Yu, T. Rodinger and R. Pomes, *J. Mol. Biol.*, 2009, **387**, 1165.
55. R. M. Henry, D. Caplan, E. Fadda and R. Pomes, *J. Phys. Condens. Matter*, 2011, **23**, 234102.
56. D. M. Popovic and A. A. Stuchebrukhov, *Biochim. Biophys. Acta*, 2012, **1817**, 506.
57. P. Goyal, J. Lu, S. Yang, M. R. Gunner and Q. Cui, *Proc. Natl. Acad. Sci. U. S. A.*, 2013, **110**, 18886.
58. N. Ghosh, X. Prat-Resina, M. R. Gunner and Q. Cui, *Biochemistry*, 2009, **48**, 2468.
59. B. M. Samudio, V. Couch and A. A. Stuchebrukhov, *J. Phys. Chem. B*, 2016, **120**, 2095.
60. S. Yang and Q. Cui, *Biophys. J.*, 2011, **101**, 61.
61. A. Tuukkanen, V. R. I. Kaila, L. Laakkonen, G. Hummer and M. Wikström, *Biochim. Biophys. Acta, Bioenerg.*, 2007, **1767**, 1102.
62. J. Xu and G. A. Voth, *Proc. Natl. Acad. Sci. U. S. A.*, 2005, **102**, 6795.
63. J. Xu and G. A. Voth, *Biochim. Biophys. Acta*, 2006, **1757**, 852.
64. J. Xu, M. A. Sharpe, L. Qin, S. Ferguson-Miller and G. A. Voth, *J. Am. Chem. Soc.*, 2007, **129**, 2910.
65. J. Xu and G. A. Voth, *Biochim. Biophys. Acta*, 2008, **1777**, 196.
66. H. J. Lee, E. Svahn, J. M. J. Swanson, H. Lepp, G. A. Voth, P. Brzezinski and R. B. Gennis, *J. Am. Chem. Soc.*, 2010, **132**, 16225.
67. Y. Peng and G. A. Voth, *Biochim. Biophys. Acta*, 2012, **1817**, 518.
68. R. Liang, J. M. J. Swanson, Y. Peng, M. Wikström and G. A. Voth, *Proc. Natl. Acad. Sci. U. S. A.*, 2016, **113**, 7420.
69. M. H. Olsson, P. K. Sharma and A. Warshel, *FEBS Lett.*, 2005, **579**, 2026.
70. M. H. Olsson and A. Warshel, *Proc. Natl. Acad. Sci. U. S. A.*, 2006, **103**, 6500.
71. M. H. Olsson, P. E. Siegbahn, M. R. Blomberg and A. Warshel, *Biochim. Biophys. Acta*, 2007, **1767**, 244.
72. A. L. Johansson, S. Chakrabarty, C. L. Berthold, M. Hogbom, A. Warshel and P. Brzezinski, *Biochim. Biophys. Acta*, 2011, **1807**, 1083.
73. J. M. J. Swanson, C. M. Maupin, H. N. Chen, M. K. Petersen, J. C. Xu, Y. J. Wu and G. A. Voth, *J. Phys. Chem. B*, 2007, **111**, 4300.
74. A. Warshel and R. M. Weiss, *J. Am. Chem. Soc.*, 1980, **102**, 6218.
75. U. W. Schmitt and G. A. Voth, *J. Phys. Chem. B*, 1998, **102**, 5547.
76. U. W. Schmitt and G. A. Voth, *J. Chem. Phys.*, 1999, **111**, 9361.

77. M. Cuma, U. W. Schmitt and G. A. Voth, *J. Phys. Chem. A*, 2001, **105**, 2814.
78. T. J. F. Day, A. V. Soudackov, M. Cuma, U. W. Schmitt and G. A. Voth, *J. Chem. Phys.*, 2002, **117**, 5839.
79. Y. Wu, H. Chen, F. Wang, F. Paesani and G. A. Voth, *J. Phys. Chem. B*, 2008, **112**, 467.
80. G. A. Voth, *Acc. Chem. Res.*, 2006, **39**, 143.
81. C. Knight, G. E. Lindberg and G. A. Voth, *J. Chem. Phys.*, 2012, **137**, 22A525.
82. T. Yamashita, Y. X. Peng, C. Knight and G. A. Voth, *J. Chem. Theory Comput.*, 2012, **8**, 4863.
83. J. G. Nelson, Y. Peng, D. W. Silverstein and J. M. J. Swanson, *J. Chem. Theory Comput.*, 2014, **10**, 2729.
84. S. Lee, R. Liang, G. A. Voth and J. M. J. Swanson, *J. Chem. Theory Comput.*, 2016, **12**, 879.
85. R. Liang, H. Li, J. M. J. Swanson and G. A. Voth, *Proc. Natl. Acad. Sci. U. S. A.*, 2014, **111**, 9396.
86. R. Liang, J. M. J. Swanson, J. J. Madsen, M. Hong, W. F. DeGrado and G. A. Voth, *Proc. Natl. Acad. Sci. U. S. A.*, 2016, **113**, E6955.
87. S. Lee, J. M. J. Swanson and G. A. Voth, *Biophys. J.*, 2016, **110**, 1334.
88. S. Lee, H. B. Mayes, J. M. J. Swanson and G. A. Voth, *J. Am. Chem. Soc.*, 2016, **138**, 14923.
89. S. Riistama, G. Hummer, A. Puustinen, R. B. Dyer, W. H. Woodruff and M. Wikström, *FEBS Lett.*, 1997, **414**, 275.
90. X. Zheng, D. M. Medvedev, J. Swanson and A. A. Stuchebrukhov, *Biochim. Biophys. Acta*, 2003, **1557**, 99.
91. E. Olkhova, M. C. Hutter, M. A. Lill, V. Helms and H. Michel, *Biophys. J.*, 2004, **86**, 1873.
92. I. Hofacker and K. Schulten, *Proteins*, 1998, **30**, 100.
93. R. Pomes, G. Hummer and M. Wikström, *Biochim. Biophys. Acta*, 1998, **1365**, 255.
94. M. Svensson-Ek, J. Abramson, G. Larsson, S. Tornroth, P. Brzezinski and S. Iwata, *J. Mol. Biol.*, 2002, **321**, 329.
95. L. Qin, C. Hiser, A. Mulichak, R. M. Garavito and S. Ferguson-Miller, *Proc. Natl. Acad. Sci. U. S. A.*, 2006, **103**, 16117.
96. V. R. Kaila, M. I. Verkhovsky, G. Hummer and M. Wikström, *Proc. Natl. Acad. Sci. U. S. A.*, 2008, **105**, 6255.
97. A. L. Woelke, G. Galstyan, A. Galstyan, T. Meyer, J. Heberle and E. W. Knapp, *J. Phys. Chem. B*, 2013, **117**, 12432.
98. Y. Peng, J. M. J. Swanson, S.-G. Kang, R. Zhou and G. A. Voth, *J. Phys. Chem. B*, 2015, **119**, 9212.
99. M. Karpefors, P. Adelroth, A. Namslauer, Y. J. Zhen and P. Brzezinski, *Biochemistry*, 2000, **39**, 14664.
100. M. Wikström and M. I. Verkhovsky, *Biochim. Biophys. Acta*, 2007, **1767**, 1200.
101. H. J. Lee, L. Ojemyr, A. Vakkasoglu, P. Brzezinski and R. B. Gennis, *Biochemistry*, 2009, **48**, 7123.

102. D. M. Popovic and A. A. Stuchebrukhov, *FEBS Lett.*, 2004, **566**, 126.
103. I. Belevich, E. Gorbikova, N. P. Belevich, V. Rauhamaki, M. Wikström and M. I. Verkhovsky, *Proc. Natl. Acad. Sci. U. S. A.*, 2010, **107**, 18469.
104. M. R. Bratton, M. A. Pressler and J. P. Hosler, *Biochemistry*, 1999, **38**, 16236.
105. G. Gilderson, L. Salomonsson, A. Aagaard, J. Gray, P. Brzezinski and J. Hosler, *Biochemistry*, 2003, **42**, 7400.
106. P. Adelroth and J. Hosler, *Biochemistry*, 2006, **45**, 8308.
107. V. Sharma, P. Ala-Vannesluoma, I. Vattulainen, M. Wikström and T. Rog, *Biochim. Biophys. Acta*, 2015, **1847**, 690.
108. D. Han, A. Namslauer, A. Pawate, J. E. Morgan, S. Nagy, A. S. Vakkasoglu, P. Brzezinski and R. B. Gennis, *Biochemistry*, 2006, **45**, 14064.
109. U. Pfitzner, K. Hoffmeier, A. Harrenga, A. Kannt, H. Michel, E. Bamberg, O. M. H. Richter and B. Ludwig, *Biochemistry*, 2000, **39**, 6756.
110. A. S. Pawate, J. Morgan, A. Namslauer, D. Mills, P. Brzezinski, S. Ferguson-Miller and R. B. Gennis, *Biochemistry*, 2002, **41**, 13417.
111. A. Namslauer, A. S. Pawatet, R. Gennis and P. Brzezinski, *Proc. Natl. Acad. Sci. U. S. A.*, 2003, **100**, 15543.
112. A. S. Vakkasoglu, J. E. Morgan, D. Han, A. S. Pawate and R. B. Gennis, *FEBS Lett.*, 2006, **580**, 4613.
113. K. L. Durr, J. Koepkee, P. Hellwig, H. Muller, H. Angerere, G. Peng, E. Olkhova, O. M. H. Richter, B. Ludwig and H. Michele, *J. Mol. Biol.*, 2008, **384**, 865.
114. L. Varanasi and J. Hosler, *Biochemistry*, 2011, **50**, 2820.
115. A. L. Johansson, J. Carlsson, M. Hogbom, J. P. Hosler, R. B. Gennis and P. Brzezinski, *Biochemistry*, 2013, **52**, 827.
116. S. A. Siletsky, J. P. Zhu, R. B. Gennis and A. A. Konstantinov, *Biochemistry*, 2010, **49**, 3060.
117. J. P. Zhu, H. Z. Han, A. Pawate and R. B. Gennis, *Biochemistry*, 2010, **49**, 4476.
118. P. Brzezinski and A. L. Johansson, *Biochim. Biophys. Acta*, 2010, **1797**, 710.
119. S. Chakrabarty, I. Namslauer, P. Brzezinski and A. Warshel, *Biochim. Biophys. Acta*, 2011, **1807**, 413.
120. H. Y. Chang, S. K. Choi, A. S. Vakkasoglu, Y. Chen, J. Hemp, J. A. Fee and R. B. Gennis, *Proc. Natl. Acad. Sci. U. S. A.*, 2012, **109**, 5259.
121. L. Salomonsson, G. Branden and P. Brzezinski, *Biochim. Biophys. Acta*, 2008, **1777**, 343.

CHAPTER 11

Water Oxidation by PSII: A Quantum Chemical Approach

PER E. M. SIEGBAHN

Department of Organic Chemistry, Arrhenius Laboratory,
Stockholm University, SE-106 91 Stockholm, Sweden
Email: per.siegbahn@su.se

11.1 Introduction

Water oxidation in photosystem II is one of the most important processes in nature. Intense research during at least half a century has been devoted towards understanding the atomic details of this process. It has already been almost 50 years ago since the basic so-called Kok-cycle was set up.[1] This cycle shows that water oxidation proceeds in a sequence of four S-transitions from S_0 to S_4, in which the oxygen evolving complex (OEC) is oxidized four times. Parallel to these oxidations, two substrate water molecules are deprotonated. At the final S_4-state, the O–O bond is formed and O_2 is released. To understand water oxidation therefore means understanding all the S-state transitions involved, and finally the mechanism for the O–O bond formation. An updated version of the Kok-cycle is shown in Figure 11.1, in which all the steps of deprotonation and oxidation are included, and also the stages where two water molecules become bound to the OEC. The steps in the cycle are labeled as S_n^m, where the lower index denotes the S-state. For each S-state there are two states, labeled with the upper index as 0 or −1, where the latter one has one proton less.

A major step towards the understanding of water oxidation in PSII came when the first X-ray structure was presented in 2004 by Ferreira *et al.*[2] Even

Chemical Biology No. 5
Mechanisms of Primary Energy Transduction in Biology
Edited by Mårten Wikström

Published by the Royal Society of Chemistry, www.rsc.org

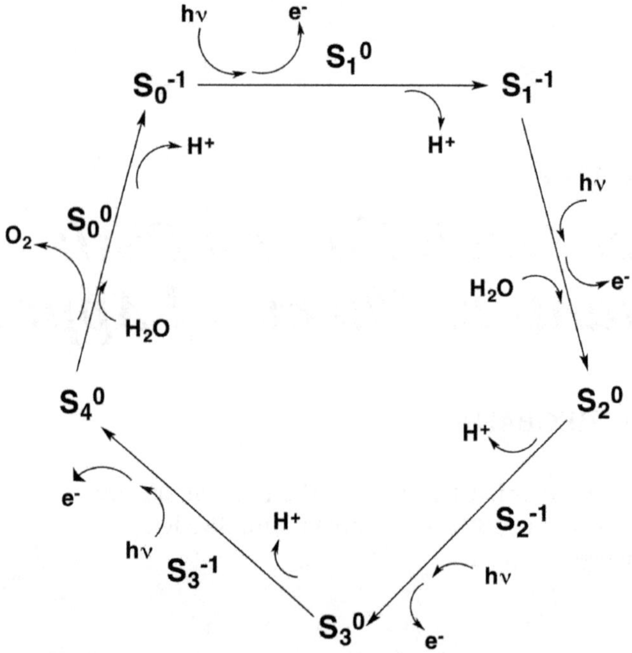

Figure 11.1 The Kok cycle S-state cycle for water oxidation updated by more recent findings.

though the resolution was rather low at 3.5 Å it was possible to identify metal positions of the OEC with three Mn atoms and one Ca in a cuboidal form, and with a fourth Mn atom outside the cube, bridging to it on an oxo group. It has later been shown (see below) that the outer Mn was misplaced by 3.2 Å with respect to the Mn_3Ca cubane part. The amino acids ligated to the cluster were also identified, but the coordinations of these ligands were unclear, as well as the locations of bridging oxo groups. The resolution was slightly improved by Loll *et al.* a year later to 2.9 Å, with a structure where the metal positions in the cube were refined,[3] but where bridging oxygens could still not be seen. The outer Mn was now instead suggested to be singly bridging with an oxo or OH ligand to one Mn-atom in the cube. It has later been shown that the outer Mn was misplaced by 1.4 Å with respect to the Mn_3Ca cubane part. The cubane part was of the closed form, see below, which later on has been shown by spectroscopy to be much less preferred than the open form. Although the resolution still did not allow it, chemically more reasonable coordinations of the amino acid ligands could also be suggested. This was the situation when quantum chemical model calculations came into the picture in 2004. In 2011, Umena *et al.* published the first high-resolution structure of PSII.[4]

In 2007–2009 a few years before the high resolution structure appeared, structures for all the S-states were suggested on the basis of density functional theory (DFT) calculations.[5,6] The X-ray study by Umena *et al.* in 2011

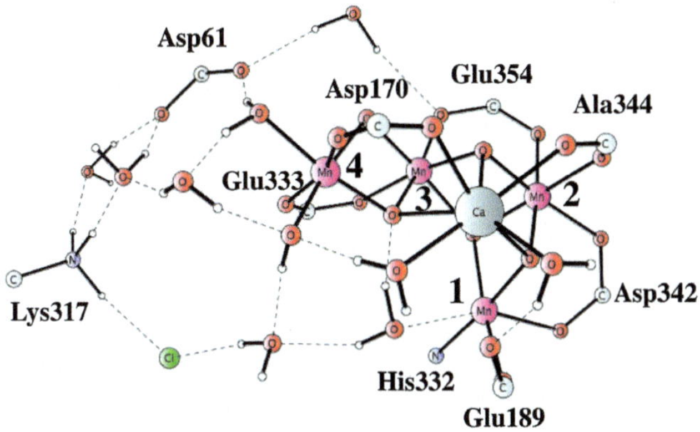

Figure 11.2 Optimized S_2^0-state. 1–4 are the numbering of the Mn-centers. Only the most essential atoms are shown.

confirmed the structure for S_1 except that Asp170 was shown to be a bridging ligand between Mn_4 and Ca, while DFT placed it as a terminal ligand on Mn_4. It is now also accepted that the X-ray structure suffered from X-ray reduction which slightly distorted the structure. In 2011, it was shown that the modification of Asp170 did not modify the suggested mechanism,[7] but new slightly modified structures were constructed, see the core structure in Figure 11.2. Very important spectroscopic studies were made in 2012, convincingly establishing a detailed structure of the S_2 state.[8] That structure is very similar to the one obtained by DFT in 2009 and slightly modified in 2011. The details confirmed by the experiments include the oxidation state of each manganese, the positions of the oxo-groups, the OH groups and the waters, as well as the position of the oxo-group that eventually forms the O–O bond in S_4, see further below. In 2014, another breakthrough appeared as a new EPR study confirmed also the details of the theoretically suggested S_3 structure.[9]

Very recently, a new structure of the S_3 state was suggested in an X-ray free electron laser (XFEL) study.[10] Most strikingly, this structure is lacking the OH-group in the center of the cluster suggested earlier by both experiments and theory. However, results of model calculations and spectroscopy make it very unlikely that the X-FEL structure can have four Mn(IV). A structure with only three Mn(IV), one Mn(III) and a Tyr_z radical is much more likely. This state should then be a precursor to the final fully oxidized S_3 state.

The leading mechanism, at present, for the O–O bond formation was suggested in 2006.[11] It can be described as an oxyl (oxygen radical) attack on a bridging oxo-ligand, and is therefore termed the oxyl–oxo mechanism. This mechanism will be discussed more in detail below. Very strong support for this mechanism came in 2012 after a set of remarkable spectroscopic studies using EPR and water exchange measurements.[8]

In the last couple of years, theoretical studies on the mechanism for water oxidation are less frequent, and most of them are in agreement with the mechanism outlined above. For theoretical work done before 2011, there has been a review where other references can be found for even earlier studies.[12] Cluster models and mechanisms have been studied by Pace *et al.*[12] They suggest lower oxidation states than the ones suggested here and in most other studies. They base their conclusions on a computational analysis of NEXAFS (Near Edge X-ray Absorption Fine-Structure) spectra using a TDDFT (Time Dependent DFT) approach. In a DFT study it has been questioned if the TDDFT method is accurate enough for the analysis, and also if NEXAFS (or XANES) is really a reliable technique for determining oxidation states.[13] The most thorough and conclusive analysis of the oxidation states in the S_2 state comes from a recent study using a combination of theory and spectroscopy.[14] That study made it very clear that the lower oxidation state suggestion by Pace *et al.* is not a viable possibility. In 2007, Kusunoki[15] suggested a structure for the S_0-state based on the low resolution structure of Loll *et al.*,[3] resulting in the same correct and incorrect aspects as that structure. The ligands are at essentially the correct place, but the outer manganese is too far out, connected to the cube by only one oxo-ligand and having three water ligands, two of which were suggested to be the substrates based on water exchange considerations. In 2011, after the high-resolution structure, the model was refined. The originally suggested mechanism was kept, now supported by EXAFS calculations. To explain the unusual position of the central oxo-ligand, it was suggested that there are two different structures for the S_1-state with equal energy.[16] In contrast to the cluster approach, the QM/MM (quantum mechanics/molecular mechanics) method has been used by Batista *et al.*[17,18] They restricted the structures investigated to be similar to the one of Ferreira *et al.*[2] For the best structure obtained they also did an analysis of EXAFS spectra,[19] and found that their suggested structure would match the experimental spectrum if it was only slightly modified. They took this information to be a confirmation of their suggested structure. After the new X-ray structure by Umena *et al.* appeared, it became clear that their suggested structure was quite different from the real structure. Later on it has been shown that, using their type of analysis, also other quite different structures match the EXAFS spectra well.[20] Furthermore, Batista *et al.* suggested an O–O bond formation mechanism that is different to the one proposed here. A nucleophilic attack from an external water (or bound to calcium) was suggested to form the bond with an oxygen radical bound to manganese. Their mechanism is the same as the one suggested earlier in several DFT studies starting 1999,[21–25] but which was demonstrated to lead to too high barriers in the study mentioned above from 2006.[11] In a comparative DFT study in 2012, it was shown that most of the earlier structures were energetically much too high in energy to be reasonable candidates for the actual structure.[26]

Some early experimental studies, made before the high-resolution X-ray structure appeared, deserve special attention. Already in 2004, Messinger speculated that the O-O bond should be formed between a terminal water ligand to Mn and an oxo-bridge between Mn and Ca.[27] The involvement of an oxo bridge was a very controversial suggestion at the time since it was known that an oxo-group in general would not exchange with water as fast as observed by water exchange experiments on PSII. However, there were also other alternatives that could fit the experimental observations. For example, in the same study it was concluded that the nucleophilic attack was an attractive possibility. The mechanism started from an S_3 state based on the Loll *et al.* structure with a closed Mn_3Ca cubane form and a singly bridged outer Mn. This form for S_3 is much higher in energy than the open form shown to be preferred later on by spectroscopy. The role of an oxyl radical for the O–O bond formation was not discussed either. Furthermore, the suggested S_3 state was considered to have an oxyl radical and one Mn(III), which later turned out to be incorrect. Still, the suggestion of the possibility that an oxo-bridge is one of the oxygens in the O–O bond formation was remarkably foresighted, although based on only weak experimental indications. In 2006, four structures were suggested by polarized EXAFS as the only possible alternatives.[28] These four suggestions could later be shown to be significantly too high energetically,[6] although one of them can be considered as a mirror image of the later X-ray structure. The structure for the S_2 state suggested by Dau *et al.* in 2008,[29] also based on EXAFS (including a molecular mechanics modeling), has a very similar core structure to the subsequent X-ray structure. It is also very similar to the 1c DFT structure from 2006,[11] but was more tightly bound. For S_3, a similar tightly bound structure with a closed form of the cubane was suggested. A significant structural change in the S_2 to S_3 transition was a key suggestion in the mechanism. In 2011, Cox *et al.*[30] did a new more extended DFT analysis and concluded that the previous DFT structure[5] matched the experimental multifrequency EPR and ^{55}Mn-ENDOR spectra best. Simultaneously, and independently, Kaupp *et al.*[31] reached the same conclusion.

In a theoretical study after the appearance of the X-ray structure in 2011,[32] Batista *et al.* analyzed the role of chloride by deleting it from the X-ray structure, and concluded that one role might be to prevent salt-bridges which would hinder proton transfer. In another study they analyzed the protonation state of the S_1-state.[33] They concluded that the X-ray structure is a mixture of oxidation states. A similar conclusion was reached also in other studies.[7,34–36]

Last year, a theoretical study by Isobe *et al.*[37] suggested that in S_3 an oxyl-radical state is in equilibrium with the one with four Mn(IV). However, that study is contrary to what has been shown by several other theoretical studies and also by experiments, and has not been confirmed by others afterwards. In some recent DFT studies the focus has been on water insertion in the S_2 to S_3 transition. The mechanisms suggested for how this occurs will be discussed below.

11.2 Methods and Models

Most of the DFT calculations discussed here were performed in the same way as described in detail in several papers.[38] The hybrid functional B3LYP*[39] was used with exact exchange reduced to 15%. Medium size basis sets were used for the geometries, and large basis sets for energies. A surrounding dielectric medium with a dielectric constant of 6.0 was added. The performance of the B3LYP functional for the present type of problems has been reviewed,[40–42] indicating a typical accuracy within 3–5 kcal mol^{-1}, normally overestimating barriers. Dispersion effects were added using the empirical D2 formula of Grimme.[43] Finally, a spin-correction of −2.8 kcal mol^{-1} was added for the S_4-state structures.[44] Computed Hessians were used to optimize transition states and to determine zero-point effects. The calculations were performed with the Jaguar program.[45]

The quantum chemical cluster model chosen for the applications mostly discussed here, is the same as the one used in the most recent studies.[38] The model is based on the high-resolution (1.9 Å) structure by Umena *et al.*,[4] which is very similar to the structure obtained from DFT calculations prior to the X-ray structure.[5,6] It is shown for the S_2-state in Figure 11.2, where only the most important atoms are shown. The full 200 atom structures have been given previously.[38] The amino acids included in the model were first the directly binding amino acids, Asp170, Glu189, His332, Glu333, Asp342, Alal344 and Glu354. The second shell residues Asp61, His337 and Arg357 and the region around the chloride were also included. This region contains, besides chloride, Lys317 and three water molecules, forming a hydrogen bonding network, as in the X-ray structure.

The energy diagrams discussed below are constructed following a scheme where only relative pK_a and redox potentials are used, see for example ref. 47. This is possible if two results from experiments are used. First, the absolute energy to remove an (H^+, e^-)-couple from the OEC and place the proton and electron at the respective acceptors, water and P_{680}^+, is chosen to fit what is known experimentally about the driving force for the entire reaction. This fixes every second energy level in the diagram from the calculated relative energy (H^+, e^-) differences. To fix also the other half of the energy levels, one additional parameter has to be chosen. How this parameter was chosen is described below. The same parameter was used for all energy levels in the diagrams.

11.3 Discussion

This discussion is divided into six sections. In the first section the DFT optimized structures of the different S-states are discussed. In the second section, the mechanism for O–O bond formation is outlined. In the third section, the energetics is discussed for all steps from the starting S_0 state to the completion of the water oxidation process. Both deprotonation and electron release steps are included. It should be mentioned that the resulting energy diagram is slightly different from the one presented before. In the fourth and fifth sections, the roles of Tyr_z and calcium are described. Finally, in the sixth section one of the few remaining problems in the mechanism is discussed. This concerns the insertion of a water molecule in the S_2 to S_3 transition.

11.3.1 S-state Structures

In 2008, a series of calculations were started to obtain structures for all the different S-states.[5] The starting point was the mechanism and the transition state (TS) obtained in 2006 for O–O bond formation in S_4.[11] This TS was placed into the Ferreira *et al.* X-ray structure. Electrons and protons were then added to come back to the other S-states. The resulting structures are very close to the high resolution structure later obtained by Umena *et al.* by ref. 4 and to the ones suggested later by EPR spectroscopy.[9] The only noticeable difference is that Asp170 was shown to be bridging between the outer Mn and Ca, rather than terminally bound to Mn as suggested by the DFT modeling.

The S_0 structure is shown in Figure 11.3. There are three short (2.8–2.9 Å) and one longer (3.2 Å) Mn–Mn distance in agreement with one suggestion based on EXAFS.[46] The oxidation states for Mn2 is IV, while they are III for the other ones. A notable feature of the structure is that the central oxygen (usually termed O5) is protonated, which is in agreement with most studies. However, the distances between this OH group and the nearest Mn-atoms is quite unusual, with three distances of 2.3 Å, see the figure, which is actually quite similar to the structure by Umena *et al.* Normally, one significantly shorter distance is expected. The reason for this is that the oxidation states for all the nearest Mn-atoms are III, and the OH binds along the Jahn–Teller axis to all of them. Furthermore, the interaction is quite ionic making the bonding less directional. The potential surface for moving the OH is very flat

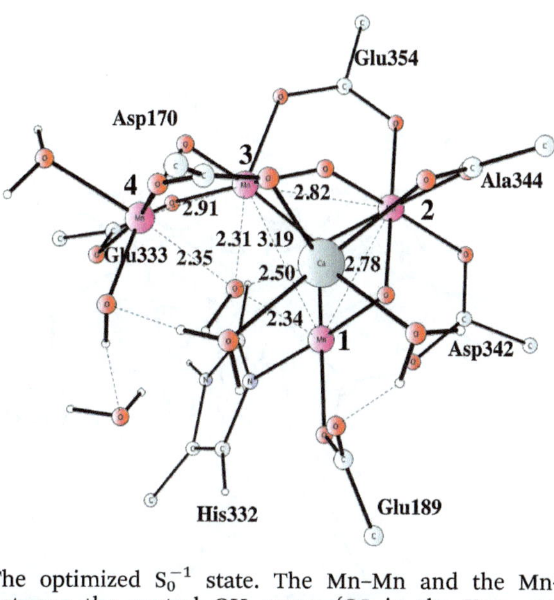

Figure 11.3 The optimized S_0^{-1} state. The Mn–Mn and the Mn–OH distances between the central OH group (O5 in the X-ray structure) to the manganese and calcium atoms are given in Å. Only the most essential atoms are shown.

and other structures are therefore close in energy. This is therefore not a very significant aspect for the mechanism, see also futher below for the S_1 state.

The S_1 state is reached by removal of one proton and one electron from S_0. The oxidation occurs for Mn3. The proton removed is the one on O5. The structure is shown in Figure 11.4. As expected, O5 now has one short bond to Mn3 and longer bonds to the other ones. The estimated pK_a for O5 is close to 7, which means that it does not matter for the mechanism if O5 is protonated or not, see further below.

In the next oxidation there is no deprotonation if O5 is already deprotonated in S_1, otherwise that proton should leave in this transition. It is clear that irrespective at what stage the O5 proton leaves, in this or the previous transition, it must leave before Mn is oxidized. The oxidized Mn is Mn4, the outer one. In S_2, see Figure 11.5, there are still three short Mn–Mn distances and one long. At this stage, Mn1 is the only Mn(III) atom. As mentioned in the introduction the details of the structure and electronic state has been confirmed by experiments.[8]

The last state observed before the O–O bond is formed is S_3, see Figure 11.6. Also this structure has been confirmed in detail by spectroscopy.[9] There was for a long time a debate whether an Mn center or a ligand was oxidized in the S_2 to S_3 transition. However, consensus has now been reached that an Mn-atom is oxidized. The one oxidized is Mn1. An important event in this transition is also that a water molecule becomes bound. Surprisingly, recent experiments indicate that this water is probably not a substrate, but this question is not yet settled, see further below.

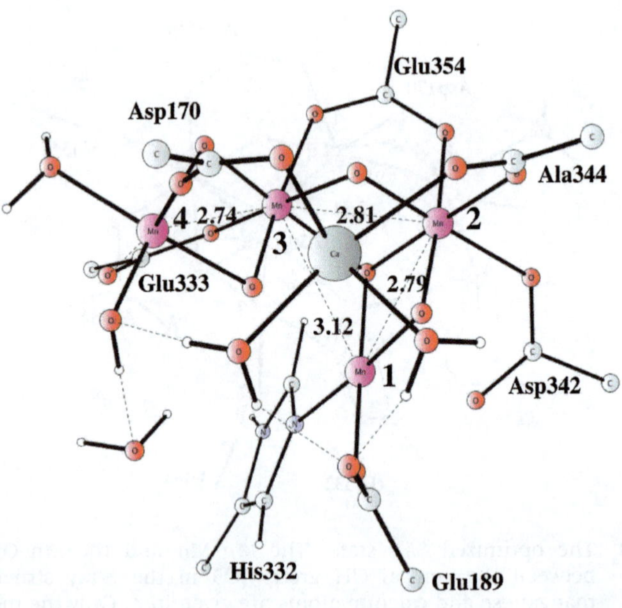

Figure 11.4 The optimized S_1^{-1} state.

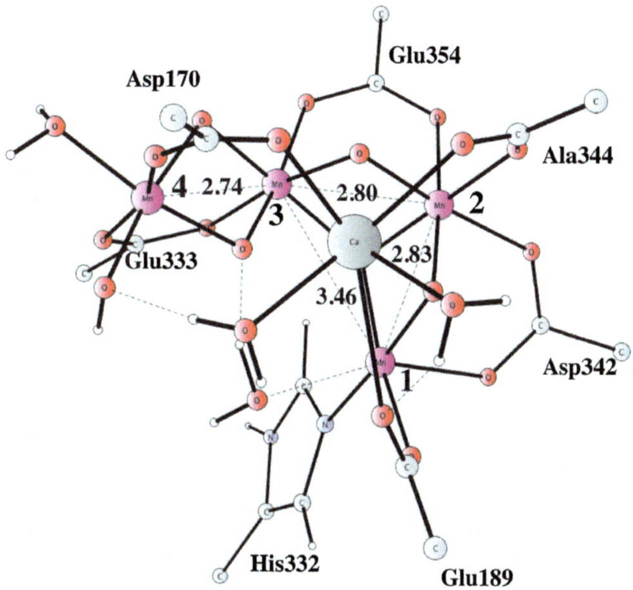

Figure 11.5 Optimized S_2^0 structure.

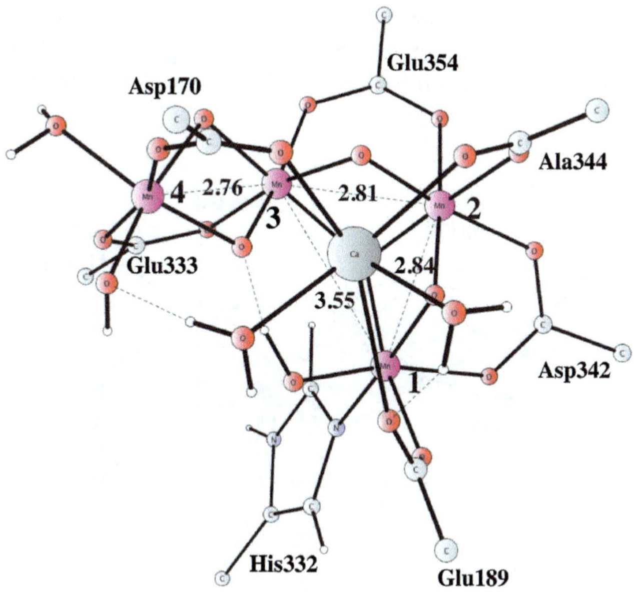

Figure 11.6 Optimized S_3^0 structure.

11.3.2 O–O Bond Formation

The present mechanism for O–O bond formation was found about 10 years ago,[11] following a tedious procedure based on the low resolution X-ray structures available at the time.[2,3] It was early realized, even before the low-resolution structures were available, that a structure with an oxyl radical bound to an Mn(IV) was required for a low barrier in S_4.[21] A reasonable state of this type was obtained, after several years of testing different models, starting with the low-resolution structures that were determined in 2003–2004. Also different types of structures not suggested before were tried. A major new finding was an additional oxo-group (later termed O4) between Mn3 and Mn4. This finding was confirmed experimentally several years later.[4] To find the mechanism for O–O bond formation, all possibilities of combining oxygens were tried, using cluster models with ferro- and antiferro-magnetic couplings between the Mn centers. One mechanism stood out with a much lower barrier than the rest of them. The most recent version of the transition state for this mechanism is shown in Figure 11.7, which has a barrier from S_4 of only 7.7 kcal mol^{-1}. The mechanism is commonly termed the oxyl–oxo mechanism, since the O–O bond is formed between a bridging oxo group (later termed O5) and a terminal oxyl ligand.

A detailed schematic mechanism for the O–O bond formation is shown in Figure 11.8. The reactant in S_4 has two low-lying states. The lowest one has the spin of the oxyl radical antiferro-magnetically coupled to the spin of its binding Mn-center. This introduces some Mn(V) character, but the best description is an Mn(IV)–oxyl state. The second state shown in the figure has the oxyl spin ferro-magnetically coupled to the spin of Mn. Typically, the energy is 8 kcal mol^{-1} higher for this state. The product in the figure has a peroxide bound to an antiferro-magnetically coupled pair of Mn-centers.

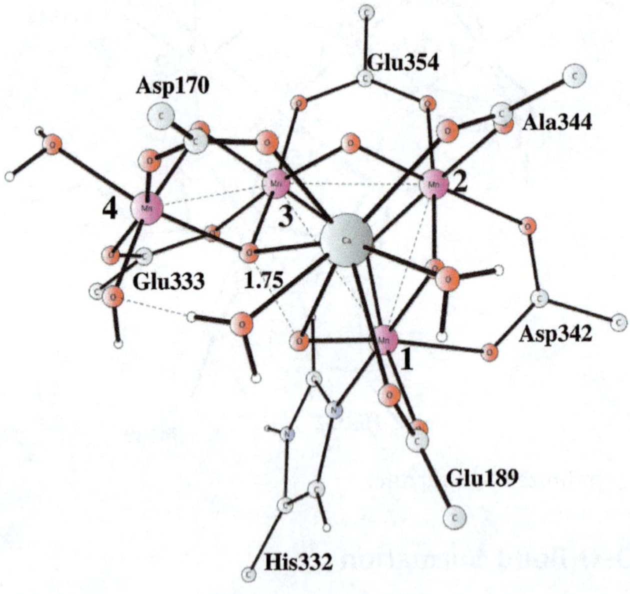

Figure 11.7 Optimized transition state for O–O bond formation in S_4. The distance between the oxygens forming the bond is also shown.

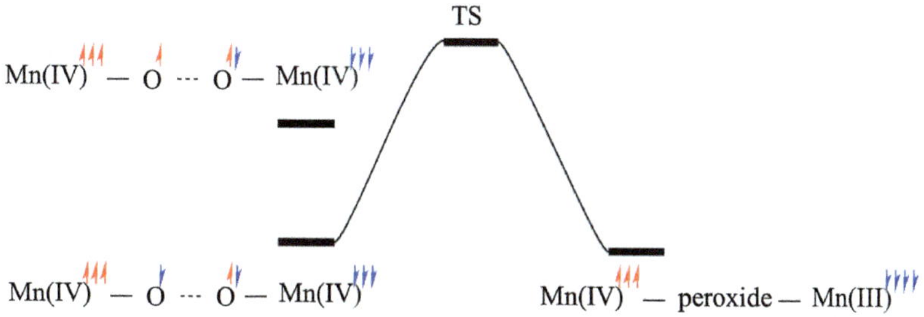

Figure 11.8 Schematic mechanism for O–O bond formation in PSII.

As seen in the figure, this type of coupling allows a smooth formation of the O–O bond between the lowest reactant and the product, without change of spin-directions, which is the key for the low barrier of the oxyl–oxo mechanism where two manganese centers have to be involved. If only one Mn-center is involved, as in a water attack mechanism, the O–O bond formation will have to start on the excited reactant to reach the product. The same is true also if cobalt is used instead of nickel, even for the oxyl-oxo mechanism.[48] There is one variant of this mechanism, where the same Mn-centers are involved but the oxyl radical is bound to Mn4 rather than to Mn1.[49,50]

A nucleophilic water attack mechanism has been essentially the only other type of mechanism suggested for O–O bond formation in PSII.[51] However, a very high barrier of 30 kcal mol^{-1} from the S_4 state, has recently been demonstrated for that mechanism.[52] One reason for this is the spin-requirement mentioned above, but an even more important reason is that the product, either O–OH or O–OH_2, is very high in energy for the OEC, probably to avoid hydrogen peroxide formation.

11.3.3 Full Energy Diagram

A mechanism for water oxidation in PSII is not complete without an energy diagram, where energies for all deprotonation and redox events are included. Since *absolute* pK_a and redox potentials are difficult to obtain accurately from computational modeling, the diagram should ideally be obtained from just the *relative* values computed. The first energy needed to obtain a diagram is the driving force for the entire reaction, which is best taken from experiments. At pH = 7 the redox potential for forming O_2 from water is 0.8 V. For P_{680}^{+} a redox potential has been measured to be 1.25 V.[53,54] This leads to a total driving force of 41.5 kcal mol^{-1}.[6] For the full diagram a single additional parameter, used for all energy levels, is needed. This parameter should also be taken from experimental information. In this case the most useful experimental results are that only an electron leaves in the S_1 to S_2 transition[56,57] and that the proton release in both S_2 and S_3 occurs only after P_{680}^{+} reduction.[58,59] These two results should ideally be reproduced simultaneously with the single parameter. Since this turned out to be possible, it supports the probability that the computed relative energies have a high accuracy. Using the experimental driving force and single parameter, a diagram based only on calculated relative values has been constructed following the description in Section II, see Figure 11.9.

A comment can be made concerning the computational accuracy of relative values as compared to absolute ones for the redox and pK_a values. Absolute values are strongly affected by charged groups even far away from the active site due to the slow decrease of the coulomb field, and these effects can therefore not be ignored for the absolute values. For the relative values these same charges have an effect only on the *difference in distance* between two redox (or deprotonation) positions. Even though the atom oxidized, or the proton released, in each transition is well localized in the OEC cluster, the subsequent charge change is smeared out over the cluster. For example, an oxidation of one particular Mn center will give rise to a change of the spin on that center of nearly one unit, but the charge will change by only an order of magnitude smaller amount. This means that the

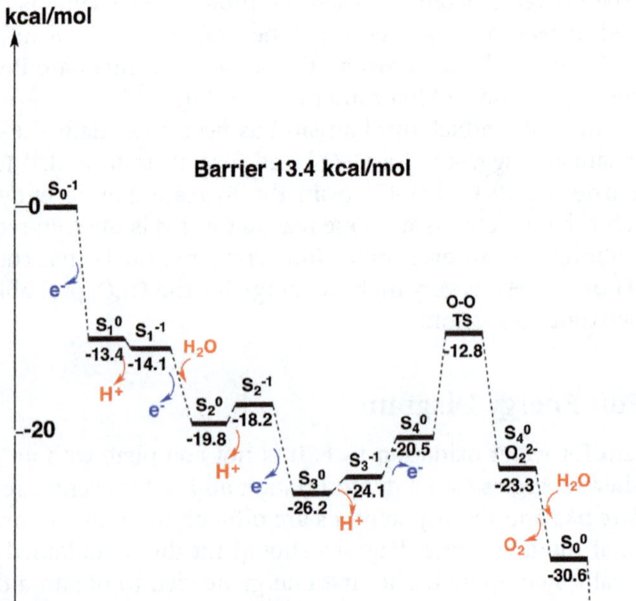

Figure 11.9 Full energy diagram for water oxidation from S_0 back to S_0.

change of charge felt by a group outside will be more or less the same independently of which center has been oxidized. It should be added that the full effect is obviously included from all charged groups within the model. The charged groups outside the model will still give rise to a small error on the computed relative values, which will be smaller the larger the model is. Experience says that a model of 200 atoms is sufficient for obtaining accurate relative values.

Figure 11.9 gives a compact description of all the calculated energies for the full cycle of water oxidation. For each level there is also an optimized structure, given earlier. A few general remarks can be given for the results in the diagram. First, before each oxidation there is an oxidation of P_{680} by sunlight. Second, as in all redox processes in enzymes studied so far, electron and proton release occur alternatingly. For each S-state there are therefore two energy levels, without and with proton release.

In the first transition from S_0 to S_1 the oxidation is quite exergonic by -13.4 kcal mol^{-1}, followed by a proton release which is very weakly exergonic by -0.7 kcal mol^{-1}. In agreement with spectroscopy, this transition therefore has both an oxidation and a proton release. There has been some recent debate whether a proton actually leaves or not in this transition. In the high-resolution structure the distances obtained indicate that the proton is still there,[4] while spectroscopy indicates that it should leave. In any case, it is likely from the calculations that the pK_a value should be close to 7. Energetically the difference is very small and it is very unlikely that this question should have any implication for the O–O bond formation mechanism occurring much later. It has also been suggested that the presence of a proton could strongly modify the redox potential and therefore also be important for the water oxidation mechanism. However, this is not true. If two cases of S_1 are considered, the first one without a proton (p$K_a = 6.5$) and the other one with a proton (p$K_a = 7.5$). The only difference in the energy diagram would be that the S_1^0 to S_1^{-1} transition would be exergonic by -0.7 kcal mol^{-1} (as in Figure 11.9) in the first case, while in the other case it would be endergonic by the same amount. For the next oxidation, it is clear that the proton would have left the OEC in both cases, so the redox potential difference would be very similar. The overall mechanistic difference between the two cases will, of course, be small just like the pK_a difference.

In the S_1 to S_2 transition an electron leaves, exergonically by -5.7 kcal mol^{-1}, but no proton in agreement with spectroscopy. This is seen by the weakly endergonic deprotonation by $+1.6$ kcal mol^{-1}. Again, this is a small energy difference that is not likely to play a significant role in water oxidation. Since the deprotonation is uphill, the next transition from S_2 to S_3 will start by an endergonic step of $+1.6$ kcal mol^{-1}, which is driven by the subsequent exergonic oxidation of -8.1 kcal mol^{-1}. The details of how this occurs is described below, where the effect of Tyr_z is discussed. There is an uncertainty whether a water enters the OEC in this transition, as in the diagram, or in the next transition. The calculations give a very weak binding of a water close to Mn1 in S_2.

All oxidations prior to S_3 have been from Mn(III) to Mn(IV). At this stage all manganese atoms are in the oxidation state IV, and the next oxidation will therefore be harder. In fact, this oxidation will be on a terminal oxygen, which then becomes a radical as it should be for the O–O bond formation mechanism discussed above. Also, the S_3 to S_4 transition will start with an endergonic deprotonation step, in this case by +2.1 kcal mol^{-1}. The subsequent oxidation of the oxygen is endergonic by +3.6 kcal mol^{-1}. Once at the S_4 state, the O–O bond formation occurs as described above with a barrier of +7.7 kcal mol^{-1}. Altogether, the barrier counted from S_3 will therefore be +13.4 kcal mol^{-1}, in agreement with a rate in milliseconds. The formation of the peroxide is exergonic by −2.8 kcal mol^{-1}. The step of O_2 release and water binding is exergonic by −7.3 kcal mol^{-1}. The final step back to S_0, completing the cycle, is a deprotonation step, exergonic by −10.9 kcal mol^{-1}. The TS for O_2 release is not shown, but a detailed mechanism has been described with a barrier similar to the one for O–O bond formation.[50]

The actual values for the redox energies and pK_a values can easily be read directly from the diagram in the following way. The reference redox potential is the one of P_{680}^+ of 1.25 V. In the first transition, the exergonicity of −13.4 kcal mol^{-1} is therefore translated to a redox potential for the OEC of $1.25 - (13.4/23.063) = 0.67$ V, using the transformation factor between kcal/mol and eV of 23.063. For the deprotonation events the reference pK_a value is that of bulk water of 7. For the next step of proton release the exergonicity of −0.7 kcal mol^{-1} will therefore correspond to a pK_a of $7 - (0.7/1.36) = 6.5$ using the transformation factor of 1.36 between kcal/mol and pK_a units.

It should finally be remarked that in one of the more recent papers,[38] the driving force for the last two transitions was increased by 5 kcal mol^{-1} in each one, leading to a total driving force of 51.5 kcal mol^{-1} instead of 41.5 kcal mol^{-1}, as used here. The reasoning behind the increase of 5 kcal mol^{-1}, has recently been realized to be incorrect. The present diagram is therefore a better representation of the energetics.

11.3.4 The Effects of Tyr$_z$

It has long been known that Tyr$_z$ is oxidized by P_{680}^+ in each transition.[60] Since Tyr$_z$ is situated in between P_{680} and the OEC this allows the OEC to be further away from the reaction center and still have a fast transfer of the electron. If the OEC had been closer to the reaction center, this would have led to destructive reduction of the OEC. In previous papers, Tyr$_z$ has not been included in the energy diagrams, but it is now included in Figures 11.10 and 11.11.

An important feature of including Tyr$_z$ in the diagram, is that the oxidation of the OEC will occur in two steps. Not to lose too much energy in the first step when Tyr$_z$ is oxidized, a redox potential as close to P_{680}^+ as possible may seem optimal. In most of the previous studies no loss of energy in this electron transfer was therefore assumed. However, this argument

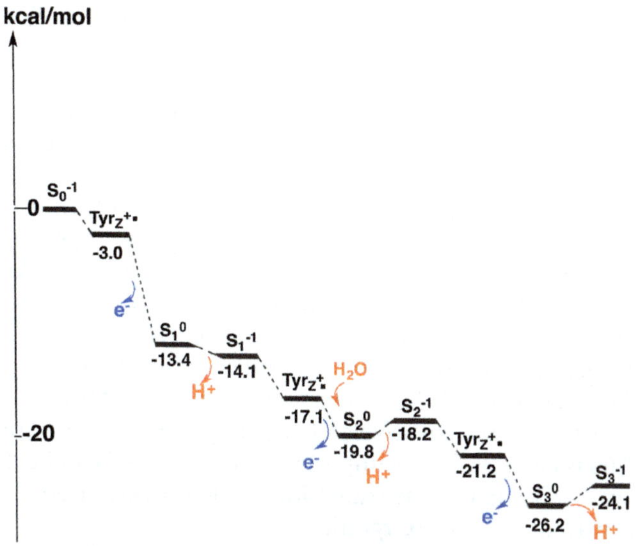

Figure 11.10 Energy diagram including Tyr_z for water oxidation from S_0 to S_3.

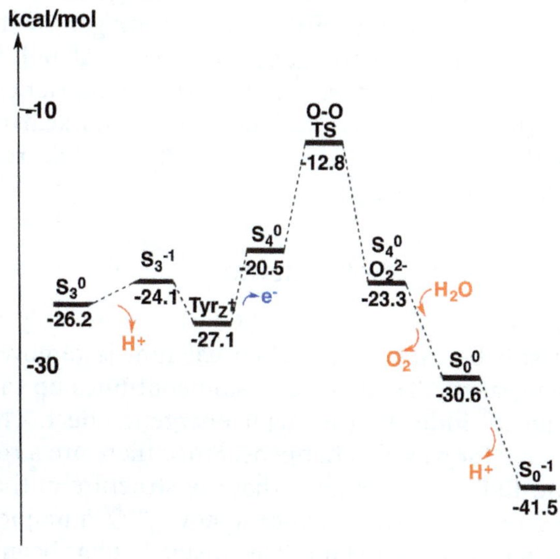

Figure 11.11 Energy diagram including Tyr_z for water oxidation from S_3 to S_0.

does not account for the effect on the charge separation in the reaction center. With a loss of energy, the charge separation would be better secured. In the present diagrams it will be assumed that there is a loss of 3 kcal mol^{-1} in the electron transfer from Tyr_z to P_{680}^+. To obtain a sufficiently high redox potential of Tyr_z for the oxidation of the OEC, there is a hydrogen bonding histidine which will keep the proton released from Tyr_z as close as possible.

The energy diagram including the Tyr_z levels from S_0 to S_3 is shown in Figure 11.10. The first step from P_{680} to Tyr_z will be very fast since the reorganisation energy is small, which is important for preserving the charge separation in the reaction center. In comparison, the next step with the OEC oxidation will be very slow since there will be a slow proton release (in most cases) and a larger reorganisation energy involved.

The transition from S_0 to S_1 is not affected in a major way by the presence of Tyr_z. The oxidation of the OEC is still strongly exergonic. In the next transition from S_1 to S_2 the effect is larger with a reduction of the exergonicity for OEC oxidation by about 50%. In S_2 to S_3 an interesting situation appears. As noted above, the initial proton transfer is endergonic by +1.6 kcal mol^{-1}. However, as Tyr_z is oxidized the proton release becomes exergonic by −1.4 kcal mol^{-1}. This has been experimentally interpreted as a repulsive effect from the positively charged Tyr_z-His$^+$ on the proton being released. This is correct but the importance of this effect can be debated. It is clear that even without Tyr_z the transition to the fully oxidized S_3 state would proceed anyway, since it is exergonic.

The beginning of the S_3 to S_4 transition is quite similar to the previous transition, see Figure 11.11. Proton release is initially endergonic but is driven by Tyr_z oxidation in an overall exergonic proton release of −0.9 kcal mol^{-1}. The oxidation of the terminal oxygen in the OEC is with the effect of Tyr_z included endergonic by +6.6 kcal mol^{-1}, while it was +5.7 kcal mol^{-1} before. This means that the overall barrier counted from S_3 will increase slightly from +13.4 kcal mol^{-1} to +14.3 kcal mol^{-1}. A higher redox potential than the one assumed here for Tyr_z will decrease the barrier, while a lower one will increase it.

11.3.5 The Importance of Calcium

Calcium is an essential part of the OEC, which can only be replaced by strontium and still be functional. When calcium is removed, it is experimentally known that the S-state progression continues up to and including S_2, but then halts,[55] indicating a major energetic effect. There have been several suggestions for why this happens. Since there are strong indications from EXAFS and EPR measurements that the structure of the OEC remains essentially the same at S_2 after calcium removal,[61,62] a major structural role for calcium has been considered unlikely. Instead, it has been suggested that calcium could bind a substrate water, making the water more nucleophilic for an attack on an electrophilic oxo-group bound to Mn(v), in the process of making the O–O bond.[51] It has also been suggested that calcium should have an effect on the H-bond network between Y_z and the manganese cluster.[62] This is a reasonable explanation for the minor effect (less than one kcal/mol on the barrier) found from calcium substitution by strontium,[63] but an effect that entirely stops the oxidation process at S_2 should require an effect at least an order of magnitude larger and therefore most probably would need a different explanation.

If calcium is removed from the OEC and nothing else is changed, there would obviously be huge effects on the redox and pK_a values due to the change of charge. However, this is a very unlikely scenario in reality. It is known from many other enzymes where metals have been removed that the loss of positive charge is compensated by adding protons. A realistic investigation of the effect of removing calcium therefore has to use an OEC-cluster with two additional protons.[64] The three structures obtained from DFT optimizations of S_1 to S_3 are shown in Figure 11.12. The best positions for the added protons were found to be one on the oxo-bridge between Mn2 and Mn3, and the second one on the previous hydroxide group on Mn4, see the figure. The oxo bridge was previously connected to calcium and is therefore a likely position for a proton. As in the natural process with calcium, the S_1 state has two Mn(III) and two Mn(IV). The Mn(III) ones are Mn1 and Mn4 just like in the OEC. The proton removed in going to S_2 is the one situated on an OH-group on Mn3 marked in the figure. The energy required

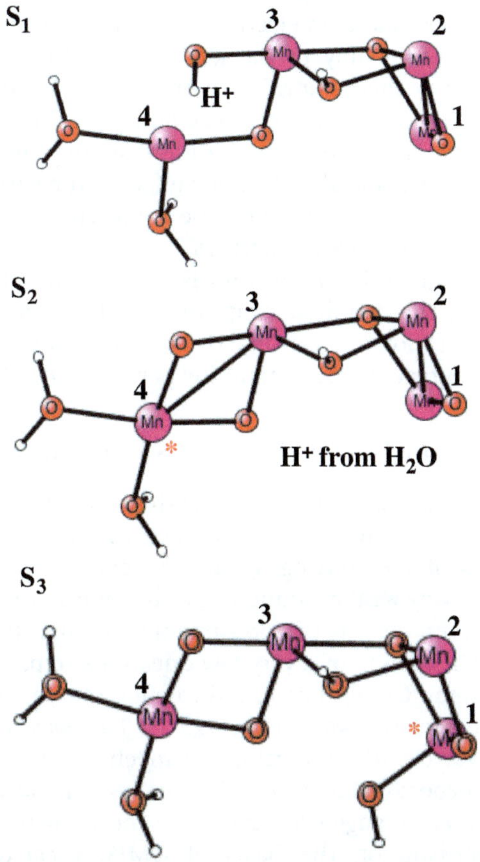

Figure 11.12 The optimal S_1 to S_3 structures obtained when calcium has been replaced by two protons. The proton removed in each transition is shown as well as the Mn-atom oxidized (asterisk) in the transition.

to remove this proton is only 2.8 kcal mol^{-1} larger than in the unmodified OEC, and should therefore occur easily. Also oxidizing the calcium depleted cluster to reach S_2 was found to be easy, actually 0.4 kcal mol^{-1} easier than for the natural OEC. The manganese oxidized is Mn4, again like in the OEC. So far, the model calculations agree with experiments that S_2 should be reached, but does not give any clue of the importance of calcium.

As mentioned above, the step from S_2 to S_3 does not occur experimentally when calcium is removed. The calculations show that the proton easiest to remove in S_2 is the one on the water that is added to Mn1 in this transition, see the figure. It turns out that this proton is easier by as much as 16.5 kcal mol^{-1} than for the natural system. Therefore, the OEC without calcium should easily reach this point in the oxidation process. However, the next electron is much harder to remove than in the natural system, by 17.3 kcal mol^{-1}. This means a substantial increase in the redox potential of 0.75 V and therefore gives a reasonable explanation for why the OEC without calcium does not reach S_3. As a point of interest, the energy required for the removal of both a proton and an electron from S_2 to S_3 is quite similar to the natural system, differing only by 0.8 kcal mol^{-1}. This is not unexpected, since the total effect of this removal is to break an O–H bond in water, with the cluster having the same charge before and after the removal. A similar removal of an (H^+, e^-)-couple from S_1 differs slightly more to the natural system but still by only 2.4 kcal mol^{-1}. The effect of the surrounding enzyme on the bond strength should therefore be minimal, which is also true in general for all bond strengths in enzymes.

In summary, a major role for calcium is thus to tune the pK_a and redox values for allowing the required oxidations. A large effect on the redox potential by incorporation of redox-inactive ions in model complexes similar to the OEC has also recently been found experimentally.[65]

11.3.6 Water Insertion in the S_2 to S_3 Transition

There is at present consensus about the fact that a water molecule is inserted during the S_2 to S_3 transition. There is also essential agreement on the detailed structures of the starting S_2 and the ending S_3 states. The water molecule is at most very weakly bound in S_2 while it is bound as an hydroxide to Mn1 in S_3. However, there is still no agreement on how the water insertion occurs in detail. This is one of a very few questions remaining in the water oxidation mechanism. Guidoni *et al.* did calculations to suggest a so called "pivot" mechanism for the water binding.[66,67] This was followed by a DFT study by Perez-Navarro *et al.* supporting this mechanism and suggesting that an intermediate 5-coordinated Mn(IV) is involved in the process.[68] In the meantime Askerka *et al.* suggested the same mechanism, now termed the "carousel" mechanism, on the basis of QM/MM calculations.[69] These mechanisms are all basically based on the Guidoni *et al.* study.

In the originally suggested mechanism for water insertion, water binds directly to the open site of Mn1 and then loses a proton in a few steps.[38] The problem with this simple mechanism became clear when a mutation experiment was done involving Val185.[70] It was shown that this valine could become a hindrance for water insertion. Since this amino acid was not included in the original DFT studies, new investigations were made, showing that Val185 indeed was in the way for a simple water binding to Mn1.[50] The calculations showed that for a water to reach Mn1, the valine needed to rotate with a rather high but not too high barrier. To study this mechanism more in detail much larger models are needed, and that type of calculation has still not been done. An advantage with this new version of the mechanism, requiring valine rotation, was that exchange of this water in S_2 became much slower. Since the same water was suggested to be a substrate water, it now led to agreement for this mechanism with water exchange studies that showed that the fast exchanging substrate water exchanges only slowly in S_2.[71,72]

In the Guidoni *et al.* mechanism, water instead binds to the outer Mn (Mn4). In order to make this possible, water binds to an excited structure of S_2, earlier termed the closed (or inner) form of S_2,[6,73] which has an open site on Mn4 instead of Mn1. After binding to Mn4, there is a circular structural change around Mn4, which was termed the "pivot" (or "carousel") mechanism, to reach a position that could eventually lead to a binding of an hydroxide on Mn1. To support this mechanism, Guidoni *et al.* performed DFT and dynamics calculations and constructed an energy diagram for the process. An immediate problem with the results is that the process was predicted to be exergonic by about 15 kcal mol^{-1}. Since this is much more than the exergonicity of the entire S_2 to S_3 transition, an alternative interpretation of the calculated results can be suggested. In that case the starting point for the process should be at an excited level of +10 kcal mol^{-1} or more above S_2. This is not unreasonable since the starting point is a 5-coordinated Mn(IV) complex,[68] which is usually rather unstable. However, this would then lead to a too high barrier from S_2 of +22 kcal mol^{-1} for the mechanism since the computed barrier was 12 kcal mol^{-1} from the starting point. The solution to the problem of the water insertion process is therefore still an open question.

11.4 Summary

During the past ten years, a very good understanding of the details of water oxidation in PSII has been reached. The first X-ray structures (2004–2005) of the enzyme made a big step towards this goal.[2,3] However, these structures were only at a low resolution, which meant that much work was still required after that. In this context theoretical modeling has played a big role. By changing the X-ray structures at key points and testing a large number of alternatives, theory could suggest both mechanism (in 2006) and structures (2006–2008) for the intermediates in the process. The first high-resolution X-ray structure appeared in 2011 and meant another big step forward.[4] This structure is in very good detailed agreement with the theoretically suggested structures, confirming the changes from the low-resolution structures that had been suggested. In the coming years, spectroscopy using mainly EPR and water exchange information could also confirm the details of the structures for S_2 and S_3.[8,9] Furthermore, these experiments could also confirm key components of the theoretically suggested mechanism, such as the controversial suggestion of a participation of a bridging oxo-group in the O–O bond formation step.

Other essential features of the water oxidation mechanism have also been well understood by now, such as the reason for the presence of a calcium in the OEC and of a redox active tyrosine (Tyr_z) in between the OEC and the oxidant P_{680}. There are obviously minor remaining questions, in particular how a water is inserted in the S_2 to S_3 transition. In this case the discovery that a valine (Val185) plays a significant role in this process[70] is important. A much larger model than so far used, including this valine and its surrounding, will be necessary to fully understand the water insertion process.

With the detailed understanding of water oxidation in PSII, the next major goal is to understand how to model this chemistry in the laboratory. Several big steps have already been taken in this direction, but much remains. There is no doubt that the detailed knowledge of PSII will be very helpful in this process.

Acknowledgements

This work was generously supported by the Knut and Alice Wallenberg Foundation and by grants from the Swedish research council. Computer time was provided by the Swedish National Infrastructure for Computing.

References

1. B. Kok, B. Forbush and M. McGloin, *Photochem. Photobiol.*, 1970, **11**, 457.
2. K. N. Ferreira, T. M. Iverson, K. Maghlaoui, J. Barber and S. Iwata, *Science*, 2004, **303**, 1831–1838.
3. B. Loll, J. Kern, W. Saenger, A. Zouni and J. Biesiadka, *Nature*, 2005, **438**, 1040–1044.
4. Y. Umena, K. Kawakami, J.-R. Shen and N. Kamiya, *Nature*, 2011, **473**, 55–60; M. Suga, F. Akita, K. Hirata, G. Ueno, H. Murakami, Y. Nakajima, T. Shimizu, K. Yamashita, M. Yamamoto, H. Ago and J.-R. Shen, *Nature*, 2015, **517**, 99–103.
5. P. E. M. Siegbahn, *Chem. Eur. J.*, 2008, **27**, 8290–8302.
6. P. E. M. Siegbahn, *Acc. Chem. Res.*, 2009, **42**, 1871–1880.
7. P. E. M. Siegbahn, *Chem. Phys. Chem.*, 2011, **12**, 3274–3280.
8. L. Rapatskiy, N. Cox, A. Savitsky, W. M. Ames, J. Sander, M. M. Nowacyzk, M. Rögner, A. Boussac, F. Neese, J. Messinger and W. Lubitz, *J. Am. Chem. Soc.*, 2012, **134**, 16619–16634.
9. N. Cox, M. Retegan, F. Neese, D. A. Pantazis, A. Boussac and W. Lubitz, *Science*, 2014, **345**, 804–808.

10. I. D. Young *et al.*, *Nature*, 2016, **540**, 7633.
11. P. E. M. Siegbahn, *Chem. Eur. J.*, 2006, **12**, 9217–9227.
12. (a) P. Gatt, R. Stranger and R. J. Pace, *J. Photochem. Photobiol. B: Biol.*, 2011, **104**, 80–93; (b) S. Petrie, P. Gatt, R. Stranger and R. J. Pace, *Phys. Chem. Chem. Phys.*, 2012, **14**, 11333–11343.
13. B. Brena, P. E. M. Siegbahn and H. J. Ågren, *Am. Chem. Soc.*, 2012, **134**, 17157.
14. V. Krewald, M. Retegan, N. Cox, J. Messinger, W. Lubitz, S. DeBeer, F. Neese and D. A. Pantazis, *Chem. Sci.*, 2015, **6**, 1676–1695.
15. M. Kusunoki, *Biochim. Biophys. Acta*, 2007, **1767**, 484–492.
16. M. Kusunoki, *J. Photochem. Photobiol. B: Biol.*, 2011, **104**, 100–110.
17. E. M. Sproviero, M. B. Newcomer, J. A. Gascon, E. R. Batista, G. W. Brudvig and V. S. Batista, *Photosynth. Res.*, 2009, **102**, 455–470.
18. E. M. Sproviero, J. A. Gascon, J. P. McEvoy, G. W. Brudvig and V. S. Batista, *J. Am. Chem. Soc.*, 2008, **130**, 3428–3442.
19. E. M. Sproviero, J. A. Gascon, J. P. McEvoy, G. W. Brudvig and V. S. Batista, *J. Am. Chem. Soc.*, 2008, **130**, 6728–6730.
20. X. Li, E. M. Sproviero, U. Ryde, V. S. Batista and G. Chen, *Inter. J. Quantum Chem.*, 2013, **113**, 474.
21. P. E. M. Siegbahn and R. H. Crabtree, *J. Am. Chem. Soc.*, 1999, **121**, 117–127.
22. P. E. M. Siegbahn, *Inorg. Chem.*, 2000, **39**, 2923–2935.
23. M. Lundberg and P. E. M. Siegbahn, *Phys. Chem. Chem. Phys.*, 2004, **6**, 4772–4780.
24. P. E. M. Siegbahn and M. Lundberg, *Photochem. Photobiol. Sci.*, 2005, **4**, 1035–1043.
25. P. E. M. Siegbahn and M. Lundberg, *J. Inorg. Biochem.*, 2006, **100**, 1035–1040.
26. P. E. M. Siegbahn, *J. Am. Chem. Soc.*, 2009, **131**, 18238–18239.
27. J. Messinger, *Phys. Chem. Phys. Chem.*, 2004, **6**, 4764–4771.
28. J. Yano, J. Kern, K. Sauer, M. J. Latimer, Y. Pushkar, J. Biesiadka, B. Loll, W. Saenger, J. Messinger, A. Zouni and V. K. Yachandra, *Science*, 2006, **314**, 821–825.
29. H. Dau, A. Grundmeier, P. Loja and M. Haumann, *Phil. Trans. R. Soc. B*, 2008, **363**, 1237–1244.
30. N. Cox, L. Rapatskiy, J.-H. Su, D. A. Pantazis, M. Sugiura, L. Kulik, P. Dorlet, A. W. Rutherford, F. Neese, A. Boussac, W. Lubitz and J. Messinger, *J. Am. Chem. Soc.*, 2011, **133**, 3635–3648.
31. S. Schinzel, J. Schraut, A. V. Arbuznikov, P. E. M. Siegbahn and M. Kaupp, *Chemistry Eur. J.*, 2010, **16**, 10424–10438.
32. I. Rivalta, M. Amin, S. Luber, S. Vassiliev, R. Pokhrel, Y. Umena, K. Kawakami, J.-R. Shen, N. Kamiya, D. Bruce, G. W. Brudvig, M. R. Gunner and V. S. Batista, *Biochemistry*, 2011, **50**, 6312–6315.
33. S. Luber, I. Rivalta, Y. Umena, K. Kawakami, J.-R. Shen, N. Kamiya, D. Bruce, G. W. Brudvig and V. S. Batista, *Biochemistry*, 2011, **50**, 6308–6311.
34. W. Ames, D. A. Pantazis, V. Krewald, N. Cox, J. Messinger, W. Lubitz and F. Neese, *J. Am. Chem. Soc.*, 2011, **133**, 19743–19757.

35. A. Grundmeier and H. Dau, *Biochim. Biophys. Acta*, 2012, **1817**, 88–105.
36. A. Galstyan, A. Robertazzi and E. W. Knapp, *J. Am. Chem. Soc.*, 2012, **134**, 7442–7449.
37. H. Isobe, M. Shoji, J.-R. Shen and K. Yamaguchi, *Inorg. Chem.*, 2015, **55**, 502–511.
38. P. E. M. Siegbahn, *Biochim. Biophys. Acta*, 2013, **1827**, 1003–1019.
39. A. D. Becke, *J. Chem. Phys.*, 1993, **98**, 5648–5652.
40. P. E. M. Siegbahn, *J. Biol. Inorg. Chem.*, 2006, **11**, 695–701.
41. P. E. M. Siegbahn and F. Himo, *J. Biol. Inorg. Chem.*, 2009, **14**, 643–651.
42. P. E. M. Siegbahn; F. Himo *Wire's Computational Molecular Science*, John Wiley & Sons, Ltd. 2011, vol. 1, 323–336.
43. S. Grimme, *J. Chem. Phys.*, 2006, **124**, 034108; T. Schwabe and S. Grimme, *Phys. Chem. Chem. Phys.*, 2007, **9**, 3397–3406.
44. L. Noodleman and D. A. Case, *Adv. Inorg. Chem.*, 1992, **38**, 423–470.
45. Jaguar 5.5, Schrödinger, L. L. C., Portland, OR, (1991–2003).
46. J. Yano, J. Kern, K. Sauer, M. J. Latimer, Y. Pushkar, J. Biesiadka, B. Loll, W. Saenger, J. Messinger, A. Zouni and V. K. Yachandra, *Science*, 2006, **314**, 821–825; J. Yano, J. Kern, Y. Pushkar, K. Sauer, P. Glatzel, U. Bergmann, J. Messinger, A. Zouni and V. K. Yachandra, *Phil. Trans. R. Soc. B*, 2008, **363**, 1139–1147.
47. P. E. M. Siegbahn, *J. Photochem. Photobiol. B: Biol.*, 2011, **104**, 94–99.
48. X. Li and P. E. M. Siegbahn, *J. Am. Chem. Soc.*, 2013, **135**, 13804–13813.
49. N. Cox and J. Messinger, *Biochim. Biophys. Acta*, 2013, **1827**, 1020–1030.
50. X. Li and P. E. M. Siegbahn, *Phys. Chem. Chem. Phys.*, 2015, **17**, 12168–12174.
51. E. M. Sproviero, J. A. Gascon, J. P. McEvoy, G. W. Brudvig and V. S. Batista, *J. Am. Chem. Soc.*, 2008, **130**, 3428–3442; E. M. Sproviero, J. A. Gascon, J. P. McEvoy, G. W. Brudvig and V. S. Batista, *J. Am. Chem. Soc.*, 2008, **130**, 6728–6730; J. P. McEvoy and G. W. Brudvig, *Chem. Rev.*, 2006, **106**, 4455–4483.
52. P. E. M. Siegbahn, *Proc. Natl. Acad. Sci. U. S. A.*, 2017, **114**, 4966–4968.
53. B. A. Diner, *Biochim. Biophys. Acta*, 2001, **1503**, 147–163.
54. F. Rappaport and J. Lavergne, *Biochim. Biophys. Acta*, 2001, **1503**, 246–259.
55. A. Boussac and A. W. Rutherford, *Biochemistry*, 1988, **27**, 3476–3483.
56. G. Renger, *Physiol. Plantarum*, 1997, **100**, 828–841.
57. V. Förster and W. Junge, *Photochem. Photobiol.*, 1985, **41**, 183–190.
58. F. Rappaport, M. Blanchard-Desce and J. Lavergne, *Biochim. Biophys. Acta*, 1994, **1184**, 178–192.
59. M. Haumann, P. Liebisch, C. Müller, M. Barra, M. Grabolle and H. Dau, *Science*, 2005, **310**, 1019–1021.
60. B. A. Barry and G. T. Babcock, *Proc. Natl. Acad. Sci. U. S. A.*, 1987, **84**, 7099–7103.
61. M. J. Latimer, V. J. DeRose, V. K. Yachandra, K. Sauer and M. P. Klein, *J. Phys. Chem. B*, 1998, **102**, 8257–8265; V. K. Yachandra and J. Yano, *J. Photochem. Photobiol. B: Biol.*, 2011, **104**, 51–59.

62. T. Lohmiller, N. Cox, J.-H. Su, J. Messinger and W. Lubitz, *J. Biol. Chem.*, 2012, **287**, 24721–24733.
63. F. Rappaport, N. Ishida, M. Sugiura and A. Boussac, *Energy Environ. Sci.*, 2011, **4**, 2520–2524.
64. P. E. M. Siegbahn, *Phys. Chem. Chem. Phys.*, 2014, **16**, 11893–11900.
65. E. Y. Tsui, J. S. Kanady and T. Agapie, *Inorg. Chem.*, 2013, **52**, 13833–13848.
66. D. Narzi, D. Bovi and L. Guidoni, *Proc. Natl. Acad. Sci. U. S. A.*, 2014, **111**, 8723–8728.
67. M. Capone, D. Bovi, D. Narzi and L. Guidoni, *Biochemistry*, 2015, **54**, 6439–6442.
68. M. Perez-Navarro, F. Neese, W. Lubitz, D. A. Pantazis and N. Cox, *Current Opinion Chem. Biol.*, 2016, **31**, 113–119.
69. M. Askerka, J. Wang, D. J. Vinyard, G. W. Brudvig and V. S. Batista, *Biochemistry*, 2016, **55**, 981–984.
70. P. L. Dilbeck, H. Bao, C. L. Neveu and L. Burnap, *Biochemistry*, 2013, **52**, 6824–6833.
71. W. Hillier and T. Wydrzynski, *Coord. Chem. Rev.*, 2008, **252**, 306–317.
72. H. Nilsson, T. Krupnik, J. Kargul and J. Messinger, *Biochim. Biophys. Acta*, 2014, **1837**, 1257–1262.
73. D. A. Pantazis, W. Ames, N. Cox, W. Lubitz and F. Neese, *Angew. Chem. Int. Ed.*, 2012, **51**, 9935–9940.

CHAPTER 12

Respiratory Supercomplexes in Mitochondria

GIORGIO LENAZ,*[a] GAIA TIOLI,[a] ANNA IDA FALASCA[b] AND MARIA LUISA GENOVA[a]

[a] Department of Biomedical and Neuromotor Sciences, Alma Mater Studiorum, University of Bologna, Via Irnerio 48, 40126 Bologna, Italy; [b] Department of Food and Drug, University of Parma, Parco Area delle Scienze 27/A, 43124 Parma, Italy
*Email: giorgio.lenaz@unibo.it

12.1 Introduction

12.1.1 The Respiratory Chain of Mitochondria

The major mechanism of energy conservation in eukaryotes is oxidative phosphorylation (OXPHOS), performed by a multi-enzyme system embedded in the inner mitochondrial membrane (IMM) and constituted by two portions: the electron transfer chain and the ATP synthase complex.

The *electron transfer chain* or *respiratory chain* is a series of enzymes that collects reducing equivalents (hydrogen atoms) released from oxidations of intermediary metabolism, mostly from oxidations occurring in the mitochondrial matrix, with some exceptions, and conveys them to molecular oxygen, reducing it to water in a four electron reduction. The free energy decrease accompanying electron transfer is exploited to create an electrochemical proton gradient ($\Delta\tilde{\mu}_{H^+}$) by proton translocation from the mitochondrial inner space, the matrix, to the space existing between the inner and outer mitochondrial membranes.[1] The proton gradient is then largely

Chemical Biology No. 5
Mechanisms of Primary Energy Transduction in Biology
Edited by Mårten Wikström

Published by the Royal Society of Chemistry, www.rsc.org

used as a source of energy to synthesise ATP from ADP and Pi by the ATP synthase complex. The ATP synthesised is moved towards the cytoplasm in exchange with ADP by the ATP/ADP translocase at further expense of the proton gradient. A model of the electron transfer chain was first described by Hatefi *et al.*[2] in David Green's laboratory at the University of Wisconsin (*cf.*, ref. 3 for a comprehensive review) as what is nowadays considered the "classical model".

Quoted from Hatefi *et al.*:[2]

> *"We have been able to isolate from mitochondria, isolated from beef heart, four primary enzyme complexes that are operative in the oxidation of succinate and [NADH]. .... These four primary enzyme systems, and certain combinations forming secondary complexes, ... are described below"* – namely NADH-Coenzyme Q reductase (Complex I, CI), succinate-Coenzyme Q reductase (Complex II, CII), ubiquinol-cytochrome *c* reductase (Complex III, CIII or cytochrome bc1 Complex) and cytochrome *c* oxidase (Complex IV, CIV).

These enzyme complexes are connected by two mobile redox-active molecules, *i.e.* Coenzyme Q (CoQ or ubiquinone), a lipophilic quinone inserted in the membrane lipid bilayer, and cytochrome *c* (cyt. *c*) a hydrophilic heme protein confined to the water phase of the mitochondrial intermembrane space, in close contact with the external surface of the inner membrane. Such a separation prevents energy-wasting direct electron exchange between the two pools of substrates despite a significant difference in redox midpoint potential of quinone and cyt. *c*. This is critically important for the efficiency of mitochondrial respiration and the energy-conserving process.[4] In fact, the respiratory chain functions with the operation of the respiratory enzyme complexes in the following sequence of redox reactions between the partner components:

$$\text{NADH} \rightarrow \text{CI} \rightarrow \text{CoQ} \rightarrow \text{CIII} \rightarrow \text{cyt. } c \rightarrow \text{CIV} \rightarrow \text{O}_2 \tag{1}$$

or

$$\text{succinate} \rightarrow \text{CII} \rightarrow \text{CoQ} \rightarrow \text{CIII} \rightarrow \text{cyt. } c \rightarrow \text{CIV} \rightarrow \text{O}_2 \tag{2}$$

The best fit unit stoichiometry between complexes in beef heart mitochondria[5] is: 1 CI: 1.3 CII: 3 CIII: 6.7 CIV.

Nevertheless, large differences in cytochrome, CoQ, and pyridine nucleotide contents in mitochondria from different species, as well as from different organs and tissues of the same species, have been reported.[6]

It is worth mentioning that the IMM contains other proteins having electron transfer activity,[6] among which are electron transfer flavoproteins capable of feeding electrons to the respiratory chain by pathways converging on CoQ while not involving CI and NADH. For example: glycerol-3-phosphate dehydrogenase involved in a shuttle of reducing equivalents from cytosol to mitochondria;[7] electron transfer flavoprotein (ETF) dehydrogenase, involved in fatty acid oxidation;[8] dihydroorotate dehydrogenase, involved in pyrimidine nucleotides biosynthesis;[9] choline dehydrogenase, important for regulation of choline and phospholipid metabolism[10] sulphide dehydrogenase involved in the disposal of sulphide.[11] In addition, a series of alternative NADH dehydrogenases exist in mitochondria from several organisms, especially plants and fungi. Alternative or branched pathways of electron transfer also occur, departing from CoQ: these are the alternative ubiquinol oxidases from bacteria and from plant and fungi mitochondria. Finally, some dehydrogenases feed electrons directly to cyt. *c* in plant mitochondria.[12] Nevertheless, sulphite oxidase, involved in sulphur metabolism, is the only enzyme capable to feed electrons directly to cyt. *c* in mammalian mitochondria.[13]

12.1.2 Organization of the Respiratory Chain: Historical Outline

As we mentioned in the previous section, Hatefi *et al.*[2] in David Green's laboratory in Wisconsin accomplished the systematic resolution and reconstitution of four respiratory complexes from mitochondria, thus leading Green[3] to postulate that the overall respiratory activity is the result of both intra-complex electron transfer in the 'solid' state of redox components (*e.g.* flavins, FeS clusters, cytochromes) having fixed steric relation and, in addition, of inter-complex electron transfer ensured by rapid diffusion of the mobile components acting as co-substrates, *i.e.* CoQ and cyt. *c*. This proposal was substantially confirmed over the following years, thanks to the basic kinetic analysis of Kröger and Klingenberg,[14,15] and led Hackenbrock *et al.*[16] to postulate the *Random Collision Model of Electron Transfer*:

> *"Electron transport is a diffusion-coupled kinetic process; Electron transport is a multicollisional, obstructed, long-range diffusional process; The rates of diffusion of the redox components have a direct influence on the overall kinetic process of electron transport and can be rate limiting, as in diffusion control.... It is concluded that mitochondrial electron transport is a diffusion-based random collision process and that diffusion has an integral and controlling effect on electron transport."*

The organization of the respiratory chain represented a major research subject in the 1970–1980s, culminating with acceptation of the random collision model by the majority of investigators in the field.

The accumulation of recent experimental evidence obtained with newly developed techniques has led to the proposal of a different model of supramolecular organisation based upon specific, though dynamic, interactions between individual components, and leading to the acquisition of new properties (substrate channelling, assembly, morphological organisation) that were unpredictable in the previous reductionist approach.

A retrospective analysis of the literature reveals however that the idea of supramolecular associations between respiratory enzymes had been present since the early times, dating from the pioneering studies of Chance and Williams,[17] who depicted the respiratory chain as a solid-state assembly of flavins and cytochromes in a protein matrix. Though aware that respiration relies on enzymes, they stated:

> *"Based upon the components that we actually observe to function in the intact mitochondria, we formulate the following respiratory chain for substrate oxidation:*
>
> $$O_2 \rightarrow a_3 \rightarrow a \rightarrow c \rightarrow b \rightarrow fp \rightarrow DPN \rightarrow substrate"$$

Evidence against a random distribution of respiratory complexes also derived from the same early investigations in Green's laboratory, reporting isolation of CI + CIII units,[18] indicating that such units may be preferentially associated in the native membrane. The authors of these studies were well aware that these preferential associations were physiological and indicated the existence of supramolecular units of electron transfer. Hatefi *et al.*[2] in their paper on the reconstitution of the electron transfer system stated that:

> *"the four primary enzyme systems may be suitably combined to form secondary units capable of catalysing the sums of the reactions catalysed by the respective subunits".*

A careful examination of these pioneering papers allows one to conclude that respiratory supercomplexes (SCs), as well as the entire DPNH oxidase (the *respirasome,* see later) had been isolated and reconstituted. The appearance of the fluid mosaic model of membranes[19] strongly influenced the researchers involved in the study of mitochondrial membranes, and this is probably the reason why the random collision model of Hackenbrock was so well and uncritically accepted, obscuring in such way Hatefi's concept of intercomplex specific associations.

There were, however, a few reports before the year 2000 on the possible presence of specific associations between respiratory complexes, either fixed[20] or dynamic.[21]

A breakthrough occurred in 2000 when Schägger applied the previously introduced technique of Blue-Native Polyacrylamide Gel Electrophoresis (BN-PAGE) to digitonin-solubilized yeast and mammalian mitochondria.[22] The newly discovered associations were considered to represent the physiological state of the respiratory complexes. In the same paper the authors also described a dimeric state for the ATP-synthase complex.

Today the structural evidence of supercomplex association of the respiratory chain is well consolidated,[23–25] although its functional role is still debated and a possible relation with a random distribution of the individual complexes is not completely clarified.[26–29]

12.2 Distribution and Composition of Respiratory Supercomplexes

All the SCs investigated exhibit highly ordered architectures, thus discarding most doubts on artificial protein–protein interactions, and supporting the idea that such interactions may also be species- or kingdom-specific.[30]

Most evidence presently favours the view that SCs are formed by the three "core" complexes of the respiratory chain, *i.e.* complexes I, III, and IV, whereas the other respiratory enzymes may be free in the lipid bilayer.[6] These "core" complexes have the common feature of carrying out proton translocation and of having subunits encoded by mitochondrial DNA.

12.2.1 Distribution in Different Organisms

Blue-Native Polyacrylamide Gel Electrophoresis and the closely related, milder, colourless native PAGE (CN-PAGE) are the preferred experimental strategy for the detection of the respiratory SCs from various biological samples and for the analysis of their protein composition.[31,32]

Respiratory SCs have been found in a wide range of living organisms; they were first discovered in yeast and mammalian mitochondria[5,22] and subsequently revealed and characterised in several mammalian tissues and in many other organisms, also including plants, fungi and bacteria (Table 12.1).

Although the classical SCs containing CI are absent in most bacteria, probably due to absence in CI of the supernumerary subunits that mediate the contacts between CI and CIII,[33] SCs formed by CIII/IV are often present in these organisms. The exception of SC I + III + IV of *P. denitrificans*[34] is explained, differently from other bacteria, with the presence in CI of three supernumerary subunits.[35]

12.2.2 Composition of Respiratory Supercomplexes

The fundamental features of the supramolecular structural organisation of the standard respiratory complexes I, III, and IV are conserved in all higher eukaryotes. The SC $I_1III_2IV_{1-4}$ contains all the redox enzymes required for the complete pathway of electron transfer from NADH to molecular oxygen, and for this reason it was called "*respirasome*". However, SCs containing only two partners are also present in detectable quantities, as the SC I_1III_2 in which CIV is not present, as well as respiratory assemblies comprising only CIII and CIV. In particular, BN-PAGE in digitonin-solubilized mitochondria of *S. cerevisiae*, which possesses no CI, revealed two bands with apparent masses of ~750 and 1000 kDa containing the subunits of complexes III and IV, as assigned after two dimensional SDS-PAGE followed by N-terminal protein sequencing.[22] The smaller SC (III_2IV_1) consisted of a CIII dimer and a CIV monomer while the larger SC (III_2IV_2) is comprised of a CIII dimer associated with two CIV monomers.[36–38]

Table 12.1 Overview of the literature about the supercomplex organization of the respiratory chain.

Study	Ref.	Species/tissue	Type of supercomplex[a]
Schägger 2002	168	Bacteria	$I_1III_4IV_4$
Sousa *et al.* 2013	167	Bacteria	$[(bc)_2:(caa_3)_4]_2$; $(bc)_2:(caa_3)_4$; $(bc)_4:(caa_3)_2$; $(SQR)_2$:NAR[b]
Stroh *et al.* 2004	34	Bacteria	$I_1III_4IV_4$; $III_4IV_{2/4}$
Krause 2006	39	Bovine heart	$I_1III_2IV_{0-4}$; $I_{1-x}III_yIV_z$; I_1IV_1; III_2IV_{1-2}
Lenaz *et al.* 2010	146	Bovine heart	I + III + IV
Schäfer *et al.* 2006	65	Bovine heart	I_1III_2; $I_1III_2IV_1$
Schäfer *et al.* 2007	144	Bovine heart	I_1III_2IV
Schägger 2002	168	Bovine heart	I_1III_2; $I_1III_2IV_{0-4}$
Schägger and Pfeiffer 2001	5	Bovine heart	I_1III_2; $I_1III_2IV_n$
Seelert *et al.* 2009	169	Bovine heart	I_1III_2; $I_1III_2IV_1$
Sousa *et al.* 2016	69	Bovine heart	$I_1III_1IV_1$
Yang *et al.* 2014	170	*Caenorhabditis elegans*	I + III; I + III + IV
Le Pécheur *et al.* 2016	171	*Drosophila melanogaster*	I_1III_2; $I_1III_2IV_{1-3}$
Hartman *et al.* 2011	172	Fish	I_1III_1
Cabrera-Orefice *et al.* 2014	173	Fungi	I_1IV_1; I_1III_2; $I_1III_2IV_{1-4}$
Davies *et al.* 2011	174	Fungi	I_1III_2
Krause *et al.* 2006	175	Fungi	$I_1III_2IV_{0-2}$; $I_{1-2}III_2$
Krause *et al.* 2004b	40	Fungi	$I_1III_2IV_{0-4}$; I_2; I_2III_2
Maas *et al.* 2009	176	Fungi	I_1IV_1; III_2IV_{1-2}
Marques *et al.* 2007	42	Fungi	I_1III_2; I_1IV_1; III_2IV_{1-2}; $I_nIII_nIV_n$
Nübel *et al.* 2009	43	Fungi	$I_1III_2IV_{0-2}$; I_2III_2; III_2IV_{1-2}
De Los Rios Castillo *et al.* 2011	177	Human syncytiotrophoblast	I_1III_2; $I_1III_2IV_{1-3}$; $I_1III_{1-2}IV_{1-4}$
Genova *et al.* 2008	80	Human papillary thyroid tumour cells (TPC1)	I + III
Greggio *et al.* 2017	178	Human skeletal muscle	I_1III_2; $I_1III_2IV_1$; III_2IV_2
Kovářová *et al.* 2012	179	Human skin fibroblasts	I_1III_2; $I_1III_2IV_1$
Lopez-Fabuel *et al.* 2017a	180	Human fibroblasts	I_2III_2; $I_1III_2IV_1$
Mejia *et al.* 2016	181	Human lymphoblasts	I_1III_2; $I_1III_2IV_1$; $I_1III_2IV_2$
Miyake *et al.* 2013	182	Human skin fibroblasts	$I_1III_2IV_n$
Bohovych *et al.* 2016	183	Mouse embryonic fibroblasts	$I_1III_2IV_1$; $I_1III_2IV_2$
Jang *et al.* 2016	184	Mouse heart	$I_1III_2IV_1$ + others
Lopez-Fabuel *et al.* 2017b	185	Mouse neurons	I_2III_2; $I_1III_2IV_1$
Sun *et al.* 2016	186	Mouse and human cell lines	I_nIII_n; $I_nIII_nIV_n$

Table 12.1 (*Continued*)

Study	Ref.	Species/tissue	Type of supercomplex[a]
Eubel *et al.* 2004	187	Plants	I_1III_2; I_2III_4; III_2IV_{1-2}; $I_1III_2IV_{1-4}$
Kakizaki *et al.* 2012	188	Plants	I_1III_2
Krause *et al.* 2004a	189	Plants – spinach green leaves	I_1III_2; III_2IV_1; $I_1III_2IV_{1-4}$
Lenaz *et al.* 2010	146	Plants – potato tuber	I + III + IV
Petereit *et al.* 2017	190	Plants – *A. thaliana*	I_1III_2
Peters *et al.* 2008	191	Plants	I_1III_2
Gómez *et al.* 2009	161	Rat heart	$I_1III_2IV_1$; III_2IV_1
Lenaz *et al.* 2010	146	Rat liver	I + III + IV
Lombardi *et al.* 2009	192	Rat skeletal muscle	I_1III_2; $I_1III_2IV_{1-3}$
Lopez-Fabuel *et al.* 2017b	185	Rat astrocytes and neurons	I_1III_2; I_2III_2; $I_1III_2IV_1$
Wernicke *et al.* 2010	193	Rat brain	I_1III_2; $I_1III_2IV_{1-3}$
Matus-Ortega *et al.* 2015	194	*Saccharomyces cerevisiae*	SCs containing NDI[c]

[a]Subscript indices indicate the number of monomers included in the supercomplex; n = variable copy numbers.
[b]bc, Complex III; caa3, Complex IV; SQR, succinate : quinone oxidoreductase, Complex II; NAR, nitrate reductase.
[c]NDI, internal alternative NADH dehydrogenase.

In bovine heart mitochondria, only 14–16% of total CI was found in free form in the presence of digitonin;[5] thus, it seems likely that all CI is bound to CIII in the absence of detergents.

Although CI is usually a monomer, a putative dimeric CI has also been described in mitochondria from *Neurospora crassa, Yarrowia lipolytica, Podospora anserina* and *Solanum tuberosum*, where it may be involved in forming string-like and patch-like megacomplexes composed of repeated units of respiratory SCs.[39–43]

A higher level of complexity in the organisation state of SCs is due to the presence of further interaction partners such as the ADP/ATP carrier, some chaperones for cytochrome *c* oxidase assembly and the carbonic anhydrase subunits, which co-assemble with the OXPHOS SCs.[44–46] Several additional proteins are probably able to interact with cytochrome *c* oxidase, and some examples such as monocarboxylate transporters, mitochondrial nitric oxide synthase, and the large-conductance Ca^{2+}-activated K^+ channel have already been identified by immune-precipitation.[47–49] Recently it was reported that mitochondrial superoxide dismutases associate with a documented respirasome.[50]

Wang *et al.*[51] provided evidence for a multifunctional fatty acid β-oxidation (FAO) complex within mitochondria that is physically associated with SCs. However, Schönfeld *et al.*[52] could not find such an association. Nsiah-Sefaa and McKenzie in a recent review[53] have analysed defects in fatty acid oxidation and OXPHOS pathways. The pathology underlying many aspects of these diseases is not well understood; for example, it is not clear why some patients with primary FAO deficiencies exhibit secondary OXPHOS defects. According to the same authors these metabolic interrelations support the view that physical interactions exist between OXPHOS proteins and FAO, and that these interactions are critical for both functions.

It has been also observed that several mitochondrial soluble NAD^+-dependent dehydrogenases such as malate dehydrogenase, β-hydroxyacyl-CoA dehydrogenase, α-ketoglutarate dehydrogenase complex, and the pyruvate dehydrogenase complex specifically associate with CI.[54–56] However, the functional role of such associations remains controversial.[57]

The other respiratory enzymes not comprising the "core" of the proton translocation machinery appear not to be associated in SCs.[6] CII has been investigated in some detail and its association with other complexes of the respiratory chain has not been identified thus far by BN-PAGE, except in a few cases,[52,58] although it was recently found that digitonin-solubilised CII quantitatively forms high molecular weight structures that can be resolved by CN-PAGE.[59] These structures differ in electrophoretic mobility between tissues (500-over 1000 kDa) and cultured cells (400–670 kDa) and are enzymatically active as shown by in-gel activity assay.

Mitochondrial glycerol phosphate dehydrogenase has been suggested not to be associated in a SC on the basis of functional studies.[60] This absence has been confirmed in the respirasome, although the enzyme was found to polymerize in the form of homo-oligomers.[61]

Fang *et al.*[62] observed that dihydroorotate dehydrogenase deficiency hampers mitochondrial function; the enzyme was shown to physically interact with respiratory complexes II and III by immunoprecipitation and BN/SDS-PAGE analysis.

According to Hildebrandt,[63] sulphide-quinone oxidoreductase and sulphite oxidase, the two membrane bound enzymes of the pathway catalysing sulphide oxidation, which transfer electrons into the respiratory chain, might also be associated with SCs containing CIV *in vivo*.

12.3 Supercomplex Association Provides a Kinetic Advantage

Immediately after the discovery of respiratory SCs it was proposed that the natural consequence of such assemblies is substrate channelling or enhanced catalysis in inter-complex electron transfer. Substrate channelling is the direct transfer of an intermediate between the active sites of two enzymes catalysing consecutive reactions;[64] in the case of electron transfer, this means direct transfer of electrons between two consecutive enzymes by successive reduction and re-oxidation of the intermediate without its diffusion in the bulk medium. In such a case, inter-complex electron transfer becomes indistinguishable from intra-complex electron transfer, so that ubiquinone and cyt. *c* (*i.e.* the so-called mobile intermediates predicted to exhibit substrate-like behaviour in the classic view of the random collision model[16]) would rather be buried with restraints between two consecutive components of the SC. However, the view that SCs enhance mitochondrial respiration by channelling the intermediate substrates is still challenged by some debated structural and functional evidence, as we discuss in the following sections.

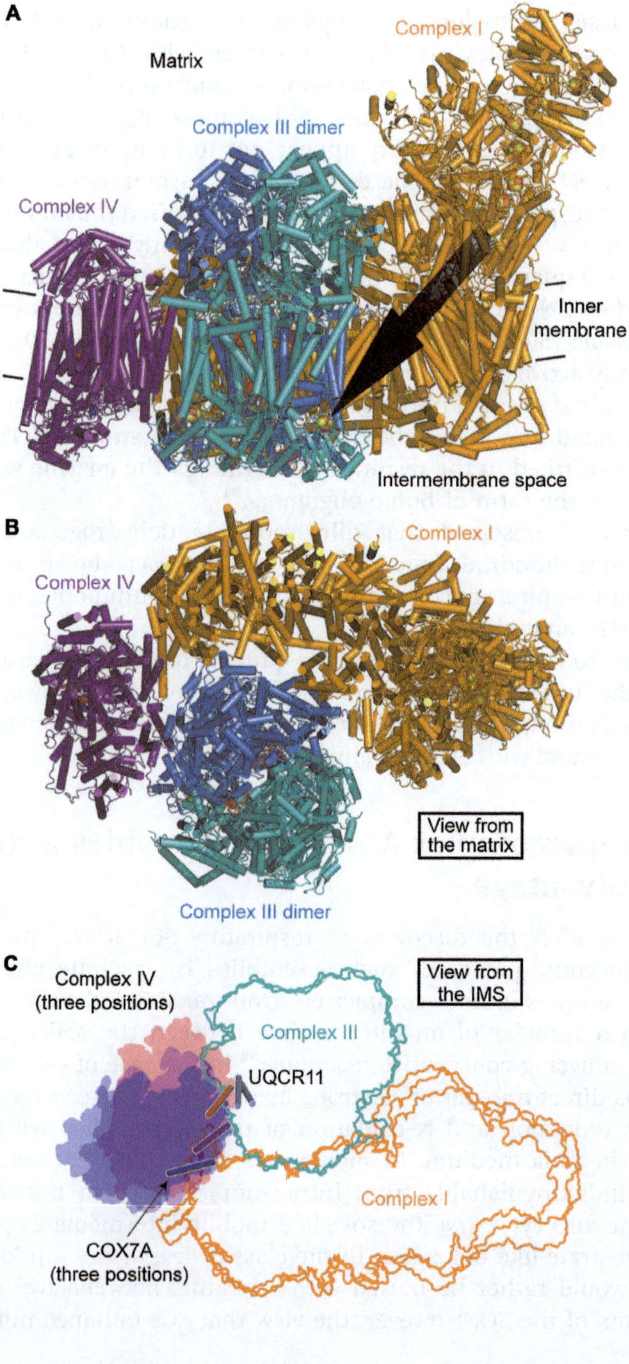

12.3.1 Structural Evidence

12.3.1.1 Molecular Structure of Supercomplexes

If channelling occurs between CI and CIII, the redox groups involved in CoQ reduction by CI and $CoQH_2$ re-oxidation in CIII must be located in such way to provide a pre-defined conducting pathway that contains CoQ itself. To ascertain this condition, it is required to have a high resolution knowledge of the molecular structure of the SCs.

At first, some SCs have been purified and analysed by negative-stain electron microscopy[65] and single-particle cryo-EM.[37,66,67] Immediately afterwards, pseudo-atomic models of the 1.7 MDa bovine respirasome consisting of one copy of CI, one CIII dimer and one CIV monomer were produced by fitting the X-ray structures of the component complexes to the 3D maps of the respirasome (Figure 12.1).

Some early evidence for possible channelling came from the observation of a unique arrangement of the three component complexes showing that the CoQ-binding sites in CI and in CIII face each other and are separated by a gap within the membrane core of the SC. In 2011, Althoff *et al.*[66] proposed that CoQ is likely to run a trajectory through such a 13 nm-long gap, which lies within the membrane imbedded domains of the SC and is therefore most likely filled with lipids (Figure 12.1).

However, understanding the mechanism of channelling still requires a more detailed knowledge of the molecular structure of the interacting sites. The very recent high resolution studies by cryo-EM of mammalian mitochondrial SCs[33,68,69] allow new predictions to be made on the pathway taken by CoQ between CI and CIII. In the architecture shown in Figure 12.1, representative of all the mammalian structures described so far, the arrangement of CI and CIII is highly conserved, with the membrane arm of CI curving around the CIII dimer, whereas the position of CIV rolls on the toe of CI, interacting with both CI and CIII or gradually dissociating from

Figure 12.1 Structures of the respirasome. (A) The arrangement of CI (orange), the CIII dimer (the monomers are shown in blue and cyan), and CIV (magenta) in the porcine respirasome, viewed along the membrane plane. The black arrow schematically indicates the distance between the CoQ binding sites in CI and CIII (*cf.* text for details). (B) The porcine respirasome viewed from the matrix. (C) Comparison of the porcine and ovine respirasome structures. The structures were superimposed using the structure of the CIII dimer, to reveal the different relative positions of CI and CIV. CIV is shown as transparent overlapping shapes (pink, porcine respirasome; magenta, tight ovine respirasome; blue, loose ovine respirasome). The transmembrane helix in CIV COX7A is highlighted to show how it moves relative to CIII UQCR11, which has been modeled in two slightly different positions.
Figure created by Milenkovic *et al.*[29] using 5GPN.pdb, 5J4Z.pdb and 5J7Y.pdb. Reprinted from *Cell Metabolism*, Volume 25, D. Milenkovic, J. N. Blaza, N.-G. Larsson, J. Hirst, The Enigma of the Respiratory Chain Supercomplex, 765–776, Copyright 2017, with permission from Elsevier.

CIII and finally from the respirasome completely, as in the smaller SC I_1III_2 of approx. 1.5 MDa. As also discussed by Milenkovic *et al.*,[29] these different structures may depict the respirasome in various stages of disassembly, either following its extraction from the membrane or resembling different physiological conditions with respirasomes constantly dissociating, reforming and reorganizing in the membrane.

In theory, within the SCs we may have the extremes from close docking of the active sites with real interprotein tunnelling of substrates, up to relatively long distances that may be covered either by important conformation changes of the protein itself or by restricted diffusion (*microdiffusion*) of the substrates within the space between the two active sites; all of these alternatives have in common obligate channelling between two fixed sites of the same SC, so that even the last situation, microdiffusion, would be kinetically quite distinguishable from random collisions as exploited through bulk diffusion, *i.e.* where the interaction of the substrate molecules may stochastically occur with a great number of possible sites in distinct respiratory complexes/SCs reached by random diffusion in the membrane bilayer. However, steady state kinetic analysis cannot distinguish among the above mentioned three possible mechanisms of channelling.

Notably, the respirasome structure recently published by Letts *et al.*[33] provides specific information about the protein contacts in the SC, which incidentally are dominated by the supernumerary subunits of CI rather than by its core subunits, but also indicates that the CoQ reduction site in CI and the $CoQH_2$ oxidation site in CIII are separated and both easily accessible to the membrane, thus the same authors interpreted that no limit to free diffusion of CoQ in the bilayer exists. On the contrary, the group of Kuhlbrandt[69] recently took into account the dimeric structure of CIII by showing that the CoQ reduction site of CI in the SC is separated by 11 nm from the proximal CIII monomer, and 18 nm from the distal one, thus suggesting that the latter is functionally inactive whereas the linear arrangement of CI with the active monomer of CIII and also with CIV is strongly in favour of channelling for electron transfer between complexes by substrate microdiffusion.

In fact, information about the molecular structure of the SC must be compatible with the recognized mechanisms of CoQ reduction by CI and its re-oxidation by dimeric CIII; in particular the latter occurs *via* a complicated mechanism known as the *Q-cycle*.[4,70] In the Q-cycle there are two sites in CIII where CoQ interacts, one located near the intermembrane space (Qo) where ubiquinol is sequentially oxidized by the Rieske iron–sulphur protein and by the b_L heme, and one located near the matrix space (Qi) where ubiquinone is reduced. Apparently if CoQ channelling occurs in the SC by microdiffusion, there should exist at least two CoQ molecules trapped in the interstices between CI and CIII in order to fulfil the Q-cycle mechanism.

However, we do not have to leave out the fact that regulatory interactions may occur in dimeric CIII, which is incidentally the dominant form in SCs, such that half of the sites reactivity (*i.e.* heme b_L-b_L bridge) allows electrons to rapidly equilibrate between the Qi sites of the two monomers.[4,71,72] A basic operational principle of this system is that it enables any connection between catalytic sites of the opposite i/o sides of the membrane to be enzymatically competent. The physiological meaning of this design of the CIII homodimer remains an open issue. With our current knowledge on the SC I_1III_2 structure taken into account, we may reason that both one molecule of ubiquinol formed by CI (*i.e.* CoQ confined in the SC) and the bulk CoQ pool of the membrane may co-participate in the complete electron pathway of the Q-cycle by cross-reacting respectively at the Qo- and Qi-site of the corresponding monomers of dimeric CIII facing the central cavity of the SC or the surrounding membrane environment.

12.3.1.2 Dynamic Nature of Supercomplexes: The Plasticity Model

In 2003, when we first provided kinetic evidence by flux control analysis in favour of metabolic channelling in the CoQ region, we also postulated[92] that the SC I_1III_2 may be in equilibrium with free complexes, depending on metabolic conditions, to explain the presence of NADH-cyt. *c* oxidoreductase activity both in the absence and presence of channelling, depending on the protein to phospholipid ratio of the examined proteoliposome samples.

When Acín-Pérez, *et al.*[58] demonstrated that purified respirasomes (*i.e.* SC $I_1III_2IV_1$) are able to respire in a Clark's electrode, they also observed other types of associations, as I + III or III + IV, and concluded that a variety of associations between respiratory complexes likely co-exists *in vivo* with free complexes. In this context they proposed an integrated model, *the plasticity model*, for the organization of the mitochondrial electron transport chain. The previous opposed models, solid *vs* fluid, would be two allowed extreme and functional situations of a dynamic range of molecular associations between respiratory complexes.[58] The stoichiometry of the complexes and the variable stability of supramolecular assemblies under different physiological conditions would determine a variety of different respiratory architectures. Intrinsic to its dynamic nature, the plasticity model proposes that segregation of both CoQ and cyt. *c* into distinct sub-domains is incomplete and that their respective functional pools are promiscuous (*i.e.* the intermediate substrates are both trapped into SCs and freely diffusible). According to Enriquez[24] this promiscuity can be generated either by the dissociation of SCs or by the escape of CoQ or cyt. *c* from the supramolecular assembly.

The plasticity model and the dynamics of mitochondrial SCs are widely discussed by Acín-Pérez and Enriquez,[26] but still debated in the literature. Possible factors involved in association/dissociation of SCs, such as membrane potential and post-translational changes, are reviewed by Genova and Lenaz.[28] However, despite the popularity of the model and the indirect evidence supporting it, no direct demonstration exists that SCs are cyclically formed and dissociated during physiological events.

12.3.1.3 The Role of Lipids: Cardiolipin in Supercomplexes

Supercomplexes have 2–5 nm gaps at the transmembrane interfaces of the individual complexes (Figure 12.1). These gaps lie within the membrane imbedded domains of the SC and are therefore most likely filled with lipids. The mammalian SC has larger gaps than the yeast SC.[37,66,67] The recent article by Wu *et al.*[68] on the high resolution structure of the purified SC $I_1III_2IV_1$ from porcine heart also localizes a large number of phospholipid molecules involved in the protein–protein interactions.

In Barth syndrome patients, where cardiolipin (CL) remodelling is altered due to the mutation of the gene Tafazzin, SCs are unstable, leading to the mitochondrial functional impairment that underlies the disease.[73] More recently, Gonzalvez *et al.*[74] confirmed that in immortalized lymphoblasts from Barth's syndrome patients the amount of SCs is decreased, as well as the amount of individual complexes I and IV; these changes were compensated by increasing mitochondrial mass.

Direct involvement of CL in the formation of SC III + IV was demonstrated in genetically manipulated strains of *S. cerevisiae* in which the CL content can be regulated *in vivo*.[75] BN-PAGE of digitonin extracts of mitochondria revealed that yeast mutants completely lacking CL did not form the stable SC III_2IV_2 as observed in the wild type parental strain, whereas the total amount of individual respiratory complexes was not affected by the lack of CL.[75,76] However, no experimental evidences about CI-containing SCs could be obtained in these studies, since *S. cerevisiae* lacks CI.[73]

Desmurs *et al.*[77] demonstrated that C11orf83, a mammalian mitochondrial inner membrane protein facing the intermembrane space, is specifically associated with the bc1 complex of the electron transport chain and involved in the early stages of its assembly by stabilizing the bc1 core complex. The same authors showed that C11orf83 is also involved in the stabilization the SC III_2IV due to direct binding of CL to its α-helices 2 and 3.

In contrast with CL depletion that destabilises SCs, it was reported that phosphatidylethanolamine (PE) depletion tends to favour the formation of larger SCs between CIII and CIV in *S. cerevisiae* mitochondria.[78] The reason why CL and PE, both non-bilayer-forming phospholipids, behave in an opposite way on supercomplex stability has been ascribed to the different charge, the former being an anionic phospholipids and the latter zwitterionic.[78] Using yeast mutants of PE and PC biosynthetic pathways, Baker *et al.*[79] showed a specific requirement for mitochondrial PE in CIII and CIV activities but not for their formation, whereas loss of PC does not affect either their function or formation. Unlike CL, mitochondrial PE or phosphatidylcholine (PC) is not required for respiratory supercomplex formation, emphasizing the specific requirement of CL in supercomplex assembly.

The importance of lipid composition in the stabilization of CI-containing SCs is supported by our results showing that reconstitution of binary CI/CIII proteoliposomes from bovine heart at high lipid to protein ratio (30 : 1 w : w) hampers the assembly of SC I_1III_2.[80,81] On the contrary, SC I_1III_2 and efficient NADH-cyt. *c* reductase activity may be preserved when similar high-lipid proteoliposomes are enriched with 20% CL (w : w), resembling the percent content of CL in the mitochondrial membrane (M. Kopuz, Y. Birinci,

S. Nesci, G. Lenaz and M.L. Genova, unpublished data). A likely explanation is that detrimental dilution of native protein-bound CL in the excess exogenous lipids is reversed by increasing the CL content and thus shifting again the equilibrium to CL binding to protein.

It is well documented that exposure of mitochondria to reactive oxygen species (ROS) can affect the respiratory activity *via* oxidative damage of CL, which is required for the optimal functioning of the enzyme complexes,[82–84] reviewed by Paradies *et al.*[85] Our results[80] demonstrated by flux control analysis that the maintenance of the SC I_1III_2 in proteoliposomes is abolished if lipid peroxidation is induced before protein reconstitution in the lipid vesicles. Evidently, the distortion of the lipid bilayer induced by peroxidation and the alteration of the phospholipid annulus originally present in the purified SC I_1III_2 determine its dissociation.

Recently SCs from bovine heart mitochondria were separated by sucrose density gradient centrifugation.[86] The SC sample did not contain cyt. *c* but did contain complexes I, III, and IV at a ratio of 1:2:1, six molecules of CoQ_{10}, and 623 atoms of phosphorus indicating the presence of phospholipids. When exogenous cyt. *c* was added, the SC exhibited KCN-sensitive NADH oxidation, confirming the previous studies of Acín-Pérez *et al.*[58] on respiratory activity of SCs isolated from BN-PAGE bands and the studies of Stroh *et al.*[34] on the respirasome from *P. denitrificans*.

12.3.1.4 Standing Uncertainties

Despite recent experimental advances in our capability to characterize and describe redox proteins in respiratory SCs, two major challenges need to be met in order to disclose the catalytic mechanisms of their enzyme function, namely, determine how the intermolecular redox reactions are coordinated with intramolecular redox reactions and how redox reactions are coordinated with other chemical events such as proton transfer or substrate binding and release (*cf.* ref. 87 for recent review).

Some authors[29] claimed that SCs have no specific function and only derive from the necessity to maintain a high protein concentration in the inner membrane avoiding unspecific aggregation. It is certainly true that SC formation depends on the lipid concentration, being favoured at high protein to lipid ratio;[80] on the other hand, if excluding that electron transfer by channelling may actually occur, it is difficult to turn to alternative reasons for why the active sites of the interacting respiratory complexes should have evolved to be located in such strict vicinity as to face off in a stereospecific and non-random manner. The problem of substrate channelling in SCs has been addressed by kinetic analysis and appears to quite differ in the CoQ region with respect to the cyt. *c* region since these two "mobile" components have different locations with respect to the membrane and are differently connected with reducing and oxidizing partner enzymes. For this reason, we will discuss the subject in the following separate sections.

12.3.2 Evidence for Channelling in the Coenzyme Q Region

12.3.2.1 Rate Advantage in the Coenzyme Q Region

The overall rate in a diffusion-coupled pathway is always less than that of the rate-limiting step and only approaches the latter when it is widely different (*i.e.* much slower) compared to the rate of the other step(s) in the pathway (*cf.* ref. 14 and 89). On the contrary, the overall rate is expected to be always equal to the rate of the limiting step if channelling of the intermediate substrate(s) occurs.

In 1978, Ragan and Heron[88] provided evidence that purified CI and CIII, when mixed as concentrated solutions in detergent and then co-dialysed, combine reversibly in a 1 : 1 molar ratio to form a SC I + III unit (NADH-cyt. *c* oxidoreductase); electron transfer between such a unit reconstituted in proteoliposomes and any extra molecules of complexes I or III freely dispersed in the lipid bilayer does not contribute significantly to the overall rate of cyt. *c* reduction. In reconstituted mixtures of CI and CIII the reduction of cytochrome b by NADH is biphasic and the extents of the fast and slow phases of reduction are determined by the amount of CIII specifically associated with CI, the extent of the slow phase increasing when CIII is in excess over the SC. These studies[88] were able for the first time to demonstrate the existence of an active SC formed by complexes I and III (*i.e.* stoichiometric behaviour for the activity of NADH-cyt. *c* reductase ascribable to the formation of the SC I + III), and that electron transfer within the SC is fast, whereas electron transfer from the SC to free CIII *via* the CoQ pool is still possible but extremely slow.

However, prevailing CoQ-pool behaviour could be induced and CI and CIII could be made to operate independently of each other (*i.e.* dissociation of the SC I + III into "free" complexes) by raising the amount of phospholipid and ubiquinone (approx. a 2-fold and a 6-fold increase, respectively) in the concentrated mixture.[88] Under this condition, electron transfer from CI to CIII is still ensured by CI + CIII units that however dissociate and reform at rates exceeding the rates of electron transfer in the individual complexes.

Heron and co-workers[89] proposed that relative mobility of CI and CIII is lost and the complexes are frozen in their SC assembly when phospholipid in excess of that needed to form an annulus is absent, thus favouring a stable orientation of the site of reduction of ubiquinone by CI with respect to the site of oxidation by CIII.

Heron *et al.*[89] also reported that endogenous CoQ_{10} leaks out of the SC I + III unit when extra phospholipid is present in the proteoliposomes, causing a decrease in activity that could be alleviated by adding more ubiquinone. It is likely that the function of the large amount of ubiquinone in the natural membrane may be, therefore, to maintain the CoQ_{10} content in the SC unit when it is formed.

A more direct comparison of the effect of channelling with respect to CoQ-pool behaviour was performed in a simple experimental condition in our laboratory.[90] A proteoliposome system obtained by fusing a crude mitochondrial fraction (R4B),[91] enriched in CI and CIII and also containing traces of CII, with different amounts of phospholipids and CoQ_{10} was used to discriminate whether the reconstituted protein fraction behaves as individual enzymes connected by random diffusion of the intermediate substrate or as assembled SC, depending on the amount of phospholipids used in the reconstitution. The comparison of the experimentally determined NADH-cyt. *c* reductase activity with the values expected by theoretical calculation applying the pool equation of Kroger and Klingenberg[14] showed overlapping results at phospholipid dilutions (w/w) from 1 : 10 to 1 : 40 protein to lipid ratios.[90] On the contrary, pool behaviour was not effective and the observed rates of NADH-cyt. *c* reductase were higher than the theoretical values[80,90,92] at low protein : lipid dilution (1 : 1 w/w), resembling the mean nearest neighbour distance between respiratory complexes in mitochondria.[93–95]

Moreover when the same proteoliposomes at 1 : 1 protein : lipid ratio were treated with n-dodecyl-β-D-maltoside (DDM) to destroy the supercomplex organization, the NADH-cyt. *c* reductase activity fell dramatically, whereas both CI and CIII individual activities were unchanged;[81] an analogous behaviour was detected by treating bovine heart mitochondria with the same detergent.[25]

These studies suggest that electron transfer between CI and CIII can take place both by CoQ channelling within the SC I_1III_2 and by a less efficient collision-based behaviour, depending on the experimental conditions.

12.3.2.2 *Evidence for Channelling by Metabolic Flux Control Analysis*

By exploiting the flux control analysis[96] using specific inhibitors in order to define the extent of metabolic control exerted by each individual complex over the entire NADH-dependent respiration in isolated bovine heart mitochondria, our laboratory achieved the first demonstration that CI and CIII behave as a single enzymatic unit (*e.g.* SC I_1III_2) and that electron transfer through CoQ is accomplished by channelling between the two redox enzymes.[97] Using the same method for succinate oxidation we found that CII, but not CIII, is rate-limiting, supporting the notion that CII does not form SCs[97] and that the oxidation of succinate follows pool behaviour (*cf.* below).

Ten years later, Blaza *et al.*[98] criticized the evidence for channelling in the respiratory chain deriving from flux control analysis:[97] they reasoned that rotenone is competitive with CoQ and therefore the extent of its inhibition of CI is affected by the additional presence of the exogenous quinone employed in CI assay but absent in the assay of NADH aerobic oxidation. Indeed, using rotenone and CoQ_1, Blaza *et al.*[98] find a flux control coefficient (FCC) for CI that is certainly not valid, since it exceeds 1. However, discrepancy with Bianchi *et al.*[97] may be ascribed to the different exogenous substrates used for CI assay in the two studies, *i.e.* CoQ_1 and decylubiquinone (DB) respectively, and also to the technical difficulty of calculating FCC from the initial slopes of the inhibition curves. In reply to the raised criticisms, we can argue that DB has much lower affinity than CoQ_1 for CI,[99] thus exerting lower competition with rotenone and lower influence on the inhibition efficiency measured by Bianchi *et al.*[97] Moreover, we note that the FCC values found by Blaza *et al.*[98] using alternative inhibitors of CI (*e.g.* piericidin and diphenyleneiodonium, not competitive with CoQ) are oddly low and it is likely that such low FCC for CI is caused by the limiting amount of cyt. *c* in their samples, that would shift the major control of the chain to the cyt. *c* region. Significantly, in Blaza's study, addition of exogenous cyt. *c* to the mitochondrial samples raised the FCC from 0.19 to 0.67.

Finally, we emphasize that the major point discussed by us[97] to demonstrate the existence of a SC I + III was not so much the high FCC of CI but the high FCC of CIII in NADH oxidation (not measured, on the contrary, by Blaza *et al.*[98]), which is incompatible with a CoQ-pool model postulating CI and CIII as independent molecules in the membrane. Such high FCC value of CIII over NADH-dependent respiration, as calculated by inhibitor titration with mucidin, is clearly not suspect of artefact because the corresponding FCC of CIII measured by using the same inhibitor in succinate oxidation is low (as expected from the rate-limiting role of CII and the lack of SC II + III). In other words, the FCC of CIII can be close to 1 if CIII is assembled together with the quinone reductase that precedes in the electron pathway and, in our hands, this condition is experimentally observed only for the NADH-dependent pathway involving CI, whereas it does not occur in the case of CII. In addition, the high FCC that CIII shows over NADH-cyt. *c* oxidoreductase activity in reconstituted proteoliposomes is drastically decreased after dissociation of the SC I_1III_2[80] by reconstitution in excess phospholipids or in peroxidized phospholipids.

Very few other studies have addressed the functional aspects of SCs using metabolic control analysis;[100–102] these studies confirmed that the respiratory chain is organized in functionally relevant supramolecular structures, particularly under certain conditions like digitonin-permeabilized HepG2 cells in the presence of high membrane potential.[100]

12.3.2.3 Separate Compartments of Coenzyme Q?

The effect on respiration of the simultaneous addition of NADH and succinate was examined in an early study[103] showing incomplete additivity and partial mutual competition between succinate oxidation and NADH oxidation; this effect was considered unlikely to be due to competition of the dehydrogenases for CoQ; also on the basis of other observations,[104] a compartmentalization of the CoQ pool was suggested, in which two sub-pools were able to partially interact through a spill-over diffusion-mediated mechanism.

In 1985, Gutman[105] investigated the properties of the NADH and succinate oxidation in submitochondrial particles in relation to the rates of energy-dependent reverse electron transfer from succinate to NAD^+ and of forward electron transfer from NADH to fumarate, concluding that:

> *"the electron flux from succinate dehydrogenase to oxygen (forward electron transfer towards Complex III) or to NADH dehydrogenase (reverse electron*

transfer) employs the same carrier and is controlled by the same reaction" whereas "the electron transfer from NADH to oxygen does not share the same pathway through which electrons flow in the NADH-fumarate reductase".

In other words, CI and CII are linked by a different pathway with respect to CI and CIII. The non-homogeneity of the ubiquinone pool with respect to succinate and NADH oxidation may be interpreted today in terms of compartmentalization of CoQ in the SC I + III in contrast with the free CoQ molecules used for connecting complexes II and III.

Several other reports have suggested in the past that the CoQ pool is not homogeneous, raising doubts on its universal validity;[106] even Kröger and Klingenberg[14] observed that an aliquot of CoQ cannot be reduced by either NADH or succinate. Later, Jørgensen *et al.*[107] noticed that three pools of ubiquinone appeared to be present in heart mitochondria: a metabolically inactive pool consisting of reduced as well as oxidized ubiquinone, a pool coupled to oxidation of added (cytoplasmic) NADH, and the well-known pool coupled to citric acid cycle oxidations. The latter pool, however, could not distinguish NADH-dependent from succinate-dependent reduction of CoQ.

Benard *et al.*[108] described the existence of three different pools of CoQ during succinate-dependent steady-state respiration in rat liver and muscle mitochondria: one pool is directly utilised, another (approx. 8% in muscle and 23% in liver) is mobilized as a reserve in case of a perturbation to maintain the energy fluxes at normal values (*e.g.* due to inhibition of the respiratory complexes or in case of mitochondrial diseases), and a third one (approx. 79% in muscle and 21% in liver) cannot be mobilized at all. These results are compatible with CoQ compartmentalization, although similar results with NADH oxidation were not provided in order to functionally prove that the fraction of CoQ that is not utilisable for succinate oxidation is channelled within SC I + III.

Lapuente-Brun *et al.* demonstrated[109] that the physical assembly between complexes I and III determines a preferential pathway for electrons mediated by a dedicated subset of CoQ molecules. According to their results, this compartmentalization prevents significant cross talk between NADH oxidation (CI-dependent) and succinate oxidation (dependent on CII) or other flavoenzyme-dependent oxidations. Moreover, CIII molecules bound in the SC I + III and free CIII are respectively and exclusively dedicated to oxidation of NADH (the former) and other substrates using the free CoQ pool (the latter, *e.g.* succinate, glycerol-3-phosphate and choline, as also suggested by other authors,[110,111] *cf.* also Section 12.3.2.4.2.).

On the contrary, the previously quoted paper by Blaza *et al.*[98] showed that the steady-state rates of aerobic NADH and succinate oxidation were not additive in bovine heart submitochondrial particles; moreover, the extents of cytochromes b_H, b_L, c and c_1 reduction in the same cyanide-inhibited particles were similar if the reductant was either NADH or succinate or a mixture of the two substrates. Blaza *et al.* interpreted the results as a demonstration that a single homogeneous pool of CoQ molecules exists that receives electrons indifferently from CI and CII. Consequently, free CIII and SC-bound CIII are able to equally receive electrons from the CoQ pool.[98] It is worth noting that, according to the pool equation,[14] the combined rate of NADH and succinate oxidation should be slightly higher than the single activities under the conditions of Blaza *et al.*,[98] as indeed was the case in their report. However, on the basis of those mere data, which considered the whole pathway of electrons from ubiquinol to oxygen and therefore comprised the redox cycling of cyt. *c* and CIV, it is not possible to discriminate whether the pool behavior is due to the homogeneity of the CoQ pool or of the cyt. *c* pool.

In fact, under such conditions, cyt. *c* may be rate-limiting, provoking a bottleneck step both in NADH and in succinate oxidation when electrons compete for a common pool of cyt. *c* and CIV molecules in free form. Indeed, most cyt. *c* and CIV are free and able to receive electrons from both NADH and succinate with only a small portion of CIV dedicated to NADH (*i.e.* forming the respirasome I_1III_2IV), as demonstrated by our flux control analysis data[97] that showed no channelling in the cyt. *c* region (*cf.* Section 12.3.3).

To overcome this problem, we performed a study of succinate and NADH oxidation by excess exogenous cyt. *c* in KCN-inhibited mitochondria;[25] this experimental approach shortens the electron transfer pathway for both oxidation reactions by including only CoQ and CIII as redox partners. Under such condition, we found that NADH and succinate oxidation by cyt. *c* are completely or almost completely additive and close to the theoretical summation of the two single activities (Table 12.2), suggesting the existence of two separate CoQ-compartments.

Interestingly, similar results were also obtained in preliminary experiments from our laboratory (Tioli G, Falasca AI, Lenaz G and Genova ML, data communicated at EBEC2016-Session P4) using R4B proteoliposomes where the additivity of NADH and succinate oxidation decreased when CIII was progressively inhibited with mucidin. This is in accordance with the idea that the increasing number of inactive CIII units forced CoQ reduced by succinate and CII to access residual active units of CIII even if linked to the

Table 12.2 NADH and succinate oxidation by exogenous cytochrome *c* in cyanide-inhibited BHM.[a]

Substrate as electron donor	Rate of cytochrome *c* reduction[b]
NADH	0.356 ± 0.031 (5)
Succinate	0.591 ± 0.060 (5)
NADH + Succinate	0.883 ± 0.071 (5)

[a]NADH-cytochrome *c* oxidoreductase activity and succinate-cytochrome *c* oxidoreductase activity were assayed spectrophotometrically in frozen and thawed BHM at 30 °C by monitoring the reduction of 50 μM cytochrome *c* in the presence of 1 mM KCN and 75 μM NADH or/and 26 mM succinate, as described in Lenaz *et al.*[25]

[b]Values are referred to two-electron transfer activity (*μmoles electron couple/min/mg protein*). Data are mean values ± standard deviation of separate experiments, as indicated by the numbers in round brackets. Reprinted from *Biochimica et Biophysica Acta (BBA) – Bioenergetics*, Volume **1857**, G. Lenaz, G. Tioli, A. I. Falasca, M. L. Genova, Complex I function in mitochondrial supercomplexes, 991–1000, Copyright 2016, with permission from Elsevier.

CI-containing SC, suggesting that the CoQ molecules in the SC I + III are in a dissociation equilibrium with the molecules in the pool,[112] *cf.* next section. The recent structural model of Letts *et al.*[33] strongly supports our kinetic observations, since the neighbourhood of the CoQ sites in CI and CIII to one another is consistent with preferred channelling of reduced CoQ within the same SC I_1III_2, while at the same time the exposure of the same sites to the membrane lipids supports diffusion in the pool. To this purpose it is suggestive to consider that in the bound CIII dimer only one monomer is active for NADH oxidation due to the neighbourhood of the active sites[69] whereas the other monomer not having such constrictions might be easily available to receive electrons from the CoQ pool.

It is intuitive that channelling is largely predominant during high turnover of the respiratory chain, whereas at low turnover CoQ diffusion in the pool would preferentially take place.

12.3.2.4 The Function of the Coenzyme Q Pool

There is no doubt that a mobile pool of CoQ in the inner mitochondrial membrane coexists with protein-bound CoQ. Is this pool just a reservoir of an excess of CoQ molecules without a specific function or is the pool necessary for functioning of the respiratory chain and/or for additional functions?

12.3.2.4.1 Dissociation Equilibrium of Bound Coenzyme Q. As described in the previous sections of this chapter, CI is almost totally associated in a SC with CIII, and CoQ channelling is likely to occur in the boundary between the two complexes. However, this does not exclude that free CoQ in the pool is also necessary for proper channelling. In fact, there is evidence suggesting that the bound inter-complex quinone that allows electron flow directly from CI to CIII is in dissociation equilibrium with the CoQ pool, so that the amount of bound CoQ, at steady state, would be dictated by the size of the pool itself. In particular, the existence of this equilibrium is suggested by the saturation kinetics for total ubiquinone exhibited by the integrated activity of CI and CIII (NADH-cyt. *c* oxidoreductase)[113] and by the decrease of respiratory activity in mitochondria fused with phospholipids causing subsequent dilution of the CoQ pool.[114] To be in agreement with the experimental observations in favour of channelling, this proposition requires that the dissociation rate constants (k_{off}) of bound CoQ be considerably slower than the rates of inter-complex electron transfer *via* the same bound quinone molecules.

In this way, free CoQ acts as the diffusible substrate of various redox enzymes not involved in respiratory supercomplex organization (*cf.* next section) and at the same time as a reservoir for binding to the SC $I_1III_2IV_n$; in addition, free CoQ may be a reservoir for other functions believed to require CoQ binding to specific proteins, such as uncoupling proteins[115] and the permeability transition pore.[116]

A different question is whether electron transfer between CI and CIII can occur *via* the CoQ pool when in absence of any supercomplex organization. Studies of respiration in pathological conditions[117,118] showed that electron transfer in absence of supercomplex organization is lost even if activity of the individual complexes is normal. Reconstitution studies,[80,88,89] however, had indicated that electron transfer is possible in both modes. In a proteoliposome system where CI was reconstituted together with an alternative oxidase and CoQ_{10}, Jones *et al.*[119] found high rates of NADH oxidation in absence of a CI-containing SC, which demonstrates that CoQ_{10} was able to shuttle electrons from CI by following pool behaviour.

12.3.2.4.2 Electron Transfer Between Individual Complexes Not Involved in Supercomplex Organization. The CoQ pool is required for electron transfer from CII to CIII: indeed CII kinetically follows pool behaviour after extraction and reconstitution[14,15] and in intact mitochondria[120] in accordance with the lack of SCs found by both BN-PAGE and flux control analysis (see previous sections).

Other enzymes such as glycerol-3-phosphate dehydrogenase, ETF dehydrogenase, dihydroorotate dehydrogenase, choline dehydrogenase, sulphide dehydrogenase, that are likely to be in minor amounts and strongly rate-limiting in integrated electron transfer, are probably inserted in the respiratory chain by interaction through the CoQ pool.[6] A study addressing this problem[121] demonstrated that in brown adipose tissue (BAT) mitochondria the inhibition curve of glycerol phosphate-cyt. *c* reductase is sigmoidal in the presence of myxothiazol and antimycin, suggesting the presence of a homogeneous CoQ pool between glycerol phosphate dehydrogenase (mtGPDH) and CIII.

More recently, it was shown that the delivery of electrons from mtGPDH to CIII in human neutrophil mitochondria takes place in absence of supercomplex organisation and of NAD-linked respiration,[117] in line with the notion that mtGPDH operates in mitochondria through the CoQ pool. Preliminary studies by BN-PAGE (M.L. Genova and H. Rauchova, unpublished) show that mtGPDH does not appear linked to any of the respiratory complexes. Accordingly, Mráček *et al.*[61] demonstrated that mtGPDH associates into homooligomers (presumably as dimer, trimer and tetramer of this rather hydrophobic dehydrogenase) as well as high molecular weight SCs of more than 1000 kDa of yet unknown composition, better evidenced by CN-PAGE than by BN-PAGE, but none of them associated with CI, CIII or CIV (*i.e.* OXPHOS complexes that may share common electron transfer pathway with mtGPDH and would therefore make kinetic sense). The mtGPDH homooligomers are endowed with in-gel activity and may represent native, though relatively labile (*i.e.* weak electrostatic interactions that are easily dissociated after Coomassie dye addition), forms of mtGPDH in the membrane of BAT mitochondria.

Schönfeld *et al.*[52] observed that oxidation of carnitine esters of medium- and long-chain fatty acids by rat heart mitochondria is not accompanied by reverse electron transfer to NAD^+ through CI although it produces the same or higher energization of mitochondria as compared to succinate oxidation by CII. No association could be found by BN-PAGE analysis between CI or other OXPHOS complexes and the ETF that participates in fatty acid oxidation. It must be remarked, however, that membrane-bound ETF dehydrogenase was not included in Schönfeld's study.

Since no clear association was demonstrated between CI and CII and most authors agree that CII is not a significant part of SCs, also reverse electron transfer from succinate to NAD^+, involving sequential interaction of complexes II and I by means of CoQ, must take place by collisional interactions in the CoQ pool. The hyperbolic relation experimentally found by Gutman[105] between the rate of reverse electron transfer and succinate oxidase is in complete accordance with the pool equation.

This observation poses the puzzling question[112] about how ubiquinol reduced in the membrane pool by CII interacts with the CoQ binding site in CI, since most CI is engaged in the SC with CIII, and the interaction of CoQ in the pool with the quinone-binding site in common between the two enzymes is necessarily slow. In view of the recent progress in the knowledge of the detailed atomic structure of CI,[122,123] the previous view that two different routes may exist for forward and reverse electron transfer within CI[124] is no longer tenable as such, unless we consider two different conformations, of which the one present during reverse electron transfer makes the CoQ site more accessible to the pool. It must be noted that the ATP-driven reverse electron transfer from succinate to NAD^+ occurs in the presence of a high mitochondrial transmembrane protonmotive force that, according to Piccoli *et al.*,[125] might be the physiological signal causing the structural reorganization of the respiratory complexes and this would no longer limit the access from the CoQ pool to the binding site in Complex I. The already mentioned analysis of SCs by cryo-EM[33,69] provides strong support for the accessibility of the CoQ site in CI to the free CoQ in the membrane. Such accessibility, however, would be no proof against the existence of channelling between CI and CIII.

12.3.2.5 Concluding Evidence About Channelling in the Coenzyme Q Region

As we described in the previous sections of this chapter, the major observations supporting the notion that supercomplex association determines channelling in the CoQ region are the following: (a) the rate advantage of NADH-cyt. *c* reductase when CI-containing SCs are present; (b) both CI and CIII are rate-limiting as measured by flux control analysis; (c) there is evidence for two compartments of CoQ in experiments of competition of NADH and succinate oxidation.

The dynamic character of CoQ bound within the SC, which is in dissociation equilibrium with the free pool of CoQ in the membrane, is one main point favouring controversy. According to our hypothesis, some CoQ molecules are trapped in a lipid micro-domain within the SC $I_1III_2IV_n$ and are channelled from CI to CIII during electron transfer at steady state; however, the relevance of CoQ dissociation from the SC to the pool becomes significant when electron transfer in the respiratory chain is slow or blocked by an inhibitor. Structural evidence pertaining the relevant sites where CoQ is reduced and re-oxidized strongly supports this idea. To this respect, the plasticity model[4] would be a functional rather than structural feature of the respiratory chain, at least in the CoQ region, a possibility advanced by Enriquez himself.[24] In other words, the SCs are stable supramolecular assemblies that do not readily dissociate, whereas CoQ behaves in a highly dynamic fashion in which channelling and diffusion take place in proportions depending on the turnover rates of the respiratory chain. It is desirable that studies on the dissociation constants of CoQ from SC will give an answer to this scientific dilemma in the future.

12.3.3 Electron Transfer Through Cytochrome *c*

The metabolic flux control studies in our laboratory clearly established that CI and CIII behave as a functional SC in bovine heart and rat liver mitochondria, whereas CIV does not since it has a low flux control coefficient, despite the BN-PAGE showing that a fraction of CIV is physically associated in the respirasome in the same mitochondrial samples.[80,97,146]

The reason for this discrepancy is not clear. It is true that most CIV appears to be free in the BN-gels (*cf.* Section 12.2), and one might consider that the cytochrome *c* oxidase activity in the respirasome is masked by the large excess of active enzyme randomly distributed in the membrane; however, the problem is rather difficult to examine critically. In fact, the experiments of Bianchi *et al.*[97] show that the NADH oxidase activity responds to cyanide titration with an inhibitor sensitivity lower than that of the cytochrome *c* oxidase specific activity. Assuming that (i) all CIV units, either free or bound in the respirasome, have the same sensitivity to CN^- and that (ii) the probability of the inhibitor to bind the enzyme is the same in both assay conditions, then the percent decrease of activity would be the same for NADH oxidase and for cytochrome *c* oxidase if electron transfer occurred by cyt. *c* channelling in the respirasome, giving a flux control coefficient close to unity. Since this is not the case, we must conclude that either one of the given assumptions is not correct or that cyt. *c* is not trapped in the respirasome and relies on an excess of free CIV units which can be randomly reduced. The latter means that the free molecules of CIV are also involved in electron transfer from NADH, and implies that the molecules of CIV assembled in the respirasomes do not participate in the channelling of cyt. *c*.[27,28] Of course, we are aware that branched pathways are difficult to approach by metabolic control analysis and that measurements in isolated mitochondria by inhibitor titration of the relevant enzymes inevitably involve using assay conditions (*i.e.* substrate concentrations) that are not necessarily intended to reproduce *in vivo* conditions.

However, the non-artifactual nature of the lack of functional channelling in the cyt. *c* region of mammalian mitochondria, as shown by the flux control coefficients, is supported by the fact that we conversely showed by the same assay method that the respirasome of potato tuber mitochondria is completely functional in cyt. *c* channelling.[80] The reason for the difference may be in a tighter binding of cyt. *c* in the potato respirasome.

The structural evidence of purified SCs by cryo-EM adds structural reasons on the looseness of CIII–CIV interactions in mammalian mitochondria. In ovine mitochondria, Letts *et al.*[33] identified two distinct arrangements of SC $I_1III_2IV_1$: a major 'tight' form and a minor 'loose' form (resolved at the resolution of 5.8 Å and 6.7 Å, respectively). In both respirasome structures, the density for CIV is weaker relative to CI and CIII, indicating the greater conformational flexibility of CIV. The clearest density for CIV can be seen in the tight respirasome in which CIV contacts both CI and CIII, whereas CIV in the loose respirasome has well-defined contacts only with CI. Tight and loose architectures may represent independent structural entities or may interconvert, indicating different stages of assembly or disassembly. The presence of two forms of the respirasome having distinct CIV linkages was also detected by Sousa *et al.*[69] and ascribed to instability of the purified SC.

Using a different approach, Lapuente-Brun *et al.*[109] demonstrated that at least part of CIV forms a functional SC with channelling of cyt. *c*. In particular, they showed that when CIV is allowed to participate in SCs by the presence of SCAF1, a significant proportion of CIV activity is not utilized by glucose-driven cell respiration. Moreover, they define three CIV populations: one dedicated exclusively to receive electrons from NADH oxidation (forming SC I + III + IV), another dedicated to receive electrons from FAD-dependent enzymes (forming SC III + IV) and a third major one that is in free form and that is able to receive electrons from both NADH and $FADH_2$ oxidization. On the contrary, if CIV is maintained permanently detached from SCs by elimination of SCAF1, Lapuente-Brun *et al.*[109] observed that the maximum respiration activity of intact cells under glucose-rich medium parallels the maximum respiration activity obtained in the same cells by directly feeding electrons to CIV by the specific electron donor tetramethylphenylenediamine (TMPD). In this case, since all CIV is in free form and electron transfer takes place *via* diffusion of a single pool of cyt. *c*, this experiment demonstrates that substrate pool behaviour is also very efficient.

Functional evidence for cyt. *c* channelling was also found in *S. cerevisiae* mitochondria;[126] accordingly, single particle cryo-EM revealed that the SC consisting of a III_2IV_2 unit (there is no CI in this yeast species) is arranged in such a way that the distance between the binding sites of cyt. *c*, *i.e.* cytochrome c_1 of CIII and the CuA-subunit II of CIV, is considerably shorter than in bovine mitochondria.[37]

Very recently, Rydström Lundin *et al.*[127] studied the kinetics of quinol oxidation-oxygen-reduction in *S. cerevisiae* mitochondria as a function of the respiratory SC factors Rcf1 and Rcf2 that mediate supramolecular interactions between CIII and CIV forming SCs III_2IV_{1-2}; they demonstrated that

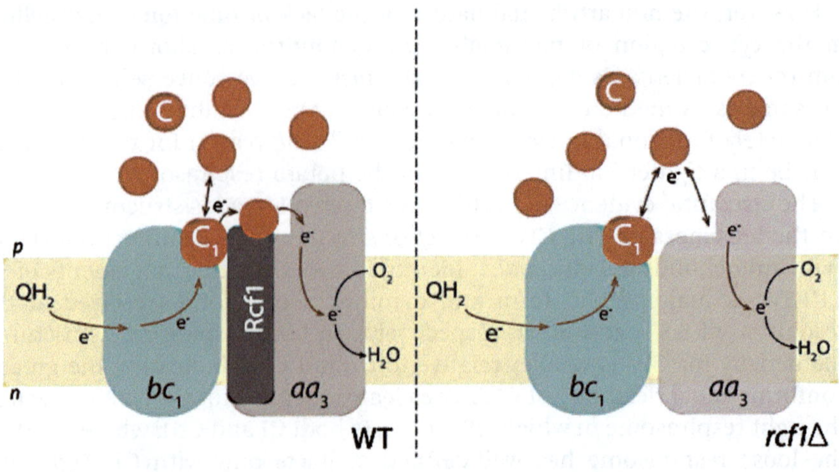

Figure 12.2 Schematic illustration of a suggested mechanism by which the Rcf1 polypeptide regulates the activity of the respiratory chain in *S. cerevisiae*. (*Left*) In the wild-type mitochondria and in the presence of added *S. cerevisiae* cytochrome *c* (C), electrons delivered by ubiquinol (QH_2) are transferred from cytochrome c_1 of CIII (the cyt. bc_1 complex) *via* a cytochrome *c* molecule that is attached at CIV (the cyt. aa_3 complex) or channelled in the interface between the two complexes, possibly bound at Rcf1. Under steady-state conditions, the direct electron transfer, *via* the bound cytochrome *c*, is faster than the equilibration of electrons with the cytochrome *c* pool (during this direct electron transfer, the cytochrome *c* pool is still slowly reduced because the pool is in equilibrium with the bound cytochrome *c*). (*Right*) In the absence of Rcf1, the direct electron transfer does not take place therefore electrons are only transferred *via* the cytochrome *c* pool: first from CIII to the pool and then from the pool to CIV.
Reprinted with permission from C. Rydström Lundin, C. von Ballmoos, M. Ott, P. Ädelroth, P. Brzezinski, *Proc Natl Acad Sci USA*, 2016, 113, E4476.

the Rcf1 polypeptide mediates formation of an electron-transfer bridge from CIII to CIV *via* a tightly bound cyt. *c* (Figure 12.2). In fact, in the wild-type mitochondria under steady-state conditions in the presence of added homologous cyt. *c*, the direct electron transfer, *via* the bound cyt. *c (i.e.* cyt. *c* channelling*)*, is faster than the equilibration of electrons with the cyt. *c* pool (noteworthy during this direct electron transfer, the cyt. *c* pool is still slowly reduced because the pool is in equilibrium with the bound cyt. *c*). Interestingly, when using heterologous cyt. *c* (from horse heart), there is no pre-bound cyt. *c* and electron transfer between CIII and fully active CIV occurs only *via* the cyt. *c* pool. Moreover, results from recent studies indicate that in the absence of Rcf1, interactions required for stability of the SC III + IV are disrupted.[128] Consistently, Rydström Lundin *et al.*[127] demonstrated that direct electron transfer does not take place in the absence of Rcf1 and electrons are only transferred *via* the cyt. *c* pool in the Rcf1Δ strain.

To further confuse the issue, it is also worth noting that Trouillard *et al.*[129] demonstrated that the time-resolved oxidation of cyt. *c* by CIV in yeast mitochondria is a random-collision process not involving substrate compartmentalization.

At the same time, however, since the purified SC $I_1III_2IV_1$ obtained from bovine heart[66] and some putative respirasomes isolated from rat liver mitochondria[58] was also shown to contain traces of bound cyt. *c*, we cannot definitively exclude that substrate channelling on this level occurs also in mammalian mitochondria.

We will discuss in Section 12.5.1. the possible physiological reasons why there is no major channelling of cyt. *c* in mammalian mitochondria.

12.4 Supercomplexes and Reactive Oxygen Species

The capacity of mitochondria to produce ROS depends on the type of tissue as well as on the metabolic conditions of the cell.[130,131] For example, succinate is important for ROS production in brain, heart, kidney and skeletal muscle, while fatty acids are major generators of ROS in kidney and liver.[132]

The redox potential of the NAD^+/NADH couple and the proton-motive force act as powerful regulators of the steady-state concentration of redox species responsible for electron leaking and ROS generation.[133,134] In turn, these forces are regulated by the redox supply to the respiratory chain, by the degree of coupling and/or by physio-pathological constraints to electron transfer, such as enzyme phosphorylation, cyt. *c* removal, CIV inhibition and oxygen concentration *etc.*

The "Redox-Optimized ROS Balance hypothesis" (R-ORB) described by Aon *et al.*[135] attempts to explain at a mechanistic level the link between mitochondrial respiration and ROS emission. The hypothesis is based on the observation that ROS levels (as the net result of production and scavenging) attain a minimum when mitochondria maximize their energetic output (*i.e.* maximal state-3 respiration) at intermediate values of the redox environment between fully oxidized and fully reduced redox couples such as NADH/NAD^+ and GSH/GSSG. These redox carriers are more oxidized than the corresponding values in state-4 respiration. On the other hand, ROS overflow will occur at both highly reduced or highly oxidized redox environments, albeit governed by the extent of proton leak and by the compromised scavenging capacity, respectively. By assessing mitochondrial respiration (*i.e.* forward electron transport), ROS emission and redox environment in isolated guinea pig heart mitochondria, Cortassa *et al.*[136] confirmed that mitochondria are able to keep ROS emission to a minimum likely compatible with signalling, while maximizing their energetic output.

In addition, mitochondrial ROS release is modulated by a series of nuclear encoded proteins such as by p53, p66Shc, the Bcl-2 family and Romo-1[137] and in response to external stimuli, such as TNFα,[138] hypoxia,[139] serum deprivation,[140] oxidative stress (the so-called ROS-induced ROS release[141]).

ROS production in mitochondria is a byproduct of core metabolism of the cell that cannot be easily down-regulated without under-mining cell function.[142]

An implication of supercomplex organization as the missing link between oxidative stress and energy failure was first suggested by our speculation[112] that dissociation of SC I_1III_2 occurs under conditions of oxidative stress, with loss of facilitated CoQ channelling, thus causing electron transfer to depend upon diffusive collisional encounters of the free ubiquinone molecules with their binding partner complexes, which is a random process and therefore much less efficient.

As a consequence, the alteration of electron transfer may elicit high reducing pressure and further induction of ROS generation. Following this line of thought, the different susceptibility of different types of cells and tissues to ROS damage[143] may be interpreted in terms of (i) extent and tightness of supercomplex organization in their respiratory chains and, related to the former, (ii) capability of the respiratory enzymes to hinder reaction of their auto-oxidizable prosthetic groups with oxygen. Disorganization of the supramolecular architecture of the respiratory chain would both decrease NAD-linked respiration and ATP synthesis and increase the capacity of producing superoxide by destabilised CI and CIII.

On the basis of their studies on rat brain mitochondria oxidizing different substrates, Panov *et al.*[130] suggested that the assembly of CI into SCs prevents excessive superoxide production during oxidation of NAD-linked substrates because efficient CoQ channelling helps maintaining the chain in the oxidized state (see also below).

In a recent study, we[81] obtained a direct demonstration that loss of SC I_1III_2 organization causes a dramatic enhancement of ROS generation by CI. In our model system of reconstituted binary CI/CIII at high lipid to protein ratio (30 : 1), where formation of the SC I_1III_2 is prevented, the generation of superoxide is several-fold higher than in the same binary system reconstituted at a 1 : 1 ratio, which is conversely rich in SC I_1III_2 (Figure 12.3). In agreement with this finding, dissociation of the SC I_1III_2 after treatment with the detergent DDM induces a strong enhancement of ROS generation both in proteoliposomes and in mitochondrial membranes. It is worth noting that, in all these experiments,[81] the production of ROS is investigated in the presence of inhibitors (mucidin and rotenone) that prevent electron transfer to any respiratory intermediate substrate, therefore the redox centres in CI are maximally reduced both in the presence and in the absence of the supercomplex assembly. Consequently, the above mentioned reasoning by Panov *et al.*,[130] giving emphasis to the relief of the redox pressure in the super-assembled respiratory chain, cannot be taken as the only explanation of the role of SC I_1III_2 in regulating ROS formation by CI.

Analysis of the 3D architecture of the SC $I_1III_2IV_1$ from bovine heart[144] suggested a slightly different conformation of bound CI showing a higher bending of the matrix arm towards the membrane (and presumably CIII), compared to free CI. Since, however, subsequent studies at higher resolution have failed to confirm such conformational changes,[66,67] further studies

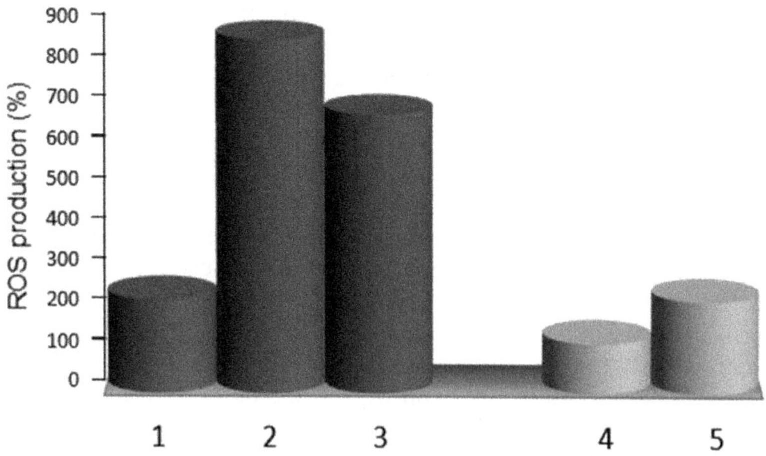

Sample ID	Relative amount of bound Complex I (% of total)	ROS production (%)
1 proteoliposomes	92	224±83
2 proteoliposomes DIL	24	851±243
3 proteoliposomes +DDM	21	672±257
4 BHM	87	121±20
5 BHM +DDM	2	223±86

Figure 12.3 Production of ROS by mitochondrial CI in different situations where supercomplexes are maintained or disassembled. The percentage value of ROS production measured in all the samples listed in the table (*lower panel*) is plotted in the graph (*upper panel*). As described in Maranzana *et al.*,[81] the ratio of bound CI versus total CI was determined by densitometric analysis of immunoblots obtained after 2D BN/SDS-PAGE. The NADH-stimulated production of ROS was measured as the relative fluorescence intensity of dichlorofluorescein in the presence of 1.8 μM mucidin and 4 μM rotenone, and expressed as percentage value of the corresponding reference samples in the absence of rotenone. In the case of BHM, the existence of endogenous systems operating to reduce ROS levels in the mitochondrial sample might have counteracted the dramatic effects of the complete dissociation of CI, thus leading to a twofold only increase of the measured ROS production. Proteoliposomes, R4B 1 : 1 w : w (*cf.* text for details); DIL, dilution at high lipid to protein ratio (30 : 1 w : w); DDM, dodecyl maltoside. Reprinted from *Biochimica et Biophysica Acta (BBA) – Bioenergetics*, Volume **1857**, G. Lenaz, G. Tioli, A. I. Falasca, M. L. Genova, Complex I function in mitochondrial supercomplexes, 991–1000, Copyright 2016, with permission from Elsevier.

are needed to solve this issue. Nevertheless, the observed destabilization of CI in the absence of SC may render the 51 kDa subunit containing the FMN more prone to interact with oxygen. According to Milenkovic *et al.*[29] the localization of the 51 kDa subunit in the matrix, far from the protein interactions at the basis of SC association, seems incompatible with this assumption.

The established architecture of respiratory SCs may also help in controlling ROS production because, due to the high oxygen-consuming activity of CIV, the local concentration of O_2 would be low near the cavity in CIII adjacent to CIV, an ideal condition for shielding a reactive semiquinone intermediate ($Q^{\bullet}$) and thus preventing ROS generation from CIII.[33]

Several observations in cellular and animal models link together SC dissociation and enhanced ROS production.

A strong decrease of high molecular weight SC correlating with higher ROS generation was observed in mouse fibroblasts expressing the activated form of the k-ras oncogene, in comparison with wild type fibroblasts.[145,146]

Diaz *et al.*[147] showed that diminished stability of SC is associated with increased levels of ROS in mouse lung fibroblasts lacking the Rieske iron–sulphur protein of CIII and hence devoid of CI-containing SC.

The cardiolipin defect in Barth syndrome, a cardio-skeletal myopathy with neutropenia which is characterized by respiratory chain dysfunction, results in destabilization of SC by weakening the interactions between CIII and CIV. Remarkably, hydroethidine staining revealed higher basal levels of superoxide production in lymphoblasts from patients, compared to control cells.[73,74] The availability of cardiolipin-deficient yeast mutants, although lacking CI, provided the opportunity to demonstrate alterations in the stabilization of SC similar to those found in Barth syndrome and exhibited increased protein carbonylation, an indicator of ROS.[148] The increase in ROS is most likely not due to defective oxidant defence systems, since the CL mutants do not display sensitivity to other oxidants like paraquat, menadione or hydrogen peroxide (H_2O_2).

The proposal that CI-containing SCs may physiologically exist in dynamic equilibrium with isolated CI (Section 12.3.1.2) raises the puzzling suspicion that the production of ROS may consequently be subjected to physiological oscillations. It is tempting to suggest that these changes may be aimed at controlling ROS levels in the cell, in view of their well-documented role in cellular redox signalling (*cf.* Section 12.5.2).

12.5 Physiological and Pathological Implications

12.5.1 Supercomplexes and Regulation of Metabolic Fluxes

Does CoQ channelling occur under physiological conditions? As discussed in the previous sections, a possible bottleneck around cyt. *c* might induce interaction of the NADH pathway (*via* channelled CoQ in SC I_1III_2IVn) with the succinate pathway (CoQ pool behaviour); however, if a rate-limiting step is situated upstream, *i.e.* in or before the dehydrogenases, the reducing pressure of CoQ on its partner oxidases (particularly in the case of CIII assembled in the SC I_1III_2IVn) may not be present and the two routes would take place independently. When oxidative metabolism proceeds mainly *via* the glycolytic pathway and the Krebs cycle, there is a prevalence of oxidation of NAD-linked substrates over succinate oxidation, allowing an undisturbed electron flux through SC I_1III_2. Strong oxidation of FAD-linked substrates as in fatty acid oxidation (where FAD-linked ETF dehydrogenase is most probably not forming SCs) might however induce strong interaction of the highly reduced CoQ pool with the SC. As postulated by Lapuente-Brun *et al.*,[109] at the metabolic level, fast kinetic adjustments of the mitochondrial respiratory activity would be followed by gene expression changes of the level of individual respiratory complexes.

The existence of two different functional CoQ compartments has profound implications for the understanding of physiological responses such as metabolic adaptation of liver mitochondria to fasting, which forces fat mobilization and fatty acid oxidation.[149] The use of different fuel molecules generates different proportions of NADH and $FADH_2$, that require an optimal equilibrium between the corresponding routes of electron transfer in the respiratory chain. In fact, an increase of the $FADH_2$-dependent activity, as in fatty acid oxidation, induces saturation of the CoQ pool oxidation capacity and reverse electron transport from ubiquinol to CI;[150] the resulting local generation of superoxide triggers protein degradation of CI subunits by oxidative damage, and consequent disintegration of the complex, as well as lipid peroxidation, disassembly of SCs and further instability of their enzyme components. Thus, we can infer that CoQ redox status acts as a metabolic sensor that fine-tunes the configuration and supramolecular organization of the respiratory chain in order to match the prevailing substrate profile.[151] In order to avoid such saturation of the CoQ pool oxidation capacity, fasting conditions require proper disassembly of SCs to achieve adjustment of respiratory supercomplex proportions favouring the FAD-linked route. Similar adaptations are expected to be necessary when there is an increase in fatty acid oxidation in response to high-fat diet. Impairment of this adaptation may be relevant to pathological processes associated to obesity.

Why does no appreciable channelling exist at the level of cyt. *c* in mammalian mitochondria? Mammalian mitochondria have several dehydrogenases directing electrons to CoQ, however they usually have only one oxidase (CIV) receiving electrons from CIII *via* cyt. *c*. For this reason, it may be worth speculating that, whilst it may be useful to separate the major NADH-dependent flux from CI from those departing from succinate, fatty acid oxidation and other metabolic pathways by separating the CoQ compartments, there is no such need for cyt. *c* that is by and large receiving electrons univocally from CIII. This assumption is reinforced by the fact that, contrary to what is found in mammalian mitochondria. CIII and CIV of plant mitochondria are functionally operating as a SC and, at the same time, cyt. *c* is tightly bound in the SC (*cf.* Section 12.3.3), as revealed by flux control analysis[97] and BN-PAGE.[80] Not surprisingly, plant mitochondria are characterized by a high branching of the electron transfer pathways feeding electrons directly to cyt. *c*,[12] which requires adjustment of the different routes as a response to physiological needs, as happens in mammalian mitochondria at the CoQ level. In plant mitochondria, segmentation might be achieved by regulating different compartments of free and bound cyt. *c*.

12.5.2 Supercomplexes and ROS Signalling

The role of mitochondrial ROS in cell signalling has been the subject of recent excellent reviews (*cf.* ref. 152, 153 and 154). Here we deal with a possible role of the supramolecular organization of the respiratory chain on ROS signalling.

As ROS are involved in cell signalling, it is expected that their generation is subjected to tight control.

The factors controlling mitochondrial ROS levels are linked to their rate of generation and of removal; these have been considered in Section 12.4. of this chapter. In this scenario, a series of nuclear encoded protein factors[137] as well as the forces directly associated with respiratory activity, which are the redox potential of the NAD^+/NADH couple and the proton-motive force, are powerful regulators of the steady-state concentration of the redox species responsible for electron leaking and ROS production.[155] Hoffman and Brookes[156] have investigated the ROS generation by rat liver mitochondria under different substrate and inhibitor conditions and different oxygen tensions, establishing the oxygen sensitivity of the different O_2-reacting sites: from such data, the apparent K_m for O_2 was lowest for CI during forward flow, followed by CI backflow, CIII Qo site, and highest for ETF dehydrogenase. They conclude that at physiological O_2 concentration, only CI may be a significant source of ROS.

We first speculated[112] and then obtained experimental demonstration[80] that dissociation of SC I_1III_2 occurs under conditions of oxidative stress, with loss of facilitated electron channelling in the CoQ region and resumption of a less efficient random diffusional behaviour. This condition forces electron-transfer to depend upon the collisional encounters of the free ubiquinone molecules with the partner complexes and may elicit high reducing pressure and further induction of ROS generation.

In fact, supercomplex disorganization eventually leading to destabilization of CI and CIII decreases NAD-linked respiration and ATP synthesis and also increases the capacity of these enzymes of producing superoxide (*cf.* Section 12.4. for experimental evidence pertaining to this topic).

The study of Guaras *et al.*[151] pinpoints another aspect of SCs in relation to ROS formation. Hyper-reduction of the CoQ pool by $ETFH_2$ oxidation during extensive fatty acid β-oxidation induces reverse electron transfer with a rise of ROS production by CI. Thus, shifting metabolic fuels from NADH-dependent to $FADH_2$-dependent substrates may adjust ROS generation by way of the specific supramolecular assembly of the respiratory complexes involved.[150]

12.5.3 Supercomplexes in Pathology and Aging

Alterations (dissociation) of the SCs often accompany pathological changes, however no direct proof exists that SC dissociation is the direct cause of the pathology.

Even so, circumstantial evidence points to a vicious circle (*cf.* ref. 157 for review).

The notion that the respiratory chain is mainly controlled at the level of CI suggests that the main alterations due to ageing must be found at the level of this enzyme,[158,159] Actually, flux control analysis in aerobic respiration in coupled liver mitochondria[160] showed that CI has little control in young rats but very high control in the old animals, meaning that aging induces a profound alteration of CI that is reflected on the entire OXPHOS.

Analysis of the occurrence of respiratory SCs (*cf.* Section 12.2.1) reveals age-related variations, suggesting that destabilization of the supramolecular organization of the respiratory chain may be crucial for the development of the aging-phenotype.[143,161,162]

In mitochondria of rat cortex, Frenzel *et al.*[163] quantified profound age-associated changes in the proportion of supramolecular assemblies of the respiratory complexes as well as of the FoF1 ATP synthase. Notably, the overall decline with age (−40%) in the CI-containing SCs is caused in large extent by the pronounced decline (−58%) of abundance of the SC I_1III_2.

The progeroid profile of the mutator mice, a remarkable model of premature aging having a defect of the proof-reading function of mtDNA polymerase-γ (Pol-γ) and hence exhibiting multiple mtDNA mutations, is accompanied by decreased respiration and impeded assembly of respiratory complexes.[164] Despite the fact that only CIV subunits were significantly reduced, the steady-state level and activity of CI was also affected and strongly decreased, in line with the idea that a decreased assembly of CIV may secondarily affect the stability of CI through disruption of the supercomplex organization.[165]

The model hypothesis that we propose poses supercomplex dissociation amongst a double influence of ROS: on one hand ROS contribute to dissociate SCs, on the other hand supercomplex dissociation enhances ROS generation (Figure 12.4). This means that if these events are not tightly controlled, they may lead to initiation of a vicious cycle of ROS generation. Is this the series of events that is conducive to aging?

Taken together, the observations collected in this chapter locate supercomplex dissociation in a physiological signalling network that can be easily altered and leads to a catastrophic event when the generation of ROS loses control.

Under physiological conditions we may envisage supercomplex association/dissociation to occur, according to the plasticity model,[58] under such stimuli as mitochondrial membrane potential and protein phosphorylation/dephosphorylation of the respiratory complexes; the ensuing changes in ROS generation modulate the signalling pathways that are initiated by ROS.

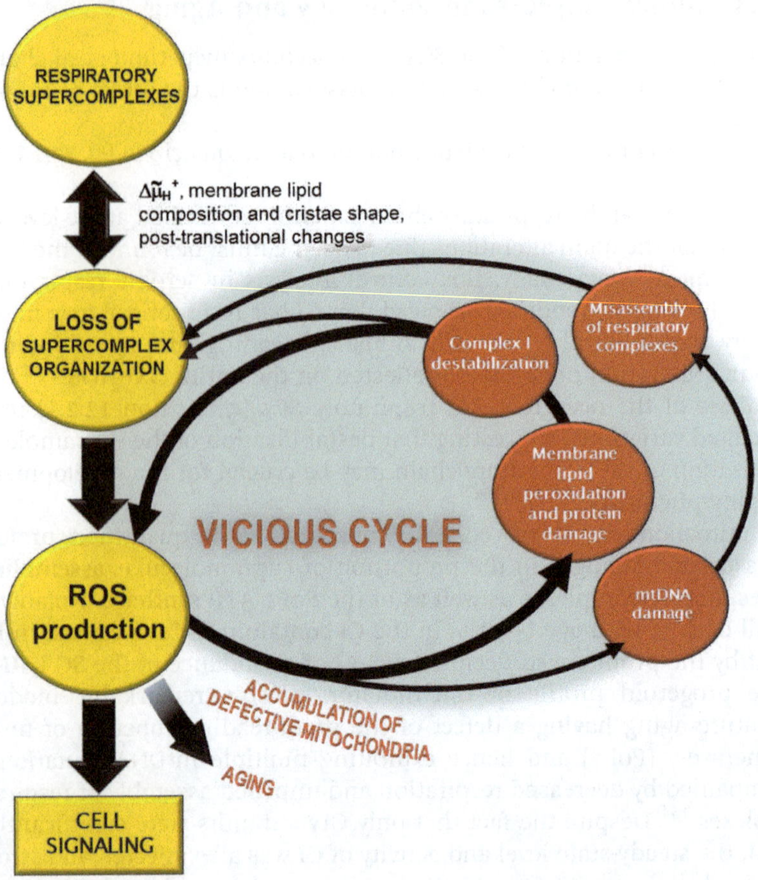

Figure 12.4 Scheme showing how the loss of supercomplex organization may be involved in a *vicious cycle* of oxidative stress and energy failure. ROS production by CI is enhanced as a consequence of respiratory supercomplex disassembly. Membrane phospholipid peroxidation, mtDNA damage and consequent further loss of supercomplex organisation may occur due to enhanced mitochondrial oxidative stress, thus perpetuating a *vicious cycle*. Depending on the amount produced, ROS can also operate as signaling molecules from mitochondria to the cell.

These changes are reversible and are kept under strict control from changes in the starting conditions.

We propose that the primary event responsible for aging is the structural damage induced by ROS in mitochondria, as predicted by the original mitochondrial theory of aging.[166]

A possible series of events might be the following. Progressive damage is induced by ROS to the mitochondrial membrane lipids and proteins. MtDNA mutations, although present, may not necessarily be an early phenomenon in the aging process.

The level of ROS may be affected by such factors as the nutrition state and the activity of the mTOR and insulin/IGF pathways. Direct protein damage and increased cardiolipin peroxidation hamper supercomplex association,[80] thus leading to further increase of ROS generation. ROS at low concentration may induce retrograde signals evoking compensatory mechanisms that attempt to counteract the ROS generation and consequent risk of damage (see Section 12.4), but at higher concentration they may induce further damage with loss of coordination of the signalling pathways (see Section 12.5.2). Mutations in mtDNA at a later stage would make the overall process irreversible and lead to the final aging phenotype.

References

1. D. G. Nicholls, S. J. Feguson, *Bioenergetics*, Academic, New York, 4th edn, 2013.
2. Y. Hatefi, A. G. Haavik, L. R. Fowler and D. E. Griffiths, *J. Biol. Chem.*, 1962, **237**, 2661.
3. D. E. Green and A. Tzagoloff, *Arch. Biochem. Biophys.*, 1966, **116**, 293.
4. M. Sarewicz and A. Osyczka, *Physiol. Rev.*, 2015, **95**, 219.
5. H. Schägger and K. Pfeiffer, *J. Biol. Chem.*, 2001, **276**, 37861.
6. G. Lenaz and M. L. Genova, *Antioxid. Redox Signaling*, 2010, **12**, 961.
7. J. Houstek, B. Cannon and O. Lindberg, *Eur. J. Biochem.*, 1975, **54**, 11.
8. J. D. Beckmann and F. E. Frerman, *Biochemistry*, 1985, **24**, 3922.
9. D. R. Evans and H. I. Guy, *J. Biol. Chem.*, 2004, **279**, 33035.
10. F. Salvi and G. Gadda, *Arch. Biochem. Biophys.*, 2013, **537**, 243.
11. L. B. Poole, *Free Radic. Biol. Med.*, 2015, **80**, 148.
12. P. Schertl and H. P. Braun, *Front. Plant Sci.*, 2014, **5**, 163.
13. U. Kappler and J. H. Enemark, *J. Biol. Inorg. Chem.*, 2015, **20**, 253.
14. A. Kröger and M. Klingenberg, *Eur. J. Biochem.*, 1973, **34**, 358.
15. A. Kröger and M. Klingenberg, *Eur. J. Biochem.*, 1973, **39**, 313.
16. C. R. Hackenbrock, B. Chazotte and S. S. Gupte, *J. Bioenerg. Biomembr*, 1986, **18**, 331.
17. B. Chance and G. R. Williams, *Nature*, 1955, **176**, 250.
18. Y. Hatefi, A. G. Haavik and D. E. Griffiths, *J. Biol. Chem.*, 1962, **237**, 1676.
19. S. J. Singer and G. L. Nicolson, *Science*, 1972, **175**, 720.
20. T. Ozawa, M. Nishikimi, H. Suzuki, M. Tanaka, Y. Shimomura, Bioenergetics: Structure and Function of Energy-Transducing Systems, ed. T. Ozawa and S. Papa, Japan Sci Soc Press, Tokyo, 1987, *Structure and Assembly of Mitochondrial Electron-transfer Complexes*, 101–119.
21. J. Hochman, S. Ferguson-Miller and M. Schindler, *Biochemistry*, 1985, **24**, 2509.
22. H. Schägger and K. Pfeiffer, *EMBO J.*, 2000, **19**, 1777.
23. I. Wittig and H. Schägger, *Proteomics*, 2005, **5**, 4338.
24. J. A. Enríquez, *Annu. Rev. Physiol.*, 2016, **78**, 533.

25. G. Lenaz, G. Tioli, A. I. Falasca and M. L. Genova, *Biochim. Biophys. Acta*, 2016, **1857**, 991.
26. R. Acín-Peréz and J. A. Enríquez, *Biochim. Biophys. Acta*, 2014, **1837**, 444.
27. M. L. Genova and G. Lenaz, *Biol. Chem.*, 2013, **394**, 631.
28. M. L. Genova and G. Lenaz, *Biochim. Biophys. Acta*, 2014, **1837**, 427.
29. D. Milenkovic, J. N. Blaza, N. G. Larsson and J. Hirst, *Cell Metab.*, 2017, **25**, 765.
30. J. Vonck and E. Schäfer, *Biochim. Biophys. Acta*, 2009, **1793**, 117.
31. I. Wittig, H. P. Braun and H. Schägger, *Nat. Protoc.*, 2006, **1**, 418.
32. F. Krause and H. Seelert, *Curr. Protoc. Protein Sci.*, 2008, **54**, 19.18.1.
33. J. A. Letts, K. Fiedorczuk and L. A. Sazanov, *Nature*, 2016, **537**, 644.
34. A. Stroh, O. Anderka, K. Pfeiffer, T. Yagi, M. Finel, B. Ludwig and H. Schägger, *J. Biol. Chem.*, 2004, **279**, 5000.
35. C. Y. Yip, M. E. Harbour, K. Jayawardena, I. M. Fearnley and L. A. Sazanov, *J. Biol. Chem.*, 2011, **286**, 5023.
36. R. A. Stuart, *J. Bioenerg. Biomembr.*, 2008, **40**, 411.
37. E. Mileykovskaya, P. A. Penczek, J. Fang, V. K. P. S. Mallampalli, G. C. Sparagna and W. Dowhan, *J. Biol. Chem.*, 2012, **287**, 23095.
38. K. Römpler, T. Müller, L. Juris, M. Wissel, M. Vukotic, K. Hofmann and M. Deckers, *J. Biol. Chem.*, 2016, **291**, 23769.
39. F. Krause, *Electrophoresis*, 2006, **27**, 2759.
40. F. Krause, C. Q. Scheckhuber, A. Werner, S. Rexroth, N. H. Reifschneider, N. A. Dencher and H. D. Osiewacz, *J. Biol. Chem.*, 2004b, **279**, 26453.
41. J. B. Bultema, H. P. Braun, E. J. Boekema and R. Kouril, *Biochim. Biophys. Acta*, 2009, **1787**, 60.
42. I. Marques, N. A. Dencher, A. Videira and F. Krause, *Eucaryotic Cell*, 2007, **6**, 2391.
43. E. Nübel, I. Witting, S. Kerscher, U. Brandt and H. Schägger, *Proteomics*, 2009, **9**, 2408.
44. R. A. Stuart, *Methods Enzymol.*, 2009, **456**, 191.
45. D. U. Mick, K. Wagner, M. van der Laan, A. E. Frazier, I. Perschil, M. Pawlas, H. E. Meyer, B. Warscheid and P. Rehling, *EMBO J.*, 2007, **26**, 4347.
46. S. Sunderhaus, N. V. Dudkina, L. Jänsch, J. Klodmann, J. Heinemeyer, M. Perales, E. Zabaleta, E. J. Boekema and H. P. Braun, *J. Biol. Chem.*, 2006, **281**, 6482.
47. T. Hashimoto, R. Hussien, H. S. Cho, D. Kaufer and G. A. Brooks, *PLoS One*, 2008, **3**, e2915.
48. T. Persichini, V. Mazzone, F. Polticelli, S. Moreno, G. Venturini, E. Clementi and M. Colasanti, *Neurosci Lett.*, 2005, **384**, 254.
49. S. Ohya, Y. Kuwata, K. Sakamoto, K. Muraki and Y. Imaizumi, *Am. J. Physiol. Heart Circ. Physiol.*, 2005, **289**, H1635.
50. W. Suthammarak, Y. Y. Yang, P. G. Morgan and M. M. Sedensky, *J. Biol. Chem.*, 2009, **284**, 6425.

51. Y. Wang, A. W. Mohsen, S. J. Mihalik, E. S. Goetzman and J. Vockley, *J. Biol. Chem.*, 2010, **285**, 29834.
52. P. Schönfeld, M. R. Wieckowski, M. Lebiedzińska and L. Wojtczak, *Biochim. Biophys. Acta*, 2010, **1797**, 929.
53. A. Nsiah-Sefaa and M. McKenzie, *Biosci. Rep.*, 2016, **36**, pi:e00313.
54. B. Sumegi and P. A. Srere, *J. Biol. Chem.*, 1984, **259**, 15040.
55. J. Ovádi, Y. Huang and H. O. Spivey, *J. Mol. Recognit.*, 1994, **7**, 265.
56. T. Fukushima, R. V. Decker, W. M. Anderson and H. O. Spivey, *J. Biol. Chem.*, 1989, **264**, 16483.
57. A. B. Kotlyar, E. Maklashina and G. Cecchini, *Biochem. Biophys. Res. Commun.*, 2004, **18**, 987.
58. R. Acín-Pérez, P. Fernández-Silva, M. L. Peleato, A. Pérez-Martos and J. A. Enriquez, *Mol. Cell*, 2008, **32**, 529.
59. N. Kovářová, T. Mráček, H. Nůsková, E. Holzerová, M. Vrbacký, P. Pecina, K. Hejzlarová, K. Kľučková, J. Rohlena, J. Neuzil and J. Houštěk, *PLoS One*, 2013, **8**, e71869.
60. H. Rauchová, M. Battino, R. Fato, G. Lenaz and Z. Drahota, *J. Bioenerg. Biomembr.*, 1992, **24**, 235.
61. T. Mráček, E. Holzerová, Z. Drahota, N. Kovářová, M. Vrbacký, P. Ješina and J. Houštěk, *Biochim. Biophys. Acta*, 2014, **1837**, 98.
62. J. Fang, T. Uchiumi, M. Yagi, S. Matsumoto, R. Amamoto, S. Takazaki, H. Yamaza, K. Nonaka and D. Kang, *Biosci. Rep.*, 2013, **33**, e00021.
63. T. M. Hildebrandt, *Biochim. Biophys. Acta*, 2011, **1807**, 1206.
64. J. Ovádi, *J. Theor. Biol.*, 1991, **152**, 135.
65. E. Schäfer, H. Seelert, N. H. Reifschneider, F. Krause, N. A. Dencher and J. Vonck, *J. Biol. Chem.*, 2006, **281**, 15370.
66. T. Althoff, D. J. Mills, J. L. Popot and W. Kühlbrandt, *EMBO J.*, 2011, **30**, 4652.
67. N. V. Dudkina, M. Kudryashev, H. Stahlberg and E. J. Boekema, *Proc. Natl. Acad. Sci. U. S. A.*, 2011, **108**, 15196.
68. M. Wu, J. Gu, R. Guo, Y. Huang and M. Yang, *Cell*, 2016, **167**, 1598.
69. J. S. Sousa, D. J. Mills, J. Vonck and W. Külbrandt, *eLife*, 2016, **5**, e21290.
70. P. Mitchell, *FEBS Lett.*, 1975, **59**, 137.
71. R. Covian and B. L. Trumpower, *Biochim. Biophys. Acta*, 2008, **1777**, 1079.
72. R. Covian and B. L. Trumpower, *J. Biol. Chem.*, 2005, **280**, 22732.
73. M. McKenzie, M. Lazarou, D. R. Thorburn and M. T. Ryan, *J. Mol. Biol.*, 2006, **361**, 462.
74. F. Gonzalvez, M. D'Aurelio, M. Boutant, A. Moustapha, J. P. Puech, T. Landes, L. Arnauné-Pelloquin, G. Vial, N. Taleux, C. Slomianny, R. J. Wanders, R. H. Houtkooper, P. Bellenguer, I. M. Møller, E. Gottlieb, F. M. Vaz, G. Manfredi and P. X. Petit, *Biochim. Biophys. Acta*, 2013, **1832**, 1194.
75. M. Zhang, E. Mileykovskaya and W. Dowhan, *J. Biol. Chem.*, 2002, **277**, 43553.

76. K. Pfeiffer, V. Gohil, R. A. Stuart, C. Hunte, U. Brandt, M. L. Greenberg and H. Schägger, *J. Biol. Chem.*, 2003, **278**, 52873.
77. M. Desmurs, M. Foti, E. Raemy, F. M. Vaz, J. C. Martinou, A. Bairoch and L. Lane, *Mol. Cell Biol.*, 2015, **35**, 1139.
78. L. Böttinger, S. E. Horvath, T. Kleinschroth, C. Hunte, G. Daum, N. Pfanner and T. Becker, *J. Mol. Biol.*, 2012, **423**, 677.
79. C. D. Baker, W. Basu Ball, E. N. Pryce and V. M. Gohil, *Mol. Biol. Cell*, 2016, **27**, 2161.
80. M. L. Genova, A. Baracca, A. Biondi, G. Casalena, M. Faccioli, A. I. Falasca, G. Formiggini, G. Sgarbi, G. Solaini and G. Lenaz, *Biochim. Biophys. Acta*, 2008, **1777**, 740.
81. E. Maranzana, G. Barbero, A. I. Falasca, G. Lenaz and M. L. Genova, *Antioxid. Redox Signaling*, 2013, **19**, 1469.
82. G. Paradies, G. Petrosillo, M. Pistolese and F. M. Ruggiero, *FEBS Lett.*, 2000, **466**, 323.
83. G. Paradies, G. Petrosillo, M. Pistolese and F. M. Ruggiero, *Gene*, 2002, **286**, 135.
84. G. Petrosillo, F. M. Ruggiero, N. Di Venosa and G. Paradies, *Faseb J.*, 2003, **17**, 714.
85. G. Paradies, G. Petrosillo, V. Paradies and F. M. Ruggiero, *Free Radic. Biol. Med.*, 2010, **48**, 1286.
86. K. Shinzawa-Itoh, H. Shimomura, S. Yanagisawa, S. Shimada, R. Takahashi, M. Oosaki, T. Ogura and T. Tsukihara, *J. Biol. Chem.*, 2016, **291**, 4178.
87. Various Authors, Redox Proteins in Supercomplexes and Signalosomes, Eds. R. O. Louro and I. Diaz-Moreno, CRC Press, Boca Raton, FL, USA, 2015.
88. C. I. Ragan and C. Heron, *Biochem. J.*, 1978, **174**, 783.
89. C. Heron, C. I. Ragan and B. L. Trumpower, *Biochem. J.*, 1978, **174**, 791.
90. G. Lenaz, R. Fato, S. Di Bernardo, D. Jarreta, A. Costa, M. L. Genova and G. Parenti Castelli, *Biofactors*, 1999, **9**, 87.
91. J. S. Rieske, *Methods Enzymol.*, 1967, **10**, 239.
92. C. Bianchi, R. Fato, M. L. Genova, G. Parenti Castelli and G. Lenaz, *BioFactors*, 2003, **18**, 3.
93. G. Vanderkooi, Molecular Biology of Membranes, ed. S. Fleischer, Y. Hatefi, D. MacLennan and A. Tzagoloff, *Organization of Protein and Lipid Components in Membranes*, Plenum Publishing Corp, New York, NY, 1978, 25–55.
94. K. Schwerzmann, L. M. Cruz-Orive, R. Eggman, A. Sänger and E. R. Weibel, *J. Cell Biol.*, 1986, **102**, 97.
95. G. Lenaz, *J. Membr. Biol.*, 1988, **104**, 193.
96. N. B. Kholodenko and H. V. Westerhoff, *FEBS Lett.*, 1993, **320**, 71.
97. C. Bianchi, M. L. Genova, G. Parenti Castelli and G. Lenaz, *J. Biol. Chem.*, 2004, **279**, 36562.
98. J. N. Blaza, R. Serrelli, A. J. Jones, K. Mohammed and J. Hirst, *Proc. Natl. Acad. Sci. U. S. A.*, 2014, **111**, 15735.

99. R. Fato, E. Estornell, S. Di Bernardo, F. Pallotti, G. Parenti Castelli and G. Lenaz, *Biochemistry*, 1996, **35**, 2705.
100. G. Quarato, C. Piccoli, R. Scrima and N. Capitanio, *Biochim. Biophys. Acta*, 2011, **1807**, 1114.
101. T. Kaambre, V. Chekulayev, I. Shevchuk, M. Karu-Varikmaa, N. Timohhina, K. Tepp, J. Bogovskaja, R. Kütner, V. Valvere and V. Saks, *J. Bioenerg. Biomembr.*, 2012, **44**, 539.
102. T. Kaambre, V. Chekulayev, I. Shevchuk, K. Tepp, N. Timohhina, M. Varikmaa, R. Bagur, A. Klepinin, T. Anmann, A. Koit, A. Kaldma, R. Guzun, V. Valvere and V. Saks, *Front. Physiol.*, 2013, **4**, 151.
103. M. Gutman and N. Silman, *FEBS Lett.*, 1972, **26**, 207.
104. M. Gutman, E. B. Kearney and T. P. Singer, *Biochemistry*, 1971, **10**, 4763.
105. M. Gutman, Coenzyme Q, ed. G. Lenaz, *Kinetic Analysis of Electron Flux Through the Quinones in the Mitochondrial System*, Wiley, Chichester, UK, 1985, 215–234.
106. C. I. Ragan and I. R. Cottingham, *Biochim. Biophys. Acta*, 1985, **811**, 13.
107. B. M. Jørgensen, H. N. Rasmussen and U. F. Rasmussen, *Biochem. J.*, 1985, **229**, 621.
108. G. Benard, B. Faustin, A. Galinier, C. Rocher, N. Bellance, K. Smolkova, L. Casteilla, R. Rossignol and T. Letellier, *Int. J. Biochem. Cell Biol.*, 2008, **40**, 1543.
109. E. Lapuente-Brun, R. Moreno-Loshuertos, R. Acín-Peréz, A. Latorre-Pellicer, C. Colás, E. Balsa, E. Perales-Clemente, P. M. Quirós, E. Calvo, M. A. Rodríguez-Hernández, P. Navas, R. Cruz, A. Carracedo, C. López-Otin, A. Pérez-Martos, P. Fernández-Silva, E. Fernández-Vizarra and J. A. Enriquez, *Science*, 2013, **340**, 1567.
110. M. R. Jackman and W. T. Willis, *Am. J. Physiol.*, 1996, **270**, C673.
111. T. Kimura, T. P. Singer and C. J. Lusty, *Biochim. Biophys. Acta*, 1960, **44**, 284.
112. G. Lenaz and M. L. Genova, *Am. J. Physiol. Cell Physiol.*, 2007, **292**, C1221.
113. E. Estornell, R. Fato, C. Castelluccio, M. Cavazzoni, G. Parenti Castelli and G. Lenaz, *FEBS Lett.*, 1992, **311**, 107.
114. H. Schneider, J. J. Lemasters and C. R. Hackenbrock, *J. Biol. Chem.*, 1982, **257**, 10789.
115. K. S. Echtay, E. Winkler and M. Klingenberg, *Nature*, 2000, **408**, 609.
116. L. Walter, H. Miyoshi, X. Leverve, P. Bernardi and E. Fontaine, *Free Radic. Res.*, 2002, **36**, 405.
117. B. J. Van Raam, W. Sluiter, E. de Wit, D. Roos, A. J. Verhoeven and T. W. Kuijpers, *PLoS One*, 2008, **3**, e2013.
118. M. G. Rosca, E. J. Vazquez, J. Kerner, W. Parland, M. P. Chandler, W. Stanley, H. N. Sabbah and C. L. Hoppel, *Cardiovasc. Res.*, 2008, **80**, 30.
119. A. J. Jones, J. N. Blaza, H. R. Bridges, B. May, A. L. Moore and J. Hirst, *Angew Chem. Int. Ed. Engl.*, 2016, **55**, 728.
120. C. D. Stoner, *J. Bioenerg. Biomembr.*, 1984, **16**, 115.

121. H. Rauchová, R. Fato, Z. Drahota and G. Lenaz, *Arch. Biochem. Biophys.*, 1997, **344**, 235.
122. R. G. Efremov and L. A. Sazanov, *Nature*, 2011, **476**, 414.
123. R. Baradaran, J. M. Berrisford, G. S. Minhas and L. A. Sazanov, *Nature*, 2013, **494**, 443.
124. V. G. Grivennikova, R. Roth, N. V. Zakharova, C. Hagerhall and A. D. Vinogradov, *Biochim. Biophys. Acta*, 2003, **1607**, 79.
125. C. Piccoli, R. Scrima, D. Boffoli and N. Capitanio, *Biochem. J.*, 2006, **396**, 573.
126. H. Boumans, L. A. Grivell and J. A. Berden, *J. Biol. Chem.*, 1998, **273**, 4872.
127. C. Rydström Lundin, C. von Ballmoos, M. Ott, P. Ädelroth and P. Brzezinski, *Proc. Natl. Acad. Sci. U. S. A.*, 2016, **113**, E4476.
128. M. Vukotic, S. Oeljeklaus, S. Wiese, F. N. Vögtle, C. Meisinger, H. E. Meyer, A. Zieseniss, D. M. Katschinski, D. C. Jans, S. Jakobs, B. Warscheid, P. Rehling and M. Deckers, *Cell Metab.*, 2012, **15**, 336.
129. M. Trouillard, B. Meunier and F. Rappaport, *Proc. Natl. Acad. Sci. U. S. A.*, 2011, **108**, E1027.
130. A. Panov, S. Dikalov, N. Shalbuyeva, R. Hemendinger, J. T. Greenamyre and J. Rosenfeld, *Am. J. Physiol. Cell Physiol.*, 2007, **292**, C708.
131. E. B. Tahara, F. D. Navarete and A. J. Kowaltowski, *Free Radic. Biol. Med.*, 2009, **46**, 1283.
132. E. B. Tahara, M. H. Barros, G. A. Oliveira, L. E. S. Netto and A. J. Kowaltowski, *FASEB J.*, 2007, **21**, 274.
133. P. Jezek and L. Hlavata, *Int. J. Biochem. Cell Biol.*, 2005, **37**, 2478.
134. G. Lenaz, *IUBMB Life*, 2001, **52**, 159.
135. M. A. Aon, S. Cortassa and B. O'Rourke, *Biochim. Biophys. Acta*, 2010, **1797**, 865.
136. S. Cortassa, B. O'Rourke and M. A. Aon, *Biochim. Biophys. Acta*, 2014, **1837**, 287.
137. Y. S. Bae, H. Oh, S. G. Rhe and Y. D. Yoo, *Mol. Cells*, 2011, **32**, 491.
138. J. J. Kim, S. B. Lee, J. K. Park and Y. D. Yoo, *Cell Death Differ*, 2010, **17**, 1420.
139. R. D. Guzy and P. T. Schumacker, *Exp. Physiol.*, 2006, **91**, 807.
140. S. B. Lee, J. J. Kim, T. W. Kim, B. S. Kim, M. S. Lee and Y. D. Yoo, *Apoptosis*, 2010, **15**, 204.
141. D. B. Zorov, M. Juhaszova and S. J. Sollott, *Biochim. Biophys. Acta*, 2006, **1757**, 509.
142. A. Y. Andreyev, Y. E. Kushnareva, A. N. Murphy and A. A. Starkov, *Biochemistry*, 2015, **80**, 517.
143. N. A. Dencher, M. Frenzel, N. H. Reifschneider, M. Sugawa and F. Krause, *Ann. NY Acad. Sci. U. S. A.*, 2007, **1100**, 291.
144. E. Schäfer, N. A. Dencher, J. Vonck and D. N. Parcej, *Biochemistry*, 2007, **46**, 12579.
145. A. Baracca, F. Chiaradonna, G. Sgarbi, G. Solaini, L. Alberghina and G. Lenaz, *Biochim. Biophys. Acta*, 2010, **1797**, 314.

146. G. Lenaz, A. Baracca, G. Barbero, C. Bergamini, M. E. Dalmonte, M. Del Sole, M. Faccioli, A. Falasca, R. Fato, M. L. Genova, G. Sgarbi and G. Solaini, Biochim. Biophys. Act*a*, 2010, **1797**, 633.
147. F. Diaz, J. A. Enriquez and C. T. Moraes, *Mol. Cell Biol.*, 2012, **32**, 415.
148. S. Chen, Q. He and M. L. Greenberg, *Mol. Microbiol.*, 2008, **68**, 1061.
149. R. Moreno-Loshuertos and J. A. Enríquez, *Free Radic. Biol. Med.*, 2016, **100**, 5.
150. D. Speijer, *Bioessays*, 2011, **33**, 88.
151. A. Guarás, E. Perales-Clemente, E. Calvo, R. Acín-Pérez, M. Loureiro-Lopez, C. Pujol, I. Martínez-Carrascoso, E. Nuñez, F. García-Marqués, M. A. Rodríguez-Hernández, A. Cortés, F. Diaz, A. Pérez-Martos, C. T. Moraes, P. Fernández-Silva, A. Trifunovic, P. Navas, J. Vazquez and J. A. Enríquez, *Cell Rep.*, 2016, **15**, 197.
152. C. R. Reczek and N. S. Chandel, *Curr. Opin. Cell Biol.*, 2015, **33**, 8.
153. G. S. Shadel and T. L. Horvath, *Cell*, 2015, **163**, 560.
154. M. D. Brand, *Free Radic. Biol. Med.*, 2016, **100**, 14.
155. M. Rigoulet, E. D. Yoboue and A. Devin, *Antioxid. Redox Signaling*, 2011, **14**, 459.
156. D. L. Hoffman and P. S. Brookes, *J. Biol. Chem.*, 2009, **284**, 16236.
157. M. L. Genova and G. Lenaz, *Antioxid. Redox Signaling*, 2015, **23**, 208.
158. A. Barrientos and C. T. Moraes, *J. Biol. Chem.*, 1999, **274**, 16188.
159. G. Lenaz, M. D'Aurelio, M. Merlo Pich, M. L. Genova, B. Ventura, C. Bovina, G. Formaggini and G. Parenti Castelli, *Biochim. Biophys. Acta*, 2000, **1459**, 397.
160. B. Ventura, M. L. Genova, C. Bovina, G. Formiggini and G. Lenaz, *Biochim. Biophys. Acta*, 2002, **1553**, 249.
161. L. A. Gómez, J. S. Monette, J. D. Chavez, C. S. Maier and T. M. Hagen, *Arch. Biochem. Biophys.*, 2009, **490**, 30.
162. L. A. Gómez and T. M. Hagen, *Semin. Cell Dev. Biol.*, 2012, **23**, 758.
163. M. Frenzel, H. Rommelspacherr, M. D. Sugawa and N. A. Dencher, *Exp. Gerontol.*, 2010, **45**, 563.
164. D. Edgar, I. Shabalina, Y. Camara, A. Wredenberg, M. A. Calvaruso, L. Nijtmans, J. Nedergaard, B. Cannon, N. G. Larsson and A. Trifunovic, *Cell Metab.*, 2009, **10**, 131.
165. M. D'Aurelio, C. D. Gajewski, G. Lenaz and G. Manfredi, *Hum. Mol. Genet.*, 2006, **15**, 2157.
166. J. Miquel, A. C. Economos, J. Fleming and J. E. Johnson Jr., *Exp. Gerontol.*, 1980, **15**, 575.
167. P. M. Sousa, M. A. Videira, F. A. Santos, B. L. Hood, T. P. Conrads and A. M. Melo, *Arch. Biochem. Biophys.*, 2013, **537**, 153.
168. H. Schägger, *Biochim. Biophys. Acta*, 2002, **1555**, 154.
169. E. Schäfer, M. Frenzel, A. Poetsch, S. Rexroth, H. J. Schwassmann, T. Suhai, J. Vonck and N. A. Dencher, *Biochim. Biophys. Acta*, 2009, **1787**, 657.
170. F. Yang, Y. Gao, Z. Li, L. Chen, Z. Xia, T. Xu and Y. Qin, *Biochim. Biophys. Acta*, 2014, **1837**, 1674.

171. G. Morrow, M. Le Pécheur and R. M. Tanguay, *Biogerontology*, 2016, **17**, 61.
172. N. Hartmann, K. Reichwald, I. Wittig and C. Englert, *Aging Cell*, 2011, **10**, 824.
173. A. Cabrera-Orefice, N. Chiquete-Félix, J. Espinosa-Jaramillo, M. Rosas-Lemus, S. Guerrero-Castillo, A. Peña and S. Uribe-Carvajal, *Biochim. Biophys. Acta*, 2014, **1837**, 73.
174. K. M. Davies, M. Strauss, B. Daum, J. H. Kief, H. D. Osiewacz, H. Rycovska, V. Zichermann and W. Kühlbrandt, *Proc. Natl. Acad. Sci.*, 2011, **108**, 14121.
175. F. Krause, C. Q. Scheckhuber, A. Werner, S. Rexroth, N. H. Reifschneider, N. A. Dencher and H. D. Osiewacz, *Ann. NY Acad. Sci. U. S. A.*, 2006, **1067**, 106.
176. M. F. Maas, F. Krause, N. A. Dencher and A. Sainsard-Chanet, *J. Mol. Biol.*, 2009, **387**, 259.
177. D. De Los Rios Castillo, M. Zarco-Zavala, S. Olvera-Sanchez, J. P. Pardo, O. Juarez, F. Martinez, G. Mendoza-Hernandez, J. J. Garcia-Trejo and O. Flores-Herrera, *J. Biol. Chem.*, 2011, **286**, 23911.
178. C. Greggio, P. Jha, S. S. Kulkarni, S. Lagarrigue, N. T. Broskey, M. Boutant, X. Wang, S. Conde Alonso, E. Ofori, J. Auwerx, C. Cantó and F. Amati, *Cell Metab.*, 2017, **25**, 301.
179. N. Kovářová, A. Ċížková Vrbacká, P. Pecina, V. Stránecký, E. Pronicka, S. Kmoch and J. Houštěk, *Biochim. Biophys. Acta*, 2012, **1822**, 1114.
180. I. Lopez-Fabuel, L. Martin-Martin, M. Resch-Beusher, G. Azkona, R. Sanchez-Pernaute and J. P. Bolaños, *Neurochem. Int.*, 2017a, pii:S0197.
181. E. M. Mejia, S. Chau, G. C. Sparagna, S. Sipione and G. M. Hatch, *Lipids*, 2016, **51**, 561.
182. N. Miyake, S. Yano, C. Sakai, H. Hatakeyama, Y. Matsushima, M. Shiina, Y. Watanabe, J. Bartley, J. E. Abdenur, R. Y. Wang, R. Chang, Y. Tsurusaki, H. Doi, M. Nakashima, H. Saitsu, K. Ogata, Y. Goto and N. Matsumoto, *Hum. Mutat.*, 2013, **34**, 446.
183. I. Bohovych, M. R. Fernandez, J. J. Rahn, K. D. Stackley, J. F. Bestan, A. Anandhan, R. Franco, S. M. Claypool, R. E. Lewis, S. S. Chan and O. Khalimonchuk, *Sci. Rep.*, 2015, **5**, 13989.
184. S. Jang, T. S. Lewis, C. Powers, Z. Khuchua, C. P. Baines, P. Wipf and S. Javadov, *Antioxid. Redox Signaling*, 2017, **27**, 57.
185. I. Lopez-Fabuel, M. Resch-Beusher, M. Carabias-Carrasco, A. Almeida and J. P. Bolaños, *Neurochem. Res.*, 2017b, **42**, 1676.
186. D. Sun, B. Li, R. Qiu, H. Fang and J. Lyu, *Int. J. Mol. Sci.*, 2016, **17**, pii: E926.
187. H. Eubel, J. Heinemeyer, S. Sunderhaus and H. P. Braun, *Plant Physiol. Biochem.*, 2004, **42**, 937.
188. Y. Kakizaki, A. L. Moore and K. Ito, *Biochem. J.*, 2012, **445**, 237.
189. F. Krause, N. H. Reifschneider, D. Vocke, H. Seelert, S. Rexroth and N. A. Dencher, *J. Biol. Chem.*, 2004a, **279**, 48369.

190. J. Petereit, K. Katayama, C. Lorenz, L. Ewert, P. Schertl, A. Kitsche, H. Wada, M. Frentzen, H. P. Braun and H. Eubel, *Front. Plant. Sci.*, 2017, **8**, 72.
191. K. Peters, N. V. Dudkina, L. Jänsch, H. P. Braun and E. J. Boekema, *Biochim. Biophys. Acta*, 2008, **1777**, 84.
192. A. Lombardi, E. Silvestri, F. Cioffi, R. Senese, A. Lanni, F. Goglia, P. de Lange and M. Moreno, *J. Proteomics*, 2009, **72**, 708.
193. C. Wernicke, J. Hellmann, B. Zieba, K. Kuter, K. Ossowska, M. Frenzel, N. A. Dencher and H. Rommelspacher, *Pharmacol. Rep.*, 2010, **62**, 35.
194. M. G. Matus-Ortega, C. A. Cárdenas-Monroy, O. Flores-Herrera, G. Mendoza-Hernández, M. Miranda, B. González-Pedrajo, H. Vázquez-Meza and J. P. Pardo, *Yeast*, 2015, **32**, 629.

CHAPTER 13

Structure, Mechanism and Regulation of ATP Synthases

JOHN E. WALKER

The Medical Research Council Mitochondrial Biology Unit, University of Cambridge, Cambridge Biomedical Campus, Hills Road, Cambridge CB2 0XY, UK
Email: walker@mrc-mbu.cam.ac.uk

13.1 Introduction

The ATP synthases, also known as F-ATPases or F_1F_o-ATPases, are multi-subunit enzyme complexes found in energy transducing membranes in eubacteria, chloroplasts and mitochondria.[1,2] They make ATP from ADP and phosphate under aerobic conditions using a proton-motive force, Δp, generated by respiration or photosynthesis, as a source of energy. The ATP hydrolase activities of the enzymes from mitochondria, chloroplasts and some eubacteria are inhibited, and, under physiological conditions, these enzymes can only synthesize ATP. However, under anaerobic conditions, other eubacterial enzymes can hydrolyze ATP, made by glycolysis, to generate the Δp that is required for other essential cellular functions, such as chemotaxis and transmembrane transport processes. Some eubacteria generate a sodium ion motive force that is coupled to ATP synthesis,[3] and the structures and functions of these ATP synthases are broadly similar to those that are driven by Δp.

Chemical Biology No. 5
Mechanisms of Primary Energy Transduction in Biology
Edited by Mårten Wikström

Published by the Royal Society of Chemistry, www.rsc.org

The proton motive force, Δp, has two components, $\Delta\Psi$ and ΔpH, the differences, respectively, in electrical potential and pH across a biological membrane. At 25 °C, they are related by the expression:[4]

$$\Delta p \text{ (in mV)} = \Delta\Psi - 59\Delta pH.$$

In mitochondria, the value of ΔpH across the inner membrane is about 0.5 units, and $\Delta\Psi$ is the dominant component of Δp. In chloroplasts, the pH of the stroma is slightly basic, whereas that of the interior of the thylakoids is about 4.0, and therefore ΔpH is the dominant term. All bacteria require Δp to be maintained across their cytoplasmic membranes, in order to grow and remain viable under conditions where the bacteria replicate, and under other conditions where they exist in a non-replicating persistent state.[5] Bacterial habitats are extremely diverse, with pH values extending from 0.5–10.5. In bacteria that thrive at acidic pH, ΔpH can be in excess of 4 units ($alkaline_{in}/acid_{out}$), and ATP synthesis is coupled mainly to the ΔpH component of Δp. At extreme acidity, intracellular acidification by proton entry is counteracted by electrogenic uptake of K^+ leading to reversal of $\Delta\Psi$. Therefore, the ATP synthase has to work against $\Delta\Psi$. At neutral pH values, $\Delta\Psi$ is the dominant component of Δp in both aerobic and anaerobic bacteria, and it drives ATP synthesis in both proton- and sodium-coupled enzymes. In alkaliphilic bacteria, the high value of $\Delta\Psi$ is counteracted by inverted ΔpH values ($acid_{in}/alkaline_{out}$) in excess of 2 units. Thus, their ATP synthases operate at a low overall value of Δp.

13.2 Structure of ATP Synthases

13.2.1 Subunit Compositions

The eubacterial and chloroplast enzymes are assemblies of eight or nine kinds of subunits, and they are the simplest and smallest ATP synthases[6–13] (Figure 13.1). Those with eight kinds of subunits have two identical b-subunits; those with nine subunits, represented by enzymes from photosynthetic bacteria and chloroplasts, have two non-identical, but related, subunits, b and b′. The number of c-subunits in the membrane domains of bacterial and chloroplast ATP synthases varies from 9–15, depending on the species (see Section 13.3.5).[14–21] The molecular weights of these enzymes range from about 500–550 kDa. In contrast, the bovine enzyme, as an example of a mitochondrial ATP synthase, has 29 subunits of 18 types (including the regulatory protein, IF_1) with a combined molecular weight of about 650 kDa.[8,22–27] They include orthologs of the eight or nine subunits that form the bacterial and chloroplast enzymes, with one exception. In the mitochondrial complex, the two bacterial b-subunits (or the single b and b′-subunits) are replaced by the mitochondrial b-subunit plus subunits d and F_6, which, although they have related functions to bacterial b- and b′-subunits in the peripheral stalk, are unrelated in sequence.[24,25,28] The membrane domains of ATP synthases from vertebrates contain eight c-subunits,

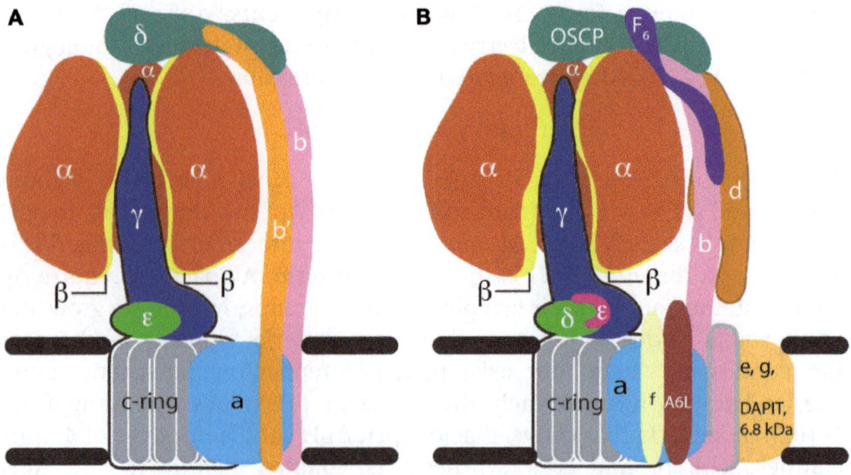

Figure 13.1 The organization of protein subunits in ATP synthases. (A), the eubacterial and chloroplast ATP synthases; (B), the mitochondrial ATP synthase, exemplified by the bovine enzyme. The upper part of each model contains the subunits in the F_1 catalytic domain. One of the three α-subunits (red) has been removed to expose the elongated α-helical structure in the γ-subunit (dark blue), which lies approximately along the central axis of the $\alpha_3\beta_3$ domain. The γ-subunit (and associated subunits) is in contact with the F_o membrane domain, which contains the c-ring (grey) and the associated a-subunit (blue). The number of c-subunits in the c-ring differs between species. The rotor of the enzyme consists of the ensemble of the c-ring and the γ-subunit (and associated subunits). The pathway for protons through the F_o domain is in the vicinity of the interface between the c-ring and a-subunit. The peripheral stalk is on the right of each model. In some eubacterial enzymes, it consists of the δ-subunit (sea green) and two identical b-subunits (pink). In other eubacterial enzymes, and in the chloroplast enzyme, the two b subunits are replaced by single copies of homologous, but non-identical, subunits b (pink) and b′ (orange). The N-terminal domain of the δ-subunit probably binds to the N-terminal regions of the three α-subunits, and the b- and b′-subunits interact with the a-subunit *via* their N-terminal transmembrane α-helices. In the mitochondrial enzyme, the peripheral stalk consists of single copies of subunits OSCP, b, d and F_6. The OSCP is the homologue of the bacterial δ-subunit, and has been demonstrated to interact with the N-terminal regions of the three α-subunits. The sequences of mitochondrial subunits b, d and F_6 are not evidently related to those of the bacterial b and b′ subunits, and their structures differ also, although they are both dominated by α-helices, except for the C-terminal domain of the OSCP (and presumably the eubacterial and chloroplast δ-subunit), which contain β-structures. The membrane domains of the mitochondrial enzyme contain a number of membrane subunits with single transmembrane spanning α-helices that are not found in eubacteria and chloroplasts. The positions of supernumerary subunits e, f, g, A6L (ATP8), DAPIT and the 6.8 kD proteolipid are indicated. Apart from A6L, they have no known roles in the generation of ATP. The same colour scheme for subunits is used consistently in subsequent Figures where the structure of the enzyme or its parts are discussed.

and this number probably persists throughout metazoans.[29] The ATP synthases from *Saccharomyces cerevisiae* and other characterized fungal enzymes have ten c-subunits.[30–32] In addition, depending on the species, the mitochondrial enzymes have up to six supernumerary subunits associated with their membrane domains.[23–27,33] With the exception of the ATP8-subunit (also known as the A6L-subunit in mammals, or the Aap1-subunit in fungi), the supernumerary subunits have no known direct role in the synthesis or hydrolysis of ATP.

Irrespective of their source, the arrangements of the constituent subunits of all ATP synthase complexes that have been characterized biochemically and structurally are closely related. As summarized in Figure 13.1, the subunits are organised in two major domains representing the membrane extrinsic and the membrane intrinsic sectors of the enzyme. The membrane extrinsic sector, known as the F_1-domain, is the catalytic part where ATP is formed from ADP and inorganic phosphate. This F_1-domain can be detached experimentally from the membrane domain in an intact state, and studied in isolation from the rest of the enzyme. Known as the F_1-ATPase, the mitochondrial F_1-domain, and similar domains from some bacteria retain the capacity to hydrolyse ATP. However, F_1-domains from other bacteria and chloroplasts have little or no ATP hydrolase activity, unless activated artificially.[34–36] The membrane intrinsic sector of ATP synthases, sometimes called F_o, contains a rotary motor driven by Δp, and connected to the extrinsic domain by a central stalk[30,37,38] and a peripheral stalk.[25,32,39–44] The central stalk, in close association with the ring of c-subunits in the membrane domain, constitutes the enzyme's rotor, and *via* its rotary action, it transmits energy from the membrane domain to the catalytic domain to drive the formation of ATP from ADP and phosphate. The rest of the enzyme constitutes the enzyme's stator against which the rotor turns. The peripheral stalk is a key component of the stator. It is an elongated and largely α-helical structure linking the topmost extremity of the $\alpha_3\beta_3$-component of the F_1-domain to the a-subunit in the membrane domain.[21,31,32,40,42–44] The a-subunit is also known as ATPase-6, or ATP6 in mitochondria and subunit IV in chloroplasts; here both subunit a and ATP6 are employed. At the top of the F_1-domain of mitochondrial enzymes, the N-terminal domain of the OSCP (oligomycin sensitivity conferral protein) binds to the N-terminal α-helical regions of the three α-subunits.[32,42,45,46] It is likely that the δ-subunit in bacterial and chloroplast ATP synthases is bound in a similar way.[21,44] In the membrane domain, the N-terminal regions of the b-subunit (or subunits) bind to the a-subunit, and hold it against the rotating c-ring in the active enzyme.[21,31,32,43,44] The two extremities are linked by a largely α-helical and probably mostly rigid section. The integrity of the peripheral stalk is essential for keeping Δp coupled to ATP synthesis, and, in enzymes where the action of the enzyme can be reversed, for maintaining the coupling of ATP hydrolysis to the generation of Δp. In mitochondrial enzymes, the ATP8-subunit is also part of the peripheral stalk.[23,25,31,32,43,47] It is anchored in the membrane by an N-terminal α-helix, and the rest of the subunit

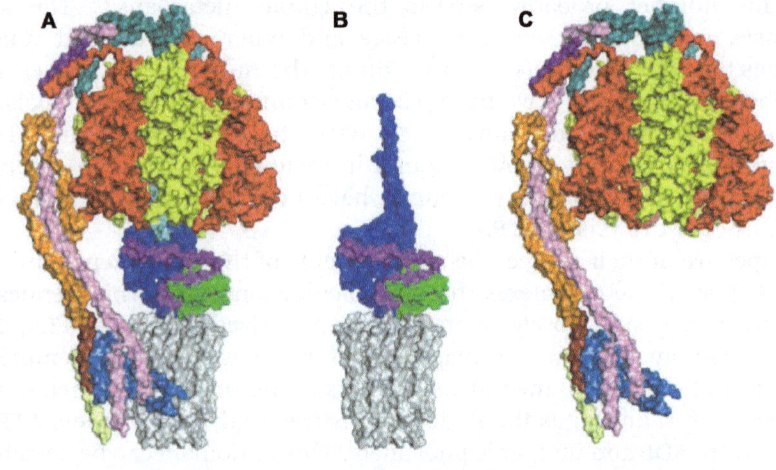

Figure 13.2 The structure of the ATP synthase from *P. angusta* resolved into its rotor and stator.[32] (A), the intact enzyme; (B), the rotor, consisting of the central stalk (subunits γ, δ and ε, shown in blue, green and magenta, respectively), and the tightly associated the c_{10}-ring (grey) in the membrane domain; (C), the stator, made of the catalytic $\alpha_3\beta_3$-domain (red and yellow subunits, respectively), and the associated peripheral stalk made of subunits OSCP, b, d and F_6 (sea-green, pink, orange and purple, respectively). The b-subunit extends into the membrane domain where it binds to the a-subunit (corn-flower blue) *via* two trans-membrane α-helices. The ATP8 subunit (brick-red) also binds to the a-subunit, and extends up into the peripheral stalk. The yellow protein is the supernumerary subunit f. Its N-terminal region may also extend into the peripheral stalk. In the intact and fully coupled enzyme in (A), the a-subunit is held against the rotating c-ring by the b and ATP8 subunits, and the trans-membrane pathway for protons is in the vicinity of the interface between subunit a and the c-ring.

extends into the membrane extrinsic region of the peripheral stalk. This subunit is essential for a fully functioning mitochondrial enzyme, and its role appears to be to provide a brace to augment the role of the b-subunit of maintaining contact between the c-ring and the a-subunit.[32]

An example of the structures of the rotor and the stator, and their interactions with each other in a mitochondrial ATP synthase, is shown in Figure 13.2.

13.2.2 High Resolution Structure

Our current knowledge of the high resolution three-dimensional structures of ATP synthases is based predominantly on the detailed analysis, largely by X-ray crystallography, of separate parts of the ATP synthase from mitochondria, and especially from bovine mitochondria. Starting in 1994 with the description of the structure of the F_1-catalytic domain,[37] over a period of more than twenty years, the almost complete structure of the bovine enzyme has emerged by building on the initial structure (Figure 13.3). Moreover, numerous structures of the catalytic F_1-domain have been solved with the enzyme inhibited with various small molecules,[48–52] and the natural inhibitor protein, IF_1.[53–57] Also, with the aim of elucidating details of the enzyme's catalytic cycle, other structures have been determined with bound substrates and substrate analogues.[58–64] Eventually, by 2012, a mosaic structure of the bovine enzyme had been constructed within the constraints of a structure of the entire complex determined at 17 Å resolution by electron cryo-microscopy (cryo-EM) (Figure 13.3).[38,40,42,55,65] In this mosaic structure, approximately 85% of the entire enzyme complex had been described at

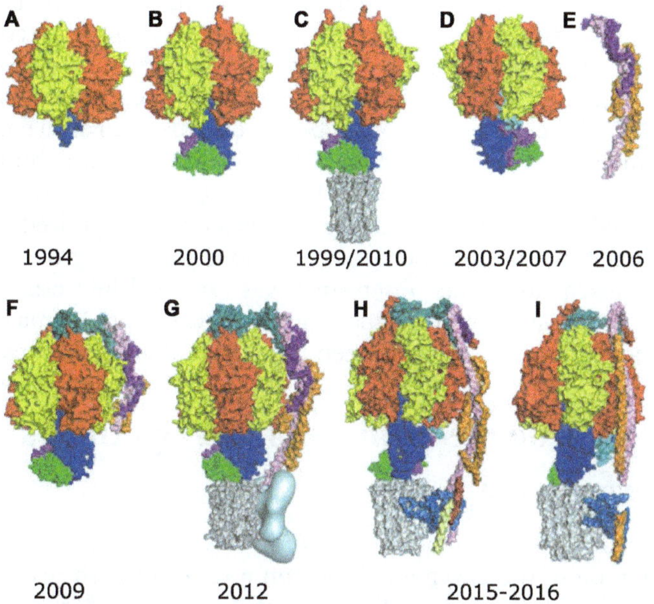

Figure 13.3 A brief history of the determination of the structure of the mitochondrial ATP synthase complex. (A), bovine F_1-ATPase complex (1BMF),[37] lacking the δ- and ε-subunits and half of the γ-subunit; (B), complete structure of bovine F_1-ATPase (1E79);[51] (C), structures of the sub-complex of the yeast and bovine F_1-ATPases with attached c10 and c8 rings, respectively (2XOK and 2XND);[30,38] (D), bovine F_1-ATPase inhibited by residues 1–60 of the natural inhibitor protein IF_1 (2V7Q);[55] (E), the sub-complex of part of the membrane extrinsic region of the peripheral stalk made from the membrane extrinsic region of subunit b plus subunits d and F_6 (2CLY);[40] (F), the complex of bovine F_1-ATPase with most of the membrane extrinsic region of the peripheral stalk (2WSS);[42] (G), mosaic structure of bovine ATP synthase built from the structures of the sub-complexes in parts C, E and F within a 17 Å cryo-EM map of the bovine enzyme;[65] (H), one of the three states of the ATP synthase from *P. angusta* at 6 Å determined by cryo-EM (5LQ1);[32] (I), for comparison the ATP synthase from *P. denitrificans* determined by X-ray crystallography at 4 Å resolution (5DN6).[44] The protruding region of the ζ-subunit is light blue.

atomic resolution. The poorly resolved 15% of the mosaic structure was in the membrane domain of the enzyme, and it included the a-subunit and the membrane region of the b-subunit (or subunits), as well as the supernumerary subunits in mitochondrial enzymes. During the period from 1999 onwards, structures of component domains of ATP synthases from other species were also described. They include the F_1-domain and the F_1-c_{10} sub-complex from *Saccharomyces cerevisiae*,[30,56,66,67] c-rings from various bacterial species,[16,20,68–70] and from chloroplasts,[18,71] catalytic F_1-domains from spinach chloroplasts,[72,73] *Escherichia coli*,[74,75] *Geobacillus stearothermophilus*[76,77] and *Caldalkalibacillus thermarum*,[78,79] and subunits from various sources[40,80,81,82]

Most recently, progress towards resolving the "missing" 15% in the membrane domains of ATP synthases have arisen from two independent advances. First, the entire ATP synthase complex from the α-proteobacterium, *Paracoccus denitrificans*, was crystallized and its structure solved by X-ray crystallography at 4 Å resolution (Figure 13.3I).[44] Second, developments in cryo-EM[83–85] allowed structures of the bovine enzyme,[43] an algal enzyme,[86] the enzyme from a fungus, *Pichia angusta*,[32] and the ATP synthase from *E. coli*[21] to be solved to about 6–8 Å resolution. In 2005, it was demonstrated that the mitochondrial ATP synthase in the alga, *Polytomella*, is organized in dimers,[87] and the dimers themselves are arranged in long rows along the edges of the cristae.[88] Subsequently, this observation was extended into other eukaryotic species.[89,90] Thus, a structure of the dimeric enzyme from *Yarrowia lipolytica*, also determined by cryo-EM, has been described, at about 6 Å resolution.[31]

13.3 Catalytic Mechanism of ATP Synthases

13.3.1 ATP Hydrolysis

Our understanding of the mechanism of how ATP is hydrolysed in the catalytic F_1 domain of the enzyme is based on a series of high resolution structures determined by X-ray crystallography of the bovine[37,38,42,48–55,57,59–64] and yeast[30,56,58,66,67] enzymes, and on biophysical "single molecule" experiments conducted mostly in bacterial enzymes,[91–94] but recently also in enzymes from eukarya.[95,96] Many of the structures have been interpreted as representing either the ground state in the catalytic cycle of ATP hydrolysis or synthesis, or of the closely related ADP inhibited state. One such structure is summarized in Figure 13.4.[51] In these structures, the three non-catalytic α-subunits and the three catalytic β-subunits of the enzyme are arranged in alternation around an asymmetric α-helical structure in the single γ-subunit. The α- and β-subunits are similar, each consisting of an N-terminal domain with a six-stranded β-barrel, a central nucleotide binding domain made of both α-helices and β-strands and an α-helical C-terminal domain containing six α-helices in β-subunits and seven in α-subunits. Because the γ-subunit is asymmetrical, F_1-ATPase is an inherently asymmetrical enzyme, and a structure of yeast F_1-ATPase without bound nucleotides demonstrated that the asymmetry is derived from the protein assembly itself.[67] When the central stalk is removed

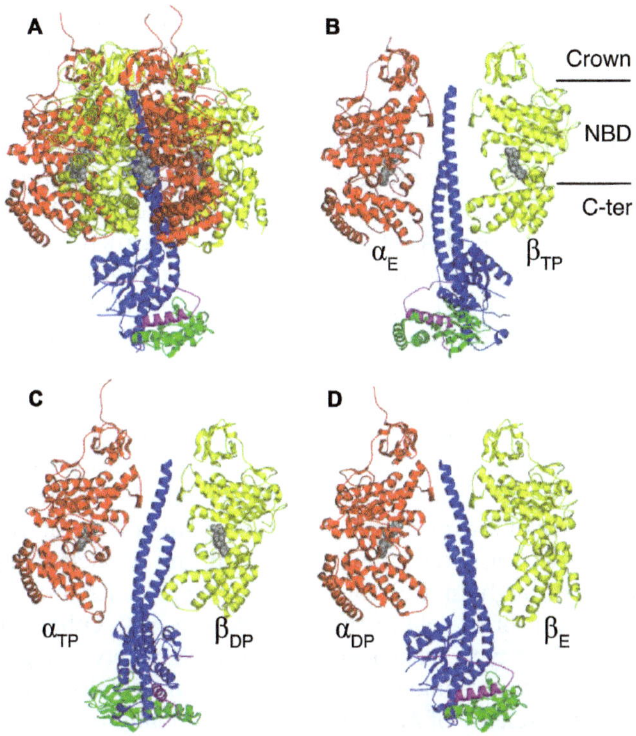

Figure 13.4 The structure of the F_1 catalytic domain bovine ATP synthase in the ground state of the catalytic cycle.[51] The subunits of the protein are depicted in ribbon representation. The α-, β-, γ-, δ- and ε-subunits are red, yellow, blue green and magenta, respectively, and nucleotides are shown in grey. (A); the complete F_1 domain with three α- and three β-subunits and single copies of γ-, δ- and ε-subunits. In the intact ATP synthase, the rotor of the enzyme consists of the central stalk interacting *via* its foot with loop regions between the N- and C-terminal α-helices of the eight c-subunits in the Fo membrane domain. (B, C and D); the three different conformations of the catalytic β-subunits that are present in the structure in (A) are shown, together with the central stalk (γ-, δ- and ε-subunits) and an α-subunit, for reference. Each β-subunit has three domains, the N-terminal domain (Crown) is made of a β-barrel, the nucleotide binding domain (NBD) is a mixed α-β structure, and the C-terminal domain (C-ter) is a bundle of six α-helices as shown in (B). Each of the three β-subunits has been obliged by the asymmetry of the α-helical coiled-coil region of the γ-subunit to adopt a different conformation, denoted β_{TP}, β_{DP} and β_E, with different affinities for nucleotides, the β_{TP} and β_{DP} subunits can each bind either ADP or ATP, and the β_E-subunit has a low affinity for a nucleotide. During the catalytic cycle of ATP synthesis or hydrolysis, the rotation of the central stalk takes each β-subunit through each of the three conformations. As the rotor turns during ATP synthesis, this action leads to the binding and entrapment of ADP and phosphate at the β_E-site as it closes, ATP formation at the β_{DP}-site and ATP release from the β_{TP} site as it opens and converts back to the β_E site. Each 360° rotation of the central stalk (clockwise as viewed from beneath) leads to the formation of an ATP molecule from each of the three β-subunits. ATP hydrolysis is presumed to proceed *via* a reversal of this process generating rotation of the central stalk in the opposite direction to the synthetic direction.

the resulting $\alpha_3\beta_3$-domain has three-fold symmetry.[76] The asymmetry of the single γ-subunit obliges each of the three catalytic β-subunits to adopt a different conformation with different nucleotide binding and catalytic properties. Two of them have similar conformations, but in one structure, now recognized as the ADP inhibited state of the enzyme (see Section 13.4), one subunit designated as β_{DP}, contains bound ADP-Mg, and the second subunit, β_{TP}, has bound AMP-PNP-Mg (adenylylimidodiphosphate (AMP-PNP), is a non-hydrolysable analog of ATP). However, in other structures, which represent the ground state of the active catalytic cycle, two molecules of AMP-PNP-Mg, ADP-BeF-Mg or ADP-Mg are bound to both β_{TP}- and β_{DP}-subunits in the same structure. Hence, the β_{DP}- and β_{TP}-subunits can bind either ADP or ATP. In all of these structures, the third subunit, the β_E-subunit adopts a radically different conformation, in which part of the nucleotide binding domain and the attached C-terminal domain have hinged outwards together, in response to the curvature of the γ-subunit. In these structures, this β-subunit has no bound nucleotide, and so it is known as the "empty" or "open" state, designated as β_E. None of the α-subunits opens in a similar way. All three of them remain closed, each binding a magnesium ion and a nucleotide, and they remain bound throughout the catalytic cycle. The nucleotides bound to α-subunits have no direct role in the formation of ATP. Other intermediate states in the catalytic cycle have been resolved by X-ray crystallography, most notably a transition state analog structure,[59] where the catalytic cycle has been arrested before the release of ADP, Mg^{2+} and phosphate from the β_E-subunit.

In order to explain the interconversion of catalytic sites through "tight", "loose" and "open" states required by a binding change mechanism of catalysis of ATP hydrolysis by F_1-ATPase,[97] it was proposed that the interconversion of sites was effected by a mechanical rotation of the γ-subunit, each 360° rotation taking each β-subunit through the three states represented by the β_{TP}-, β_{DP}- and β_E-subunits thereby hydrolyzing three ATP molecules.[98]

13.3.2 Structural Description of the Rotary Catalytic Cycle

Following the description of the structure of the bovine F_1-ATPase, and the proposal of a molecular basis for a rotary mechanism,[37] the rotation of the central stalk was observed directly in the $\alpha_3\beta_3\gamma$-complex from *G. stearothermophilus*.[91] This rotation was dependent on the hydrolysis of ATP. These experiments established that each 360° rotary cycle could be resolved at high concentrations of ATP (2 mM) into three 120° steps (Figure 13.5A).[92] The three intervening pauses are referred to as the "catalytic dwells", where the enzyme is considered to be poised to carry out, or to be carrying out, the hydrolysis of ATP. At lower concentrations of ATP (2 μM), a second pause was observed 40° after the catalytic dwell.[94,99] This is the "ATP binding dwell", where the enzyme is awaiting the binding of the substrate, ATP. The 360° rotary cycle of the human F_1-ATPase (Figure 13.5B) also contains three catalytic dwells separated by 120°, and, 30° after each catalytic dwell, three

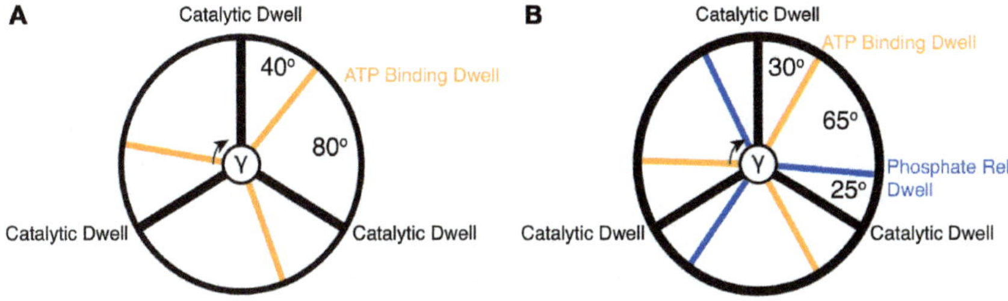

Figure 13.5 Rotary cycles in the catalytic domain of ATP synthases. The hydrolytic cycle is shown in each case. (A), bacterial ATP synthases. The cycle is divided into three 120° steps.[94] The catalytic dwells are pauses, indicated by black radii, where the enzyme is poised to hydrolyse ATP at one of the three catalytic sites. Sub-steps (orange radii) 40° later are observed at limiting concentrations of ATP, when the enzyme is waiting for ATP to bind.[94,99] (B), the mammalian enzyme has a similar rotary cycle insofar as the three main catalytic dwells are separated by rotations of 120°.[64,95] However, in these enzymes, the ATP binding dwell occurs 30° after the catalytic dwell, and a second sub-step (blue radii), corresponding to the release of phosphate, and not observed in bacterial enzyme, occurs 65° later.

ATP binding dwells.[95] However, unlike the bacterial enzyme, both dwells in the human enzyme were observed at both high (4 mM) and low (50 μM) concentrations of ATP.[95] Inhibition of the human enzyme with a monomeric version of human regulatory protein, IF_1, halted the rotary cycle at a position corresponding to the catalytic dwell.[95] Thus, the structure of the closely related bovine enzyme inhibited by a monomeric version of the inhibitory region of the bovine, IF_1, provided a structural representation of the catalytic dwell.[55] Moreover, upon addition of the ATP analogue AMP-PNP and the phosphate analogue, monothiophosphate (referred to as "thiophosphate"), the rotating human enzyme (but not the bacterial enzyme) stalls at an additional intermediate position 25° before the catalytic dwell.[95] This intermediate stall position is known as the "phosphate release dwell", but it differs from the catalytic and ATP binding dwells in that it has not been observed during each 120° step in an active catalytic cycle, and it is manifest only when rotation of the enzyme has been stopped with the inhibitors AMP-PNP and thiophosphate.[95] It is reasonable to assume that this state corresponds to a point in catalysis after the cleavage of the bond between the β- and γ-phosphates of ATP, where phosphate is about to be released. In the structure of the bovine F_1-ATPase inhibited with thiophosphate, a thiophosphate ion is bound in the $\alpha_E\beta_E$-catalytic interface approximately 7.7 Å from where the γ-phosphate of an ATP molecule would be bound in the $\alpha_{DP}\beta_{DP}$- or $\alpha_{TP}\beta_{TP}$-catalytic interface.[64]

These structural descriptions of the phosphate release dwell[64] and the catalytic dwell[53,55–57,63,64,66,67] have provided a molecular explanation of the 25° rotary sub-step accompanying release of phosphate from the β_E-subunit of the active enzyme. Release of phosphate is accompanied by a subtle rearrangement of the structure of its binding site that prevents the released phosphate from rebinding. Associated with this re-arrangement, a loop in the β_E-subunit disrupts interactions in the $\alpha_E\beta_E$-catalytic interface and opens it fully. Other rearrangements disrupt interactions between the γ-subunit and the C-terminal domain of the α_E-subunit. In order to restore most of these interactions, and to make compensatory new ones, the γ-subunit has to rotate through the required 25–30° of the sub-step.[64] As yet, there are no clear molecular explanations of the other sub-steps in the 120° steps of the rotary cycle.

13.3.3 Generation of Rotation in the Membrane Domain

The c-subunits forming the c-ring in the membrane domain of the rotor are simple proteins folded into two canonical transmembrane α-helices, with the intervening loop regions exposed in the head group region of the membrane, on the same side as the F_1 domain. Some of them help *via* protein–protein interactions to cement the c-ring and the foot of the central stalk together (Figure 13.1). The inner core of the c-ring consists of an annulus of N-terminal α-helices,[30] and the inner ring is occupied by lipids.[30] The external region of the c-ring consists of a second concentric annulus of C-terminal α-helices.[30] Each of them has an acidic amino acid residue (aspartate or glutamate depending on the species) exposed on the external surface of the ring near to its mid-point in the membrane. This acidic amino acid is an essential component in the mechanism of generation of rotation of the rotor from the proton-motive force by translocation of protons through the membrane domain.[100,101]

During ATP synthesis, the rotation of the c-ring is driven by Δp in a clockwise direction (as viewed from the membrane domain towards the F_1 domain), with an estimated rotary speed of about 100–150 revolutions per second, depending on the species.[102] In the ATP synthases that are able to hydrolyse ATP under anaerobic conditions, the energy to propel rotation is provided by the hydrolysis of ATP, which drives the rotor in a counter-clockwise direction,[93] and protons are pumped across the energy transducing membrane in the outward direction (away from the F_1 domain of the enzyme), thereby generating Δp.

As depicted schematically in Figure 13.6, part of the external surface of the c-ring is in contact with a surface in the a-subunit, and the c-ring rotates against this surface. In this model,[103] the a-subunit provides a pathway *via* a channel or two half channels for protons to access the negatively charged carboxylate residues on the C-terminal α-helices of c-subunits on the external surface of the c-ring in the rotor, and, following neutralization of the successive carboxylates, rotation of the ring, and successive reionization of the carboxylates, to allow the protons to be released on the opposite side of the F_o domain. The side-chain of a conserved arginine in subunit a (residues 210 and 179, respectively, in *E. coli* and *P. angusta*) participates in the process of proton translocation.[104]

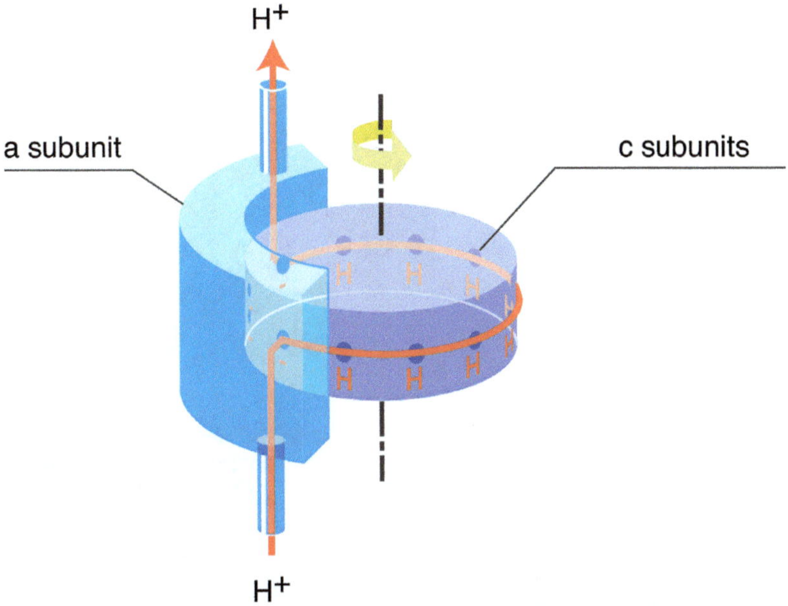

Figure 13.6 The generation of rotation in the F_o membrane domain of F-ATPases.[103] The a-subunit (blue) is shown in contact with the cylindrical ring of c-subunits (purple). The C-terminal α-helix of each c-subunit contains a carboxyl (in the side chain of an aspartate or glutamate residue; dark blue circles) exposed on the external circumference of the cylinder. From the bottom of the Figure, protons from the side of the membrane distal from the F_1 domain of the F-ATPase are shown entering a half-channel and neutralizing a negatively charged carboxylate in the interface between the a-subunit and the c-ring. Once neutralized, this residue moves by Brownian motion to the more hydrophobic environment of the lipid bilayer generating a rotation in the direction indicated. This rotational step brings another negatively charged carboxylate into the lower half-channel allowing another proton to be consumed, generating another rotational sub-step. Thus, as more negative charges are neutralized by protons, the protons are carried around on the external surface of the ring by the neutralized carboxyl, until they reach a second site in contact with the a-subunit, where the local environment reionizes the carboxyl, releasing the proton through a second half channel on the opposite of the membrane from which it entered. The direction and magnitude of Δp (−180 mV inwards in mitochondria) ensures that the rotation is unidirectional, and there is no requirement for a physical ratchet mechanism. The number of protons required to generate each 360° rotation of the c-ring corresponds to the number of c-subunits that form the ring. The γ-subunit (and associated subunits) are attached firmly to the ring, and each 360° rotation of this part of the rotor provides the energy to generate three molecules of ATP from the F1 domain (see below).

Recent structures of intact ATP synthases[21,31,32,43,44] are in accord with this model, and for the purposes of discussion, the example provided in Figure 13.7 shows the structure of the membrane domain of the ATP

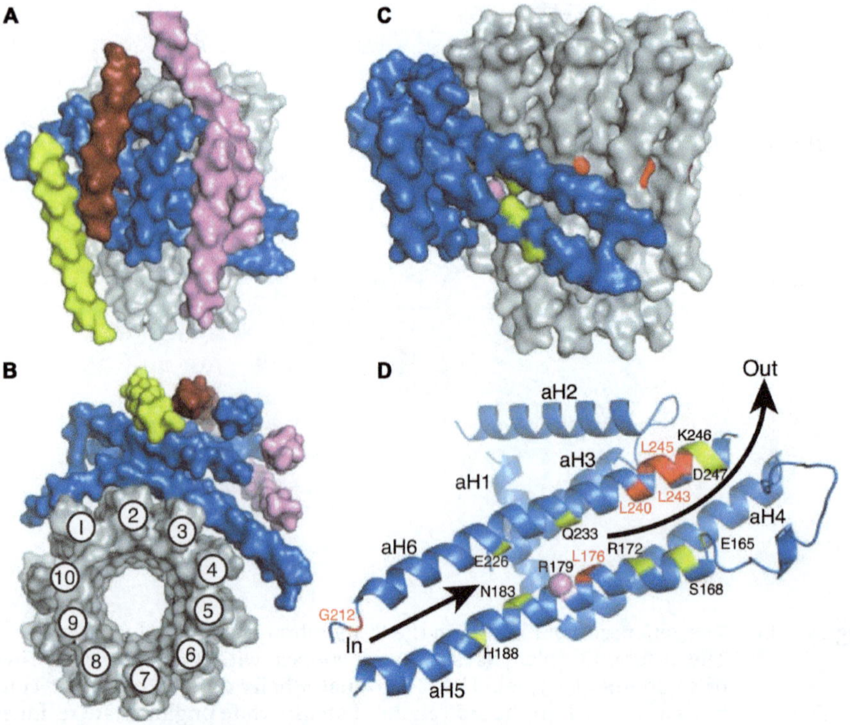

Figure 13.7 The membrane domain of the ATP synthase from *P. angusta*.[32] In (A–D), the a-subunit is corn-flower blue. (A and B), views in solid representation of the molecular model from the side, and below the membrane domain. The c_{10}-ring is grey, the b-subunit (upper part not shown) is pink, and the pale yellow, brick-red and corn-flower blue segments are transmembrane α-helices, Ch1–Ch4 assigned to subunit f, ATP8 and aH1, respectively. In the c-ring, 1–10 indicate the transmembrane C-terminal; α-helices 1–4 are in contact with subunit a. (C and D), views of the a-subunit in solid and cartoon representation viewed from outside and looking out from the interface with the c-ring, respectively. Conserved polar residues are yellow, the positions of human mutations associated with pathologies (see Table S1) are red. The pink sphere denotes the conserved Arg-179 in aH5 that is essential for proton translocation. The lower arrow indicates the inlet pathway for protons, which transfer to Glu-59 in the C-terminal α-helix-II of the c-ring. They are carried around the ring by anticlockwise rotation as viewed from above, until they arrive at Arg-179 where they enter the exit pathway, denoted by the upper arrow.

synthase from *P. angusta*.[32] In this structure, the c_{10}-ring in the rotor is intimately associated with a bundle of four α-helices in the a-subunit (aH3-aH6 in Figure 13.7), and the most striking feature of this arrangement is that these four α-helices are inclined at 30° to the plane of the membrane. α-Helix, aH1, has the more usual canonical trans-membrane topography, and lies on the outside of the bundle, distal from the c-ring; aH2 lies in the head group region of the membrane on the matrix side. In the bundle, α-helices aH5 and aH6 are in intimate contact with four of the ten C-terminal α-helices that provide the outer surface of the ring, with aH3 and aH4 packed closely behind aH5 and aH6. aH5 contains the strictly conserved Arg-179, which is known to be involved in proton translocation, and, in the structure, it is found in close proximity to the Glu-59 residue in α-helix 3 (see Figure 13.7B) in the c-ring. The proposed transmembrane pathway for protons involves both the a-subunit and the c-ring, as explained in Figure 13.7. Similar features are found in the interfaces between c-rings and a-subunits in other species. However, the ring sizes in these structures vary from 8–12, and will require structural adjustments in the a-subunit that are not readily apparent at the current levels of resolution. These initial descriptions of the proton pathway represent a highly significant advance in our understanding of the generation of rotation from Δp, but they are rudimentary. A more complete understanding of the generation of rotation from Δp will require minimally the definition of the positions of amino acid side-chains, and any water molecules that participate in proton translocation, and a grasp of the role, if any, of cardiolipin in that process (see Section 13.3.4).

13.3.4 Role of Cardiolipin

Cardiolipin is an essential component of active ATP synthases isolated from mitochondria,[105–109] and under physiological conditions each molecule of the lipid carries two negative charges.[110–112] It has been suggested that cardiolipin could act to stabilize and lubricate the rotating c-ring, or aid in proton transfer.[38] Coarse grain molecular dynamics simulations conducted with c-rings and a mixture of phospholipids plus cardiolipin have shown that cardiolipin molecules are attracted repeatedly and selectively to two sites, each involving a lysine residue and other adjacent polar residues, in the c-subunit (Figure 13.8).[112] Both sites are found in the head-group region of the lipid bilayer, one on each side of the membrane. However, although the attraction of cardiolipin molecules is selective, the residence times are brief and are of the order of about 400–500 nanosecs, sufficient time for a ring rotating in the active enzyme at 100 Hz to turn about 0.02°. This unusual behaviour fits with a possible role for cardiolipin in lubricating the ring or in bringing protons to the entry and taking them away from the transmembrane proton channel. The latter mechanism could be advantageous in ATP synthases driven primarily by $\Delta\Psi$, such as those in mitochondria and some bacteria. As the chloroplast enzyme is driven mainly by ΔpH, it would have no requirement for such a mechanism, in keeping with the lack of cardiolipin in chloroplast membranes.

13.3.5 Bioenergetic Cost of Making an ATP Molecule

Knowledge of the main architectural features of the ATP synthases and their functions, allows a fundamental deduction to be made about energy transduction in the organisms in which these enzymes are found, concerning the

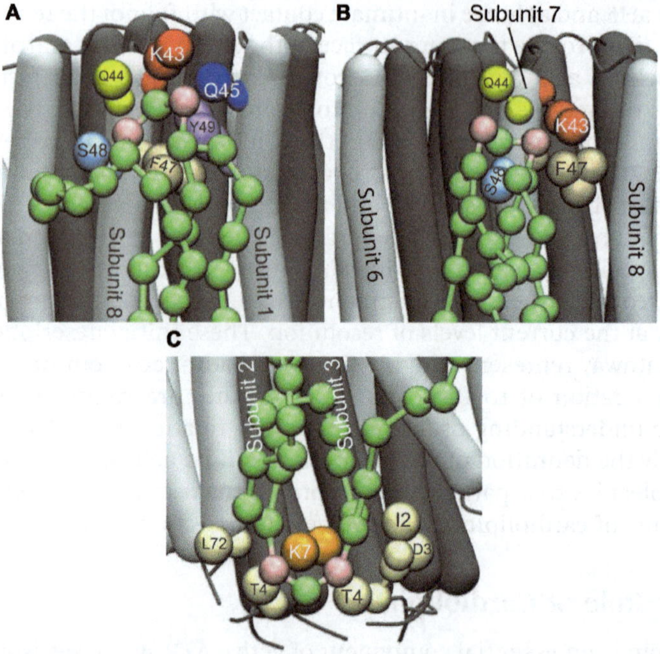

Figure 13.8 Modes of binding of cardiolipin to the c_8-rings found, for example, in human and bovine mitochondria.[112] The N- and C-terminal α-helices of c-subunits are dark and light grey, respectively. The large coloured spheres represent the coarse grain beads for specific amino acids, as indicated, lying within 0.7 nm of the phosphate beads of cardiolipin. Residue K43 (red) is completely trimethylated; residue K7 (orange) is unmodified. Cardiolipin molecules are green, with pink phosphate groups. (A) and (B), the head-group region of cardiolipin in the inner leaflet of the membrane bound, respectively, to two adjacent c-subunits (subunits 8 and 1), and to a single c-subunit; (C), a cardiolipin molecule in the outer leaflet of the membrane bound to c-subunits in the outer leaflet.

number of protons that are required to be translocated through the energy transducing membrane for each ATP molecule that is synthesized by the ATP synthase from ADP and inorganic phosphate. This parameter is the bioenergetic cost to the organism of making each ATP molecule. Since each 360° rotation of the rotor of the ATP synthase produces three ATP molecules from the F_1 domain of the enzyme, and each 360° rotation of the c-ring requires the translocation of the same number of protons as there are c-subunits in the ring, the bioenergetic cost (protons/ATP) to the ATP synthase is the number of c-subunits in the c-ring divided by 3. Vertebrates (about 50 000 species) have eight c-subunits in their c-rings, and invertebrates (about 2 million species) probably have c_8-rings also.[29,38] Therefore, the c_8-ring is probably a characteristic feature of ATP synthases throughout the multi-cellular biological world. It is the smallest c-ring to have been observed hitherto, and the associated bioenergetic cost to these ATP synthases is 2.7 protons/ATP. In contrast, unicellular organisms, including fungi and eubacteria (and also chloroplasts from green plants), have a variety of different numbers of c-subunits in their c-rings, and values of 9, 10, 11, 12, 13, 14 and 15 have been reported.[16,18,20,21,30,44,68,71] The bioenergetic cost to these ATP synthases will be 3.0, 3.3, 3.7, 4.0, 4.3, 4.7 and 5.0 protons/ATP, respectively. In mitochondria, ATP is produced in the matrix of the organelle, and it is made available as a source of cellular energy by exchanging it for external ADP. Inorganic phosphate is also actively transported back into the matrix of mitochondria. The combination of the electrogenic exchange of an internal ATP for an external ADP and the electroneutral symport of phosphate and a proton adds one proton to the total required to provide ATP to the cellular cytoplasm, and so the cost to mitochondria containing an ATP synthase with a c_8-ring will be 3.7 protons. In mammalian mitochondria for each two electrons transferred to oxygen from NADH or succinate, 10 or 6 protons, respectively, are translocated out of the matrix. Therefore, the number of moles of ADP phosphorylated to ATP per two electrons transferred to oxygen, known as the P/O ratio, will be 10/3.7 and 6/3.7 or 2.7 and 1.6, respectively for NADH and succinate, similar to the experimental values of 2.5 and 1.5.[113,114] P/O ratios have been less studied in other species, and there is uncertainty about the proton stoichiometries of their electron transfer complexes. Nonetheless, similar considerations apply to their ATP synthases, except that in eubacteria there is no requirement to translocate ATP, ADP and phosphate across the energy transducing membrane.

13.4 Regulatory Mechanisms

Different types of ATP synthases have disparate mechanisms of inactivation and reactivation of ATP synthesis under conditions of low and increasing available energy. In mitochondria and chloroplasts, when Δp is low, ADP-Mg (without inorganic phosphate) remains bound to one of the three catalytic sites of the enzyme (E) forming an inactive complex, E-ADP. In chloroplasts, it is thought that the inactive E-ADP state is stabilized during the hours of darkness by formation of an intramolecular disulfide linkage in the γ-subunit of the ATP synthase. When daylight is restored, the enzyme is reactivated by thioredoxin regulated reduction of the disulfide, and energy dependent dissociation of the ADP-Mg.[115]

13.4.1 Mitochondrial ATP Synthases

The role of the bound ADP-Mg in the physiological regulation of the mitochondrial enzyme is less clear, and regulation by the inhibitor protein IF_1 is probably more important physiologically. At a pH of about 6.7 or below, and in the absence of Δp, the mitochondrial ATP synthase forms a 1 : 1 complex with an inhibitory protein, IF_1.[116,117] The formation of the inhibited complex requires the hydrolysis of two ATP molecules, and in the inhibited complex, the inhibitor occupies a complex binding site at a catalytic interface between the α_{DP}- and β_{DP}-subunits.[53,55,57] The inhibition is relieved by restoration of an adequate Δp, allowing the enzyme to turn the rotor in the synthetic direction, with the ejection of the inhibitor from its binding site, and the resumption of the synthesis of ATP.

Active bovine IF_1 is a homodimer, in which the C-terminal regions from residues 44–84 form an antiparallel α-helical coiled-coil (Figure 13.9A).[118] In this dimeric state, the two N-terminal inhibitory regions extend in approximately opposite directions, allowing the dimer to bind to two F_1-ATPase complexes simultaneously *in vitro*. However, *in vivo* the membrane bound ATP synthase is dimerised by an independent mechanism involving protein–protein interactions between specific subunits in the F_o membrane sector of the enzyme,[31,119–121] and the catalytic sites of the two constituent F_1-domains in the dimeric enzyme are too far apart for the dimeric IF_1 to bind to both catalytic components of the dimer simultaneously. However, it cannot be excluded that the dimeric IF_1 can bind to two monomeric ATP synthases in adjacent dimers in mitochondrial membranes.

In crystals of the dimeric IF_1 alone, the dimers form fibres involving extensive intra-dimer interactions.[118] In these crystals, the α-helical C-terminal region extends in an N-terminal direction as a continuous α-helix into the inhibitory region, so that the structure of each IF_1 molecule consists of a single α-helix from residues 18–80 (Figure 13.9A). The dimeric inhibitor is active at pH values of 6.5–7.5.[122] At higher pH values, the dimers of bovine IF_1 form dimers of dimers and higher aggregates, occluding the inhibitory portion of the protein and rendering the inhibitor inactive.[122] Disaggregation of these oligomers into dimers appears to be controlled by the ionic state of a histidine residue, H49, which is thought to provide a pH-sensitive switch between inactive and active states.[122] Deletion of the C-terminal region from residues 61–84 removes the capacity to dimerise, and produces a potent monomeric inhibitor that binds to a single F_1-ATPase moiety.[55] In some species, for example *S. cerevisiae*, IF_1 lacks the C-terminal dimerization region, and the protein is naturally monomeric.[56] A fragment of the bovine protein consisting of residues 1–60 is monomeric and active, and has been used extensively in biochemical and structural studies of the inhibitory activity.[29,32,33,43,55,57,64,123]

In the structure of the complex of bovine F_1-ATPase inhibited by the monomeric version of IF_1, residues 8–50 are resolved with residues 21–50 forming a single α-helix. The structures of the bovine[55] and yeast F_1-IF_1 complexes, each with one bound inhibitor protein, show that IF_1 occupies a complex binding site at a catalytic interface between the α_{DP}- and β_{DP}-subunits (Figure 13.9B).[55,56] In the bovine complex, the inhibitor protein is in a deep groove lined with α-helices in the C-terminal domains of the α_{DP}- and β_{DP}-subunits, and its N-terminal region interacts with the coiled-coil region of the γ-subunit *via* a short α-helix, and extends into the central aqueous cavity of F_1-ATPase. Yeast IF_1 binds to yeast F_1-ATPase in a similar

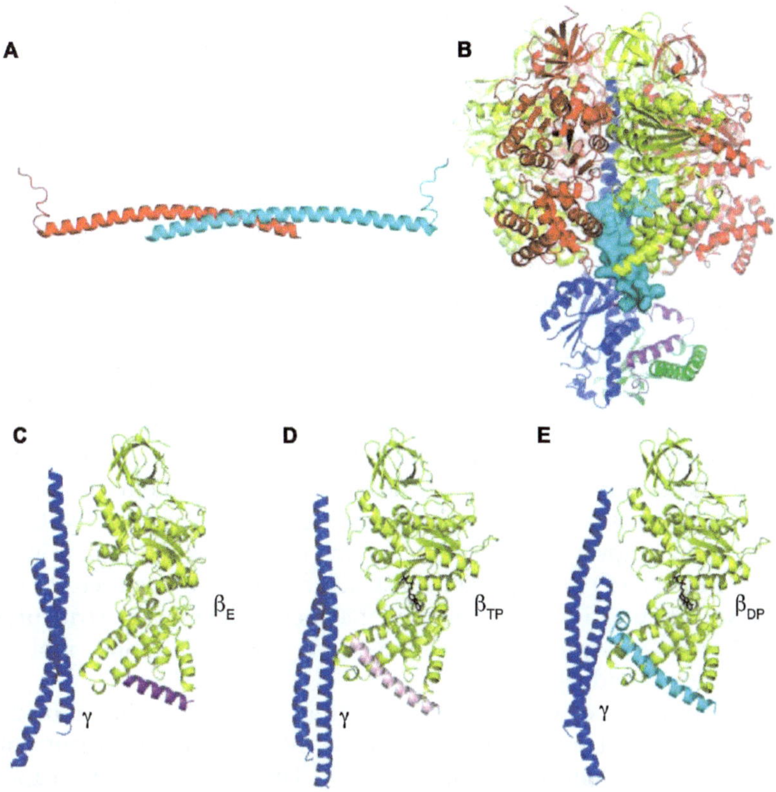

Figure 13.9 Structure and mode of binding of the regulatory protein IF_1 to mitochondrial ATP synthase. (A), Crystal structure of the dimeric bovine IF_1.[118] In crystals of the isolated protein, the two identical monomers (red and blue) forming the dimer are almost entirely α-helical, except for residues 1–18, which are disordered. The proteins have dimerised *via* an α-helical coiled-coil from residues 49–81; (B), in a structure of bovine F_1-ATPase inhibited by residues 1–60 of IF_1,[55] the truncated and monomeric inhibitor protein (light blue, in solid representation) is bound in a largely α-helical conformation from residues 21–50 in the $\alpha_{DP}\beta_{DP}$-catalytic interface, with residues 14–18 forming a second short α-helix, interacting with the γ-subunit in the central stalk. Residues 1–10 are disordered and 11–13 form an extended structure found in the central cavity of the F_1-domain. In solution, residues 1–43 are unfolded, whereas residues 44–84 form the α-helical coiled-coil observed in the crystal structure of intact IF_1 in isolation. (C–E), the unstructured region becomes structured progressively as the unfolded protein binds initially to the F_1-domain and two ATP molecules are hydrolysed.[57] (C), the initial interactions of the unfolded region of IF_1 are mainly with the β_E-subunit in the $\alpha_E\beta_E$-catalytic interface, when residues 32–49 form an α-helix; (D), upon hydrolysis of an ATP molecule, the β_E-subunit is converted to the β_{TP}-subunit, which closes the catalytic interface, and residues 25–50 are now α-helical; (E), upon hydrolysis of a second ATP molecule, the β_{TP}-subunit is converted to the β_{DP}-subunit, and the inhibitor reaches its most fully folded extent, as observed in (B).

manner.[56] However, the yeast inhibitor arrests the catalytic cycle at an earlier stage than in the bovine complex. In the yeast complex, the Mg^{2+} and phosphate have been released following hydrolysis of ATP, but ADP still remains bound to the β_E-subunit; in the bovine inhibited complex, the β_E-subunit is not occupied by ADP.[56] The residues in bovine IF_1 that provide its binding energy have been identified by systematic mutational analysis and measurement of kinetic constants.[124] Most of the residues are involved in hydrophobic interactions with the C-terminal domain of the β_{DP}-and β_{TP}-subunits. Additional binding energy is provided by an ionic interaction involving glutamate residue Glu-30 of IF_1 and arginine residue Arg-408 in the β_{DP}-subunit.

In solution, a fragment consisting of residues 44–84 of bovine IF_1 forms a dimeric antiparallel α-helical coiled-coil, as in the crystal structure.[125] In contrast, the fragment of bovine IF_1 consisting of residues 10–48 is largely or completely unstructured in solution.[125] Thus, the inhibitory region of free bovine IF_1 is an example of an intrinsically disordered protein that becomes structured progressively as it binds to the F_1-domain of the ATP synthase.[57] A model of the pathway leading from the unstructured N-terminal inhibitory region of free IF_1, to the largely folded structure observed in structures of the enzyme with bound inhibitor proteins, has been provided by structures of bovine F_1-ATPase with one, two and three bound inhibitor molecules (Figure 13.9C, D and E).[57]

The monomeric bovine IF_1 with C-terminal affinity tags has provided a simple and efficient way of purifying ATP synthase complexes from mitochondria in a single chromatographic step.[123] This method has been applied successfully to enzymes from a wide range of metazoan and fungal species. However, the method is restricted to eukarya because, for reasons that are not understood, mitochondrial IF_1 has no inhibitory effect on the eubacterial ATP synthases that have been tested, despite the similarities between the structures and sequences of subunits of the F_1 domains of mitochondrial and eubacterial enzymes.

13.4.2 Bacterial ATP Synthases

Bacterial ATP synthases fall into two broad classes. Some enzymes, such as those in *E. coli* and *G. stearothermophilus* can synthesise ATP in the presence of Δp, and, in its absence, for example under anaerobic conditions, they can hydrolyze ATP and use the energy of hydrolysis to generate Δp. Others, for example the enzymes from *P. denitrificans*,[44,126,127] *C. thermarum*[35] and *M. tuberculosis*[34,36] are capable of carrying out ATP synthesis under appropriate conditions, but are inhibited in the hydrolytic direction. The enzymes from *E. coli*[128] and *G. stearothermophilus*[129] can be inhibited *in vitro* by the ε-subunit. This subunit is bound both to the γ-subunit in the central stalk of the F_1-domain and to the c-ring in the membrane domain of the enzyme (Figure 13.1). The structure of the isolated *E. coli* ε-subunit has two domains.[80,130] The N-terminal domain is folded into a 10-stranded β-sandwich, and the C-terminal region consists of two side-by-side α-helices. In the intact enzyme, the β-sandwich domain is involved in binding the ε-subunit to the γ-subunit and the c-ring, and the α-helical C-terminal region has been proposed to adopt two different conformations, "down" and "up". In the "down" conformation, the two α-helices are held associated closely with the β-sandwich by an ATP molecule bound between the two domains (Figure 13.10A).[80] When the two α-helices assume the "up" position, it has been proposed that they interact with the $\alpha_3\beta_3$ complex and inhibit ATP hydrolysis. In the *E. coli* enzyme, this "up" position has been captured in a high-resolution structure of the intact F_1-domain (Figure 13.10B),[74,75] and in a cryo-EM structure of the intact ATP synthase at 6 Å resolution.[21] In the "up" position, the two α-helices at the C-terminus of ε-subunit penetrate into the catalytic domain along the axis of the coiled-coil of the N- and C-terminal α-helices of the γ-subunit, and the N-terminal domain of the ε-subunit has no bound ATP molecule. From these observations, it has been proposed that when Δp and the concentration of ATP are low, the ATP molecule leaves the ε-subunit, allowing the two α-helices to dissociate from the β-domain, forming the inhibitory "up" conformation.[74] Other eubacterial ATP synthases, exemplified by the enzymes from *C. thermarum* and *M. tuberculosis*, synthesize ATP in the usual way under aerobic conditions.[34–36] However, they are not able to hydrolyze ATP under physiological conditions, and their latent ATP hydrolase activities can be activated only under non-physiological conditions.[34–36] In the inhibited F_1-ATPase from *C. thermarum*, the ε-subunit subunit is "down", with an ATP molecule and an accompanying Mg^{2+} ion bound to it (Figure 13.10C).[79] However, when the capacity of this ε-subunit to bind an ATP molecule was removed by the introduction of specific mutations in the ATP binding site, the ε-subunit remained in the "down" position, and the ATP hydrolytic activity of the enzyme was still inhibited.[79] Therefore, in this organism, it is unlikely that the ε-subunit contributes to the inhibitory mechanism. A more likely explanation is that the ATP hydrolytic activity of this enzyme is inhibited by the failure to release the products of ATP hydrolysis, ADP and phosphate, which, in the structure of the enzyme, are observed bound in the β_E-subunit,[79] although the reason why they remain bound awaits an explanation. Currently, the structure of the F_1-ATPase from *M. tuberculosis* is not known, and as yet there is no adequate molecular explanation for the inhibition of ATP hydrolysis in this enzyme. It is worth noting that the ε-subunit from *C. thermarum* and the bovine δ-subunit have closely similar folds (Figure 13.12D).[79] However, in the bovine protein, the groove equivalent to the groove where ATP is bound in the *C. thermarum* ε-subunit, is occupied by the bovine ε-subunit which, despite its name, has no bacterial equivalent (Figure 13.10D).[8,22,51]

The molecular basis for the inhibition of the ATP hydrolytic activity of the eubacterial ATP synthase from *P. denitrificans* and other α-proteobacteria is somewhat unexpected. The F_1-domains of these enzymes bind an additional subunit known as the ζ-subunit that is involved in the regulation of ATP

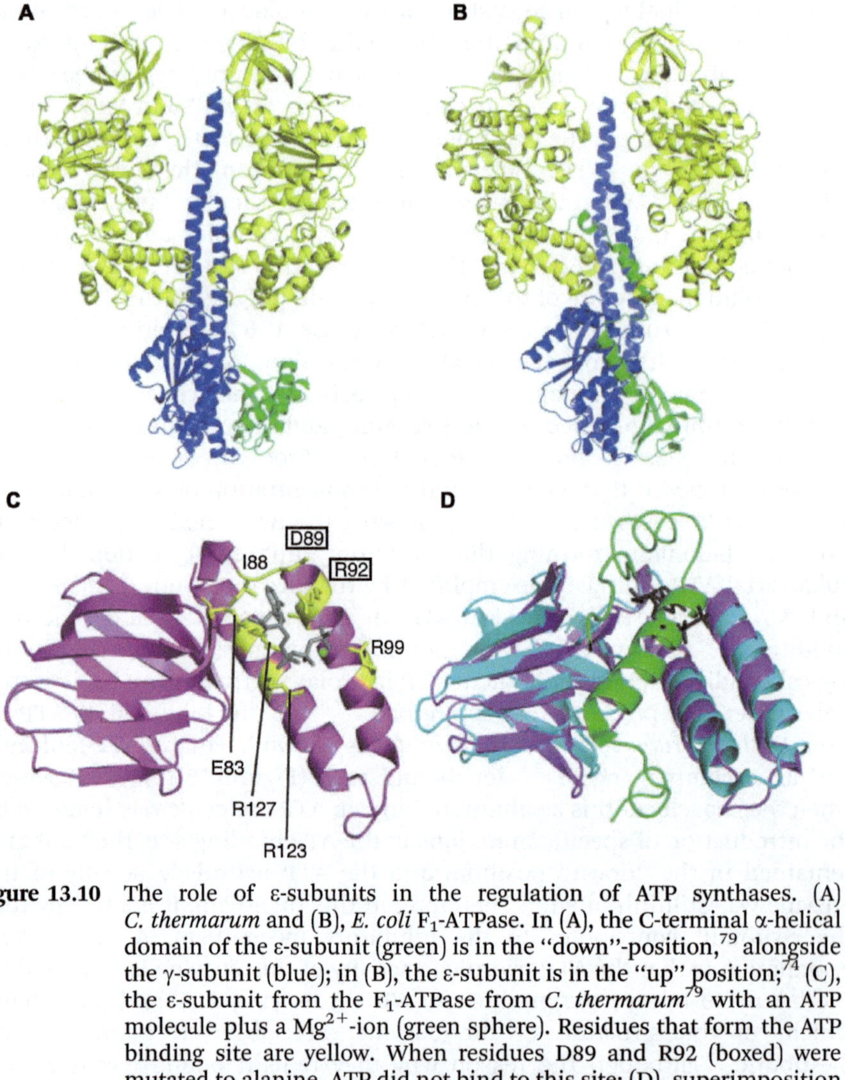

Figure 13.10 The role of ε-subunits in the regulation of ATP synthases. (A) *C. thermarum* and (B), *E. coli* F_1-ATPase. In (A), the C-terminal α-helical domain of the ε-subunit (green) is in the "down"-position,[79] alongside the γ-subunit (blue); in (B), the ε-subunit is in the "up" position;[74] (C), the ε-subunit from the F_1-ATPase from *C. thermarum*[79] with an ATP molecule plus a Mg^{2+}-ion (green sphere). Residues that form the ATP binding site are yellow. When residues D89 and R92 (boxed) were mutated to alanine, ATP did not bind to this site; (D), superimposition of the structures of the ε-subunit from *C. thermarum*[79] and the equivalent δ-subunit from bovine mitochondrial enzyme (blue).[51] In the bovine F_1-ATPase, the groove between the C-terminal α-helices is occupied by the ε-subunit (green), which has no bacterial equivalent.

hydrolysis.[44,126,127] In the structure of the intact ATP synthase complex from *P. denitrificans*, the ζ-subunit is bound to the F_1-domain *via* residues 1–19 of its N-terminal α-helix, which occupy a cleft in the lower region of the $\alpha_{DP}\beta_{DP}$-catalytic interface (Figure 13.3I), with the rest of the α-helix (residues 20–32) extending from the surface of the enzyme.[44] Residues 1 and 2 are close to the N-terminal α-helix of the γ-subunit, along the central axis of the

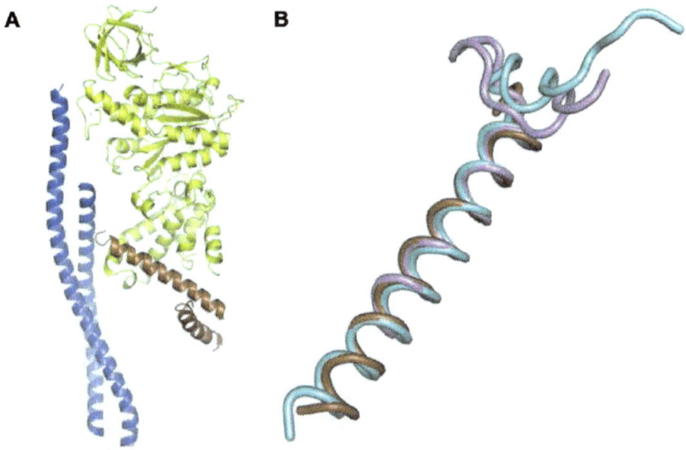

Figure 13.11 Comparison between the ζ-subunit from *P. denitrificans* with mitochondrial IF_1. (A), mode of binding of the ζ-subunit to the C-terminal domain of the β_{TP}-subunit of the *P. denitrificans* ATP synthase[44] (compare with Figure 13.9E); (B), superposition of the N-terminal region of the ζ-subunit (brown)[44] with the corresponding inhibitory regions of IF_1 from bovine[55] and yeast[56] mitochondria (cyan and pink, respectively). The α-helical regions are residues 3–24, 21–50, and, 16–36 respectively.

F_1-domain. Residues 3–19 probably form polar and hydrophobic interactions with other residues in α-helices in the C-terminal domains of the α_{DP}- and β_{DP}-subunits. The N-terminal α-helix of the ζ-subunit and the inhibitory regions of bovine and yeast IF_1 are bound to their cognate F_1-ATPases in a similar way.[44] The sequences of the N-terminal α-helix of the ζ-subunit and the N-terminal inhibitory regions of bovine and yeast IF_1 are related weakly, and their structures are very similar (Figure 13.11), and all three inhibitors occupy equivalent positions in the $\alpha_{DP}\beta_{DP}$-catalytic interface, interacting with α-helices in the lower regions of their C-terminal domains. Thus, they appear to inhibit their cognate F_1-ATPases in a similar manner. Moreover, the inhibitory regions of all three proteins are intrinsically unfolded, and therefore their mechanisms of binding to the F_1-domains are likely to be similar also.

13.5 Perspectives

13.5.1 Determination of Structures of ATP Synthases

In 2012, the residual "missing" 15% of the structure of the ATP synthase from bovine mitochondria represented the most difficult part of the enzyme complex for structural analysis (Figure 13.3). Since then, a description of the missing region has emerged in ATP synthases from both bacteria and mitochondria, and current models have a resolution of 4–7 Å.[21,31,32,43,44]

As described above, these models are providing new insights into the generation of rotation from Δp in the various enzyme complexes, although a deeper molecular understanding of the generation from Δp will require structural information in the membrane domains of the enzymes at higher resolution than at present, as well as information from other sources. This recent progress has been aided by two technical advances. One was the finding that, despite pessimistic prognostications, it has proved to be possible to crystallise an intact ATP synthase, allowing the specific conformation of the enzyme caught in the crystals to be solved by X-ray crystallography. The second, with much wider consequences for structural biology in general, has been the recent advances in electron cryo-microscopy (cryo-EM),[83–85] which has provided the means of solving the structures of many protein complexes, and even individual proteins, at atomic resolution by averaging images of single particles of the proteins and protein complexes, embedded in vitreous ice.[131] The ATP synthase complex is a particularly challenging topic for study by this means, as the enzyme can adopt many different conformations arising from a combination of at least three factors. First, in each catalytic domain, the structures of the three catalytic sites differ from each other at any moment in time, and the enzyme is a mixture of complexes where any one of these three different sites can be proximal to the peripheral stalk. The classification of the images of mitochondrial enzymes has been aided by studying ATP synthase complexes inhibited with a single copy of monomeric IF_1, which protrudes from the F_1-domain and marks the $\alpha_{DP}\beta_{DP}$-catalytic interface, allowing the images to be sorted into three major classes characterized by the position of this interface relative to the peripheral stalk.[32,43] Second, in the membrane domain of the enzyme, the rotation of the c-ring can be arrested at a range of different positions between each c-subunit relative to a fixed point in the stator. It may be possible to reduce heterogeneity from this source by inhibiting rotation of the c-ring, for example by covalent modification of the c-subunit with dicyclohexylcarbodiimide,[100,132] or by using an inhibitor of rotation, such as oligomycin.[133,134] Third, the central and peripheral stalks can bend, allowing the F_1-domain to adopt a range of conformations relative to the F_o-domain. This source of heterogeneity can be eliminated by removal of the peripheral stalk and F_1-domains biochemically, and studying the F_o-domain in isolation.[135] This approach has been applied to the V-ATPase,[136] where, in contrast to F_o, V_o can be formed *in vivo*. Nonetheless, the combination of these factors leads to an unusually high level of structural heterogeneity in samples of biochemically pure ATP synthases. This heterogeneity is a hindrance when the immediate aim is to describe a unique structure of the complex in order to understand its underlying architecture. However, providing that images of a sufficiently large number of particles can be collected, then, in principle, it is highly advantageous as it should allow the structures of the individual conformations of the enzyme to be resolved, and provide unparalleled detailed insights into the workings of the ATP synthases. So far, the conformational complexity of the enzymes, and the number of images of particles that it has been possible to record and resolve into the various structural classes, have limited the resolution of the cryo-EM structures of ATP synthases to about 6–8 Å. Nonetheless, structures of ATP synthases at about 6–10 Å are emerging from an increasingly wide range of species. Some of them are being studied by analysis of single particles,[21,31,32,43,86] and others are being investigated *in situ* in mitochondrial membranes by tomography.[89,90,137–140] So far, the main benefit is that these structures confirm the wide conservation of the key features of the enzyme involved in transmembrane proton translocation, and catalysis. One possible exception is the structure of the enzyme from the mitochondria of *Trypanosoma brucei*.[140] On the basis of tomographic imaging of the mitochondrial membranes at 18 Å resolution, it has been proposed that the catalytic part of its ATP synthase has a non-canonical structure and catalytic mechanism.[140] It remains to be seen whether these proposals will be sustained by more detailed investigation.

It has long been known that the subunits that form the peripheral stalks in bacterial and mitochondrial enzymes are the least conserved parts of the enzyme,[28] but the new studies are uncovering unexpected complexity in the peripheral stalks of algal enzymes,[86,138] for example.

13.5.2 The Catalytic Cycle and Rotary Mechanism

The problem of reconciling structures of the catalytic F_1-domain with various bound substrates, substrate analogues and inhibitors with data from rotational experiments, and of providing a molecular description of the substeps in the rotary cycle, has been discussed before.[2] As described in Section 13.3.2, currently, in the mammalian enzyme, there are convincing molecular descriptions of the phosphate release dwell[64,95] and the catalytic dwell, and of the 30° sub-step between them, but not of the ATP binding dwell, and the two sub-steps between the catalytic and phosphate binding dwells, and the catalytic and the ATP binding dwells (see Figure 13.5). An associated uncertainty is that the order of release from the enzyme of the products hydrolysis, ADP, Mg^{2+} and phosphate has not been established unambiguously. Based on a bacterial F_1-ATPase modified genetically to promote early release of phosphate, the release of ADP is impaired, suggesting that ADP is released before phosphate.[141] On the basis of molecular dynamics calculations based on bovine structural data[62] and bacterial rotational data[94] it has been proposed that phosphate is released from the β_E-subunit after ADP.[142,143] However, the structural data, derived almost entirely from eukaryotic enzymes, are ambiguous. The structure of bovine F_1-ATPase inhibited with thiophosphate (F_1-ThioP) and a structure of yeast F_1-ATPase[66] can be interpreted as being consistent with the release of ADP and the magnesium ion before the release of phosphate from the β_E-subunit. However, the structure of bovine F_1-ATPase crystallized in the presence of the magnesium ion chelator, phosphonate, contains an ADP molecule only without an accompanying magnesium ion or phosphate, bound to a fully open β_E-subunit,[63] suggesting that the release of both phosphate and magnesium (in an unspecified order) precedes that of ADP. This order is consistent with the order of product release from other P-loop containing nucleotide hydrolases such as kinesin,[144] G proteins[145] and myosin.[146] In the structure of the bovine enzyme inhibited with phosphonate (known as F_1-PH),[64] the γ-subunit is rotated by 30.5° relative to the γ-subunit in F_1-ThioP, to a similar position to that of the γ-subunit in a structure of bovine F_1-ATPase inhibited by thiophosphate and with three truncated IF_1 molecules. Therefore, it appears from this structure that ADP can remain bound to bovine F_1-ATPase after the release of phosphate. Further experiments are required to determine whether thiophosphate will bind to F_1-PH, and whether ADP will bind to F_1-ThioP.

Another issue relating to product release from F_1-ATPase, where there are conflicting data, concerns the identity of the subunit from which phosphate is released. The structure of bovine F_1-ATPase inhibited by thiophosphate, together with rotational studies on the human F_1-ATPase,[95] show that phosphate is released from the β_E-subunit. In an alternative catalytic mechanism, it is proposed that phosphate is released from the β_{DP}-subunit of the bacterial enzyme immediately following the hydrolysis of ATP and before the release of ADP.[147] It is now apparent from rotational experiments that the catalytic cycles of human and bacterial F_1-ATPases differ significantly, as indeed they differ from the catalytic cycle of the related V-type ATPases, where there are three 120° steps per cycle with no observable substeps.[148] These differences are likely to underlie the difficulties encountered hitherto in reconciling structural data from the bovine enzyme with rotational data determined on bacterial enzymes. The structures of F_1-ThioP and of F_1-ATPase inhibited by IF_1 demonstrate that the structural and rotational data from the same or closely related species are compatible. The remaining challenge is to provide molecular explanations of the generation of the rest of the 120° step in the catalytic cycle.

13.5.3 The Peripheral Stalk

Recent structural advances[21,31,32,43] have helped to add substance to the earlier proposal that the role of the peripheral stalk is to hold the static a-subunit against the rotating c-ring, thereby maintaining the coupling between Δp and the synthesis of ATP. The bacterial peripheral stalk is made mostly of two long α-helices in b- and b′-subunits, packed in parallel, which lie along most of the external surface of the $\alpha_3\beta_3$-domain.[21,44] The α-helices extend unbroken into the membrane domain, where the two parallel N-terminal α-helical regions are packed against the a-subunit. The central part of the membrane extrinsic region of the mitochondrial peripheral stalk is also made of parallel α-helices in the b-, d- and F_6-subunits,[32,40,42,43] packed against each other, but the mitochondrial peripheral stalk differs from the bacterial version in the membrane domain, where the N-terminal region of the b-subunits forms two antiparallel canonical transmembrane α-helices packed against the a-subunit, augmented by the ATP8 subunit and possibly subunit f.[31,32,43] In both bacterial and mitochondrial enzymes, this extensive α-helical region of the peripheral stalk appears to be stiff and rather inflexible. At the upper extremity, the N-terminal domain of the δ-subunit in bacterial and chloroplast enzymes, or the mitochondrial ortholog, the OSCP, provides the connection of the peripheral stalk to the $\alpha_3\beta_3$-domain (Figure 13.12). The C-terminal domain connects *via* interactions of α-helices OH7 and OH8 with the C-terminal and long α-helices of the b-subunit, respectively. Part of this region remains incompletely resolved in all current structures of ATP synthases. It is probably flexible and the region appears to act like a pivot or hinge. In a structure of the *E. coli* enzyme, it is claimed that the structure of

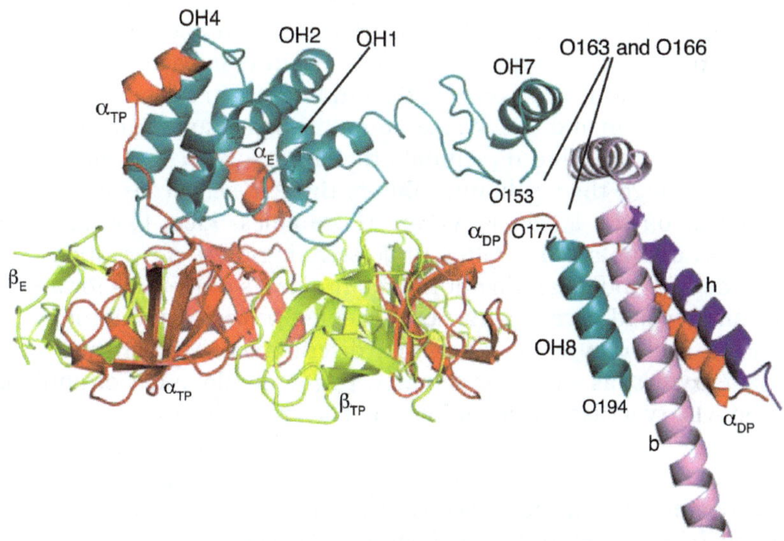

Figure 13.12 The pivot region in the peripheral stalk of the ATP synthase from *P. angusta*.[32] The OSCP is sea-green, the three α-subunits are red, the three β-subunits are yellow, the b-subunit is pink and the h-subunit is purple (the h-subunit is the fungal orthologue of subunit F_6 in mammalian enzymes). In the OSCP, the positions of α-helices OH1, OH2, OH4, OH7 and OH8 are indicated. The link between OH7 and OH8 (residues 154–176) was not resolved, but the positions of residues O163 and O166, inferred from a structure of the bovine F_1-ATPase with a truncated peripheral stalk containing selenomethionine residues, are indicated. The N-terminal regions of the α- and β-subunits are designated α_{TP}, α_{DP} and α_E, and β_{TP}, β_{DP} and β_E. N-Terminal α-helices extend from the crown region of each α-subunit. Two of them, in the α_{TP}- and α_E-subunits, interact in different ways with the N-terminal domain of the OSCP. The third, in the α_{DP}-subunit, binds to the long α-helix in the b-subunit. The N-terminal α-helix of the b-subunit lies alongside OH7. This region is proposed to form a pivot in the peripheral stalk, allowing it to accommodate lateral rocking movements of the $\alpha_3\beta_3$-domain during catalysis.

the peripheral stalk is "complete",[21] but, in reality, the density in the pivot region has not been interpreted correctly, and, as in other structures, it cannot be resolved definitively. It is possible that in addition to providing a pivot, the region could also have spring-like properties to contribute to holding the stator and rotor in contact. A full understanding of how the enzyme works requires further elucidation of the structure and properties of the pivot region.

13.5.4 The Mitochondrial Inhibitor Protein

The mechanism of binding of bovine and yeast IF_1 to their cognate F_1-ATPases, and how they prevent ATP hydrolysis, are well understood.[55–57,123,124] Also, the yeast IF_1 inhibits the bovine enzyme and *vice versa*,[33,123] and the bovine IF_1, inhibits F-ATPases from a wide range of metazoans and from several fungi.[29] Yet, bovine IF_1 has no effect, for example, on the bacterial F_1-ATPase from *E. coli*, despite the high conservation in bacterial enzymes of the amino acid residues that provide the binding site in the bovine and yeast enzymes (J. V. Bason and J. E. Walker, unpublished results), although it has been reported that bovine IF_1 inhibits the enzyme from *P. denitrificans*.[149] Similarly, although the ζ-subunit, inhibits the *P. denitrificans* F_1-ATPase, it has no effect on the *E. coli* F_1-ATPase (E. Morales- Rios, J. V. Bason and J. E. Walker, unpublished results). Understanding these observations could lead to a deeper understanding of the requirements of features for an effective IF_1, and it could have the practical benefit of providing inhibitor proteins and related inhibitory compounds that discriminate between human and bacterial enzymes, as would be required in the development of antibiotics to inhibit ATP synthases from pathogenic bacteria (see Section 13.5.5).

Another area where there is need for further investigation is the physiological role of IF_1. It has been clearly established in experiments with the purified bovine ATP synthase reconstituted into phospholipid vesicles that the action of IF_1 is unidirectional, inhibiting ATP hydrolysis and not its synthesis.[123] Yet, it has been proposed that *in vivo*, IF_1 inhibits both synthesis and hydrolysis.[150] As yet, there is no plausible molecular explanation for the apparently different behaviour of IF_1 *in vitro* and in a cellular context.

13.5.5 ATP Synthase as a Drug Target

Somewhat unexpectedly, a new antibiotic, bedaquiline,[151–153] introduced to combat tuberculosis, has been proved to act by inhibiting the ATP synthase from the causative bacterium, *M. tuberculosis*, and it does so by binding to the c-ring of the enzyme's rotor.[20] Bedaquiline appears to have no significant effect on the ATP synthase in human mitochondria. These observations provide proof of the principle that, despite the many similarities between the human and bacterial enzymes, the ATP synthases in bacterial pathogens and human mitochondria differ sufficiently to provide potential therapeutic targets, against which new antibiotics to combat bacterial infections can be

Table 13.1 Conservation of sequences of subunits of human and bovine ATP synthases.

Subunit	Identity (%)
α	98.4
β	98.6
γ	91.6
δ	92.5
ε	94.0
OSCP	87.4
F_6	82.9
a	77.9
b	86.4
c	100.0
d	88.1
e	84.3
f	73.4
g	92.2
ATP8	50.0
DAPIT	94.8
6.8 kDa proteolipid	73.3

directed. One approach to developing new drugs is by rational design of inhibitory compounds based upon the structures and properties of the bacterial enzymes. This approach requires knowledge of the ATP synthase in the pathogen, and also of the human ATP synthase. The structure of the human enzyme has not been studied, but it is known to have the same subunit composition as the bovine enzyme,[29] and the sequences of human and bovine subunits are closely related (Table 13.1). Therefore, the structure of the bovine enzyme provides an excellent surrogate for that of the human enzyme, and from it a model of the human enzyme can be calculated easily. At the moment, the structures of ATP synthases from mycobacteria, and other bacterial pathogens have not been described, and the mechanisms of regulation of their ATP hydrolytic reactions are not understood either.

13.5.6 The Permeability Transition Pore

It has been suggested that a non-specific pore, known as the permeability transition pore (PTP),[154,155] is associated with the ATP synthase complex in human mitochondria.[156] The permeability transition refers to the opening of a non-specific channel, the PTP, in the inner membrane.[154,155] Opening of the pore has been associated with necrotic cell death, and has been linked to human diseases including cardiac ischemia, neurodegeneration, cancer and muscle dystrophy.[157] Thus, knowledge of the proteins that form the pore has considerable medical relevance. Opening of the pore is triggered by uptake of cytoplasmic calcium ions *via* the mitochondrial calcium uniporter, leading to swelling of the organelle, disruption of the inner membrane and

ATP synthesis, followed by cell death. The proposal that has been made is that the pore may be provided by proteins in the interface region between monomers in the dimeric assembly of the ATP synthase complex,[158] or by the ring of c-subunits in the enzyme's rotor.[159,160] In humans, identical copies of the mature c-protein with different N-terminal mitochondrial import sequences are encoded by three nuclear genes, ATP5G1, ATP5G2 and ATP5G3.[161,162] Disruption of these three genes produced cells devoid of the c-subunit, but with the characteristic properties of the PTP preserved.[163] Therefore, the c-ring does not provide the PTP. Similarly, human ρ^0 cells, which lack mitochondrial DNA, and therefore cannot make subunits ATP6 and ATP8, also retain the PTP.[163] Therefore, none of the membrane subunits of the ATP synthase that are involved directly in transmembrane proton translocation participates in forming the PTP. It remains to be tested whether any or all of the remaining membrane subunits, b, e, f, g, DAPIT and 6.8PL participate in the formation of the pore.

Note Added in Proof

Recently, it has been demonstrated that the separate disruption of the human genes for the OSCP and subunit b, results in a vestigial F-ATPase completely devoid of any of the stator subunits, including the peripheral stalk subunits (OSCP, F_6, b and d), subunit a and the supernumerary subunits ATP8, e, f, g, DAPIT and 6.8PL. Yet, the cells retain the characteristic properties of PTP.[164] It is increasingly unlikely that the PTP is associated with the ATP synthase complex.

Acknowledgements

I am grateful to G. M. Cook and A. G. W. Leslie for their comments and advice. I thank M. G. Montgomery for producing the Figures and for making the reference list. This work was funded by the Medical Research Council, UK by Programme grant [MR/M009858/1] to the author.

References

1. J. E. Walker, *Angew. Chem., Int. Ed.*, 1998, **37**, 2308.
2. J. E. Walker, *Biochem. Soc. Trans.*, 2013, **41**, 1.
3. W. Laubinger and P. Dimroth, *Eur. J. Biochem.*, 1987, **168**, 475.
4. D. G. Nicholls and S. J. Ferguson, *Bioenergetics 4*, Academic Press, Cambridge, Massachusetts, USA, 2013.
5. A. Harms, E. Maisonneuve and K. Gerdes, *Science*, 2016, 354.
6. D. L. Foster and R. H. Fillingame, *J. Biol. Chem.*, 1979, **254**, 8230.
7. D. L. Foster and R. H. Fillingame, *J. Biol. Chem.*, 1982, **257**, 2009.
8. J. E. Walker, M. J. Runswick and M. Saraste, *FEBS Lett.*, 1982, **146**, 393.
9. J. E. Walker, M. Saraste and N. J. Gay, *Biochim. Biophys. Acta*, 1984, **768**, 164.

10. G. Falk, A. Hampe and J. E. Walker, *Biochem. J.*, 1985, **228**, 391.
11. A. L. Cozens and J. E. Walker, *J. Mol. Biol.*, 1987, **194**, 359.
12. G. Falk and J. E. Walker, *Biochem. J.*, 1988, **254**, 109.
13. H. S. Van Walraven, R. Lutter and J. E. Walker, *Biochem. J.*, 1993, **294**, 239.
14. J. Hermolin and R. H. Fillingame, *J. Biol. Chem.*, 1989, **264**, 3896.
15. W. Jiang, J. Hermolin and R. H. Fillingame, *Proc. Natl. Acad. Sci. U. S. A.*, 2001, **98**, 4966.
16. T. Meier, P. Polzer, K. Diederichs, W. Welte and P. Dimroth, *Science*, 2005, **308**, 659.
17. D. Pogoryelov, J. Yu, T. Meier, J. Vonck, P. Dimroth and D. J. Muller, *EMBO Rep.*, 2005, **6**, 1040.
18. M. Vollmar, D. Schlieper, M. Winn, C. Büchner and G. Groth, *J. Biol. Chem.*, 2009, **284**, 18228.
19. D. Matthies, L. Preiss, A. L. Klyszejko, D. J. Muller, G. M. Cook, J. Vonck and T. Meier, *J. Mol. Biol.*, 2009, **388**, 611.
20. L. Preiss, J. D. Langer, O. Yildiz, L. Eckhardt-Strelau, J. E. G. Guillemont, A. Koul and T. Meier, *Sci. Adv.*, 2015, **1**, e1500106.
21. M. Sobti, C. Smits, A. S. Wong, R. Ishmukhametov, D. Stock, S. Sandin and A. G. Stewart, *Elife*, 2016, 5.
22. J. E. Walker, I. M. Fearnley, N. J. Gay, B. W. Gibson, F. D. Northrop, S. J. Powell, M. J. Runswick, M. Saraste and V. L. Tybulewicz, *J. Mol. Biol.*, 1985, **184**, 677.
23. I. M. Fearnley and J. E. Walker, *EMBO J.*, 1986, **5**, 2003.
24. J. E. Walker, R. Lutter, A. Dupuis and M. J. Runswick, *Biochemistry*, 1991, **30**, 5369.
25. I. R. Collinson, M. J. van Raaij, M. J. Runswick, I. M. Fearnley, J. M. Skehel, G. L. Orriss, B. Miroux and J. E. Walker, *J. Mol. Biol.*, 1994, **242**, 408.
26. R. Chen, M. J. Runswick, J. Carroll, I. M. Fearnley and J. E. Walker, *FEBS Lett*, 2007, **581**, 3145.
27. B. Meyer, I. Wittig, E. Trifilieff, M. Karas and H. Schägger, *Mol. Cell. Proteomics*, 2007, **6**, 1690.
28. J. E. Walker, M. J. Runswick and L. Poulter, *J. Mol. Biol.*, 1987, **197**, 89.
29. T. B. Walpole, D. N. Palmer, H. Jiang, S. Ding, I. M. Fearnley and J. E. Walker, *Mol. Cell. Proteomics*, 2015, **14**, 828.
30. D. Stock, A. G. W. Leslie and J. E. Walker, *Science*, 1999, **286**, 1700.
31. A. Hahn, K. Parey, M. Bublitz, D. J. Mills, V. Zickermann, J. Vonck, W. Kühlbrandt and T. Meier, *Mol. Cell*, 2016, **63**, 445.
32. K. R. Vinothkumar, M. G. Montgomery, S. Liu and J. E. Walker, *Proc. Natl. Acad. Sci. U. S. A.*, 2016, **113**, 12709.
33. S. Liu, T. J. Charlesworth, J. V. Bason, M. G. Montgomery, M. E. Harbour, I. M. Fearnley and J. E. Walker, *Biochem. J.*, 2015, **468**, 167.
34. H. Nakagawa, S. H. Lee, V. K. Kalra and A. F. Brodie, *J. Biol. Chem.*, 1977, **252**, 8229.

35. G. M. Cook, S. Keis, H. W. Morgan, C. von Ballmoos, U. Matthey, G. Kaim and P. Dimroth, *J. Bacteriol.*, 2003, **185**, 4442.
36. A. C. Haagsma, N. N. Driessen, M. M. Hahn, H. Lill and D. Bald, *FEMS Microbiol. Lett.*, 2010, **313**, 68.
37. J. P. Abrahams, A. G. W. Leslie, R. Lutter and J. E. Walker, *Nature*, 1994, **370**, 621.
38. I. N. Watt, M. G. Montgomery, M. J. Runswick, A. G. W. Leslie and J. E. Walker, *Proc. Natl. Acad. Sci. U. S. A.*, 2010, **107**, 16823.
39. S. Karrasch and J. E. Walker, *J. Mol. Biol.*, 1999, **290**, 379.
40. V. K. Dickson, J. A. Silvester, I. M. Fearnley, A. G. W. Leslie and J. E. Walker, *EMBO J.*, 2006, **25**, 2911.
41. J. E. Walker and V. K. Dickson, *Biochim. Biophys. Acta*, 2006, **1757**, 286.
42. D. M. Rees, A. G. W. Leslie and J. E. Walker, *Proc. Natl. Acad. Sci. U. S. A.*, 2009, **106**, 21597.
43. A. Zhou, A. Rohou, D. G. Schep, J. V. Bason, M. G. Montgomery, J. E. Walker, N. Grigorieff and J. L. Rubinstein, *Elife*, 2015, **4**, e10180.
44. E. Morales-Rios, M. G. Montgomery, A. G. W. Leslie and J. E. Walker, *Proc. Natl. Acad. Sci. U. S. A.*, 2015, **112**, 13231.
45. R. J. Carbajo, F. A. Kellas, M. J. Runswick, M. G. Montgomery, J. E. Walker and D. Neuhaus, *J. Mol. Biol.*, 2005, **351**, 824.
46. R. J. Carbajo, F. A. Kellas, J. C. Yang, M. J. Runswick, M. G. Montgomery, J. E. Walker and D. Neuhaus, *J. Mol. Biol.*, 2007, **368**, 310.
47. J. Lee, S. Ding, T. B. Walpole, A. N. Holding, M. G. Montgomery, I. M. Fearnley and J. E. Walker, *J. Biol. Chem.*, 2015, **290**, 13308.
48. M. J. van Raaij, J. P. Abrahams, A. G. W. Leslie and J. E. Walker, *Proc. Natl. Acad. Sci. U. S. A.*, 1996, **93**, 6913.
49. J. P. Abrahams, S. K. Buchanan, M. J. Van Raaij, I. M. Fearnley, A. G. W. Leslie and J. E. Walker, *Proc. Natl. Acad. Sci. U. S. A.*, 1996, **93**, 9420.
50. G. L. Orriss, A. G. W. Leslie, K. Braig and J. E. Walker, *Structure*, 1998, **6**, 831.
51. C. Gibbons, M. G. Montgomery, A. G. W. Leslie and J. E. Walker, *Nat. Struct. Biol.*, 2000, **7**, 1055.
52. J. R. Gledhill, M. G. Montgomery, A. G. W. Leslie and J. E. Walker, *Proc. Natl. Acad. Sci. U. S. A.*, 2007, **104**, 13632.
53. E. Cabezón, M. G. Montgomery, A. G. W. Leslie and J. E. Walker, *Nat. Struct. Biol.*, 2003, **10**, 744.
54. M. W. Bowler, M. G. Montgomery, A. G. W. Leslie and J. E. Walker, *Proc. Natl. Acad. Sci. U. S. A.*, 2006, **103**, 8646.
55. J. R. Gledhill, M. G. Montgomery, A. G. W. Leslie and J. E. Walker, *Proc. Natl. Acad. Sci. U. S. A.*, 2007, **104**, 15671.
56. G. C. Robinson, J. V. Bason, M. G. Montgomery, I. M. Fearnley, D. M. Mueller, A. G. W. Leslie and J. E. Walker, *Open Biol.*, 2013, **3**, 120164.
57. J. V. Bason, M. G. Montgomery, A. G. W. Leslie and J. E. Walker, *Proc. Natl. Acad. Sci. U. S. A.*, 2014, **111**, 11305.

58. K. Braig, R. I. Menz, M. G. Montgomery, A. G. W. Leslie and J. E. Walker, *Structure*, 2000, **8**, 567.
59. R. I. Menz, J. E. Walker and A. G. W. Leslie, *Cell*, 2001, **106**, 331.
60. R. I. Menz, A. G. W. Leslie and J. E. Walker, *FEBS Lett*, 2001, **494**, 11.
61. R. Kagawa, M. G. Montgomery, K. Braig, A. G. W. Leslie and J. E. Walker, *EMBO J.*, 2004, **23**, 2734.
62. M. W. Bowler, M. G. Montgomery, A. G. W. Leslie and J. E. Walker, *J. Biol. Chem.*, 2007, **282**, 14238.
63. D. M. Rees, M. G. Montgomery, A. G. W. Leslie and J. E. Walker, *Proc. Natl. Acad. Sci. U .S. A.*, 2012, **109**, 11139.
64. J. V. Bason, M. G. Montgomery, A. G. W. Leslie and J. E. Walker, *Proc. Natl. Acad. Sci. U. S. A.*, 2015, **112**, 6009.
65. L. A. Baker, I. N. Watt, M. J. Runswick, J. E. Walker and J. L. Rubinstein, *Proc. Natl. Acad. Sci. U. S. A.*, 2012, **109**, 11675.
66. V. Kabaleeswaran, N. Puri, J. E. Walker, A. G. W. Leslie and D. M. Mueller, *EMBO J.*, 2006, **25**, 5433.
67. V. Kabaleeswaran, H. Shen, J. Symersky, J. E. Walker, A. G. W. Leslie and D. M. Mueller, *J. Biol. Chem.*, 2009, **284**, 10546.
68. L. Preiss, A. L. Klyszejko, D. B. Hicks, J. Liu, O. J. Fackelmayer, Ö. Yildiz, T. A. Krulwich and T. Meier, *Proc. Natl. Acad. Sci. U. S. A.*, 2013, **110**, 7874.
69. D. Matthies, W. Zhou, A. L. Klyszejko, C. Anselmi, O. Yildiz, K. Brandt, V. Muller, J. D. Faraldo-Gomez and T. Meier, *Nat. Commun.*, 2014, **5**, 5286.
70. L. Preiss, J. D. Langer, D. B. Hicks, J. Liu, O. Yildiz, T. A. Krulwich and T. Meier, *Mol. Microbiol.*, 2014, **92**, 973.
71. D. Pogoryelov, O. Yildiz, J. D. Faraldo-Gómez and T. Meier, *Nat. Struct. Mol. Biol.*, 2009, **16**, 1068.
72. G. Groth and E. Pohl, *J. Biol. Chem.*, 2001, **276**, 1345.
73. G. Groth, *Proc. Natl. Acad. Sci. U. S. A.*, 2002, **99**, 3464.
74. G. Cingolani and T. M. Duncan, *Nat. Struct. Mol. Biol.*, 2011, **18**, 701.
75. A. Roy, M. L. Hutcheon, T. M. Duncan and G. Cingolani, *Acta Cryst. F*, 2012, **68**, 1229.
76. Y. Shirakihara, A. G. W. Leslie, J. P. Abrahams, J. E. Walker, T. Ueda, Y. Sekimoto, M. Kambara, K. Saika, Y. Kagawa and M. Yoshida, *Structure*, 1997, **5**, 825.
77. Y. Shirakihara, A. Shiratori, H. Tanikawa, M. Nakasako, M. Yoshida and T. Suzuki, *FEBS J.*, 2015, **282**, 2895.
78. A. Stocker, S. Keis, J. Vonck, G. M. Cook and P. Dimroth, *Structure*, 2007, **15**, 904.
79. S. A. Ferguson, G. M. Cook, M. G. Montgomery, A. G. W. Leslie and J. E. Walker, *Proc. Natl. Acad. Sci. U. S. A.*, 2016, **113**, 10860.
80. U. Uhlin, G. B. Cox and J. M. Guss, *Structure*, 1997, **5**, 1219.
81. H. Yagi, N. Kajiwara, H. Tanaka, T. Tsukihara, Y. Kato-Yamada, M. Yoshida and H. Akutsu, *Proc. Natl. Acad. Sci. U. S. A.*, 2007, **104**, 11233.

82. E. Morales-Ríos, M. G. Montgomery, A. G. W. Leslie, J. J. García-Trejo and J. E. Walker, *Acta Crystallogr. F Struct. Biol. Commun.*, 2015, **71**, 1309.
83. M. T. Smith and J. L. Rubinstein, *Science*, 2014, **345**, 617.
84. K. R. Vinothkumar and R. Henderson, *Q. Rev. Biophys.*, 2016, **49**, e13.
85. S. Subramaniam, L. A. Earl, V. Falconieri, J. L. Milne and E. H. Egelman, *Curr. Opin. Struct. Biol.*, 2016, **41**, 194.
86. M. Allegretti, N. Klusch, D. J. Mills, J. Vonck, W. Kühlbrandt and K. M. Davies, *Nature*, 2015, **521**, 237.
87. N. V. Dudkina, J. Heinemeyer, W. Keegstra, E. J. Boekema and H. P. Braun, *FEBS Lett.*, 2005, **579**, 5769.
88. N. V. Dudkina, G. T. Oostergetel, D. Lewejohann, H. P. Braun and E. J. Boekema, *Biochim. Biophys. Acta*, 2010, **1797**, 272.
89. M. Strauss, G. Hofhaus, R. R. Schröder and W. Kühlbrandt, *EMBO J.*, 2008, **27**, 1154.
90. K. M. Davies, M. Strauss, B. Daum, J. H. Kief, H. D. Osiewacz, A. Rycovska, V. Zickermann and W. Kuhlbrandt, *Proc. Natl. Acad. Sci. U. S. A.*, 2011, **108**, 14121.
91. H. Noji, R. Yasuda, M. Yoshida and K. Kinosita Jr., *Nature*, 1997, **386**, 299.
92. R. Yasuda, H. Noji, K. Kinosita and M. Yoshida, *Cell*, 1998, **93**, 1117.
93. H. Noji, K. Häsler, W. Junge, K. Kinosita, M. Yoshida and S. Engelbrecht, *Biochem. Biophys. Res. Commun.*, 1999, **260**, 597.
94. R. Yasuda, H. Noji, M. Yoshida, K. Kinosita and H. Itoh, *Nature*, 2001, **410**, 898.
95. T. Suzuki, K. Tanaka, C. Wakabayashi, E. Saita and M. Yoshida, *Nat. Chem. Biol.*, 2014, **10**, 930.
96. B. C. Steel, A. L. Nord, Y. Wang, V. Pagadala, D. M. Mueller and R. M. Berry, *Sci. Rep.*, 2015, **5**, 8773.
97. P. D. Boyer, *FEBS Lett*, 1975, **58**, 1.
98. P. D. Boyer, *Biochim. Biophys. Acta*, 1993, **1140**, 215.
99. T. Masaike, F. Koyama-Horibe, K. Oiwa, M. Yoshida and T. Nishizaka, *Nat. Struct. Mol. Biol.*, 2008, **15**, 1326.
100. J. Hoppe and W. Sebald, *Eur. J. Biochem.*, 1980, **107**, 57.
101. M. J. Miller, M. Oldenburg and R. H. Fillingame, *Proc. Natl. Acad. Sci. U. S. A.*, 1990, **87**, 4900.
102. C. Etzold, G. Deckers-Hebestreit and K. Altendorf, *Eur. J. Biochem.*, 1997, **243**, 336.
103. W. Junge, *Proc. Natl. Acad. Sci. U. S. A.*, 1999, **96**, 4735.
104. R. N. Lightowlers, S. M. Howitt, L. Hatch, F. Gibson and G. B. Cox, *Biochim. Biophys. Acta*, 1987, **894**, 399.
105. Y. Kagawa and E. Racker, *J. Biol. Chem.*, 1966, **241**, 2467.
106. K. Kopaczyk, J. Asai, D. W. Allmann, T. Oda and D. E. Green, *Arch. Biochem. Biophys.*, 1968, **123**, 602.
107. A. Pitotti, A. R. Contessa, F. Dabbeni-Sala and A. Bruni, *Biochim. Biophys. Acta*, 1972, **274**, 528.

108. E. Santiago, N. López-Moratalla and J. F. Segovia, *Biochem. Biophys. Res. Commun.*, 1973, **53**, 439.
109. D. M. Laird, J. W. Parce, R. I. Montgomery and C. C. Cunningham, *J. Biol. Chem.*, 1986, **261**, 14851.
110. G. Olofsson and E. Sparr, *PLoS One*, 2013, **8**, e73040.
111. M. Sathappa and N. N. Alder, *Biochim. Biophys. Acta*, 2016, **1858**, 1362.
112. A. L. Duncan, A. J. Robinson and J. E. Walker, *Proc. Natl. Acad. Sci. U. S. A.*, 2016, **113**, 8687.
113. B. Chance and G. R. Williams, *J. Biol. Chem.*, 1955, **217**, 383.
114. P. C. Hinkle, *Biochim. Biophys. Acta*, 2005, **1706**, 1.
115. T. Hisabori, E. Sunamura, Y. Kim and H. Konno, *Antioxid. Redox. Signalling*, 2013, **19**, 1846.
116. D. A. Harris and A. M. Das, *Biochem. J.*, 1991, **280**, 561.
117. J. E. Walker, *Curr. Opin. Struct. Biol.*, 1994, **4**, 912.
118. E. Cabezón, M. J. Runswick, A. G. W. Leslie and J. E. Walker, *EMBO J.*, 2001, **20**, 6990.
119. I. Arnold, K. Pfeiffer, W. Neupert, R. A. Stuart and H. Schagger, *EMBO J.*, 1998, **17**, 7170.
120. G. Arselin, M. F. Giraud, A. Dautant, J. Vaillier, D. Brèthes, B. Coulary-Salin, J. Schaeffer and J. Velours, *Eur. J. Biochem.*, 2003, **270**, 1875.
121. D. M. Bustos and J. Velours, *J. Biol. Chem.*, 2005, **280**, 29004.
122. E. Cabezon, P. J. Butler, M. J. Runswick and J. E. Walker, *J. Biol. Chem.*, 2000, **275**, 25460.
123. M. J. Runswick, J. V. Bason, M. G. Montgomery, G. C. Robinson, I. M. Fearnley and J. E. Walker, *Open Biol.*, 2013, **3**, 120160.
124. J. V. Bason, M. J. Runswick, I. M. Fearnley and J. E. Walker, *J. Mol. Biol.*, 2011, **406**, 443.
125. D. J. Gordon-Smith, R. J. Carbajo, J. C. Yang, H. Videler, M. J. Runswick, J. E. Walker and D. Neuhaus, *J. Mol. Biol.*, 2001, **308**, 325.
126. E. Morales-Ríos, F. de la Rosa-Morales, G. Mendoza-Hernández, J. S. Rodríguez-Zavala, H. Celis, M. Zarco-Zavala and J. J. García-Trejo, *FASEB J.*, 2010, **24**, 599.
127. P. Serrano, M. Geralt, B. Mohanty and K. Wüthrich, *J. Mol. Biol.*, 2014, **426**, 2547.
128. P. P. Laget and J. B. Smith, *Arch. Biochem. Biophys.*, 1979, **197**, 83.
129. Y. Kato-Yamada, D. Bald, M. Koike, K. Motohashi, T. Hisabori and M. Yoshida, *J. Biol. Chem.*, 1999, **274**, 33991.
130. S. Wilkens, F. W. Dahlquist, L. P. McIntosh, L. W. Donaldson and R. A. Capaldi, *Nat. Struct. Biol.*, 1995, **2**, 961.
131. I. Schmidt-Krey and J. L. Rubinstein, *Micron*, 2011, **42**, 107.
132. J. Symersky, V. Pagadala, D. Osowski, A. Krah, T. Meier, J. D. Faraldo-Gómez and D. M. Mueller, *Nat. Struct. Mol. Biol.*, 2012, **19**, 485.
133. H. A. Lardy, D. Johnson and W. C. McMurray, *Arch. Biochem. Biophys.*, 1958, **78**, 587.

134. J. Symersky, D. Osowski, D. E. Walters and D. M. Mueller, *Proc. Natl. Acad. Sci. U. S. A.*, 2012, **109**, 13961.
135. I. R. Collinson, M. J. Runswick, S. K. Buchanan, I. M. Fearnley, J. M. Skehel, M. J. van Raaij, D. E. Griffiths and J. E. Walker, *Biochemistry*, 1994, **33**, 7971.
136. M. T. Mazhab-Jafari, A. Rohou, C. Schmidt, S. A. Bueler, S. Benlekbir, C. V. Robinson and J. L. Rubinstein, *Nature*, 2016, **539**, 118.
137. K. M. Davies, C. Anselmi, I. Wittig, J. D. Faraldo-Gómez and W. Kühlbrandt, *Proc. Natl. Acad. Sci. U. S. A.*, 2012, **109**, 13602.
138. K. N. Yadav, H. V. Miranda-Astudillo, L. Colina-Tenorio, F. Bouillenne, H. Degand, P. Morsomme, D. González-Halphen, E. J. Boekema and P. Cardol, *Biochim. Biophys. Acta*, 2017, **1858**, 267.
139. A. W. Mühleip, F. Joos, C. Wigge, A. S. Frangakis, W. Kühlbrandt and K. M. Davies, *Proc. Natl. Acad. Sci. U. S. A.*, 2016, **113**, 8442.
140. A. W. Mühleip, C. E. Dewar, A. Schnaufer, W. Kühlbrandt and K. M. Davies, *Proc. Natl. Acad. Sci. U. S. A.*, 2017, **114**, 992.
141. R. Watanabe and H. Noji, *Nat. Commun.*, 2014, **5**, 3486.
142. K. Nam, J. Pu and M. Karplus, *Proc. Natl. Acad. Sci. U. S. A.*, 2014, **111**, 17851.
143. K. Okazaki and G. Hummer, *Proc. Natl. Acad. Sci. U. S. A.*, 2013, **110**, 16468.
144. B. Milic, J. O. Andreasson, W. O. Hancock and S. M. Block, *Proc. Natl. Acad. Sci. U. S. A.*, 2014, **111**, 14136.
145. A. Wittinghofer and I. R. Vetter, *Annu. Rev. Biochem.*, 2011, **80**, 943.
146. A. M. Gulick and I. Rayment, *BioEssays*, 1997, **19**, 561.
147. W. Junge, H. Sielaff and S. Engelbrecht, *Nature*, 2009, **459**, 364.
148. Y. Minagawa, H. Ueno, M. Hara, Y. Ishizuka-Katsura, N. Ohsawa, T. Terada, M. Shirouzu, S. Yokoyama, I. Yamato, E. Muneyuki, H. Noji, T. Murata and R. Iino, *J. Biol. Chem.*, 2013, **288**, 32700.
149. J. J. García-Trejo, M. Zarco-Zavala, F. Mendoza-Hoffmann, E. Hernández-Luna, R. Ortega and G. Mendoza-Hernández, *J. Biol. Chem.*, 2016, **291**, 538.
150. L. Formentini, M. Sánchez-Aragó, L. Sánchez-Cenizo and J. M. Cuezva, *Mol. Cell*, 2012, **45**, 731.
151. A. C. Haagsma, R. Abdillahi-Ibrahim, M. J. Wagner, K. Krab, K. Vergauwen, J. Guillemont, K. Andries, H. Lill, A. Koul and D. Bald, *Antimicrob. Agents Chemother.*, 2009, **53**, 1290.
152. A. Koul, N. Dendouga, K. Vergauwen, B. Molenberghs, L. Vranckx, R. Willebrords, Z. Ristic, H. Lill, I. Dorange, J. Guillemont, D. Bald and K. Andries, *Nat. Chem. Biol.*, 2007, **3**, 323.
153. K. Hards, J. R. Robson, M. Berney, L. Shaw, D. Bald, A. Koul, K. Andries and G. M. Cook, *J. Antimicrob. Chemother.*, 2015, **70**, 2028.
154. R. A. Haworth and D. R. Hunter, *Arch. Biochem. Biophys.*, 1979, **195**, 460.
155. M. Zoratti and I. Szabò, *Biochim. Biophys. Acta*, 1995, **1241**, 139.
156. V. Giorgio, S. von Stockum, M. Antoniel, A. Fabbro, F. Fogolari, M. Forte, G. D. Glick, V. Petronilli, M. Zoratti, I. Szabó, G. Lippe and P. Bernardi, *Proc. Natl. Acad. Sci. U. S. A.*, 2013, **110**, 5887.

157. A. Rasola and P. Bernardi, *Apoptosis*, 2007, **12**, 815.
158. M. Carraro, V. Giorgio, J. Šileikytė, G. Sartori, M. Forte, G. Lippe, M. Zoratti, I. Szabò and P. Bernardi, *J. Biol. Chem.*, 2014, **289**, 15980.
159. T. Azarashvili, I. Odinokova, A. Bakunts, V. Ternovsky, O. Krestinina, J. Tyynelä and N. E. Saris, *Cell Calcium*, 2014, **55**, 69.
160. K. N. Alavian, G. Beutner, E. Lazrove, S. Sacchetti, H. A. Park, P. Licznerski, H. Li, P. Nabili, K. Hockensmith, M. Graham, G. A. Porter and E. A. Jonas, *Proc. Natl. Acad. Sci. U. S. A.*, 2014, **111**, 10580.
161. M. R. Dyer and J. E. Walker, *Biochem. J.*, 1993, **293**, 51.
162. W. L. Yan, T. J. Lerner, J. L. Haines and J. F. Gusella, *Genomics*, 1994, **24**, 375.
163. J. He, H. C. Ford, J. Carroll, S. Ding, I. M. Fearnley and J. E. Walker, *Proc. Natl. Acad. Sci. U. S. A.*, 2017, **114**, 3409.
164. J. He, J. Carroll, S. Ding, I. M. Fearnley and J. E. Walker, *Proc. Natl. Acad. Sci. U. S. A.*, 2017, **114**, 9086.

Subject Index